STRUCTURAL CONTROL

managing editor
DONALD E. GRIERSON

technical editors
PAM McCUAIG
LINDA STROUTH

STRUCTURAL CONTROL

Proceedings of the International IUTAM Symposium on Structural Control held at the University of Waterloo,
Ontario, Canada, June 4-7, 1979

Edited by
H. H. E. LEIPHOLZ
University of Waterloo,
Ontario, Canada

1980

NORTH-HOLLAND PUBLISHING COMPANY
AMSTERDAM • NEW YORK • OXFORD
SM PUBLICATIONS

© International Union of Theoretical and Applied Mechanics, 1980

All rights reserved. No part of this publication may be reproduced, stored in a retrieval system, or transmitted, in any form or by any means, electronic, mechanical, photocopying, recording or otherwise, without the prior permission of the copyright owner.

ISBN: 0 444 85485 1

Published by:
North-Holland Publishing Company
Amsterdam • New York • Oxford

Sole distributors for the U.S.A. and Canada:
Elsevier North-Holland, Inc.
52 Vanderbilt Avenue
New York, N.Y. 10017

Library of Congress Cataloging in Publication Data

International IUTAM Symposium on Structural Control,
　　University of Waterloo, 1979.
　　Structural control.

　　1. Structural engineering--Congresses.　2. Control
theory--Congresses.　3. System analysis--Congresses.
I. Leipholz, Horst.　II. International Union of
Theoretical and Applied Mechanics.　III. Title.
TA630.I545　1979　　　624.1'7　　　80-12950
ISBN 0-444-85485-1

Printed in The Netherlands

Editor's Preface

At the General Assembly meeting of IUTAM in Delft, September, 1976, a proposal of the Canadian delegation for a symposium on Structural Control was discussed and taken into further consideration. At a following meeting of the Bureau of IUTAM in Vienna, September, 1977, a Study Group was established consisting of:

H.H.E. Leipholz	-	(Canada), Chairman,
A. Bryson	-	(U.S.A.)
G. Maier	-	(Italy)
Yu.N. Rabotnov	-	(U.S.S.R.)
A. Sawczuk	-	(Poland)
M. Shinozuka	-	(U.S.A.)
J.T.P. Yao	-	(U.S.A.)

This Study Group recommended to the Bureau that the symposium should be held in June, 1979, at the University of Waterloo, Waterloo, Ontario, Canada. The recommendation was approved at the General Assembly meeting of IUTAM in Herrenalb, September, 1978, and the above mentioned Group was appointed the Scientific Committee of the symposium.

After thorough preparations and a careful selection of participants, the symposium took place from June 4 to 7, 1979, in Waterloo under the sponsorship of the University of Waterloo and it's Solid Mechanics Division. Members of this Division, Professor J. Roorda, Professor N.C. Lind, and Professor R. Schuster, served on the Local Organizing Committee. The scientific sessions were visited by numerous members of the Faculty of Engineering of the host University and by 45 delegates from 10 countries (3 continents).

Aim of the Symposium was to give three groups of people an opportunity for contacts and for a fruitful exchange of experience, data, and information:

(a) researchers concerned with the fundamentals of control and optimization theory;
(b) researchers and engineers involved in the application of control and optimization theory to industrial processes and to vehicles as well as structures in the area of aero- and space mechanics;
(c) researchers and engineers interested and active in the application of passive and active control to civil engineering structures.

The idea was that people in group (a) should recognize a new direction of applications, a new set of basic problems, and a new challenge for providing the analytical and methodological material needed to solve the upcoming problems in civil engineering. People in group (b) were supposed to share their experience with the other participants in the Symposium by reporting on applications of optimal control to already well understood and accomplished engineering systems. Finally, people in group (c) were supposed to report on their attempts and results concerning an application of optimal control specifically to civil engineering structures, thus, identifying new types of problems and approaches and indicating the advantageous and beneficial possibilities involved in optimal control with respect to civil engineering problems.

The main results of the Symposium may be summarized as follows: it came clearly to light that there are immediate, useful, and relevant applications of optimal control to civil engineering structures in the following cases:

(i) control of bridges against excessive deformations and accelerations due to moving loads and wind forces;
(ii) control of tall buildings, masts, and towers against wind forces;
(iii) control of buildings against earthquake effects;
(iv) control of building foundations against changing soil conditions.

Editor's Preface

Although impressive examples of applications of control systems in these cases are already known, it became obvious that a number of questions are still open. In the first place, control theory involving distributed parameters and large degree of freedom systems must be made available for civil engineers and developed further for their purposes. However, the most important finding was that a new approach to the design of civil engineering structures should be found. Instead of applying traditional control to traditional buildings, the "kinetic structure" as Professor Zuk put it, should be developed. Such a structure is supposed to be controllable by its very nature. In a changing environment and under changing functional demands, the kinetic structure should be designed with the aim in mind of making this structure due to its inherent control devices adjustable in a natural way to changing requirements. Of course, new building codes have to be developed for such futuristic structures.

Special thanks are due to the Bureau of IUTAM for constant assistance, advice, and a financial contribution to the travel expenses of younger scientists, thus enabling them to participate in the event. I am also greatly indebted to the National Research Council of Canada and the University of Waterloo for generous financial support and for assistance in the local organization of the Symposium.

H.H.E. Leipholz
Waterloo, Ontario, Canada
January, 1980

Contents

EDITOR'S PREFACE	v
PARTICIPANTS	xiii
A General Approach to Active Structural Control M. ABDEL-ROHMAN and H.H.E. LEIPHOLZ	1
Automatic Active Control of Structures M. ABDEL-ROHMAN and H.H.E. LEIPHOLZ	29
Design of Reduced-Order Observers for Structural Control Systems M. ABDEL-ROHMAN, H.H.E. LEIPHOLZ and V.H. QUINTANA	57
A Transducer System for the Identification of Railway Track Modulus Parameters J.D. APLEVICH	79
Flexure Oscillations of Bridges - Estimation of the Velocity and of the Acceleration by Using Leunberger Observers J. AULOGE	95
Active Control of Large Civil Engineering Structures: A Naïve Approach M.J. BALAS	107
Optimization of Control Devices in Base Isolation Systems for Aseismic Design M.A. BHATTI, K.S. PISTER and E. POLAK	127
Optimally Stable Structures Subjected to Follower Forces R. BOGACZ and O. MAHRENHOLTZ	139
Sur les Solutions de Problèmes Associés en Optimisation des Structures Mécaniques P. BROUSSE	159
Multilevel Control Concepts in Relation to the Structural Design Problem D.G. CARMICHAEL and D.H. CLYDE	171
The Use of Aerodynamic Appendages for Tall Building Control J.C.H. CHANG and T.T. SOONG	199

Principle Types of Optimization Problems in
the Mechanics of Deformed Solids and Their
Mathematical Models
A.A. ČYRAS — 211

On the Morphology of Controlled Systems
M.S. EL NASCHIE and S. AL ATHEL — 237

Two Extreme Cases of On-Line Control of
Structures
M. FANELLI and G. GIUSEPPETTI — 255

Control of Structural Vibration
S.F. GIRGIS — 269

Minimum Deformability Design and Control
of Constraints
J. GRABACKI — 281

Active Flutter Control in Transonic Conditions
A. GRAVELLE — 297

Critical Comparison Between Active and Passive
Control of Wind Induced Vibrations of Structures
by Means of Mechanical Devices
G. HIRSCH — 313

Finite Element Analysis of Contact Problems Based
on the Unilateral Constraints Formulation
N.D. HUNG, G. DE SAXCE and G.M.L. GLADWELL — 341

Complementarity Conditions in Engineering
Optimization
I. KANEKO — 375

Control Devices for Earthquake-Resistant Structural
Design
J.M. KELLY — 391

The Time-Optimal Control of Wind Induced Structural
Vibrations Using Active Appendages
R.E. KLEIN and H. SALHI — 415

Optimal Project of a Cylindrical Shell for
Moderately Large Deflections
J. LELLEP and A. SAWCZUK — 431

Application of the Control Theory for Optimal Design
of Nonelastic Beams under Dynamic Loading
Ü. LEPIK — 447

Contents

Active Damping of Large Structures in Winds
R. LUND — 459

On-Line Pulse Control of Tall Buildings
S.F. MASRI, G.A. BEKEY and F.E. UDWADIA — 471

Effects of Reduced Order Mathematical Models on Dynamic Response of Flexible Aircraft with Closed-Loop Control
D. MCLEAN — 493

Active Control of Structures by Modal Synthesis
L. MEIROVITCH and H. ÖZ — 505

On Optimal Force Action and Reaction on Structures
Z. MRÓZ — 523

Optimal Control of Unilateral Structural Analysis Problems
P.D. PANAGIOTOPOULOS — 545

The Prager-Shield Optimality Criterion - An Efficient Extension to Finite Element Problems
G. PAPE and G. THIERAUF — 563

Design of Large Scale Tuned Mass Dampers
N.R. PETERSEN — 581

Structural Optimization Based on the Workhardening Adaptation Concept
C. POLIZZOTTO and C. MAZZARELLA — 597

An Active System for Neutralization of Vibratory Force Interactions in Structures
J.W. ROBERTS and M.C. BORGOHAIN — 613

Experiments in Feedback Control of Structures
J. ROORDA — 629

Active Control of Large Building Structures
J.N. JUANG, S. SAE-UNG and J.N. YANG — 663

Hierarchical Control of Design Organization for Structural Optimization
A.S. SAMBURA — 677

Active Control of Flutter and Vibration of an Aircraft
O. SENSBURG, J. BECKER and H. HÖNLINGER — 693

On Optimal Control Configuration in Theory of
Modal Control
T.T. SOONG and M.I.J. CHANG   723

A Unified Approach to Optimal Control and
Mechanics of Constrained Continua
G. SZEFER   739

Identification and Control of Structural Damage
J.T.P. YAO   757

The Past and Future of Active Structural Control
Systems
W. ZUK   779

Optimal Structural Design of Flexible Beams with
Respect to Creep Rupture Time
M. ŻYCZKOWSKI and W. ŚWISTERSKI   795

Participants

ABDEL-ROHMAN, M., Dr., Department of Civil Engineering, University of Waterloo, Waterloo, Ontario, Canada N2L 3G1.

APLEVICH, J.D., Professor, Department of Electrical Engineering, University of Waterloo, Waterloo, Ontario, Canada N2L 3G1.

ARIARATNAM, S.T., Professor, Department of Civil Engineering, University of Waterloo, Waterloo, Ontario, Canada N2L 3G1.

AULOGE, Jean-Yves, Dr., Ecole Centrale de Lyon, 36, Route de Dardilly, 69130 Ecully, France.

BALAS, M.J., Professor, Electrical and Systems Department, Rensselaer Polytechnic Institute, Troy, New York 12181, U.S.A.

BECKER, J., Mr., Messerschmitt-Bölkow-Blohm GmbH, Airplane Division, P.O. Box 80 11 60, 8 München 80, West Germany.

BOGACZ, R., Dr., Institute of Fundamental Technological Research, Polish Academy of Sciences, Swietokrzyska 11/21, 00-049 Warsaw, Poland.

CARMICHAEL, D.G., Dr., Department of Civil Engineering, University of Western Australia, Nedlands 6009, Western Australia.

CHANG, J.C-H., Mr., Department of Civil Engineering, State University of New York at Buffalo, Buffalo, New York 14214, U.S.A.

CLYDE, D.H., Professor, Department of Civil Engineering, University of Western Australia, Nedlands 6009, Western Australia.

ČYRAS, A.A., Professor, Kariu Kapu 5, 232055 Vilnius, U.S.S.R.

CZERNY, L., Mr., Department of Mechanical Engineering, University of Illinois, Urbana, Illinois 61801, U.S.A.

DONE, G., Professor, Department of Applied Mechanics, The City University, Northampton Square, London EC1V 0HB, England.

GAUS, M.P., Dr., National Science Foundation, 1800 G Street, N.W., Room 1130, Washington, D.C. 20550, U.S.A.

GERMAN, D., Mr., Department of Mechanical Engineering, University of Illinois, Urbana, Illinois 61801, U.S.A.

GRAVELLE, A., Mr., O.N.E.R.A., 29 Avenue de la Division Leclerc, Châtillon-Sous-Bagneux, 02320 Châtillon, France.

HAWRANEK, R., Mr., Department of Civil Engineering, University of Toronto, Toronto, Ontario, Canada.

HIRSCH, G., Mr., Dipl.-Ing., Institut für Leichtbau, Technische Hochschule Aachen, Postfach D-5100 Aachen, West Germany.

HUNG, N.D., Dr., Institut du Génie Civil, University of Liège, 6 Quai Banning, Liège B400, Belgium.

JUANG, J.-N., Dr., Jet Propulsion Laboratory, California Institute of Technology, 4800 Oak Grove Drive, Pasadena, California 91103, U.S.A.

KANEKO, I., Professor, Department of Industrial Engineering, University of Wisconsin, 1513 University Avenue, Madison, Wisconsin 53706, U.S.A.

KELLY, J.M., Professor, Department of Civil Engineering, Division of Structural Mechanics, University of California, Berkeley, California 94720, U.S.A.

KLEIN, R.E., Dr., Department of Mechanical Engineering, University of Illinois, Urbana, Illinois 61801, U.S.A.

KOZIN, F., Professor, Department of Electrical Engineering, Polytechnic Institute of Brooklyn, Long Island Graduate Center, Route 110, Farmingdale, New York 11735, U.S.A.

LEIPHOLZ, H.H.E., Professor, Department of Civil Engineering, University of Waterloo, Waterloo, Ontario, Canada N2L 3G1.

McLEAN, D., Dr., Senior Lecturer, Department of Transport Technology, University of Technology, Loughborough, Leicestershire LE11 3TU, England.

MEIROVITCH, L., Professor, Department of Engineering Science and Mechanics, Virginia Polytechnic Institute and State University, Blacksburg, Virginia 24061, U.S.A.

MRÓZ, Z., Professor, Institute of Fundamental Technological Research, Polish Academy of Sciences, Swietokrzyska 21, 00-049 Warsaw, Poland.

PAPE, G., Mr., Gesamthochschule, Universität Essen, Fachbereich 10, Postfach 6843, 4300 Essen 1, West Germany.

PETERSEN, N.R., Mr., MTS Systems Corporation, P.O. Box 24012, Minneapolis, Minnesota 55424, U.S.A.

PISTER, K.S., Professor, Department of Civil Engineering, University of California, Berkeley, California 94720, U.S.A.

POLIZZOTTO, C., Professor, Facoltá di Architettura, University of Palermo, via Maqueda 175, Palermo I-90133, Italy.

QUINTANA, V.H., Professor, Department of Electrical Engineering, University of Waterloo, Waterloo, Ontario, Canada N2L 3G1.

ROBERTS, J.W., Professor, School of Engineering, University of Edinburgh, Edinburgh EH9 3JL, Scotland.

ROORDA, J., Professor, Department of Civil Engineering, University of Waterloo, Waterloo, Ontario, Canada N2L 3G1.

SCHUSTER, R.M., Professor, School of Architecture, University of Waterloo, Waterloo, Ontario, Canada N2L 3G1.

SOONG, T.T., Professor, Chairman, Department of Civil Engineering, State University of New York at Buffalo, Buffalo, New York 14214, U.S.A.

STUBBS, N., Dr., Department of Civil Engineering, Columbia University, 610 S.W. Mudd Building, New York, New York 10027, U.S.A.

SZEFER, G., Professor, Instytut Mechaniki i Podstaw Konstrukeji Maszyn Politechniki Krakowskiej, Warszawska 24, 31-155 Kraków, Poland.

THIERAUF, G., Professor, Gesamthochschule, Universität Essen, Fachbereich 10, Postfach 6843, 4300 Essen 1, West Germany.

UDWADIA, F.E., Professor, Department of Civil Engineering, University of Southern California, University Park, Los Angeles, California 90007, U.S.A.

YANG, J.N., Dr., Department of Civil Engineering, The George Washington University, Washington, D.C. 20052, U.S.A.

YAO, J.T.P., Professor, School of Civil Engineering, Purdue University, West Lafayette, Indiana 47907, U.S.A.

ZUK, W., Professor, Director, School of Architecture, University of Virginia, Campbell Hall, Charlottesville, Virginia 22903, U.S.A.

ŻYCZKOWSKI, M., Professor, Instytut Mechaniki i Podstaw Konstrukeji Maszyn Politechniki Krakowskiej, Warszawska 24, 31-155 Kraków, Poland.

STRUCTURAL CONTROL, H.H.E. Leipholz (ed.)
North-Holland Publishing Company & SM Publications
© IUTAM, 1980

A GENERAL APPROACH TO ACTIVE STRUCTURAL CONTROL*

M. Abdel-Rohman and H.H.E. Leipholz

Department of Civil Engineering
University of Waterloo
Waterloo, Ontario, Canada N2L 3G1

1.0 INTRODUCTION

New trends in civil engineering involving megastructures have posed some specific problems to structural engineers. These problems are: ensuring the safety of such structures in the presence of the uncertainties with respect to the type of loading; providing the desired human comfort; guaranteeing the serviceability; and observing the code limitations. Structural control is proposed as an approach to provide solutions to these problems.

An appropriate way to control a structure is to apply a set of auxiliary forces which are called control forces. To generate these forces, one may use passive control mechanism or active control mechanism. The control forces generated by passive control mechanisms are only able to control the structural response up to a certain limit due to the lack of energy needed to provide larger forces. That is why structural engineers [18, 20] have recently paid their attention to the application of active control to control structures subjected to severe loading conditions.

---

*Published with permission of the ASCE. Originally appeared in Proc. of ASCE, *J. of the Eng. Mech. Div.*, Vol. 105, No. EM6, Dec. 1979, pp. 1007-1023.

The pioneering work in this area has not yet led to a unified approach to be used in the active control of civil engineering structures. Some researchers paid attention to the implementation of control forces by using active control mechanisms [19], and designing the control forces by classical control methods [11, 12, 14, 17]. Other researchers have shown interest in applying modern control methods to design the control forces [8, 15, 16], but the implementation of these forces was left for further investigations. In a recent paper [2], it has been shown how to implement the control forces by using auxiliary structural elements, and the design of the forces by various control methods was offered in [2, 3, 4]. However, the design objectives were mainly to ensure the safety and the serviceability of the structure, disregarding the satisfaction of human comfort and feasibility of control. Providing the appropriate human comfort was recently considered by Yao and Sae-Ung [13]. The comfort control law was found by using a mechanical principle, and the implementation of this control was left for further investigation.

The purpose of this paper is to propose a unified approach to be used in the active control of structures, in order to satisfy simultaneously the safety, serviceability, human comfort, and the feasibility of control. The approach is applied to the active control of a two-story building frame as an example.

2.0 REQUIREMENTS OF THE GENERAL APPROACH

An extensive study [1] of active control of civil engineering structures has led to the following conclusions: (1) optimal control design methods [4] are most convenient because of their general applicability to the design of linear, nonlinear, time-varying, and/or time-invariant multivariable control systems. However, the numerical calculations involved for high order systems are cumbersome and time consuming. One should therefore make the order of the system as small as possible in order to apply efficiently the optimal control methods. (2) One has to strive for applying closed-loop control because of its effectiveness in the presence of uncertainties in loading and structural parameters. (3) One has to satisfy simultaneously the safety of the structure, human comfort for the users of these structures, and the feasibility of the energy needed to control the structure as desired. That can be done by imposing inequality constraints to satisfy these requirements. However, a constrained optimal control problem usually leads to an open-loop control solution [5, 9] which is not desirable in structural control. A method is proposed herein for finding a closed-loop control even in the presence of inequality constraints.

The approach is summarized in three subsections. The first one is an outline of a method aimed at making the order of the system as small as desired. The second one consists of an enumeration of the constraints needed for an active controlled structure. The third subsection describes the problem's formulation and its solution.

2.1 *Structure's State Equations*

Different methods can be used for a dynamic analysis of structures [6]. The continuous mass method [10] provides the mode equations for undamped structures. Internal damping of a certain mode can only be introduced approximately. This is done by assuming the damping to be a certain ratio of the critical damping of this mode. In any case, one may represent the structure by its first few mode equations and express them in the state form as has been done in references [2 - 4].

To account for damping in the structure's equations of motion, one may use the consistent or lumped mass methods [6]. In these methods, the equations of motion are expressed in a matrix form as

$$\underline{M}\,\underline{\ddot{D}} + \underline{C}\,\underline{\dot{D}} + \underline{K}\,\underline{D} = \underline{W} - \underline{U}, \qquad (1)$$

in which $\underline{M}$ = mass matrix of dimension n x n; $\underline{C}$ = damping matrix of dimension n x n; $\underline{K}$ = stiffness matrix of dimension n x n; $\underline{D}$ = displacement vector of dimension n x 1; $\underline{W}$ = disturbance vector; and $\underline{U}$ = control vector of dimension n x 1.

Equation (1) can only be decoupled if $\underline{C}$ satisfies the orthogonality conditions. As an example, matrix $\underline{C}$ must be proportional to matrix $\underline{M}$, and/or $\underline{K}$ [6]. In this case, equation (1) could be decoupled into damped modes and the first few dominant modes are considered in the state variable form.

The problem arises when matrix $\underline{C}$ does not satisfy the orthogonality conditions, which is generally the case in practice. A method was proposed and used in references [16, 17], in which equation (1) is completely transformed into the state form. For a structure with n-degrees-of-freedom, the state equations will be of order 2n, causing more numerical difficulties in the design by optimal control methods. Another method is offered here in the following steps:

(1) Determine the natural frequencies and mode shapes from the equations of undamped free vibration, i.e.,

$$\underline{M}\,\underline{\ddot{D}} + \underline{K}\,\underline{D} = 0. \qquad (2)$$

(2) Consider out of these n modes those ones that are dominant. The displacement vector is expressed as

$$\underline{D} = \sum_{i=1}^{r} \phi_i\, A_i(t) = \underline{\Phi}\,\underline{A}(t), \qquad (3)$$

in which $\phi_i = i^{th}$ mode shape; $A_i(t)$ = generalized coordinate of mode i; $\underline{\Phi}$ = modal matrix of dimension n x r; $\underline{A}(t)$ = generalized coordinate vector of dimension r x 1; and r = number of dominant modes.

(3) Substituting equation (3) into equation (1) yields

$$\underline{M} \, \underline{\Phi} \, \underline{\ddot{A}} + \underline{C} \, \underline{\Phi} \, \underline{\dot{A}} + \underline{K} \, \underline{\Phi} \, \underline{A} = \underline{W} - \underline{U} . \qquad (4)$$

(4) Multiplying each term in equation (4) by $\underline{\Phi}^T$ leads to

$$\underline{\Phi}^T \underline{M} \, \underline{\Phi} \, \underline{A} + \underline{\Phi}^T \underline{C} \, \underline{\Phi} \, \underline{A} + \underline{\Phi}^T \underline{K} \, \underline{\Phi} \, \underline{A} = \underline{\Phi}^T \underline{W} - \underline{\Phi}^T \underline{U} . \qquad (5)$$

(5) Using the orthogonality conditions, equation (5) can be written as

$$\underline{Z} \, \underline{\ddot{A}} + \underline{G} \, \underline{\dot{A}} + \underline{\Gamma} \, \underline{A} = \underline{\Phi}^T \underline{W} - \underline{\Phi}^T \underline{U} , \qquad (6)$$

in which $\underline{Z}$ = diagonal matrix; $\underline{\Gamma}$ = diagonal matrix; and $\underline{G}$ = a general matrix of dimensions r x r.

(6) Equation (6) is expressed in the state form as

$$\underline{\dot{X}} = \overline{\underline{A}} \, \underline{X} + \underline{B} \, \underline{U} + \underline{d} , \qquad (7)$$

in which $\underline{X} = [\underline{A} \, \vdots \, \underline{\dot{A}}]^T$, $\underline{d} = [0 \, \vdots \, \underline{Z}^{-1} \, \underline{\Phi}^T \, \underline{W}]^T$; $B = [0 \, \vdots \, -\underline{Z}^{-1} \, \underline{\Phi}^T]^T$; and $\overline{\underline{A}}$ is given by

$$\overline{\underline{A}} = \begin{bmatrix} \underline{0} & \vdots & \underline{I} \\ \hdashline -\underline{Z}^{-1} \, \underline{\Gamma} & \vdots & -\underline{Z}^{-1} \, \underline{G} \end{bmatrix} , \qquad (8)$$

2.2 Desired Constraints

The prime objective of introducing active control to civil engineering structures is to ensure their safety and to satisfy the comfort demands of people using these structures. It is attempted to achieve these objectives with a minimum cost possible. Therefore, a total of three general constraints should be imposed on the controlled structure, that is to say, safety, human comfort, and minimum cost.

Safety can be ensured by imposing constraints on the maximum allowed deflection and/or acceleration conditioned by building's code, at some critical sections of the structure. These constraints can be expressed as

$$-\varepsilon1_i \leq D_i \leq \varepsilon1_i \quad , \quad i = 1,2,\ldots,n \, , \qquad (9)$$

$$-\varepsilon2_j \leq \ddot{D}_j \leq \varepsilon2_j \quad , \quad j = 1,2,\ldots,n \, , \qquad (10)$$

in which $\varepsilon1_i$ = maximum allowed displacement at coordinate i; and $\varepsilon2_j$ = maximum allowed acceleration at coordinate j.

The human comfort can be satisfied by imposing constraints on the acceleration at the critical sections. Criteria to provide human comfort were studied in [7, 13, 20] and are expressed here in a general sense as

$$-\varepsilon3_\ell \leq \ddot{D}_\ell \leq \varepsilon3_\ell \quad , \quad \ell = 1,2,\ldots,n \, , \qquad (11)$$

in which $\varepsilon3_\ell$ = maximum allowed acceleration at coordinate ℓ, which usually is less than $\varepsilon2_j$ of equation (10).

The feasibility and cost of control can generally be expressed as a function of the magnitude of control forces applied to the structure. One may therefore impose constraints as

$$-\varepsilon4_k \leq U_k \leq \varepsilon4_k \quad , \quad k = 1,2,\ldots,n \, , \qquad (12)$$

in which $\varepsilon4_k$ = maximum allowed magnitude of the control force U_k.

2.3 Problem's Formulation and Solution

To satisfy the constraints, the problem is formulated as an optimal control problem which reads:

find $\underline{U}$

such that $J = M[\underline{X}(t_f)] + \int_{t_o}^{t_f} L(\underline{X},\underline{U},t)dt = \text{minimum}$

subject to $\underline{\dot{X}} = \overline{\underline{A}}\,\underline{X} + \underline{B}\,\underline{U} + \underline{d}$,

$$\underline{D} = \underline{\psi}\,\underline{X}, \tag{13}$$

$\zeta_i = \varepsilon 1_i \pm D_i \geq 0$,

$\zeta_m = \varepsilon 3_m \pm \ddot{D}_m \geq 0$,

$\zeta_k = \varepsilon 4_k \pm U_k \geq 0$,

in which $i,m,k = 1,2,\ldots,n$; t_o = original control time; t_f = final control time; M and L are functions notions; and $\underline{\psi}$ = matrix of appropriate dimension.

The objective function J should carefully be chosen such that the controlled response of the structure is satisfactory. It may take several trials before the optimal solution is obtained. A solution for optimal control can be obtained by converting the constrainted objective function into an unconstrained objective function and applying the optimality conditions. In the presence of inequality constraints one may include them by using penalty functions. Several penalty functions are used in the literature [9], however only the step penalty function and exponential penalty functions are mentioned here.

Using the step function, the unconstrained objective function is

$$\hat{J} = J + \int_{t_o}^{t_f} [\underline{\lambda}^T (\underline{\bar{A}}\ \underline{X} + \underline{B}\ \underline{U} + \underline{d} - \underline{\dot{X}}) + \sum_{i=1}^{j} K_i\ \zeta_i^2] dt\ , \quad (14)$$

in which $\underline{\lambda} = \underline{\lambda}(t)$ is a costate vector of dimension n x 1; K_i = weighing factor which is positive if $\zeta_i \leq 0$ and zero if $\zeta_i > 0$; and j = number of inequality constraints.

Using the exponential penalty function, the unconstrained objective function is

$$\hat{J} = J + \int_{t_o}^{t_f} [\underline{\lambda}^T (\underline{\bar{A}}\ \underline{X} + \underline{B}\ \underline{U} + \underline{d} - \underline{\dot{X}}) + \sum_{i=1}^{j} \mu_i\ e^{-\beta_i\ \zeta_i}] dt\ , \quad (15)$$

in which μ_i and β_i are positive weighing factors.

Due to the nonlinearity imposed by considering the inequality constraints as penalty terms, a solution is usually obtained in open-loop type. However, by using a quadratic objective function as performance index and step penalty function for the inequality constraints, one is able to obtain a closed-loop control by the method proposed below:

(1) Determine the two-point-boundary-value-problem from the Hamiltonian. In general, it consists of linear, nonhomogeneous, first order vector differential equations due to working with quadratic objective function and step penalty functions.

(2) A closed-loop solution is obtained by assuming that the costate variable $\lambda(t)$ is a combination of the current state $\underline{X}(t)$ and of a predetermined vector $\underline{q}(t)$ (4). That leads to a Riccati matrix differential equation and a vector differential equation determining $\underline{q}(t)$.

(3) Using the Riccati matrix $\underline{P}(t)$ and vector $\underline{q}(t)$ find the controlled state $\underline{X}(t)$ and check whether the constraints have been violated or not.

(4) In the case of violating a constraint, the corresponding weighing factor K_i is increased for the next iteration as explained in the flow chart of Figure 1.

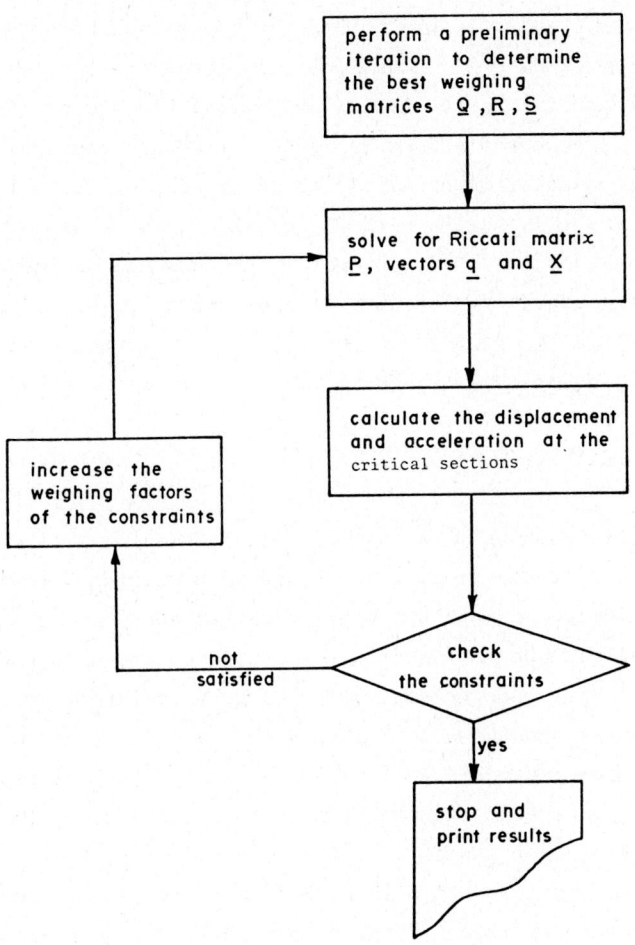

Figure 1 - Flow Chart for the General Approach

One should notice that if the problem were prescribed as finding the optimal solution for a given objective function, satisfying the imposed constraints, then the previously selected solution may not be satisfactory. However, in the suggested approach, a closed-loop control solution is sought satisfying under all circumstances the given constraints. At the same time, it is being shown with respect to which objective function the obtained solution is an optimal one. Therefore, the obtained solution by the suggested approach may be called an optimal "in a weak sense". This solution is acceptable because it satisfies all constraints. Moreover, it is usually not possible to define the proper objective function before obtaining the optimal solution. For example, in the classical case of optimal control, in the absence of inequality constraints, the objective function is obtained by trial and error, changing the weighing matrices in the objective function until one obtains a solution which would be called an optimal one.

3.0 NUMERICAL EXAMPLE

The two-story frame shown in Figure 2 is subjected to a steady state disturbance at the floor levels. It is required to control the frame's response against the applied disturbance, using the control mechanism shown in Figure 3. The tendons are arranged in X-shapes [20] in order to control the structure in either of the possible sways, as shown in Figures 4 and 5. It is noticed that although each tendon is activated separately, the two tendons, in each story, complement each other. That means when one tendon is activated, the other tendon is not activated. Therefore, one may consider the control action in each floor to be a continuous one. The mathematical model of the structure is shown in Figure 6 in which u_i is the horizontal component of u_i'.

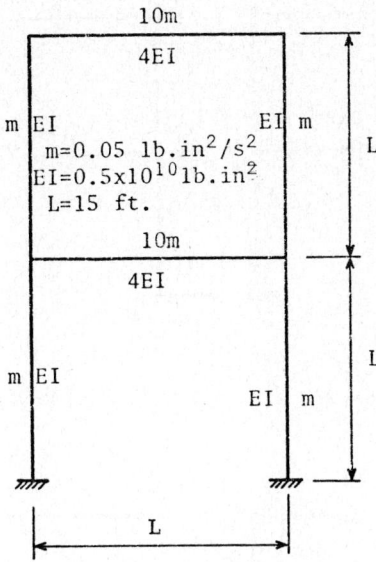

Figure 2 - Two-Story Frame

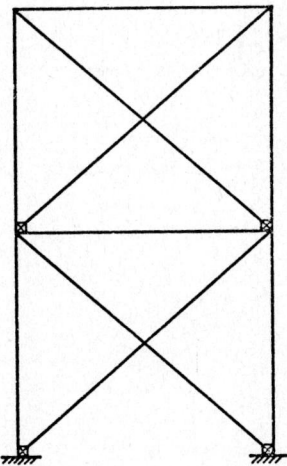

Figure 3 - Control Mechanism

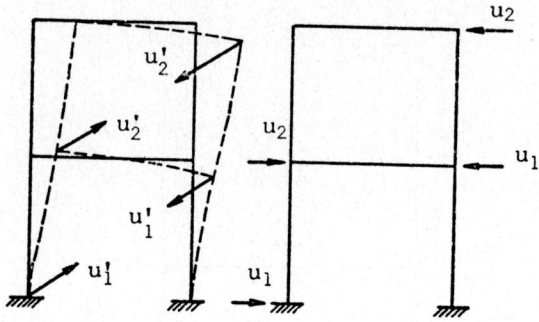

Figure 4 - Frame in Right Sway

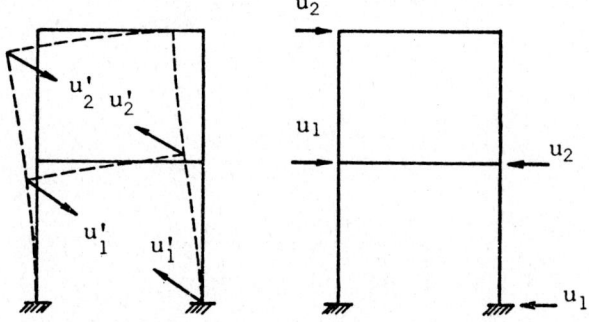

Figure 5 - Frame in Left Sway

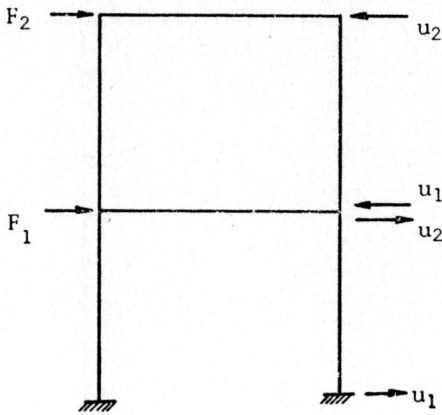

Figure 6 – Frame with the Applied Disturbance and Control

3.1 Structure's Equations of Motion

The structural coordinates are shown in Figure 7, whereas the elements' coordinates are indicated in Figure 8. The compatibility (connectivity) matrix is given by:

$$\underline{C} = \begin{bmatrix} 0 & 0 & 0 & 0 & 0 & 0 & -1 & 0 & 1 & 0 & 0 & 0 & 0 & 0 & 0 & 0 & 0 & 0 & 0 \\ 0 & 1 & 0 & 0 & 0 & 0 & 0 & 1 & 0 & 0 & 0 & 0 & 0 & 0 & 0 & 0 & 0 & 0 & 0 \\ 0 & 0 & 0 & 1 & 0 & 0 & 0 & 0 & 0 & 1 & 0 & 0 & 0 & 0 & 0 & 0 & 0 & 0 & 0 \\ 0 & 0 & 0 & 0 & -1 & 0 & 0 & 0 & 0 & 0 & -1 & 0 & 0 & 0 & 0 & -1 & 0 & 1 & 0 \\ 0 & 0 & 0 & 0 & 0 & 1 & 0 & 0 & 0 & 0 & 0 & 0 & 1 & 0 & 0 & 0 & 1 & 0 & 0 \\ 0 & 0 & 0 & 0 & 0 & 0 & 0 & 0 & 0 & 0 & 0 & 1 & 0 & 0 & 0 & 1 & 0 & 0 & 1 \end{bmatrix}^T \quad (16)$$

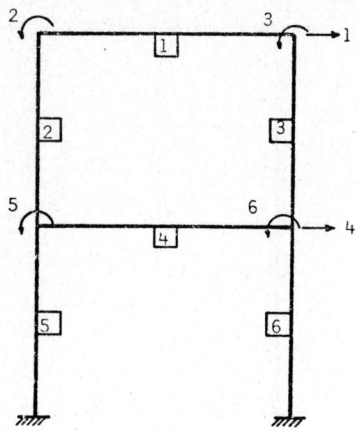

Figure 7 - Structural Coordinates

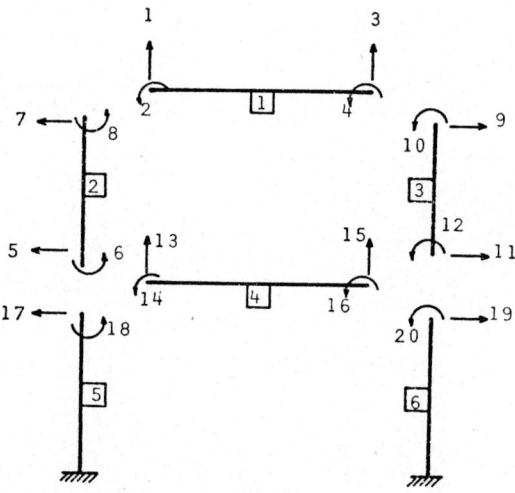

Figure 8 - Members Coordinates

Using the consistent mass method [6], the overall stiffness and mass matrices are obtained from

$$\underline{M}^* = \underline{C}^T \underline{M} \underline{C}, \tag{17}$$

$$\underline{K}^* = \underline{C}^T \underline{K} \underline{C}, \tag{18}$$

in which $\underline{M}$ = band mass matrix of dimension 20 x 20; $\underline{K}$ = band stiffness matrix of dimension 20 x 20; $\underline{M}^*$ = overall mass matrix of dimension 6 x 6; and $\underline{K}^*$ = overall stiffness matrix of dimension 6 x 6.

The structural mass matrix $\underline{M}^*$ must be modified to include the masses of the horizontal beams, since their masses contribute to the response if the structure is subjected to a horizontal acceleration. The elements which need modifications are then M*(1,1) and M*(4,4). The elements are replaced by

$$M^*(1,1) \rightarrow M^*(1,1) + 10 \text{ mL}, \tag{19}$$

$$M^*(4,4) \rightarrow M^*(4,4) + 10 \text{ mL}. \tag{20}$$

Calculating the dynamic matrix form $\underline{M}^{*-1} \underline{K}^*$, one is able to determine the natural frequencies and mode shapes which are summarized in Table 1.

Table 1 - Frequencies and Mode Shapes

Node Number	Mode Number					
	1	2	3	4	5	6
1	0.867335	0.534157	0.0	0.0	-0.77055	0.3056
2	-0.000369	-0.001616	-0.6274	-0.35303	0.25474	-0.4592
3	-0.000369	-0.001616	0.6274	0.35303	0.25474	-0.4592
4	0.497723	-0.845382	0.0	0.0	-0.30568	0.5219
5	-0.000884	-0.000472	0.326	-0.61267	-0.3025	-0.3257
6	-0.000884	-0.000472	-0.326	0.61267	-0.3025	-0.3257
ω^2	62.141	495.7171	5803.67	8997.55	59791.68	100579.26

Considering only the first two modes, and using the method previously outlined, the equations of motion are

$$\ddot{A}_1 + 62.141\, A_1 = (0.4977\, F_1 + 0.86733\, F_2 - 0.49772\, u_1 - 0.36961\, u_2)/100.367 , \tag{21}$$

$$\ddot{A}_2 + 495.7171\, A_2 = (-0.84538\, F_1 + 0.53415\, F_2 + 0.84538\, u_1 - 1.37954\, u_2)/99.454 . \tag{22}$$

3.2 Structure's State Equations

Denominating $x_1 = A_1$, $x_2 = \dot{A}_1$, $x_3 = A_2$, and $x_4 = \dot{A}_2$, equations (21) and (22) are expressed in the state form as

$$\underline{\dot{X}} = \underline{A}\,\underline{X} + \underline{B}\,\underline{U} + \underline{d} , \tag{23}$$

in which $\underline{X} = [x_1\ x_2\ x_3\ x_4]^T$; $\underline{U} = [u_1\ u_2]^T$; and

$$\underline{A} = \begin{bmatrix} 0 & 1 & 0 & 0 \\ -\omega_1^2 & 0 & 0 & 0 \\ 0 & 0 & 0 & 1 \\ 0 & 0 & -\omega_1^2 & 0 \end{bmatrix}, \quad \underline{B} = \begin{bmatrix} 0 & 1 \\ \dfrac{-0.4977}{100.367} & \dfrac{-0.3696}{100.367} \\ 0 & 0 \\ \dfrac{0.8453}{99.454} & \dfrac{-1.3785}{99.454} \end{bmatrix}$$

$$\underline{d} = [0 \ (0.4977 \ F_1 + 0.86733 \ F_2)/100.367 \ 0 \ (0.5341 \ F_2 - 0.8453 \ F_1)/99.45]^T.$$

The forcing functions are assumed as pulsating forces having the structure's frequencies. Resonance was assumed in order to show that automatic active control can be used to guarantee the safety of structures and human comfort, even under severe loading conditions. The values F_1 and F_2 are assumed as

$$F_1 = 2000 \sin 7.88 \ t + 600 \sin 22.2 \ t \ , \quad (24)$$

$$F_2 = 3000 \sin 7.88 \ t + 900 \sin 22.2 \ t \ , \quad (25)$$

in which the units are pounds.

3.3 *Constraints*

A limitation on the displacement is taken as 0.002 of the height of the structure. A constraint on the acceleration or velocity is taken as 6 inch/sec^2 or 6 inch/sec, respectively [20]. These constraints are imposed on coordinate 1 of Figure 7. Displacement and velocity at this coordinate are given by

$$D_1 = 0.86733 \ x_1 + 0.5341 \ x_3 \ , \quad (26)$$

$$\dot{D}_1 = 0.86733 \ x_2 + 0.5341 \ x_4 \ . \quad (27)$$

Therefore, the used constraints are:

$$\zeta_1 = -0.86733\ x_1 - 0.5341\ x_3 + 0.002 \times 30 \geq 0, \quad (28a)$$

$$\zeta_2 = 0.86733\ x_1 + 0.5341\ x_3 + 0.002 \times 30 \geq 0, \quad (28b)$$

$$\zeta_3 = -0.86733\ x_2 - 0.5341\ x_4 + 0.5 \geq 0, \quad (28c)$$

$$\zeta_4 = 0.86733\ x_2 + 0.5341\ x_4 + 0.5 \geq 0. \quad (28d)$$

The control energy is considered to be unlimited in this example. However, constraints on the magnitude of the control forces can also be imposed.

3.4 Problem's Formulation

The problem is formulated as

$$\left.\begin{array}{ll} \text{find:} & u_1,\ u_2 \\[6pt] \text{such that:} & J = \dfrac{1}{2}\underline{X}^T(t_f)\underline{S}\,\underline{X}(t_f) + \dfrac{1}{2}\displaystyle\int_{t_o}^{t_f}(\underline{X}^T\,\underline{Q}\,\underline{X} + \underline{U}^T\,\underline{R}\,\underline{U})dt \\[6pt] \text{subject to:} & \text{equations (23) and (28).} \end{array}\right\} \quad (29)$$

It is assumed that the design of the observer has already been completed [3, 5] and the task of the designer is then to find u_1 and u_2 satisfying equation (29).

3.5 Problem's Solution

Using step penalty functions, the unconstrained objective function is

$$\hat{J} = J + \int_{t_o}^{t_f} [\underline{\lambda}^T(\underline{A}\,\underline{X} + \underline{B}\,\underline{U} + \underline{d} - \underline{\dot{X}}) + \sum_{j=1}^{4} K_j\ \zeta_j^2]dt. \quad (30)$$

The Hamiltonian is given by

$$H = \frac{1}{2}\underline{X}^T \underline{Q}\,\underline{X} + \frac{1}{2}\underline{U}^T \underline{R}\,\underline{U} + \underline{\lambda}^T(\underline{A}\,\underline{X} + \underline{B}\,\underline{U} + \underline{d}) + \sum_{j=1}^{4} K_i \zeta_i^2 \,. \quad (31)$$

The optimality conditions [4, 5] lead to the following two-point-boundary-value-problem:

$$\underline{\dot{X}} = \underline{A}\,\underline{X} - \underline{B}\,\underline{R}^{-}\underline{B}^T\underline{\lambda} + \underline{d} \quad ; \quad \underline{X}(t_o) = \underline{X}_o , \quad (32a)$$

$$\underline{\dot{\lambda}} = -\underline{Q}\,\underline{X} - \underline{A}^T\underline{\lambda} - \sum_{j=1}^{4} 2 K_j \zeta_j \frac{\partial \zeta_j}{\partial \underline{X}} \quad ; \quad \underline{\lambda}(t_f) = \underline{S}\,\underline{X}(t_f) . \quad (32b)$$

Since equations (32) are nonhomogeneous, a solution is assumed in the form of

$$\underline{\lambda} = \underline{P}\,\underline{X} + \underline{q} , \quad (33)$$

in which $\underline{P}$ = Riccati matrix of dimension 4 x 4; and $\underline{q}$ = a vector of dimension 4 x 1.

Substituting equation (33) into equations (32), one has

$$-\underline{\dot{P}} = \underline{P}\,\underline{A} + \underline{A}^T\underline{P} + \underline{Q} + \underline{\bar{H}} - \underline{P}\,\underline{B}\,\underline{R}^{-1}\underline{B}^T\underline{P} \quad ; \quad \underline{P}(t_f) = \underline{S} , \quad (34)$$

$$-\underline{\dot{q}} = \underline{A}^T\underline{q} - \underline{P}\,\underline{B}\,\underline{R}^{-1}\underline{B}^T\underline{q} + \underline{P}\,\underline{d} + \underline{W}_c \quad ; \quad \underline{q}(t_f) = \underline{0} , \quad (35)$$

in which $\underline{\bar{H}}$ and $\underline{W}_c$ are obtained from

$$\sum_{j=1}^{4} 2 K_j \zeta_j \frac{\partial \zeta_j}{\partial \underline{X}} = \underline{\bar{H}}\,\underline{X} + \underline{W}_c . \quad (36)$$

Since the structure is subjected to a steady state disturbance, then $t_f = \infty$ and $\underline{P}(t)$ becomes a constant gain matrix. This matrix could be obtained by integrating equation (34) backward [4] for a long time. However, $t_f = 4$ seconds was already enough to obtain the constant gain matrix. Once $\underline{P}$ has been determined, $\underline{q}$ is obtained from integrating equation (35) backward. The optimal control forces are calculated from

$$\underline{U} = -\underline{R}^{-1} \underline{B}^T \underline{P} \underline{X} - \underline{R}^{-1} \underline{B}^T \underline{q} . \tag{37}$$

Substituting equation (37) into equation (23), the closed-loop system becomes

$$\underline{\dot{X}} = (\underline{A} - \underline{B} \underline{R}^{-1} \underline{B}^T \underline{P})\underline{X} + \underline{d} - \underline{B} \underline{R}^{-1} \underline{B}^T \underline{q} \; ; \; \underline{X}(t_o) = \underline{X}_o . \tag{38}$$

A satisfactory controlled measure is mainly depending on presuming appropriate weighing matrices $\underline{S}$, $\underline{Q}$, and $\underline{R}$, as well as the weighing factors K_j. The trial process has indicated that the best controlled response is obtained when $|\underline{R}|$ = 1.21186, $|\underline{S}|$ = 1, and $|\underline{Q}|$ = 10,000, which were assumed to be diagonal matrices.

Considering zero initial conditions, displacement and acceleration at coordinate 1 are plotted in Figures 9 and 10 and compared with the uncontrolled response. It is noticed that the constraints are satisfied ensuring the safety of the structure and guaranteeing the human comfort. The response of control forces (open-loop plus closed-loop) are also shown in Figures 11 and 12, whereas the open-loop control forces are plotted in Figures 13 and 14.

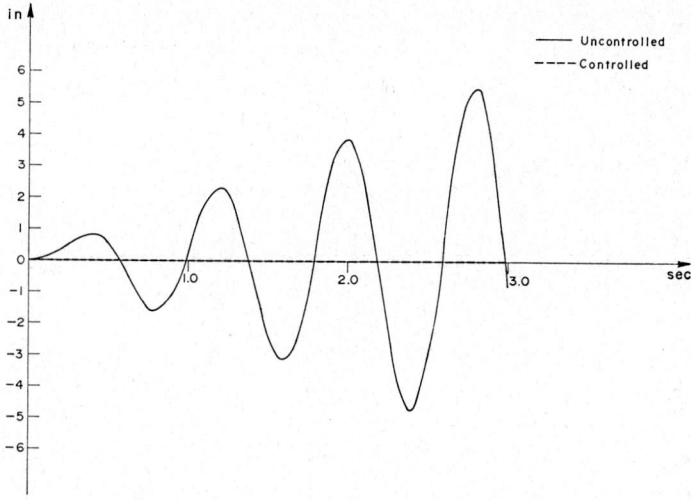

Figure 9 - Deflection Response of Second Floor

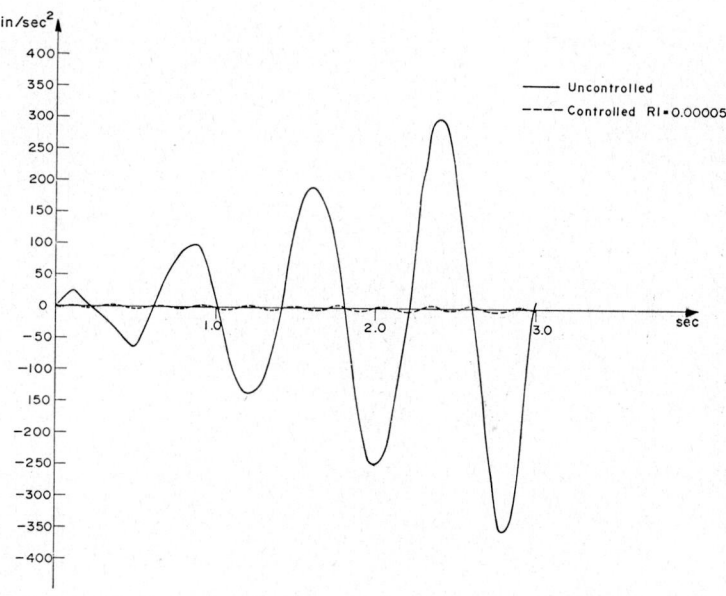

Figure 10 - Acceleration Response of Second Floor

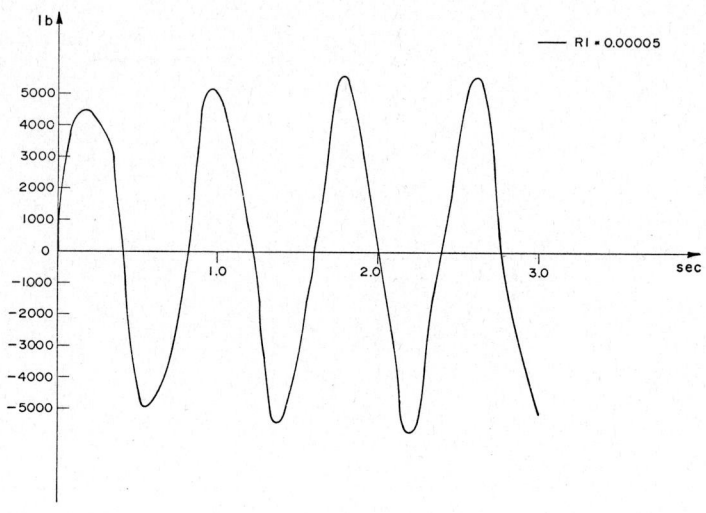

Figure 11 – Total Active Control Response at First Floor

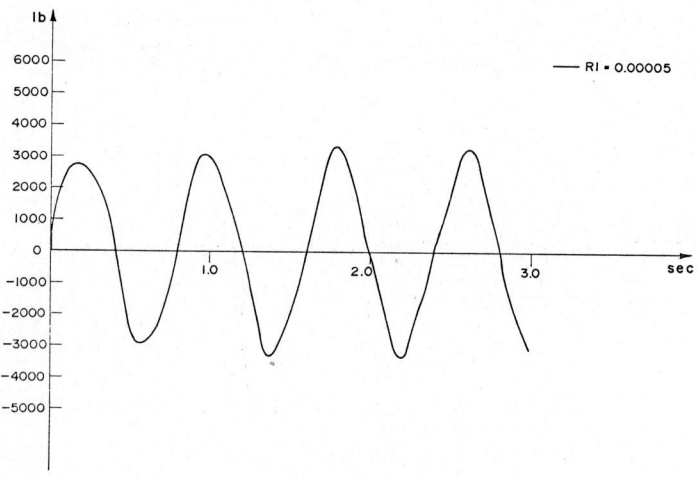

Figure 12 – Total Active Control Response at Second Floor

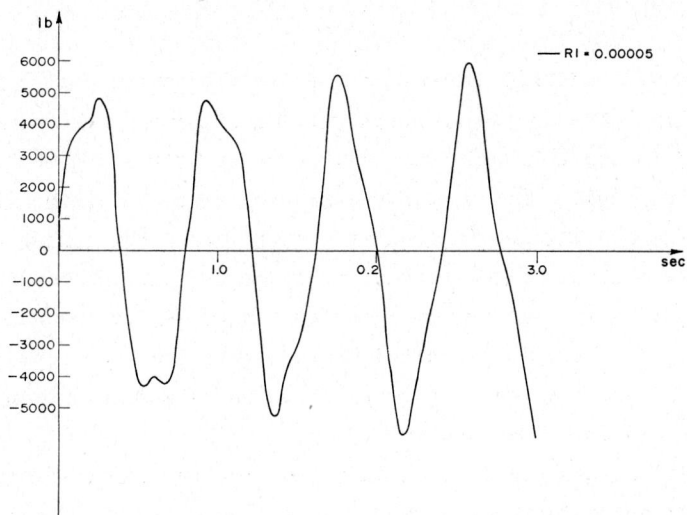

Figure 13 - External Control Response at First Floor

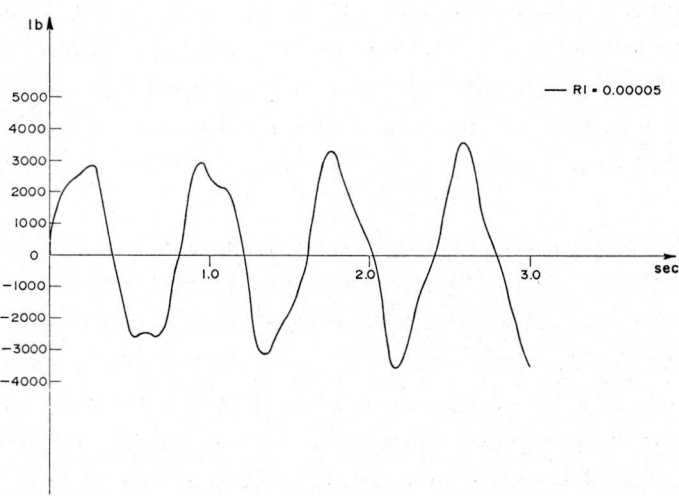

Figure 14 - External Control Response at Second Floor

4.0 CONCLUSIONS

In this paper, a general approach to be used in active control of structures was given. The approach satisfies all the requirements with regard to safety, human comfort, and cost of the control system. The problem was formulated as an optimal control problem subjected to equality and inequality constraints. For linear structural systems, using a quadratic objective function for performance indices as well as step penalty functions for inequality constraints, a closed-loop control can be iteratively obtained. The control is of the tracking type [4] which is a combination of closed-loop and open-loop control.

It has also been shown that choosing the control forces having the same character as the applied disturbance will easily provide a satisfactory controlled response. This verifies the results obtained in references [4, 17].

The proposed approach is valid as long as the disturbance is known a priori, or can be expected within a tolerance, leaving any difference between the assumed and the actual disturbance to be controlled by the closed-loop control which provides the active damping and stiffness to the structure disturbance. However, an extension including random loading is recommended for future investigation.

The application of the proposed approach to the control of high-rise buildings would be an area of worthwhile research. The authors believe this could be achieved through iteration. Starting with the design of control considering the first few dominant modes, and checking the controlled response taking into account, in turn, the neglected modes, will eventually indicate which modes should be considered in the design of the control system.

In the presented approach, since restricting costs and getting the best controlled response possible are conflicting criteria, one may not easily get an optimal solution. However, one should realize that other parameters can be introduced to the optimization problem, e.g., locations of the control forces, structural parameters, etc. In that way, greater flexibility in arriving at an optimum is obtained. Pursuing such an idea will be left for future studies.

ACKNOWLEDGEMENT

The authors are most grateful to Professor V.H. Quintana, Department of Electrical Engineering, University of Waterloo, for his valuable comments. The financial support by the National Research Council of Canada through Grant No. A7297 and a scholarship is acknowledged.

NOTATIONS

The following symbols are used in this paper:

$A_i(t)$	-	generalized coordinate of mode i
$\underline{A}$	-	open-loop matrix
$\underline{\bar{A}}$	-	open-loop matrix
$\underline{A}(t)$	-	generalized coordinate vector
$\underline{B}$	-	control matrix
$\underline{C}$	-	damping, or compatibility matrix
$\underline{D}$	-	displacement vector
$\underline{d}$	-	equivalent disturbance vector
E	-	Young's modulus
F_i	-	force
$\underline{G}$	-	equivalent damping matrix
H	-	Hamiltonian
I	-	moment of inertia
J	-	objective function

$\underline{K}$	-	stiffness matrix
$\underline{K}^*$	-	overall stiffness matrix
$\underline{K}_i$	-	weighing factor for constraint i
L	-	length of n of a member
$\underline{M}$	-	mass matrix
$\underline{M}^*$	-	overall mass matrix
m	-	mass per unit length
$\underline{P}$	-	Riccati matrix
$\underline{Q}$	-	weighing matrix
$\underline{q}$	-	vector for open-loop control
$\underline{R}$	-	weighing matrix
r	-	number of dominant modes
$\underline{S}$	-	weighing matrix
t	-	time
t_o	-	initial control time
t_f	-	final control time
$\underline{U}$	-	control force vector
$\underline{W}$	-	disturbance vector
$\underline{X}$	-	state variable vector
x_i	-	state variable no. i
$\underline{Z}$	-	equivalent mass matrix
β_i	-	weighing factor
$\underline{\psi}$	-	constant matrix
$\varepsilon 1$	-	limit on displacement
$\varepsilon 2$	-	limit on acceleration for serviceability
$\varepsilon 3$	-	limit on acceleration for human comfort
$\varepsilon 4$	-	limit on control forces
Φ	-	modal matrix
Γ	-	equivalent stiffness matrix
$\underline{\lambda}$	-	co-state variable vector
μ_i	-	weighing factor
ω	-	natural frequency
ζ_i	-	constraint no. i

REFERENCES

[1] ABDEL-ROHMAN, M., "Contributions to Automatic Active Control of Civil Engineering Structures", *Ph. D. Thesis*, Department of Civil Engineering, University of Waterloo, Waterloo, Ontario, Canada, 1978.

[2] ABDEL-ROHMAN, M. and LEIPHOLZ, H.H.E., "Active Control of Flexible Structures", *ASCE, J. of the Struc. Division*, Vol. 104, No. ST8, Proc. Paper 13964, August, 1978, pp. 1251-1266.

[3] ABDEL-ROHMAN, M. and LEIPHOLZ, H.H.E., "Structural Control by Pole Assignment Method", *ASCE, J. of the Eng. Mech. Division*, Vol. 104, No. EM5, Proc. Paper 14060, October, 1978, pp. 1159-1175.

[4] ABDEL-ROHMAN, M., QUINTANA, V.H. and LEIPHOLZ, H.H.E., "Optimal Control of Civil Engineering Structures", submitted to ASCE for publication.

[5] BRYSON, A.E. and HO, Y-C., *Applied Optimal Control*, Hemisphere Publishing Co., Washington, D.C., 1975.

[6] CLOUGH, R.A. and PENZIEN, J., *Dynamics of Structures*, McGraw-Hill, Inc., New York, 1975.

[7] FARAH, A., "Human Response: A Criterion for the Assessment of Structural Serviceability", *Ph. D. Thesis*, Department of Civil Engineering, University of Waterloo, Waterloo, Ontario, Canada, 1977.

[8] MARTIN, C.R. and SOONG, T.T., "Model Control of Multistory Structures", *ASCE, J. of the Eng. Mech. Division*, Vol. 104, No. EM4, Proc. Paper 12321, August, 1976, pp. 613-623.

[9] QUINTANA, V.H., "Some Numerical Methods for Solving Optimal Control Problems", *Ph. D. Thesis*, University of Toronto, Toronto, Ontario, Canada, 1970.

[10] OVUNC, B.A., "Dynamics of Frameworks by Continuous Mass Method", *J. of Computers and Structures*, Vol. 4, 1974, pp. 1061-1089.

[11] ROORDA, J., "Active Damping in Structures", *Report Aero No. 8*, Cranfield Institute of Technology, Cranfield, England, July, 1971.

[12] ROORDA, J., "Tendon Control in Tall Structures", *ASCE, J. of the Struc. Division*, Vol. 101, No. ST3, Proc. Paper 11168, March, 1975, pp. 505-521.

[13] SAE-UNG, S. and YAO, J.T.P., "Active Control of Building Structures", *Technical Report CE-STR-76-1*, School of Civil Engineering, Purdue University, West Lafayette, Indiana, 1976.

[14] SCHORN, G., "Feedback Control of Structures", *Ph. D. Thesis*, Department of Civil Engineering, University of Waterloo, Waterloo, Ontario, Canada, 1975.

[15] YANG, J-N. and YAO, J.T.P., "Formulation of Structural Control", *Technical Report CE-STR-74-2*, School of Civil Engineering, Purdue University, West Lafayette, Indiana, 1975.

[16] YANG, J-N., "Application of Optimal Control Theory to Civil Engineering Structures", *ASCE, J. of the Eng. Mech. Division*, Vol. 101, No. EM6, Proc. Paper 11812, December, 1975, pp. 810-838.

[17] YANG, J-N. and GIANNOPOULOS, F., "Active Tendon Control of Structures", *ASCE, J. of the Eng. Mech. Division*, Vol. 104, No. EM3, Proc. Paper 13836, June, 1978, pp. 551-568.

[18] YAO, J.T.P., "Concept of Structural Control", *ASCE, J. of the Struc. Division*, Vol. 98, No. ST7, Proc. Paper 9048, July, 1972, pp. 1567-1574.

[19] ZUK, W. and CLARK, R.H., *Kinetic Architecture*, Van Nostrand Reinhold Co., New York, New York, 1970.

[20] ZUK, W., "Kinetic Structures", *ASCE, Civil Engineering*, Vol. 39, No. 12, December, 1968, pp. 62-64.

STRUCTURAL CONTROL, H.H.E. Leipholz (ed.)
North-Holland Publishing Company & SM Publications
© IUTAM, 1980

AUTOMATIC ACTIVE CONTROL OF STRUCTURES*

M. Abdel-Rohman and H.H.E. Leipholz

Department of Civil Engineering
University of Waterloo
Waterloo, Ontario, Canada N2L 3G1

1.0 INTRODUCTION

New trends in construction technology and megastructures require new design and construction techniques. Automatic active structural control was recently recommended [14, 15] as one technique that, if used properly, would provide safety and serviceability for a megastructure, and at the same time guarantee human conform [5, 11].

Most of the applications of active control to the control of civil engineering structures [2-5, 9-13] have led to designing closed-loop control systems because of their usefulness in controlling structures subjected to uncertain and unexpected disturbances. However, because of the difficulties involved in designing such closed-loop controls for nonlinear and high order systems, active structural control was restricted to an application to simplified structural models. None of the previous studies have indicated how one should proceed when designing an optimal closed-loop control law for high order, time-varying, or even nonlinear systems. More-

---

*Published with permission of the ASCE. Originally appeared in Proc. of the ASCE, *J. of the Struc. Div.*, Vol. 106, 1980.

over, the benefits gained from using closed-loop control were not specifically illustrated by examples in the previous studies. This paper is contributing to these open questions.

The paper shows that if an optimal closed-loop control law was designed properly, however neglecting parameters which may complicate the design of the control system, then the controlled response of the actual structure, considering the previously neglected parameters and other types of disturbances, may still be acceptable. That is so, because in the closed-loop control, the control at any instant, is applied as a result of comparing the controlled response of the structures with its current response. In other words, the closed-loop control provides active damping and active stiffness [2] needed to control the structure against any unexpected disturbance and also against the influence of uncertain or even neglected structural parameters. In the application to a simple span bridge as an example, it will be shown that, indeed, closed-loop control is able to overcome the effect of actually present, but designwise neglected, parameters.

2.0 ACTIVE CONTROL OF THE SIMPLIFIED SYSTEM

Consider a one-span bridge idealized by a simply supported beam with a constant flexural rigidity EI and span L under a vehicle which is represented by a concentrated load of constant magnitude P, moving with constant speed v. The control mechanism shown in Figure 1 is used to control the bridge's vibration. Other parameters such as the normal force caused by the control mechanism, the uneveness of the bridge's deck, the inertia due to the moving loads, and different types of moving loads, will be investigated in the subsequent sections. In this section, an optimal closed-loop control will be determined assuming a smooth deck of the bridge and neglecting the normal force caused by the control mechanism as well as neglecting the inertia due to the moving load.

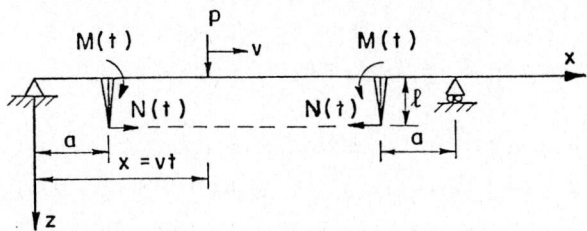

Figure 1 - Free Body Diagram

From Figure 2, one obtains the following equation of motion:

$$EI \frac{\partial^4 z}{\partial x^4} + m \frac{\partial^2 z}{\partial t^2} = P\delta(x-vt) + M(t)\delta'(x-a) - M(t)\delta'(x-L+a) \,, \quad (1)$$

in which z = lateral displacement; δ = Dirac delta function; δ' = first derivative of Dirac delta function; a = distance from the post to nearest support; and M(t) = control moment.

$$M(t) = S\ell[u(t) + \ell z'(a,t) - \ell z'(L-a,t)] \,, \quad (2)$$

in which S = stiffness of spring; ℓ = length of the post; and u(t) = active elongation or shortening of the spring, i.e., $u \lessgtr 0$.

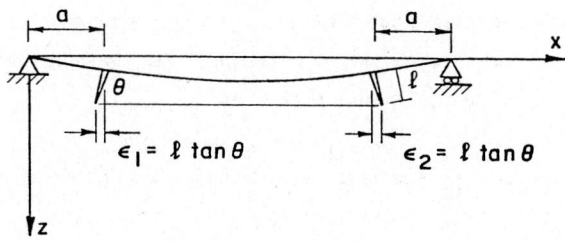

Figure 2 - Beam's Deformation

The solution of equation (1) is assumed to be

$$z(x,t) = \sum_{j=1}^{\infty} \phi_j(x) A_j(t) , \qquad (3)$$

in which A_j = generalized coordinate of mode j; and $\phi_j(x)$ = $\sin(j\pi x/L)$ in order to satisfy the boundary conditions.

Applying an integral transformation and after some manipulations, one arrives at the following mode equations:

$$\ddot{A}_1(t) + \omega_1^2 A_1(t) = \frac{2P}{mL} \sin\Omega_1 t - B_1[u(t)+\ell z'(a,t) - \ell z'(L-a,t)] , \qquad (4)$$

$$\ddot{A}_2(t) + \omega_2^2 A_2(t) = \frac{2P}{mL} \sin\Omega_2 t , \qquad (5)$$

$$\ddot{A}_3(t) + \omega_3^2 A_3(t) = \frac{2P}{mL} \sin\Omega_3 t - B_3[u(t)+\ell z'(a,t) - \ell z(L-z,t)] , \qquad (6)$$
$$\vdots$$

in which $\Omega_j = j\pi v/L$; $B_j = (4S\ell j\pi)/(LmL)\cos(j\pi a/L)$ for j odd; and $B_j = 0$ for j even.

The natural frequency ω_j is obtained from

$$\omega_j^2 = \frac{j^4 \pi^4}{L^4} \frac{EI}{m}$$

in which m = mass of the bridge per unit length; and E = Young's modulus.

Considering only the first three modes and since the even modes are uncontrolled, the simplified system [4] will be represented by the following equations:

$$\ddot{A}_1(t) + \omega_1^2 A_1(t) = \frac{2P}{mL} \sin\Omega_1 t - B_1[u(t) + C_1 A_1(t) + C_3 A_3(t)] , \qquad (8)$$

$$\ddot{A}_3(t) + \omega_3^2 A_3(t) = \frac{2P}{mL} \sin\Omega_3 t - B_3[(t) + C_1 A_1(t) + C_3 A_3(t)] , \qquad (9)$$

in which $C_j = (2\ell j\pi/L)\cos(j\pi a/L)$ for j odd; and $C_j = 0$ for j even.

Denoting $A_1(t)$, $\dot{A}_1(t)$, $A_3(t)$, and $\dot{A}_3(t)$ by the state variables $x_1(t)$, $x_2(t)$, $x_3(t)$, and $x_4(t)$, respectively, equations (8) and (9) can be expressed in the matrix state variable form by

$$\dot{\underline{X}}(t) = \underline{A}_c \underline{X}(t) + \underline{B}_c \underline{U}(t) + \underline{d}(t) , \tag{10}$$

where $\underline{A}_c$, $\underline{B}_c$, and $\underline{d}(t)$ are, respectively, given by

$$\underline{A}_c = \underline{A} - \underline{B}\,\underline{B}_j\,\underline{C}_j , \tag{11}$$

$$\underline{B}_c = - \underline{B}\,\underline{B}_j , \tag{12}$$

$$\underline{d}(t) = \underline{B}\,\underline{r}(t) . \tag{13}$$

In the above equations, $\underline{A}$, $\underline{B}$, $\underline{B}_j$, $\underline{r}(t)$, and $\underline{C}_j$ are given by

$$\underline{A} = \begin{vmatrix} 0 & 1 & 0 & 0 \\ -\omega_1^2 & 0 & 0 & 0 \\ 0 & 0 & 0 & 1 \\ 0 & 0 & -\omega_3^2 & 0 \end{vmatrix} , \quad \underline{B} = \begin{vmatrix} 0 & 0 \\ 1 & 0 \\ 0 & 0 \\ 0 & 1 \end{vmatrix} , \quad \underline{B}_j = \begin{vmatrix} B_1 \\ B_3 \end{vmatrix} ,$$

$$r(t) = \frac{2P}{mL} \begin{vmatrix} \sin \Omega_1 t \\ \sin \Omega_3 t \end{vmatrix} , \quad \underline{C}_j = \begin{vmatrix} C_1 & 0 & C_3 & 0 \end{vmatrix} ,$$

where the numerical values of the parameters in these variables can be found in Appendix II.

The rate of change of the post's rotation is supposed to be measured as an output which may be expressed as

$$\underline{Y}(t) = \underline{C}\,\underline{X} = \begin{vmatrix} 0 & \frac{C_1}{\ell} & 0 & \frac{C_3}{\ell} \end{vmatrix} \underline{X} . \tag{14}$$

The closed-loop system is represented by the block diagram shown in Figure 3. It is assumed that the design of the observer, which has to estimate the state variables from the available output, has previously been done [3, 7]. The task of the designer, then, is to design the optimal closed-loop control law, satisfying the following quadratic objective function:

$$J = \frac{1}{2} \underline{Y}^T(t_f) \underline{S}_o \underline{Y}(t_f) + \frac{1}{2} \int_{t_o}^{t_f} (\underline{Y}^T \underline{Q}_o \underline{Y} + \underline{U}^T \underline{R}\, \underline{U}) dt , \qquad (15)$$

in which t_o = initial control time; t_f = final control time; and $\underline{S}_o$, $\underline{Q}_o$, and $\underline{R}$ are weighing matrices, all of dimension 1 x 1.

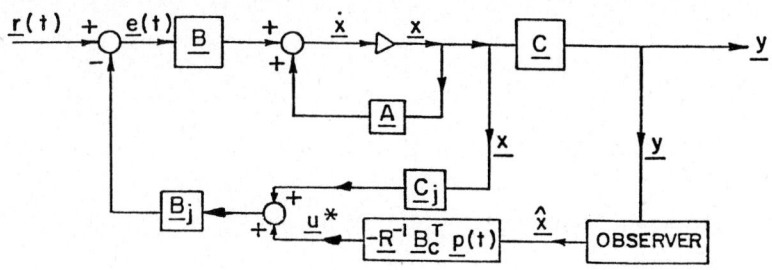

Figure 3 - Block Diagram

The optimal control problem, which consists in minimizing equation (15), subjected to conditions in equation (14) and equation (10) while considering an unspecified disturbance ($\underline{d}(t) = \underline{0}$), is called the regulator problem [4]. The optimal control law of this problem is found [7] to be

$$\underline{U}^*(t) = - \underline{R}^{-1} \underline{B}_c^T \underline{P}(t) \underline{X}(t) , \qquad (16)$$

in which $\underline{U}^*(t)$ = optimal control signal for any type of disturbance; and $\underline{P}(t)$ = the Riccati matrix of dimension 4 x 4.

Riccati matrix is obtained from solving the following differential equation:

$$-\underline{\dot{P}}(t) = \underline{P}\,\underline{A}_c + \underline{A}_c^T\underline{P} - \underline{P}\,\underline{B}_c\underline{R}^{-1}\underline{B}_c^T\underline{P} + \underline{C}^T\underline{Q}_o\underline{C} \;; \tag{17}$$

$$\underline{P}(t_f) = \underline{C}^T\underline{S}_o\underline{C} ,$$

in which $\underline{Q}_o = 1000$; $\underline{S}_o = 100$; and $\underline{R} = 1$ provide a desirable trajectory.

The controlled system is obtained by substituting equation (16) into equation (10). In order to check the controlled response of the system, the disturbance $\underline{d}(t)$ is considered in equation (10).

The controlled system is then expressed as

$$\underline{\dot{X}}(t) = [\underline{A}_c - \underline{B}_c\underline{R}^{-1}\underline{B}_c^T\underline{P}(t)]\underline{X}(t) + \underline{d}(t) . \tag{18}$$

Deflection and acceleration at mid-span of the bridge are calculated, respectively, from

$$z(L/2,t) = x_1(t) - x_3(t) , \tag{19}$$

$$\ddot{z}(L/2,t) = [x_2(t) - x_2(t-h) - x_4(t) + x_4(t-h)]/h , \tag{20}$$

in which h = time step of integration taken to be 0.01 [2].

Deflection at mid-span, and acceleration at mid-span for the simplified structure are shown, by means of a broken line, in Figures 4 and 5, respectively. The ram's displacement of the control device is also shown in Figure 6.

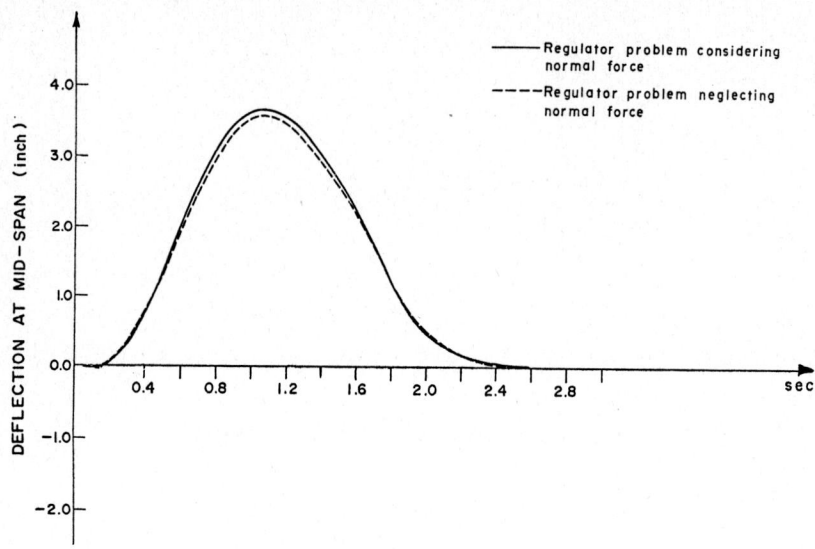

Figure 4 - *Active Controlled Response of Deflection*

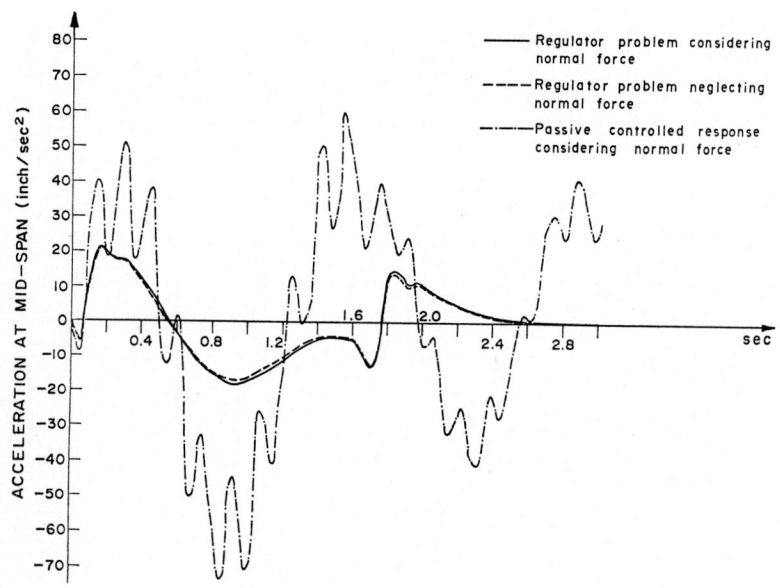

Figure 5 - *Active Controlled Response of Acceleration by Regulator Control*

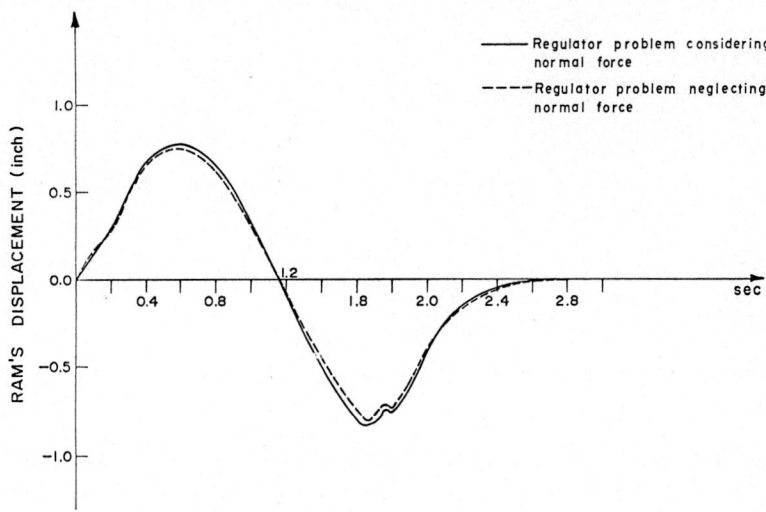

Figure 6 - Active Control Response by Regulator Control

3.0 ACTIVE CONTROL RESPONSE OF THE ACTUAL SYSTEM

In the previous section, the bridge was represented by an idealized-smooth simple beam loaded by a constant load which is moving with constant speed. The effect of a normal force caused by the control mechanism, the effect of unevenness in the bridge deck, and the effect of the inertia coming from the moving load have been neglected. Also, only one type of disturbance was tested whereas in reality the bridge is subjected to many kinds of moving loads. In the following sections, the above mentioned factors will be considered using the same control system designed previously with neglecting these effects. In order to possess good information about the influence of these factors, the effect of each factor will be studied separately.

4.0 NORMAL FORCE EFFECT

The control mechanism used in the previous section introduces the normal force N(t) between the posts as shown in Figure 1. The equation of motion for the portion between the posts becomes

$$EI \frac{\partial^4 z}{\partial x^4} + N(t) \frac{\partial^2 z}{\partial x^2} + m \frac{\partial^2 z}{\partial t^2} = P\delta(x-vt) + M(t)\delta'(x-a)$$

$$- M(t)\delta'(x-L+a) , \qquad (21)$$

in which it is assumed that N(t), at any instant, is constant between the posts.

An approximate solution for equation (21) is assumed to be in the form of equation (3), in which the function $\sin(j\pi x/L)$ is chosen to satisfy the boundary conditions as well as the compatibility conditions $x = a$ and $x = L-a$. The normal force N(t) is evaluated for three modes as

$$N(t) = S[C_1 A_1(t) + C_3 A_3(t) + U(t)] . \qquad (22)$$

Substituting equation (22) into equation (21), and applying the sine integral transformation, one obtains after some manipulations, for three modes, the following equations:

$$\ddot{A}_1 + \omega_1^2 A_1 - D_1[C_1 A_1 + C_3 A_3 + U]A_1 = \frac{2P}{mL} \sin \Omega_1 t$$

$$- B_1[C_1 A_1 + C_3 A_3 + U] , \qquad (23)$$

$$\ddot{A}_2 + \omega_2^2 A_2 - D_2[C_1 A_1 + C_3 A_3 + U]A_2 = \frac{2P}{mL} \sin \Omega_2 t , \qquad (24)$$

$$\ddot{A}_3 + \omega_3^2 A_3 - D_3[C_1 A_1 + C_3 A_3 + U]A_3 = \frac{2P}{mL} \sin \Omega_3 t$$

$$- B_3[C_1 A_1 + C_3 A_3 + U] , \qquad (25)$$

in which D_j is given by

$$D_j = \frac{j^2 \pi^2 S}{mL^2} \,. \tag{26}$$

Investigating equations (23), (24) and (25), one finds that these equations are simultaneous nonlinear differential equations. The control $U(t)$ in the above equations is substituted by the one found for the simplified system, i.e., equation (16). The active control responses of deflection and acceleration at mid-span for this case as compared with the case of neglecting the normal force are shown, respectively, in Figures 4 and 5. The ram's displacement in the two cases is also plotted in Figure 6. It is concluded that the effect of the normal force on the controlled response is trivial and hardly noticeable. This result shows that in some cases, like this one, representing a nonlinear system by an approximated linear system, in order to ease designing the control system, may give satisfactory results.

5.0 EFFECT OF UNEVENESS IN THE BRIDGE DECK

It has been assumed, previously, that the lateral displacement of the vehicle is the same as that of the bridge. In such case, the vehicle may be represented as a constant force (unsprung weight) equal to the vehicle's weight, which has been considered in the previous sections. However, for an accurate analysis, one has to view the vehicle as an independent dynamic system, which may be represented by an undamped one-degree-of-freedom system (sprung mass), as shown in Figure 7. For an uneveness expressed by $r(x)$, the load applied on the bridge is evaluated from Figure 7 as

$$P(t) = K(v_1(t) - z(\bar{x},t) - r(\bar{x})] + m_v g \,, \tag{27}$$

in which K = stiffness of the tires; m_v = mass of the vehicle; $\bar{x}$ = position of the vehicle from the left support; g = acceleration of gravity; and v_1 = lateral displacement of the vehicle.

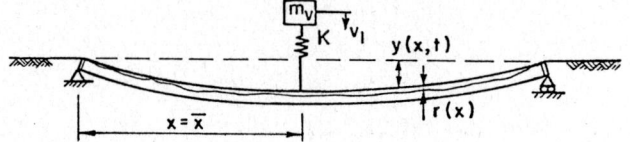

Figure 7 - Deformed Irregular Bridge's Deck

The equations of lateral motion of the vehicle and the bridge are, respectively, given by

$$m_v \ddot{v}_1(t) + K[v_1(t) - r(\overline{x}) - z(x,t)] = 0 , \qquad (28)$$

$$EI \frac{\partial^4 z}{\partial x^4} + m \frac{\partial^2 z}{\partial t^2} = K[v_1(t) - r(\overline{x}) - z(\overline{x},t)] \delta(x-\overline{x}) + m_v g \delta(x-\overline{x})$$

$$+ M(t)\delta'(x-a) - M(t)\delta'(x-L+a) . \qquad (29)$$

A solution is assumed in the form of equation (3) and by applying the sine integral transformation, one obtains, for three modes, the following relationships:

$$m_v \ddot{v}_1 + K[v_1 - r(\overline{x}) - A_1 \sin\Omega_1 t - A_2 \sin\Omega_2 t - A_3 \sin\Omega_3 t] = 0 , \qquad (30)$$

$$\ddot{A}_1 + \omega_1^2 A_1 = \left(\frac{2m_v g}{mL} - \frac{2m_v}{mL} \ddot{v}_1\right) \sin\Omega_1 t - B_1 [C_1 A_1 + C_3 A_3 + U] , \qquad (31)$$

$$\ddot{A}_2 + \omega_2^2 A_2 = \left(\frac{2m_v g}{mL} - \frac{2m_v}{mL} \ddot{v}_1\right) \sin\Omega_2 t , \qquad (32)$$

$$\ddot{A}_3 + \omega_3^2 A_3 = \left(\frac{2m_v g}{mL} - \frac{2m_v}{mL} \ddot{v}_1\right) \sin\Omega_3 t - B_3 [C_1 A_1 + C_3 A_3 + U] , \qquad (33)$$

in which $\overline{x}$ = vt = position of the vehicle measured from the left support.

Although the unevenness of the bridge deck is generally a random quantity, a simple profile will here be assumed as shown in Figure 8. This profile can be expressed as

$$r(x) = h_1\left(1-\cos\frac{2\pi x}{\ell_1}\right), \qquad (34)$$

in which $h_1 = 0.25$ inch (6.35 mm), and $\ell_1 = 4$ feet (1.22 m).

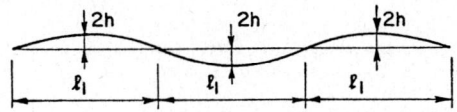

Figure 8 - Uneveness Profile of Bridge's Deck

Substituting the control found previously into equations (31) and (33), one is able to determine the controlled response of the bridge in the presence of the uneveness. The controlled responses of deflection and acceleration at mid-span as compared with the cases of neglecting uneveness, either the moving load is sprung or not, are shown, respectively, in Figures 9 and 10. Also, the ram's response is shown in Figure 11. It is concluded that there is not much difference in the deflection response between the cases of considering or neglecting the uneveness. Although the difference in acceleration response is remarkable, the controller has dampened the vibration, and the reduction in the acceleration response for the controlled response, Figure 10, as compared with the uncontrolled response, Figure 12, was significant. Yet, from the human comfort point of view, the *acceleration response* of the controlled structure is not acceptable. Therefore, it is suggested to redesign the controller considering the uneveness in order to satisfy the requirement for the human comfort. The conclusion is, that the uneveness in the bridge deck is such an important parameter that it should always be taken into account for the control system, even in the design of the bridge itself [6, 8].

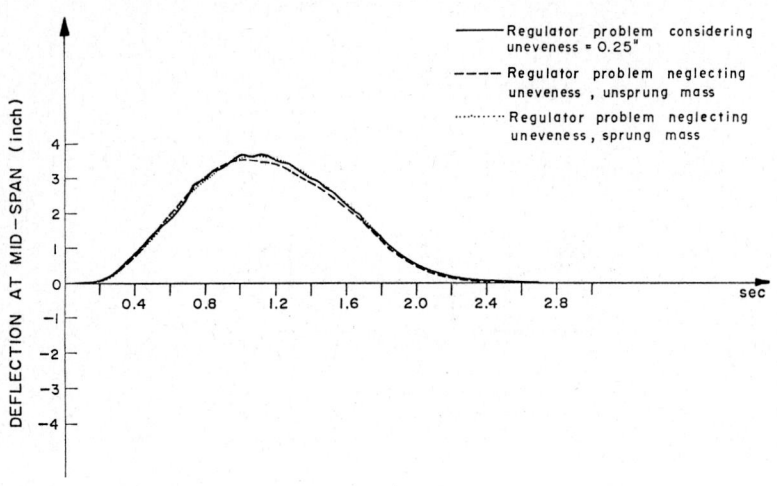

Figure 9 - Controlled Response of Deflection by Regulator Control, $h = 0.25$ in.

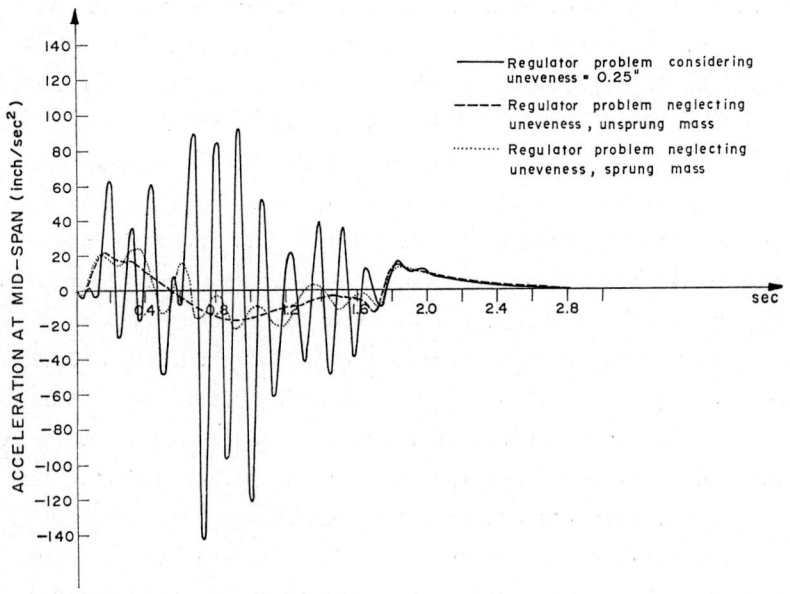

Figure 10 - Controlled Response of Acceleration by Regulator Control, $h = 0.25$ in.

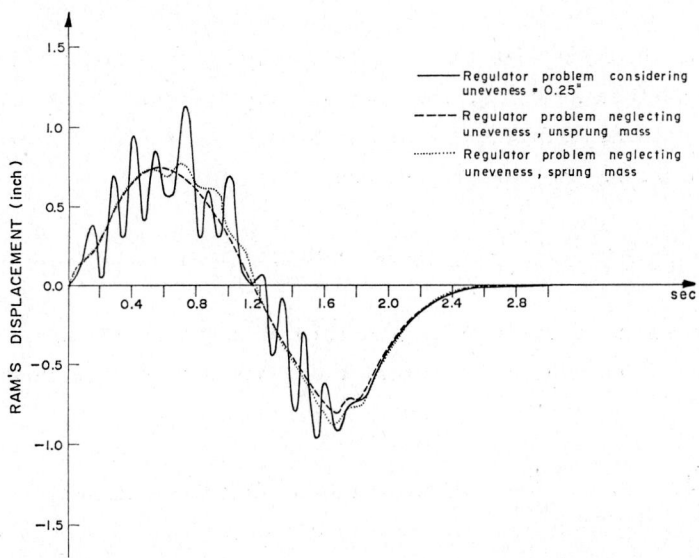

Figure 11 - *Active Control Response by Regulator Control, $h = 0.25$ in.*

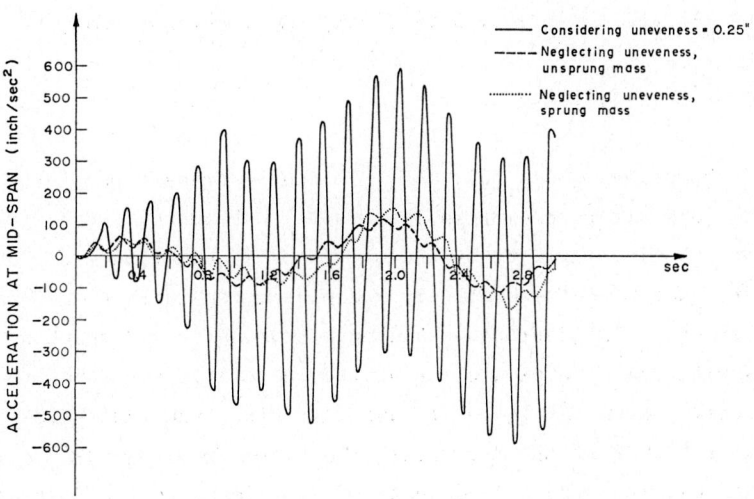

Figure 12 - *Uncontrolled Response of Acceleration, $h = 0.25$ in.*

6.0 INERTIA EFFECT DUE TO THE MOVING LOAD

Herein, it is assumed that the vehicle is always in contact with the bridge surface such that the lateral displacement of the vehicle equals the lateral displacement of the bridge. The equation of motion considering the inertia is then

$$EI \frac{\partial^4 z}{\partial x^4} + m \frac{\partial^2 z}{\partial t^2} = \left(P - \frac{P}{g}\frac{\partial^2 z}{\partial t^2}\right)\delta(x-vt) + M(t)\delta'(x-a) - M(t)\delta'(x-L+a) . \tag{35}$$

Assuming a solution in the form of equation (3) and applying the sine integral transformation, one obtains for three modes

$$\ddot{A}_1 + \omega_1^2 A_1 = \frac{2P}{mL}\sin\Omega_1 t - \frac{2P}{mLg}\sin\Omega_1 t[\ddot{A}_1\sin\Omega_1 t + \ddot{A}_2\sin\Omega_2 t + \ddot{A}_3\sin\Omega_3 t]$$

$$- B_1[C_1 A_1 + C_3 A_3 + U] , \tag{36}$$

$$\ddot{A}_2 + \omega_2^2 A_2 = \frac{2P}{ML}\sin\Omega_2 t - \frac{2P}{mLg}\sin\Omega_2 t[\ddot{A}_1\sin\Omega_1 t + \ddot{A}_2\sin\Omega_2 t + \ddot{A}_3\sin\Omega_3 t] , \tag{37}$$

$$\ddot{A}_3 + \omega_3^2 A_3 = \frac{2P}{mL}\sin\Omega_3 t - \frac{2P}{mLg}\sin\Omega_3 t[\ddot{A}_1\sin\Omega_1 t + \ddot{A}_2\sin\Omega_2 t + \ddot{A}_3\sin\Omega_3 t]$$

$$- B_3[C_1 A_1 + C_3 A_3 + U] . \tag{38}$$

Equations (36), (37) and (38) can be transformed into the state form after some manipulations [1]. From this state form, and after substituting U(t) by the one found in equation (16), one is able to determine the controlled response, considering the inertia effect. The deflection and acceleration at mid-span, as compared with the case of neglecting the inertia effect, are, respectively, shown in Figures 13 and 14. Also, the ram's displacement response of the two cases is plotted in Figure 15. From these figures, it is concluded that the difference in the response, when including or neglecting the inertia effect, is negligible.

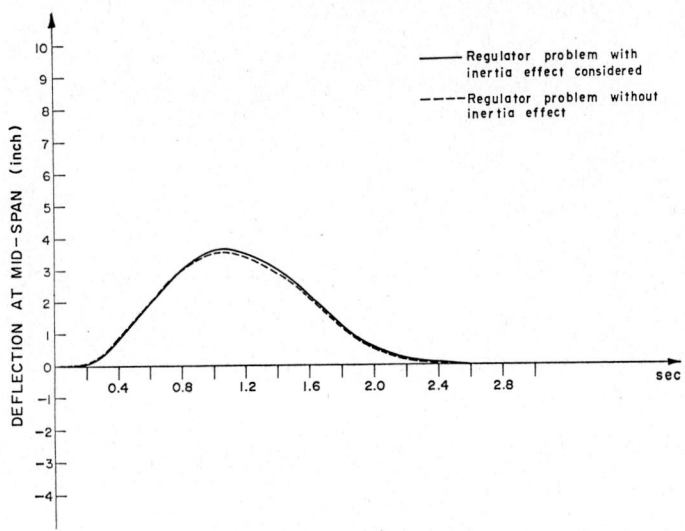

Figure 13 - Controlled Response of Deflection by Regulator Control, P = 20 Kips

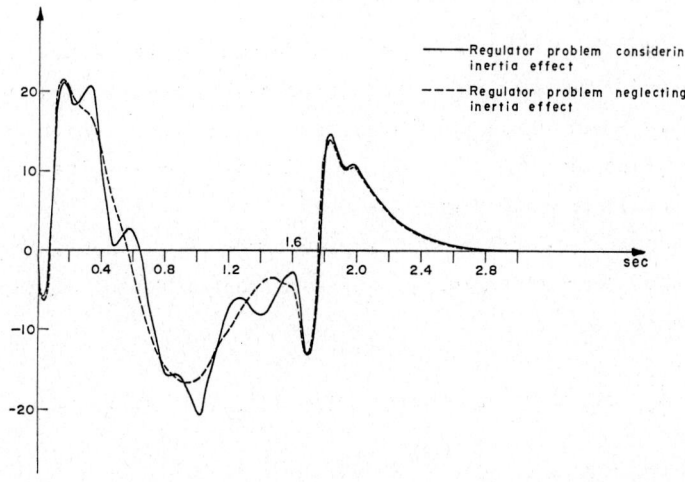

Figure 14 - Controlled Response of Acceleration by Regulator Contro, P = 20 Kips

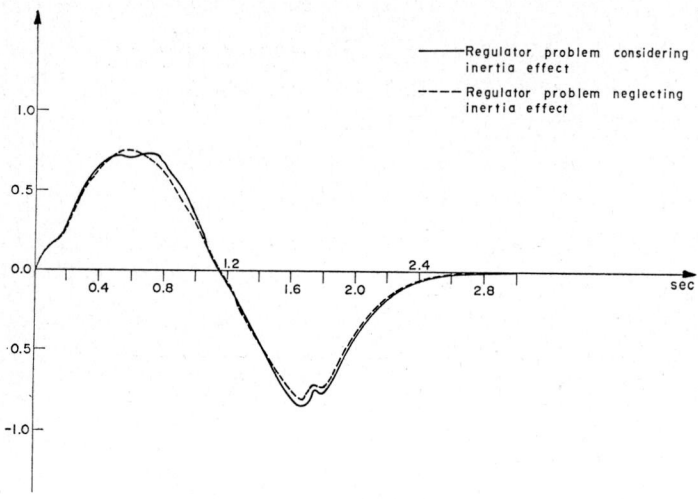

Figure 15 - Active Control Response by Regulator Control, P = 20 Kips

7.0 STUDY OF THE EFFECT OF VARIOUS MOVING LOADS

In reality, a bridge is subjected to different types of moving loads. In order to make this study as close to reality and applicable in practice as possible, some moving loads which may act on the structure, in the presence of the control system designed previously, will be investigated.

Considering first a pulsating force, $P \sin \Omega_p t$ which is moving with constant speed v, the equation of motion of the bridge becomes

$$EI \frac{\partial^4 z}{\partial x^4} + m \frac{\partial^2 z}{\partial t^2} = P \sin \Omega_p t \delta(x-vt) + M(t)\delta'(x-a) - M(t)\delta'(x-L+a) , \quad (39)$$

in which Ω_p = frequency of the pulsating force.

Resonance occurs if the frequency Ω_p equals the frequency of any vibrational mode of the bridge. Assuming that Ω_p is the same as the first mode frequency, ω_1, will enable one to investigate the controlled response of the bridge in a resonance situation. The controlled response of deflection and acceleration at mid-span, as compared with the uncontrolled response, is shown in Figures 16 and 17, respectively. The ram's displacement is also plotted in Figure 18. It is shown that although the effect of resonance is pronounced for the uncontrolled response, the designed control could stabilize the structure against the resonance effect. This is ascribed to the active damping and active stiffness introduced to the controlled structure.

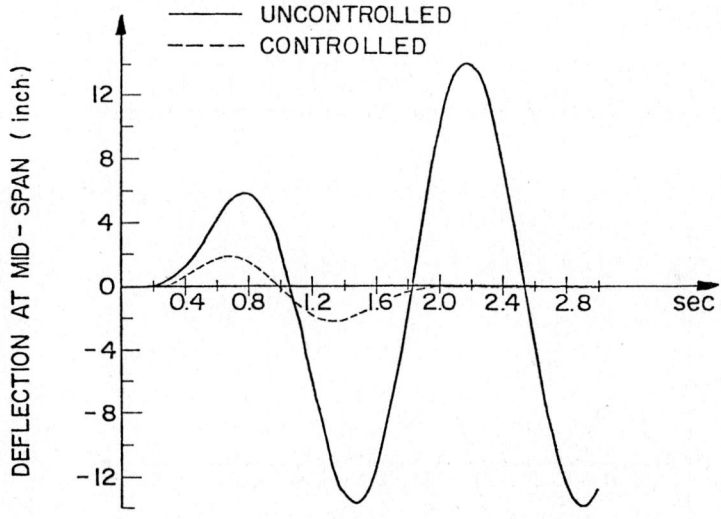

Figure 16 - *Deflection Response Due to Harmonic Loading*

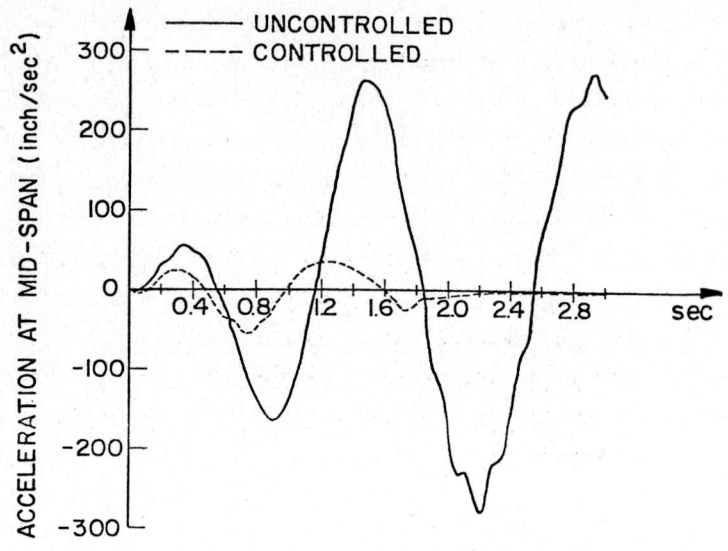

Figure 17 - *Acceleration Response Due to Harmonic Loading*

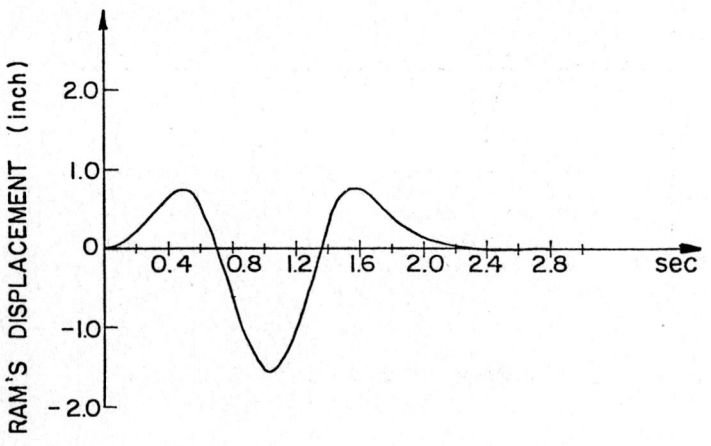

Figure 18 - *Active Control Response Due to Harmonic Loading*

Secondly, consider a train of 300 feet (91.5 m) long which is moving with a constant speed 60 feet/second (18.3 m/s). The train then takes 5 seconds to traverse the bridge, and the final control time t_f should be greater than 5 seconds in order to accomplish the control function properly. However, it was found that [1] even for t_f = 6 seconds the Riccati matrix, $\underline{P}(t)$, of equation (17) is the same as the one of t_f = 3 seconds.

The equations of motion of the forced response are divided into three portions, as shown in Figure 19. Each portion lasts one third of the five seconds. These equations for arrival, passage, and departure, are respectively given by [8].

$$EI \frac{\partial^4 z}{\partial x^4} + m \frac{\partial^2 z}{\partial t^2} = q[1-H(x-vt)]+M(t)\delta'(x-a)-M(t)\delta'(x-L+a) \quad , \tag{40}$$

$$EI \frac{\partial^4 z}{\partial x^4} + m \frac{\partial^2 z}{\partial t^2} = q+M(t)\delta'(x-a)-M(t)\delta'(x-L+a) \quad , \tag{41}$$

$$EI \frac{\partial^4 z}{x} + m \frac{\partial^2 z}{\partial t^2} = qH(x-vt)+M(t)\delta'(x-a)-M(t)\delta'(x-L+a) \quad , \tag{42}$$

in which H(x) = Heaviside unit function, which is unity for $x \geq 0$ and zero for $x < 0$; q = 50 lb/ft. (729.5 N/m) = magnitude of the applied load per unit length.

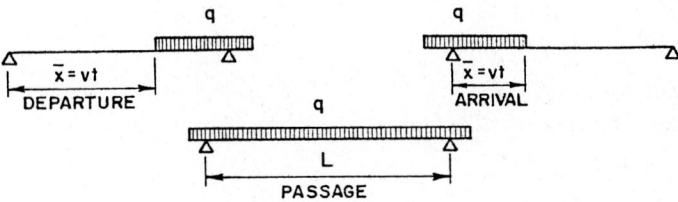

Figure 19 - Bridge Under Moving Train Load

Applying the integral transformation to each of equations (40), (41) and (42), one is able to determine the modes' equations [1]. The controlled response can be calculated after substituting the control U(t) from equation (16). A comparison between the controlled and uncontrolled response of deflection and acceleration at mid-span is shown in Figures 20 and 21. The ram's response is also plotted in Figure 22. It is concluded that the controller, in this case too, could dampen the vibration and reduce the transient responses of deflection and acceleration at mid-span of the bridge.

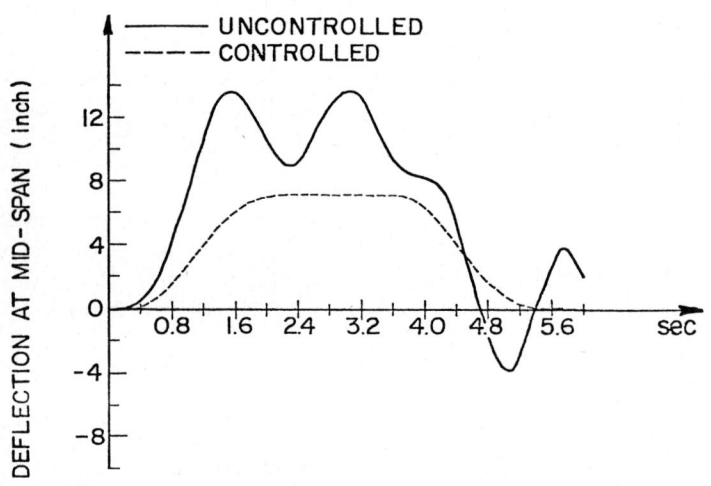

Figure 20 - *Deflection Response Due to Moving Train Load*

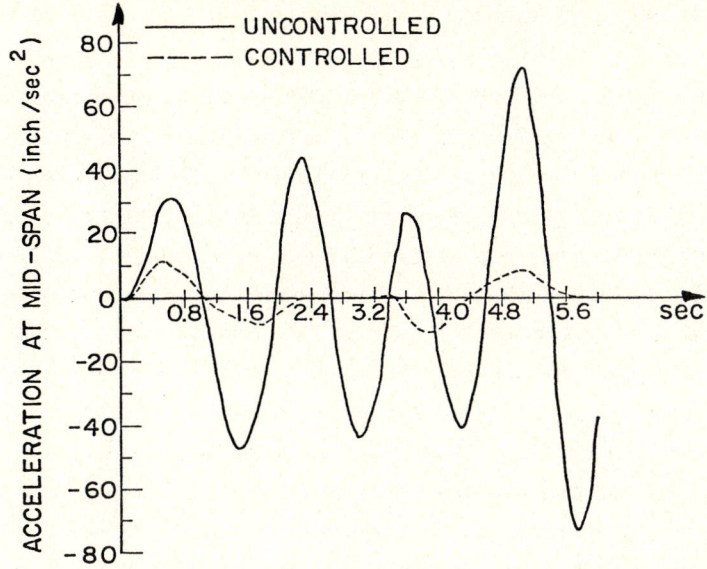

Figure 21 - *Acceleration Response Due to Moving Train Load*

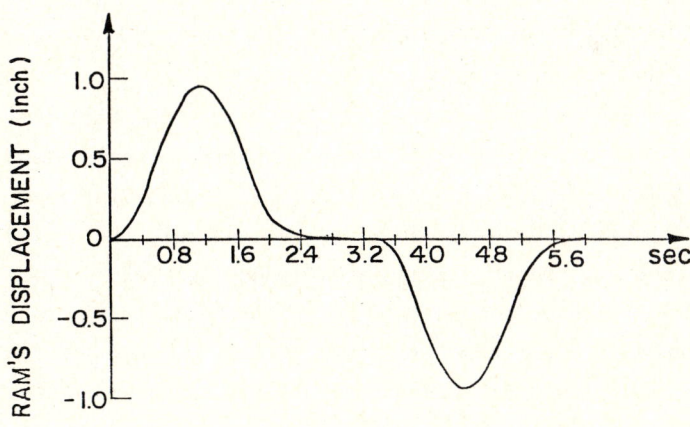

Figure 22 - *Active Control Response Due to Moving Load*

8.0 SUMMARY AND CONCLUSIONS

In this paper, some of the benefits gained from using a closed-loop control in controlling flexible civil engineering structures have been pointed out. For structures subjected to unexpected disturbances and uncertain structural parameters, it becomes obvious that the closed-loop active control is a powerful tool to guarantee the safety of these structures under very general conditions, since closed-loop active control provides the active damping and stiffness needed by any kind of current response of the structure. Realizing these facts, one is also led to the idea to design a closed-loop control system for a simplified structural model, neglecting those factors which complicate the design. The efficiency of the designed control system may then be checked afterwards by considering the factors which have been neglected previously. The design may be revised for the very rare cases in which the controlled response, when considering a neglected factor, is not acceptable. Such a procedure would be very helpful if the system considered is a high-order or nonlinear one, because of the difficulties involved in designing a closed-loop control for those kinds of systems.

In the presented example, it has been shown that the optimal closed-loop control designed for the simplified system was for instance efficient in the presence of the nonlinearity caused by the normal force generated by the control, the inertia of the moving mass, and different kinds of moving loads. This efficiency is obviously due to the fact that the closed-loop control depends on the current response of the system which may then include not only unexpected disturbances, but also the effect of neglected parameters. However, things to not always work that well. There is, for example, a remarkable difference in the controlled response of the acceleration when considering or neglecting the unevenness of the bridge's deck. Then, there is a need for redesigning the control system in order to meet the human comfort requirements, although the controlled response was very small as compared with the uncontrolled response of acceleration.

ACKNOWLEDGEMENT

The first author was supported financially with a postdoctoral fellowship from the National Research Council of Canada. The second author acknowledges financial support by the National Research Council of Canada under Grant No. A7397.

NOTATIONS

The following symbols are used in this paper:

$\underline{A}$	- uncontrolled system matrix
$\underline{A}_c$	- uncontrolled total system matrix
a	- distance from the post to nearest support
a_1	- acceleration
$\underline{B}$	- control coefficient matrix
$\underline{B}_c$	- total control coefficient matrix
B_j	- variable for mode j
$\underline{C}$	- gain matrix
C_j	- variable for mode j
$\underline{d}(t)$	- disturbance vector
E	- Young's Modulus
$\underline{e}$	- activating error vector
g	- acceleration of gravity
h	- time step of integration
h_1	- unevenness parameter
I	- moment of inertia of a cross-section
L	- span length

ℓ	- length of the post
ℓ_1	- uneveness parameter
$M(t)$	- control moment
m	- mass per unit length
$N(t)$	- normal force
$\underline{P}(t)$	- Riccati matrix
P	- magnitude of the moving load
$\underline{Q}_o$	- weighing matrix
q	- uniform moving load
$\underline{R}$	- weighing matrix
$\underline{r}(t)$	- reference vector
$r(x)$	- uneveness profile
S	- spring constant
$\underline{S}_o$	- weighing matrix
t	- time
t_o	- initial time
t_f	- final time
$\underline{U}$	- control vector
$\underline{U}^*$	- optimal control vector
v	- speed
$\underline{X}$	- state vector
$\underline{Y}$	- output vector
$z(x,t)$	- deflection at section x and time t
ω_j	- natural frequency
Ω_j	- speed parameter
Ω_p	- frequency of a moving load

APPENDIX I - REFERENCES

[1] ABDEL-ROHMAN, M., "Contributions to Automatic Active Control of Civil Engineering Structures", *Ph. D. Thesis*, Department of Civil Engineering, University of Waterloo, Waterloo, Ontario, Canada, 1979.

[2] ABDEL-ROHMAN, M. and LEIPHOLZ, H.H.E., "Active Control of Flexible Structures", *ASCE, J. of the Struc. Division*, Vol. 104, No. ST8, August, 1978, pp. 1251-1266.

[3]   ABDEL-ROHMAN, M. and LEIPHOLZ, H.H.E., "Structural Control by Pole Assignment Method", *ASCE, J. of the Eng. Mech. Division*, Vol. 104, No. EM5, October, 1978, pp. 1159-1175.

[4]   ABDEL-ROHMAN, M., QUINTANA, V.H. and LEIPHOLZ, H.H.E., "Optimal Control of Civil Engineering Structures", paper submitted to ASCE for publication.

[5]   ABDEL-ROHMAN, M. and LEIPHOLZ, H.H.E., "A General Approach to Active Structural Control", paper submitted to ASCE for publication.

[6]   BIGGS, J.M., *Introduction to Structural Dynamics*, McGraw-Hill Book Co., New York, 1964.

[7]   BRYSON, A.E. and HO, Y.C., *Applied Optimal Control*, Hemisphere Publishing Corp., Washington, D.C., 1975.

[8]   FRÝBA, L., *Vibration of Solids and Structures under Moving Loads*, Academia Publishing House of the Czechoslovak Academy of Sciences, Prague, 1972.

[9]   MARTIN, C.R. and SOONG, T.T., "Modal Control of Multi-Story Structures", *ASCE, J. of the Eng. Mech. Division*, Vol. 102, No. EM4, August, 1976, pp. 613-623.

[10] ROORDA, J., "Tendon Control in Tall Structures", *ASCE, J. of the Struc. Division*, Vol. 101, No. ST3, March, 1975, pp. 503-521.

[11] SAE-UNG, S. and YAO, J.T.P., "Active Control of Building Structures", *Technical Report No. CE-STR-76-1*, Department of Civil Engineering, Purdue University, West Lafayette, Indiana, U.S.A., 1976.

[12] YANG, J-N., "Application of Optimal Control Theory to Civil Engineering Structures", *ASCE, J. of the Eng. Mech. Division*, Vol. 101, No. EM6, December, 1975, pp. 819-838.

[13] YANG, J-N. and GIANNOPOULOS, F., "Active Tendon Control of Structures", *ASCE, J. of the Eng. Mech. Division*, Vol. 104, No. EM3, June, 1978, pp. 551-568.

[14] YAO, J.T.P., "Concept of Structural Control", *ASCE, J. of the Struc. Division*, Vol. 98, No. ST7, August, 1972, pp. 1567-1574.

[15] ZUK, W. and CLARK, R.H., *Kinetic Architecture*, Van Nostrand, Reinhold Co., 1970.

APPENDIX II - NUMERICAL DATA

L = 100 ft. (30.5 m)   a = 10 ft. (3.05 m)
EI = 12 x 10^{10} lb.in^2 ℓ = 3 ft. (0.915 m)
m = 0.3 lb.sec^2/in^2 (2.07 x 10^{-3}Ns^2/mm^2)
v = 60 ft./sec. (18.3 m/s)   P = 20 kips (89000 N)
S = 62.5 kips/inch (10949.8 N/mm)
t_f = 3 seconds

The variables involved in this paper have been computed from these parameters and are given by

ω_1^2 = 18.79   ; B_1 = 63.159   ;
ω_2^2 = 300.05   ; B_2 = 0.0   ;
ω_3^2 = 1522.02   ; B_3 = 102.51   ;

C_1 = 0.1773   ; Ω_1 = 1.885   ;
C_2 = 0.0   ; Ω_2 = 3.770   ;
C_3 = 0.2878   ; Ω_3 = 5.655   .

DESIGN OF REDUCED-ORDER OBSERVERS FOR
STRUCTURAL CONTROL SYSTEMS

M. Abdel-Rohman and H.H.E. Leipholz

Department of Civil Engineering
University of Waterloo
Waterloo, Ontario, Canada

V.H. Quintana

Department of Electrical Engineering
University of Waterloo
Waterloo, Ontario, Canada

1. INTRODUCTION

In the design of an optimal state feedback control system for a flexible structure, it is usually assumed that all system's state variables are measurable and available for control. However, the availability of measurable state variables is not the usual case in practice, whereas the measurable output variables are usually a linear combination of the state variables. In a previous paper [1], it has been shown that an output feedback is subjected to restrictions and difficulties when used for determining a stable and desirable solution. It has been recommended in reference [1] to use an observer [5, 6] in order to estimate the state feedback control system. However, the observer mentioned in [1] is a full-order observer, i.e., its order is the same as the plant's order. For high-order plants, a full-order observer becomes expensive and not easy to implement. A reduction in order is therefore desirable.

This paper presents a review of the design of reduced-order observers, used for the state feedback control of structural systems. It is shown that a reduced-order observer provides the control system with the information needed to implement the state feedback control law. It is also shown that the order of the observer can be made dependent on the number of independent output measurements. An application is presented for the control of a simple span bridge, controlled by active tendons. The response due to a constant force moving with constant speed is used as a check for the control system's performance.

2. DESIGN OF OBSERVERS

An observer is nothing more than an estimator of the system's state variables using the available outputs. Therefore, it is needed whenever the control engineer is interested in controlling a structure by state feedback control, whether the control is based on pole assignment [1], or optimal control methods [2].

The dynamic equation of an observer is described by

$$Z(t) = FZ + GY + Hv \quad , \quad Z(0) , \qquad (1)$$

in which Z = observer's state vector of dimension $p \times 1$; Y = plant's output vector of dimension $m \times 1$; v = the estimated control available of dimension $r \times 1$; F, G and H are gain matrices of appropriate dimensions to be determined.

The estimated control is given by:

$$v(t) = - \underline{D} \, Y - \underline{E} \, Z , \qquad (2)$$

where $\underline{D}$ and $\underline{E}$ are gain matrices to be determined.

Let the plant, state feedback control, and output equations of the structure described by the block diagram of Figure 1 be given, respectively, by

$$\dot{X} = AX + BU \quad , \quad X(0) \, , \tag{3}$$

$$U = -KX \, , \tag{4}$$

$$Y = CX \, , \tag{5}$$

where X = plant's state vector of dimension n × 1; U = state feedback control vector of dimension r × 1; $\underline{A}$, $\underline{B}$, $\underline{K}$ and $\underline{C}$ are given matrices of appropriate dimensions. It is assumed that $\underline{C}$ is a full rank matrix, i.e., Rank ($\underline{C}$) = m.

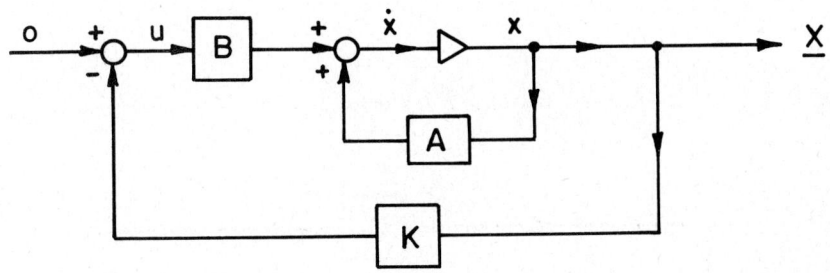

Figure 1 - State Feedback Control System

The observer should be designed in such a way that it is a fast-response stable system. The observer state vector $\underline{Z}$ must be an approximation to a linear combination of the plant's states. That is to say, the observer design is based on

$$Z = TX \, , \tag{6}$$

where T is a transformation matrix of dimension p × n. Matrix T is the identity matrix in the case of a full-order observer [1].

In order to ensure the asymptotic stability of the observer, all the eigenvalues of the observer matrix $\underline{F}$ must be in the left half of the s-plane. Furthermore, in order to minimize the time delay in $\underline{Z}$ approaching $\underline{T}\,\underline{X}$, the eigenvalues of the observer must be placed to the left of the eigenvalues of the closed-loop system of Figure 1.

The dynamic error of the observer can be written as

$$\dot{Z} - T\dot{X} = FZ + GCX + Hv - TAX - TBU . \qquad (7)$$

Assuming v equal to U, as in Figure 2, and letting $H \triangleq T B$, $F T \triangleq G C - T A$, equation (7) becomes

$$\dot{Z} - T\dot{X} = F(Z - T X) , \qquad (8)$$

where $(Z - T X)$ is the observer's error.

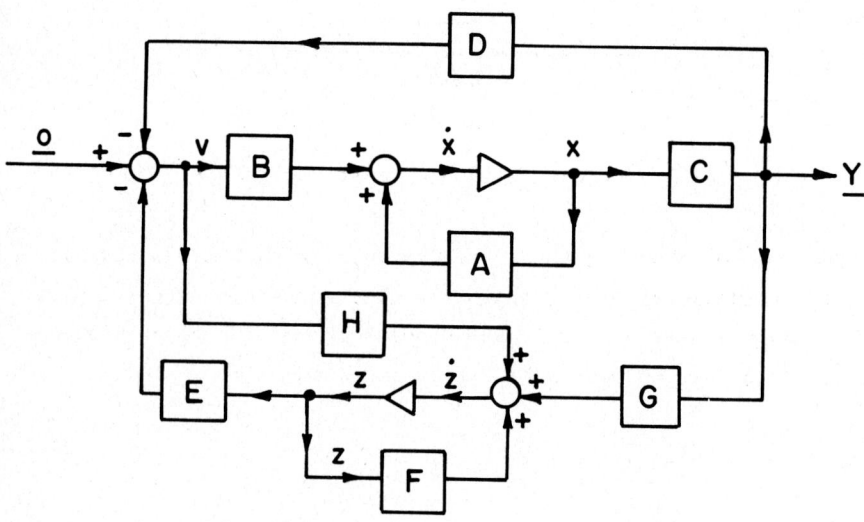

Figure 2 - State Feedback Control Using an Observer

From equation (8), it is evident that for $\underline{Z}$ to asymptotically approach $\underline{T}\,\underline{X}$ as $t \to \infty$, it is required that $\underline{F}$ has all its eigenvalues in the left-hand-side of the s-plane. Therefore, the conditions that must be satisfied in designing an observer are as follows:

$$H = T B , \tag{9}$$

$$K = D C + E T , \tag{10}$$

$$-F T = G C - T A , \tag{11}$$

$$F = \text{asymptotically stable matrix, i.e., } \mathrm{Re}\delta_i(F) < 0 , \tag{12}$$

where $\delta_i(F)$ denotes the ith eigenvalues of $\underline{F}$.

It can be observed that for a full-order observer, for which $\underline{T} = \underline{I}$, the above conditions are the same as the ones developed in reference [1]. The design problem now becomes the determination of gain matrices, $\underline{T}$, $\underline{H}$, $\underline{D}$, $\underline{E}$ and $\underline{G}$, assuming the gain matrix $\underline{F}$ having eigenvalues farther to the left of the eigenvalues of the closed-loop matrix $\hat{\underline{A}} = \underline{A} - \underline{B}\,\underline{K}$. In order to ease the design procedure for a reduced-order observer, it is recommended to assume $\underline{F}$ as a diagonal matrix whose elements (eigenvalues) are negative real numbers [11].

3. DETERMINATION OF GAIN MATRICES

To determine the transformation matrix T, equation (11) is used. This equation can be written in the following form:

$$\begin{bmatrix} T_1^r \\ T_2^r \\ \vdots \\ T_i^r \\ \vdots \\ T_p^r \end{bmatrix} \underline{A} - \begin{bmatrix} \delta_1 & & & 0 \\ & \delta_2 & & \\ & & \ddots & \\ & & & \delta_i \\ & & & & \ddots \\ 0 & & & & & \delta_p \end{bmatrix} \begin{bmatrix} T_1^r \\ T_2^r \\ \vdots \\ T_i^r \\ \vdots \\ T_p^r \end{bmatrix} = \begin{bmatrix} G_1^r \\ G_2^r \\ \vdots \\ G_i^r \\ \vdots \\ G_p^r \end{bmatrix} \underline{C} \tag{13}$$

in which T_i^r = the ith row of matrix $\underline{T}$; δ_i = the ith eigenvalue of matrix $\underline{F}$; and G_i^r = the ith row of matrix $\underline{G}$.

The ith row of equation (13) yields

$$T_i^r \underline{A} - \delta_i T_i^r = G_i^r \underline{C} , \qquad (14)$$

for which the T_i^r row can be obtained as

$$T_i^r = G_i^r \underline{C} (\underline{A} - \delta_i \underline{I})^{-1} . \qquad (15)$$

Using equation (15), one can easily evaluate T_i^r provided that G_i^r is known. It is convenient to assume that each element of G_i^T is proportional to the determinant of $(\underline{A} - \delta_i \underline{I})$. For high-order systems, for which an evaluation of the inverse and the determinant of $(\underline{A} - \delta_i \underline{I})$ becomes tedious, one may use Leverrier's algorithm [10]. This algorithm is summarized in Appendix I.

Determining the gain matrix $\underline{T}$ enables one to find the gain matrix $\underline{H}$ from equation (9). To evaluate the gain matrices $\underline{D}$ and $\underline{E}$, equation (10) is written as

$$\underline{K} = [D \ E] \begin{bmatrix} C \\ T \end{bmatrix} . \qquad (16)$$

Since the dimension of $\underline{C}$ is m × n, and the dimension of T is p × n, it becomes obvious that letting m + p = n will lead to a unique solution for the gain matrices $\underline{D}$ and $\underline{E}$. This condition can in fact be used to determine the order of the observer, which may be much less that the order of structural system. In fact, if m = n, there is no need at all for an observer since the state variables $\underline{X}$ can be determined directly from the output equation, equation (5), by

$$\underline{X} = \underline{C}^{-1} \underline{Y} , \qquad (17)$$

provided that rank (C) = n as assumed above.

Therefore, if p = n - m is satisfied, the gain matrices
$\underline{D}$ and $\underline{E}$ can be evaluated from

$$[D\ E] = K \begin{bmatrix} C \\ T \end{bmatrix}^{-1} . \qquad (18)$$

However, if p > n - m, one has to arbitrarily assign values to some elements of $\underline{D}$ and $\underline{E}$ such that the number of equations (r × n) equals the remainder of the elements of $\underline{D}$ and $\underline{E}$. The case for which p < n - m represents a numerical problem since the number of equations are greater than the number of elements of $\underline{D}$ and $\underline{E}$, i.e., rx(m+p). A solution can be obtained if it would be possible to find (n-m-p) columns of the matrix $[C\ T]^T$ to be a linear combination of the rest of the columns in this matrix [12].

In all of the previous cases, the design of the observer should be checked by considering the performance of the overall system, as that shown in Figure 2. The eigenvalues of the overall system must be the same as the eigenvalues of the closed-loop structural system, i.e., (A - BK), plus the eigenvalues of the observer. This fact is proven in Appendix II.

4. NUMERICAL EXAMPLE

Consider a simply supported beam with a constant flexural rigidity EI and span L under a concentrated force of constant magnitude P moving with a constant speed v. The control mechanism of Figure 3 is used to control the beam's vibration. Studying the free body diagram of Figure 4, the following equation of motion is obtained:

$$EI \frac{\partial^4 Z}{\partial x^4} + m \frac{\partial^2 Z}{\partial t^2} = P\delta(x-vt) + M(t)\delta'(x-a) - M(t)\delta'(x-L+a) , \qquad (20)$$

in which Z = lateral displacement; δ = Dirac delta function; δ' = first derivative of Dirac delta function; a = distance from the post to nearest support; and M(t) = control moment.

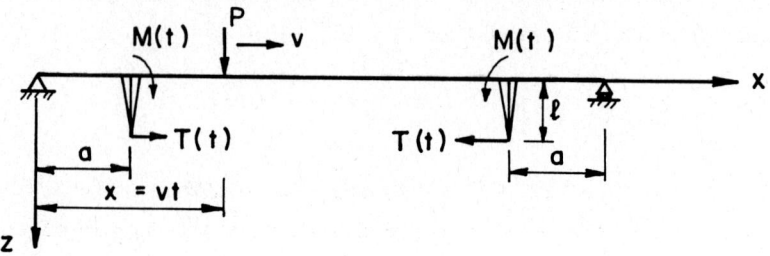

Figure 3 - *Control Mechanism*

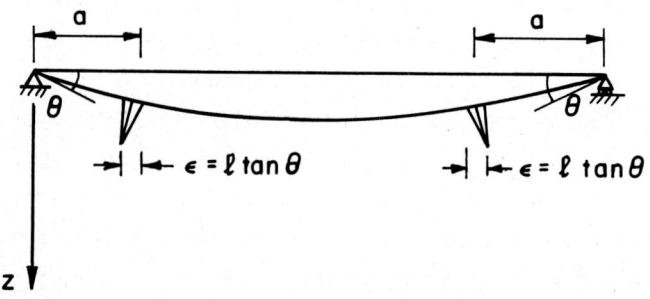

Figure 4 - *Beam's Deformation*

The control moment is given by

$$M(t) = S\ell[u(t) + \ell z'(a,t) - \ell z'(L-a,t)] , \qquad (21)$$

where S = stiffness of the spring; ℓ = length of the post; and u(t) = active elongation or shortening of the spring.

The first term in equation (21) represents the active control moment and the other two terms represent the passive control moment generated by the beam's deformation. The solution of equation (20) is assumed to be

$$Z(x,t) = \sum_{J=1}^{\infty} \phi_J(x) A_J(t) , \qquad (22)$$

where $A_J(t)$ = generalized coordinate of mode J; and the function $\phi_J(x)$ is chosen so that it satisfies the boundary conditions, that is $Z(0,t) = Z(L,t) = Z''(0,t) = Z''(L,t) = 0$. A possible choice for this function is for instance $\phi_J = \sin(J\pi x/L)$. Applying an integral transformation, after some manipulation, one arrives at the following mode equations:

$$\ddot{A}_1(t) + \omega_1^2 A_1(t) = \frac{2P}{mL} \sin\Omega_1 t - B_1[u(t) + \ell Z'(a,t) - \ell Z'(L-a,t)] , \qquad (23)$$

$$\ddot{A}_2(t) + \omega_2^2 A_2(t) = \frac{2P}{mL} \sin\Omega_2 t , \qquad (24)$$

$$\ddot{A}_3(t) + \omega_3^2 A_3(t) = \frac{2P}{mL} \sin\Omega_3 t - B_3[u(t) + \ell Z'(a,t) - \ell Z'(L-a,t)] , \qquad (25)$$

in which $B_J = (4S\ell J\pi/LmL)\cos(J\pi a/L)$ for odd modes and zero for the even modes; $\Omega_J = J\pi v/L$; and $\omega_J^2 = (\pi J/L)^4 EI/m$.

The above formulae indicate that by locating the posts symmetrically with respect to the centre of the beam, each even mode is uncontrolled. Considering three modes only, equations (23) and (25) become

$$\ddot{A}_1(t) + \omega_1^2 A_1(t) = \frac{2P}{mL} \sin\Omega_1 t - B_1[u(t) + C_1 A_1(t) + C_3 A_3(t)] , \qquad (26)$$

$$\ddot{A}_3(t) + \omega_3^2 A_3(t) = \frac{2P}{mL} \sin\Omega_3 t - B_3[u(t) + C_1 A_1(t) + C_3 A_3(t)] , \qquad (27)$$

in which $C_j = (2\ell j\pi/L)\cos(j\pi a/L)$ for odd modes and zero for the even modes.

Denominating the state variables as $x_1(t) = A_1(t)$, $x_2(t) = \dot{A}_1(t)$, $x_3(t) = A_3(t)$, and $x_4(t) = \dot{A}_3(t)$ one may write equations (25), (27) in a matrix form as

$$\underline{\dot{X}}(t) = \underline{A}_c \underline{X} + \underline{B}_c \underline{U} + \underline{d}(t) , \qquad (28)$$

in which $\underline{A}_c$, $\underline{B}_c$, and $\underline{d}(t)$ are respectively given by

$$\underline{A}_c = \begin{bmatrix} 0 & 1 & 0 & 0 \\ (-\omega_1^2 - B_1 C_1) & 0 & -B_1 C_3 & 0 \\ 0 & 0 & 0 & 1 \\ -B_3 C_1 & 0 & (-\omega_3^2 - B_3 C_3) & 0 \end{bmatrix}, \quad (29)$$

$$\underline{B}_c = \begin{bmatrix} 0 & -B_1 & 0 & -B_3 \end{bmatrix}^T, \quad (30)$$

$$\underline{d}(t) = \frac{2P}{mL} \begin{bmatrix} 0 & \sin\Omega_1 t & 0 & \sin\Omega_3 t \end{bmatrix}^T. \quad (31)$$

It is assumed that the control $\underline{U}$ was found by the optimal control method [2] and is given by

$$U = -R^{-1} B_c^T P X = -[K_1, K_2, K_3, K_4] \underline{X}. \quad (32)$$

It is also assumed that only one measurement is available, which represents the rate of change of the post's rotation and is expressed by

$$Y = C X = \begin{bmatrix} 0, \frac{C_1}{\ell}, 0, \frac{C_3}{\ell} \end{bmatrix} X. \quad (33)$$

The numerical data for all of the above variables are given in Appendix III. The objective is now to design a reduced-order observer for the above completely observable and controllable system [1]. Since n = 4 and m = 1, it is easier to consider p = 3 in order to have a unique solution for $\underline{E}$ and $\underline{D}$ of equation (10). Therefore the dimension of T becomes 3 × 4 and the dimension of $\underline{F}$ is 3 × 3. In order to assume the matrix $\underline{F}$, one has first to determine the eigenvalues of the closed-loop structural system. The closed-loop matrix $\hat{\underline{A}}$ is given by

$$\hat{\underline{A}} = \underline{A}_C - \underline{B}_C \underline{K} \,. \tag{34}$$

Considering the data given in Appendix III, the eigenvalues of the closed-loop structural system are obtained and they are tabulated in Table 1. The eigenvalues of the observer can then be assumed as in Table 2, in order to quickly reduce the observer's error to zero.

Table 1 - Eigenvalues of $\hat{A}$

Real	Imaginary
-5.7558	1.8333
-5.7558	-1.8333
-12.12	33.4525
-12.12	-33.4525

Table 2 - Chosen Eigenvalues of F

Real	Imaginary
-8.7558	0.0
-10.7558	0.0
-19.12	0.0

Using equation (15), one can determine the ith row of $\underline{T}$ after evaluating $(A_C - \delta_i I)$ and hence its inverse and determinant. The matrix T was found to be

$$T = 100 \begin{bmatrix} 29.9978 & 8.2727 & 159.2521 & 0.8018 \\ 31.3699 & 10.4104 & 217.7549 & 1.3876 \\ 40.15619 & 21.3297 & 592.3814 & 7.0503 \end{bmatrix}. \tag{35}$$

The row G_i^T was chosen to be the same as the determinant of $(A_C - \delta_i I)$. The matrix G is then given by

$$G = 10^4 [17.3321 \quad 24.2543 \quad 75.8009]^T \,. \tag{36}$$

The gain matrices D and E are obtained from equation (18), i.e.,

$$[D\ E] = K\begin{bmatrix}C\\T\end{bmatrix}^{-1} = [-2.63524 \quad -0.696\times10^{-7} \quad 0.804\times10^{-7} \quad -0.180\times10^{-7}]\ , \quad (37)$$

in which the $D = -2.63524$ and the other three elements represent E.

The calculations are checked by finding the eigenvalues of the overall system shown in Figure 5. The overall system can be expressed as

$$\dot{X} = A_C X + B_C(-DCX - EZ) + d(t) \quad , \quad X(0)\ , \quad (38)$$

$$\dot{Z} = F\ Z + GCX + H(-DCX - EZ) \quad , \quad Z(0)\ . \quad (39)$$

Therefore, the closed-loop matrix of the overall system is given by

$$\hat{A}_C = \begin{bmatrix}(A_C - B_C DC) & -B_C E\\(GC - HDC) & (F - HE)\end{bmatrix}\ . \quad (40)$$

The eigenvalues of matrix $\hat{A}_C$ were found to be as tabulated in Table 3.

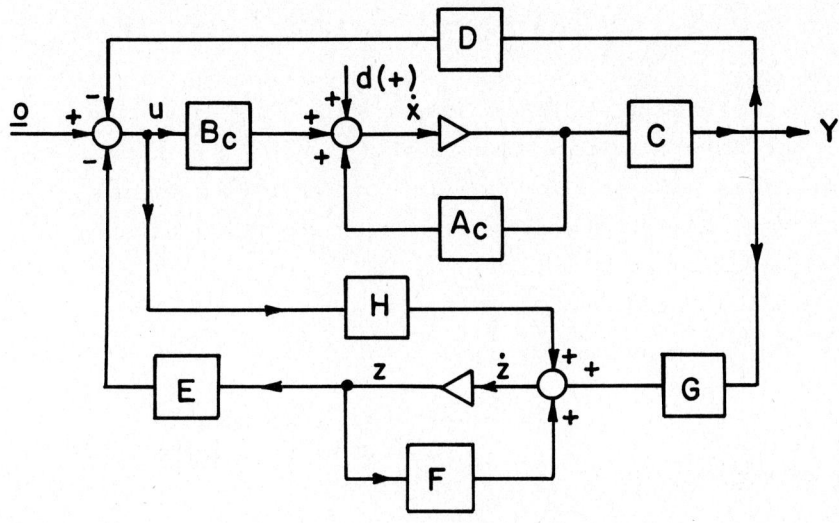

Figure 5 - Simulation Diagram of the Example

Table 3 - Eigenvalues of $\hat{A}_C$

Real	Imaginary
-5.7558	1.8333
-5.7558	-1.8333
12.12	33.4525
12.12	-33.4525
-8.7558	0.0
-10.7558	0.0
-19.12	0.0

The controlled response of deflection and acceleration at mid-span are calculated from

$$Z\left(\frac{1}{2}, t\right) = x_1(t) - x_3(t) , \tag{41}$$

$$Z\left(\frac{1}{2}, t\right) = [x_2(t) - x_4(t) - x_2(t-h) + x_4(t-h)]/h ,\quad (42)$$

where h = time step of integration, taken to be 0.01 sec.

The controlled responses of deflection and acceleration at mid-span of the beam for optimal state feedback control as compared with using a reduced order observer are shown in Figures 6 and 7, respectively.

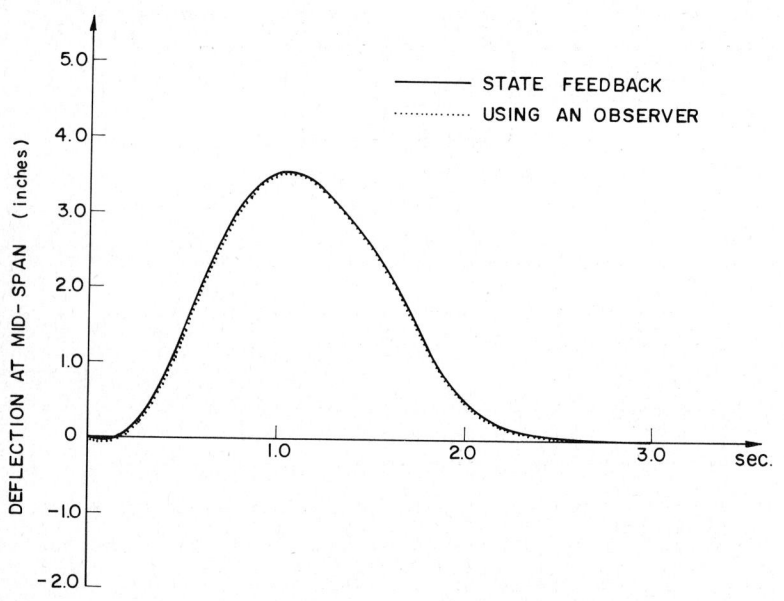

Figure 6 - Deflection Response at Mid-Span

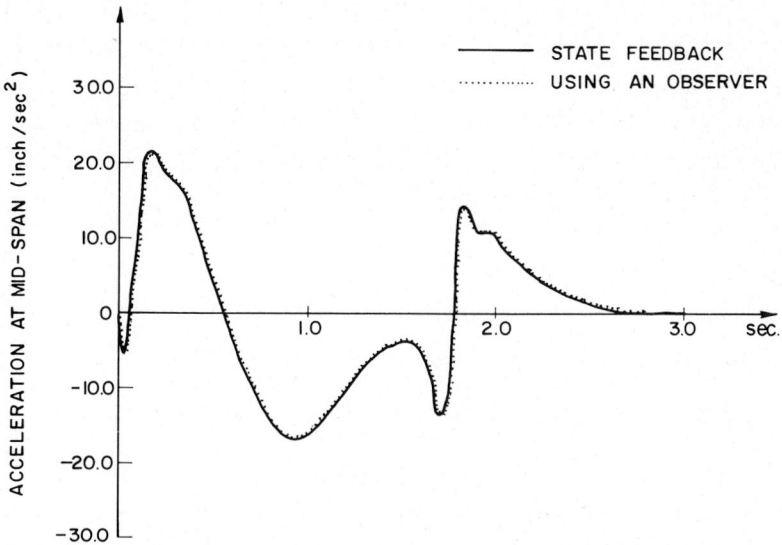

Figure 7 - *Acceleration Response at Mid-Span*

5. CONCLUSIONS

This paper has shown that in most cases, designing a full-order observer is not necessary. One should take full advantage of the available measurements of control. The order of the observer, P, can be made to be the difference between the order of the system, n, and the number of output measurements, m.

In such a case, there is no difficulty in designing the gain matrices $\underline{E}$ and $\underline{D}$, since their evaluation becomes unique, after having determined $\underline{T}$.

The most relevant result of this paper is that a reduced-order observer can be used to implement the state control of structural systems, whether the state control law may be found by the pole assignment method [1], model control method [7], or optimal control methods [2, 3, 4, 8, 9].

6. ACKNOWLEDGEMENT

The research of this paper was supported financially through a post-doctoral fellowship offered to the first writer by the National Research Council of Canada. Also, the financial support by the National Research Council of Canada under Grants No. A7397 and A8835 is acknowledged.

NOTATIONS

The following symbols are used in this paper:

$\underline{A}$ = open-loop matrix of dimension n×n
$A_j(t)$ = generalized coordinate of mode j
$\underline{A}_c$ = a matrix of dimension 4×4
$\hat{\underline{A}}$ = closed-loop matrix of dimension 4×4
$\hat{\underline{A}}_c$ = closed-loop overall matrix of dimension 7×7
a = distance between the post and nearest support
B = open-loop system matrix of dimension n×r
B_c = a matrix of dimension 4×1
B_j = coefficient for mode j relating to control
$\underline{C}$ = measurability matrix of dimension m×n
C_j = coefficient for mode j relating to output
$\underline{D}$ = gain matrix in the observer dimension r×m
$\underline{d}(t)$ = disturbance vector of dimension 4×1
$\underline{E}$ = gain matrix in the observer of dimension r×p
E = Young's modulus
$\underline{F}$ = observer's matrix of dimension p×p
$\underline{G}$ = gain matrix of the observer of dimension p×m
$\underline{G}_i^T$ = row no. i of $\underline{G}$
$\underline{H}$ = gain matrix in the observer of dimension p×r
h = time step of integration
I = moment of inertia of a section

$\underline{K}$	=	state feedback gain matrix of dimension r×n
L	=	span of the bridge
ℓ	=	length of the post
M(t)	=	control moment
m	=	mass per unit length
n	=	number of independent measurements
P	=	magnitude of the moving load
$\underline{P}$(t)	=	Riccati matrix of dimension 4×4
p	=	order of the observer
$\underline{R}$	=	weighing matrix of dimension (x)
S	=	spring constant
$\underline{T}$	=	transformation matrix of the observer of dimension p×n
$\underline{T}_i^T$	=	row no. i of the matrix T
t	=	time
$\underline{U}$	=	control vector
$\underline{V}$	=	estimated control vector
v	=	speed of the moving load
$\underline{X}$	=	state vector of the plant of dimension n×1
x	=	distance from the left support
$\underline{Y}$	=	output vector of dimension m×1
$\underline{Z}$	=	state vector of the observer of dimension p×1
z(x,t)	=	deflection at section x and time t
δ_i	=	eigenvalue no. i
$\delta(x)$	=	Dirac delta function
$\delta'(x)$	=	first derivative of Dirac delta function
$\phi_j(x)$	=	mode shape no. j
ω_j	=	natural frequency of mode j
Ω_j	=	speed parameter of mode j

REFERENCES

[1] ABDEL-ROHMAN, M. and LEIPHOLZ, H.H.E., "Structural Control by Pole Assignment Method", *ASCE, J. of the Eng. Mech. Div.*, Vol. 104, No. EM5, Proc. Paper 14060, October, 1978, pp.1159-1175.

[2] ABDEL-ROHMAN, M., QUINTANA, V.H. and LEIPHOLZ, H.H.E., "Optimal Control of Civil Engineering Structures", submitted to ASCE for possible publication on May, 1978.

[3] ABDEL-ROHMAN, M. and LEIPHOLZ, H.H.E., "A General Approach for Active Structural Control", accepted by ASCE for publication.

[4] ABDEL-ROHMAN, M. and LEIPHOLZ, H.H.E., "Automatic Active Control of Structures", accepted by ASCE for publication.

[5] BROGAN, W.L., *Modern Control Theory*, Q.P.I. Series, New York, New York, 1974.

[6] LUENBERGER, D.G., "An Introduction to Observers", *Inst. of Electrical and Electronic Engineers, Trans. on Automatic Control*, Vol. AC-16, No. 6, December, 1971, pp.796-802.

[7] MARTIN, C.R. and SOONG, T.T., "Model Control of Multistory Structures", *ASCE, J. of the Eng. Mech. Div.*, Vol. 102, No. EM4, Proc. Paper 12321, August, 1976, pp. 613-623.

[8] YANG, J-N., "Application of Optimal Control Theory to Civil Engineering Structures", *ASCE, J. of the Eng. Mech. Div.*, Vol. 101, No. EM6, Proc. Paper 11812, December, 1975, pp. 819-838.

[9] YANG, J-N. and YAO, J.T.P., "Formulation of Structural Control", *Technical Report CE-STR-74-2*, School of Civil Engineering, Purdue University, Lafayette, Indiana, 1974.

[10] KWAKERNAAK, H. and SIVAN, R., *Linear Optimal Control Systems*, Wiley Interscience, 1972.

[11] QUINTANA, V.H., *Lecture Notes on Introduction to Optimal Control*, Department of Electrical Engineering, University of Waterloo.

[12] QUINTANA, V.H. and MOHARRAM, O., "Minimal-Order Observers for Synchronous Maching Control", *Proc. of the IEEE PES 1978 Winter Meeting*, Paper A 78 134-9.

APPENDIX I - LEVERRIER'S ALGORITHM

The inverse of the matrix $(\underline{A} - \delta_i \underline{I})$ can be obtained as

$$(\underline{A} - \delta_i \underline{I})^{-1} = - \frac{Ad_j(\delta_i \underline{I} - \underline{A})}{\Delta(\delta_i)} , \qquad (I.1)$$

where $Ad_j(\delta_i \underline{I} - \underline{A})$ and $\Delta(\delta_i)$ are, respectively, given by

$$Ad_j(\delta_i \underline{I} - \underline{A}) = \delta_i^{n-1} \underline{I}_n + \delta_i^{n-2} \underline{R}_{n-2} + \cdots + \delta_i \underline{R}_1 + \underline{R}_0 , \qquad (I.2)$$

$$\Delta(\delta_i) = \delta_i^n + a_{n-1} \delta_i^{n-1} + \cdots + a_1 \delta_i + a_0 . \qquad (I.3)$$

The quantities a_i and $\underline{R}_i$ are calculated from

$$a_{n-1} = -\text{Trace}(\underline{A}) ,$$

$$\underline{R}_{n-2} = \underline{A} + a_{n-1} \underline{I}_n ,$$

$$a_{n-2} = -\frac{1}{2} \text{Trace}(\underline{A} \, \underline{R}_{n-2})$$

$$\vdots$$

$$a_{n-i} = -\frac{1}{i} \text{Trace}(\underline{A} \, \underline{R}_{n-i}) , \qquad (I.4)$$

$$\underline{R}_{n-i-1} = \underline{A} \, \underline{R}_{n-i} + a_{n-i} \underline{I}_n . \qquad (I.5)$$

APPENDIX II - PROOF OF THE SEPARATION PRINCIPLE [5]

Equations (38) and (39) can be written as

$$\begin{bmatrix} \dot{X} \\ \dot{Z} \end{bmatrix} = \begin{bmatrix} (A_C - B_C DC) & -B_C E \\ (GC - TB_C DC) & F - TB_C E \end{bmatrix} \begin{bmatrix} X \\ Z \end{bmatrix} \qquad (II.1)$$

Assume that S* is an eigenvalue of the closed-loop system given by equation (II.1), and $\underline{q} = [q_1 \vdots q_2]^T$ is the corresponding eigenvector. Then, from equation (II.1) one has

$$\begin{bmatrix} A-B_C DC & -B_C E \\ GC-TB_C DC & F-TB_C E \end{bmatrix} \begin{bmatrix} q_1 \\ q_2 \end{bmatrix} = S^* \begin{bmatrix} q_1 \\ q_2 \end{bmatrix}. \qquad (II.2)$$

From equation (II.2) one obtains

$$(A-B_C DC)q_1 - B_C E q_2 = S^* q_1, \qquad (II.3)$$

$$(GC-TB_C DC)q_1 + (F-TB_C E)q_2 = S^* q_2. \qquad (II.4)$$

Multiplying equation (II.3) by T and subtracting the result from equation (II.4), one has

$$(GC-TA)a + Fq_2 = S^*(q_2 - Tq_1), \qquad (II.5)$$

or

$$F(q_2 - Tq_1) = S^*(q_2 - Tq_1). \qquad (II.6)$$

The case that $q_2 - Tq_1 \neq 0$ indicates that S* is the eigenvalue of $\underline{F}$. However, if $q_2 - Tq_1 = 0$, then from equation (II.3) one has

$$(A - B_C DC - B_C ET)q_1 = S^* q_1, \qquad (II.7)$$

using equation (10), equation (II.7) becomes

$$(A - B_C K)q_1 = S^* q_1, \qquad (II.8)$$

which indicates that S* is an eigenvalue of $(A - B_C K)$.

Therefore, the eigenvalues of the overall closed-loop system are the eigenvalues of the closed-loop plant $(A - B_C K)$ plus the eigenvalues of the observer $(\underline{F})$.

APPENDIX III - NUMERICAL DATA

$L = 100$ ft.	$a = 10$ ft.	$EI = 12 \times 10^{10}$ lb.in^2
$\ell = 3$ ft.	$m = 0.3$ lb.sec^2/in^2	$v = 60$ ft./sec.
$P = 20$ Kips	$S = 62.5$ Kips/inch	$h = 0.01$ sec.
$\omega_1^2 = 18.79$	$B_1 = 63.159$	$C_1 = 0.1773$
$\omega_3^2 = 1522.02$	$B_3 = 102.51$	$C_3 = 0.2878$

$$K = [0 \quad -0.15574 \quad 0 \quad -0.25281] = -\underline{R}^{-1}\underline{B}_C^T \underline{P}(t) ,$$

in which $R = 1$, $\underline{P}$ = Ricatti matrix as calculated in Reference [2].

STRUCTURAL CONTROL, H.H.E. Leipholz (ed.)
North-Holland Publishing Company & SM Publications
© IUTAM, 1980

A TRANSDUCER SYSTEM FOR THE IDENTIFICATION OF
RAILWAY TRACK MODULUS PARAMETERS

J.D. Aplevich

Department of Electrical Engineering
University of Waterloo
Waterloo, Ontario, Canada

1. INTRODUCTION

It would be of considerable interest to be able to measure the coefficient of railway track stiffness, or track modulus, from a moving track recorder car without using unorthodox wheels or contacts on the rail surface. With this in mind, a feasibility study was undertaken to determine how measurements of the relative deflection of the axles under loaded 3-axle trucks could be used to estimate the modulus coefficient. The feasibility of such a measurement requires a model of the behaviour of jointed rail under load.

The classical model of rail as a beam on elastic foundation may be reformulated as a vector matrix differential equation which may then be solved as an initial-value problem without numerical difficulties. From such a solution, a proposed transducer system may be calibrated with respect to sensitivity. However, imperfections in the track profile will affect the measurement and must be filtered to improve measurement precision. A model for track profiles including deterministic and random variations thus becomes necessary, in order to evaluate the filtering process. The following sections will include a closed-form solution for deflection of jointed track under load, a model for imperfect track profiles, and a method for determining the length of track required to estimate its modulus to a given precision with specified confidence level.

2. THE DEFLECTION OF JOINTED RAIL

The rail model used in this study is essentially the conventional model of the track structure shown in Figure 1. Reference [1] contains a survey of the present state of knowledge of the analysis of track stresses and deflections, and also a brief discussion of the validity of some of the assumptions involved here. Reference [2] contains many solutions to similar problems involving beams on elastic foundations, without the reformulation used here for convenient computer solution. Referring to Figure 1, the following assumptions are made:

(1) The rail is a straight, continuously-supported beam.

(2) The foundation resists upward as well as downward rail movement, and resists rotation of the rail under load, with the moment reaction coefficient proportional to the deflection reaction coefficient.

(3) All stress-strain relations for the rail and supporting structure are linear.

(4) Kinetic effects are ignored.

(5) Interaction between parallel rails is negligible.

The above assumptions all break down under certain conditions, but are known to be reasonably accurate for loadings and precision required here.

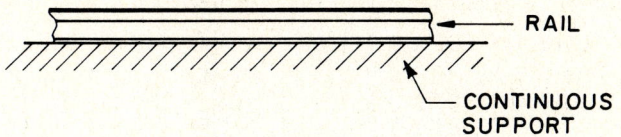

Figure 1 - Rail Foundation Model

Figure 2 is a free-body diagram of a piece of rail of infinitesimal length dt. At any point t the upward deflection is y, the moment is M, the shear is V, the vertical reaction of the foundation is -ky, the moment reaction is $-\rho \frac{dy}{dt}$, and the loading produced by the vehicle is -u(t).

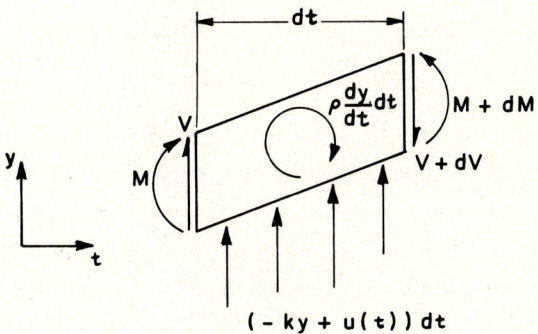

Figure 2 - Free-Body Diagram of Rail Section

Summing forces and moments to zero for constant rail stiffness EI results in the equation

$$EI \frac{d^4y}{dt^4} - \rho \frac{d^2y}{dt^2} + ky = u(t) , \qquad (1)$$

the solution of which has been published for various loadings and boundary conditions [1, 2]. For jointed track, however, the boundary conditions and nonconstant coefficients in (1) require computer solution which is rendered practical as follows. The following equation is used instead of (1):

$$\frac{d}{dt}\begin{bmatrix} y \\ y' \\ M \\ V \end{bmatrix} = \begin{bmatrix} 0 & 1 & 0 & 0 \\ 0 & 0 & 1/EI & 0 \\ 0 & \rho & 0 & 1 \\ -k & 0 & 0 & 0 \end{bmatrix} \begin{bmatrix} y \\ y' \\ M \\ V \end{bmatrix} + \begin{bmatrix} 0 \\ 0 \\ 0 \\ 1 \end{bmatrix} u(t) , \qquad (2)$$

or, in matrix notation,

$$\frac{d}{dt} X = A X + Bu(t) , \qquad (3)$$

which has the solution

$$X(t) = e^{(t-t_0)A} X(t_0) + \int_{t_0}^{t} e^{(t-\tau)A} Bu(\tau) d\tau . \qquad (4)$$

In order to use (4) it is crucial to express e^{tA} in closed form, since this matrix exponential contains both positive and negative exponential terms, which must be separated explicitly for the computed solution to have meaning at a significant distance from $t = t_0$. Use may be made of a similarity transformation matrix P so that

$$e^{tA} = P e^{t(P^{-1}AP)} P^{-1} . \qquad (5)$$

If P is chosen as the product

$$P = \begin{bmatrix} 1/EI & 0 & 0 & 0 \\ 0 & 1/EI & 0 & 0 \\ 0 & 0 & 1 & 0 \\ 0 & -\rho/EI & 0 & 1 \end{bmatrix} \begin{bmatrix} 1 & 0 & 1 & 0 \\ \alpha & \omega & -\alpha & \omega \\ \alpha^2-\omega^2 & 2\alpha\omega & \alpha^2-\omega^2 & -2\alpha\omega \\ \alpha^3-3\alpha\omega^2 & 3\alpha^2\omega-\omega^3 & 3\alpha\omega^2-\alpha^3 & 3\alpha^2\omega-\omega^3 \end{bmatrix} , \qquad (6)$$

where

$$\alpha = \sqrt{\sqrt{\frac{k}{EI}} + \frac{\rho}{4EI}} \quad , \tag{7}$$

$$\omega = \sqrt{\sqrt{\frac{k}{EI}} - \frac{\rho}{4EI}} \quad , \tag{8}$$

then

$$P^{-1}AP = \begin{bmatrix} \alpha & \omega & 0 & 0 \\ -\omega & \alpha & 0 & 0 \\ 0 & 0 & -\alpha & \omega \\ 0 & 0 & -\omega & -\alpha \end{bmatrix} \tag{9}$$

so that

$$e^{tA} = e^{\alpha t} P_1 Q(t) R_1 + e^{-\alpha t} P_2 Q(t) R_2 \quad , \tag{10}$$

where P_1 and P_2 are the leftmost and rightmost halves of P respectively, R_1 and R_2 are respectively the top and bottom halves of the explicitly obtainable matrix $P^{-1} = R$, and where

$$Q(t) = \begin{bmatrix} \cos\omega t & \sin\omega t \\ -\sin\omega t & \cos\omega t \end{bmatrix} . \tag{11}$$

If the loading u(t) is a point-load of force p at t = 0, the solution of (3) is

$$X(t) = \begin{cases} e^{tA} X(0) & t < 0 \\ e^{tA}(X(0)+Bp) \, , & t > 0 \end{cases} , \tag{12}$$

so that the initial condition vector X(0) is all that is required to obtain the solution, from which the deflection caused by any other loading can be derived using the linearity of the equations. The derivation of P in (6) above and its use for obtaining the initial conditions of (3) for the rail deflection problem is believed to be new, at least for the case of jointed rail. For infinite, that is, welded, rail the deflection and slope must approach zero at $t \to +\infty$ and $t \to -\infty$ so that, from (10) and (12),

$$\lim_{t \to -\infty} [I_2, 0] \, e^{-\alpha t} P_2 Q(t) R_2 X(0) = 0 \, , \qquad (13)$$

and

$$\lim_{t \to \infty} [I_2, 0] \, e^{\alpha t} P_1 Q(t) R_1 (X(0) + Bp) = 0 \, , \qquad (14)$$

where I_2 is a 2 by 2 unit matrix. Inspection of P_1 and P_2 shows that (12) and (13) are true if and only if

$$\begin{bmatrix} R_1 \\ R_2 \end{bmatrix} X(0) = \begin{bmatrix} -R_1 \, Bp \\ 0 \end{bmatrix} , \qquad (15)$$

which allows the solution of $X(0)$, from which the standard solution [2] is obtained, allowing for changes in reference directions. However, as limiting cases of jointed rail it was necessary to obtain solutions for free joints, for the joint considered as a hinge, and for the joint considered as an interval of reduced rail stiffness. Only the last case will be described here as illustration.

Assume that each rail has length T and that for $0 < t < t_1$ as well as for $T - t_1 < t < T$ the matrix A in (3) equals A_1 corresponding to a reduced value of EI. Otherwise A is equal to A_2, which contains normal values for EI. Then for positive t, the state-transition matrix $\Phi(t)$ will be defined as

$$\Phi(t) = \begin{cases} e^{tA_1} \, , & 0 < t \leq t_1 \\ e^{(t-t_1)A_2} e^{t_1 A_1} \, , & t_1 < t \leq T - t_1 \\ e^{(t-T+t_1)A_1} e^{(T-2t_1)A_2} e^{t_1 A_1} \, , & T - t_1 < t \leq T \end{cases} , \qquad (16)$$

using which, for positive t,

$$X(t) = \Phi(t)X(0) + G(t) , \qquad (17)$$

where $G(t)$ contains $u(t)$ but not $X(0)$. To solve for $X(0)$ for some loading between $t = 0$ and $t = T$ giving arise to a specific function $G(t)$, let the following matrices be defined:

$X_1(t)$ = top two entries of $X(t)$,

$X_2(t)$ = bottom two entries of $X(t)$,

$$\Psi = \Phi(T)^{-1} = \begin{bmatrix} \Psi_{11} & \Psi_{12} \\ \Psi_{21} & \Psi_{22} \end{bmatrix} \quad \text{where each } \Psi_{ij} \text{ is 2 by 2.}$$

Suppose that at joint nT far to the right of the load

$$X_2(nT) = S_n X_1(nT) , \qquad (18)$$

where S_n is a 2 by 2 matrix. Since $X((n-1)T) = \Psi X(nT)$,

$$\begin{bmatrix} X_1((n-1)T) \\ X_2((n-1)T) \end{bmatrix} = \begin{bmatrix} \Psi_{11} & \Psi_{12} \\ \Psi_{21} & \Psi_{22} \end{bmatrix} \begin{bmatrix} I_2 \\ S_n \end{bmatrix} X_1(nT) , \qquad (19)$$

from which

$$X_2((n-1)T) = (\Psi_{21} + \Psi_{22} S_n)(\Psi_{11} + \Psi_{12} S_n)^{-1} X_1((n-1)T)$$

$$= S_{n-1} X_1((n-1)T) , \qquad (20)$$

which is the defining equation for S_{n-1} in terms of S_n. Then S, the limit of S_{n-k} as $k \to \infty$, can be computed recursively from (20) or as the solution to the Riccati equation

$$S(-\Psi_{11}) + \Psi_{22}S - S\Psi_{12}S + \Psi_{21} = 0 , \qquad (21)$$

so that, at t = T

$$X_2(T) = SX_1(T) . \qquad (22)$$

Similarly, starting far to the left at joint -nT a matrix $\hat{S}$ may be obtained from (21), substituting submatrices of $\Phi(T)$ rather than Ψ, so that

$$X_2(0) = \hat{S}X_1(0) . \qquad (23)$$

Combining (22) and (23) with (17) evaluated at t = T, and solving for X(0),

$$X(0) = - \begin{bmatrix} I \\ \hat{S} \end{bmatrix} \left([-S,I]\Phi(T) \begin{bmatrix} I \\ \hat{S} \end{bmatrix} \right)^{-1} [-S,I]G(T) . \qquad (24)$$

Thus, X(t) at any point may be calculated by evaluating G(T), solving (24) for X(0), and then using (17).

3. A TRANSDUCER SYSTEM

The capability of solving for rail deflection with various assumptions on joint behaviour allows the calibration of a proposed transducer system. It also allows the calculation of rail profiles to be compared with actual profiles as measured by existing track recorder cars, and thereby the evaluation of any model of the random component of actual track profiles.

The vehicles of interest are two-truck, six-axle converted coaches with nominal weight of 90 tonnes (200,000 lbs.). Figure 3 shows typical deflections under a single truck for various rail weights and track modulus values. It is observed that the difference between the deflection of the centre axle and the average of the outer two is a function of track stiffness. It was proposed to measure this deflection difference since the transducer system can be robust and simple. Figure 4 shows the calibration for a recorder car of nominal weight and dimensions with a normal truck, and with a truck altered to carry half of the truck load on the central axle. Unfortunately the measurement will be perturbed by deviations in track profile, and by transient effects due to vehicle bounce and hunt as well as foundation dynamics. Ideally the transients due to motion can be eliminated by reduced train speed or filtering, but a fundamental question to be answered was the length of track to be tested in order to average out deviations in track profile.

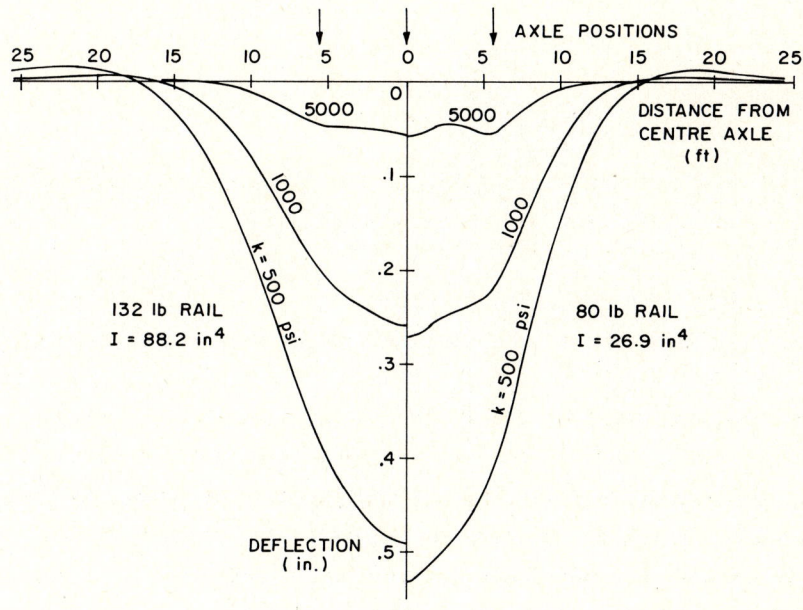

Figure 3 - Deflection under a 3-axle Truck: 16,667 lb. per axle

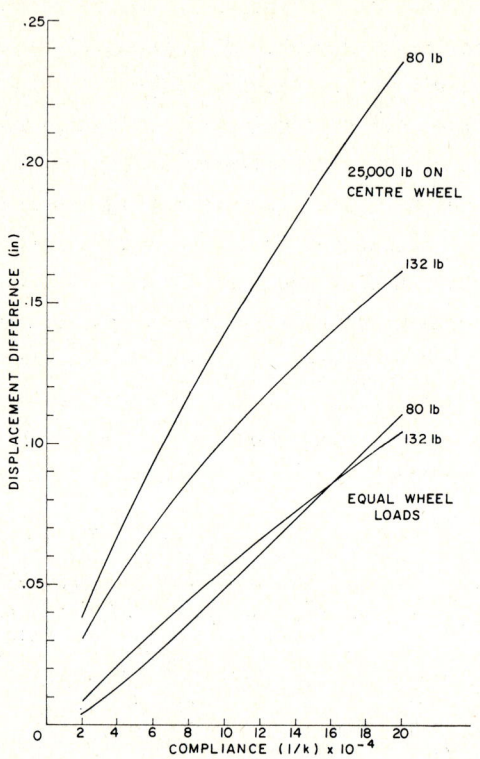

Figure 4 - *Displacement of Centre Axle with Respect to Outer Axles*

4. A MODEL OF IMPERFECT TRACK

The model of track profile used in this study is illustrated in Figure 5, in which the dynamic deflection refers to vehicle-induced rather than permanent deflections. Imperfectly-jointed track takes on a permanent deterministic set at the joints, modelled here as the profile traversed by a point load over open-jointed track. The random deflection is assumed to be caused by a random load simulating the cumulative effect of traffic loadings, differential roadbed settling, temperature fluctuations and maintenance operations. Rail surface defects, although possibly significant, were ignored. The random loading was assumed Gaussian where necessary. Real track is assigned a surface roughness (SR) number defined as

$$SR = \frac{10^5}{T} \int_0^T \left|\frac{dy}{dt}\right| dt , \qquad (25)$$

where $\frac{dy}{dt}$ is the rate of change of the vertical deflection of the leading wheel of the trailing truck, and T a suitable interval. This number is a function of the deterministic component as well as the standard deviation of the random component of the profile. Thus to arrive at realistic track models, the proportions of deterministic and random components had to be visually adjusted to give simulated leading-wheel profiles resembling actual track with the same SR number. A typical simulated track is shown in Figure 6. The proposed transducer system is sensitive only to the random component, and joints were assumed to have no effect to simplify the remainder of the study, although this simplification is not necessary in principle.

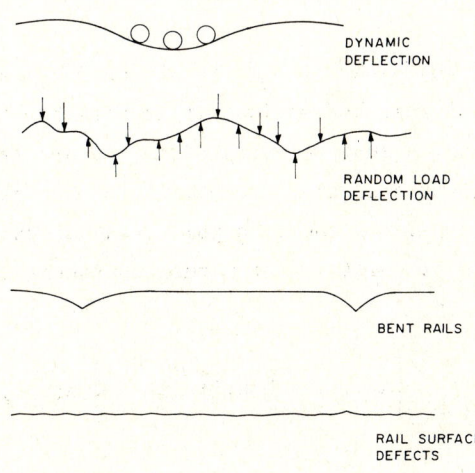

Figure 5 - Components of the Track Deflection Model

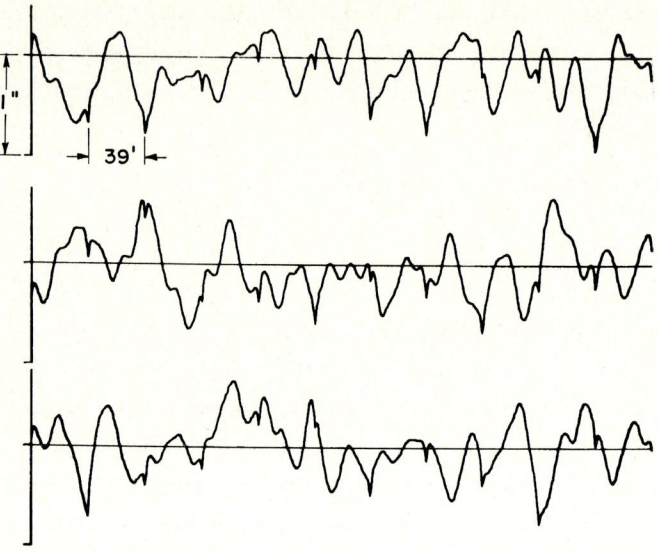

Figure 6 - Typical Simulation of Imperfect Jointed Track: Modulus = 1,000 psi, SR = 361

5. SAMPLE LENGTH AND ACCURACY: A TRADE-OFF

Given representative values of σ_N, the standard deviation of the random component of rail deflection, it is possible to estimate the track length which must be averaged to measure track compliance (1/k) to a given accuracy. Let $h(i,j)$ be the rail deflection produced at sample point i due to a point load at point j. Assuming the centre axle is at point i, the random component of the transducer measurement is

$$y(i) = \sum_{j=-\infty}^{\infty} \left[h(i,j) - \frac{1}{2} h(i+\ell,j) - \frac{1}{2} h(i-\ell,j) \right] n(j)$$

$$= \sum_{j=-\infty}^{\infty} \hat{h}(i,j) n(j) , \qquad (26)$$

where ℓ is the number of sample points between axles and $n(j)$ the random force at point j. It turns out that the standard deviation σy_N of y_N, the average of N samples $y(i)$, is given by

$$\sigma y_N = \frac{\sigma_N}{N}\left\{NR(0)+2(N-1)R(1)+2(N-2)R(2)+\cdots+2(1)R(N-1)\right\}^{1/2}, \qquad (27)$$

where

$$R(i) = \sum_{j=-\infty}^{\infty} \hat{h}(i,j)\hat{h}(j) . \qquad (28)$$

The function $\log_{10}(\sigma y_N/\sigma_N)$ is plotted in Figure 7, which shows that after about one rail length,

$$\sigma y_N = \frac{\alpha \sigma_N}{N}, \qquad (29)$$

where α depends on modulus and rail weight. The standard deviation of $1/k$ is given in terms of the slope $s(1/k)$ of the appropriate curve in Figure 4 as

$$\sigma_{1/k} = \frac{\sigma y_N}{s(1/k)}, \qquad (30)$$

so that, using the 90% confidence value of 1.649 standard deviations, the relation between sample length N and transducer system accuracy ν is

$$N\nu = 1.649 \frac{\alpha \sigma_N}{s(1/k)} . \qquad (31)$$

Using the above formula and the known precision of readily-available transducer components, it was determined that typically 4 to 6 rail lengths, or in extreme cases, 32 rail lengths, would have to be averaged to obtain reliable compliance values.

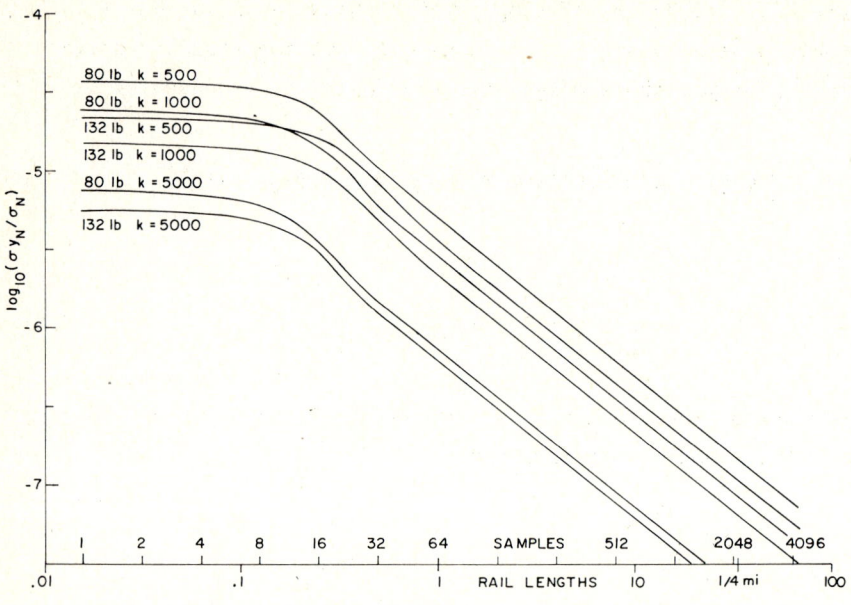

Figure 7 - Measurement Error versus Sample Size

6. CONCLUSIONS

The study summarized here was the solution of the simplest idealization of the transducer signal-processing problem. Several further possibilities exist for removing the random profile component. The measurement can be made on both sides of a truck and averaged, or better, the random component can be removed by instrumenting a second truck identically to the first except for an altered weight distribution, and subtracting the signals for each point on the track.

It is quite possible that the basic measurement system described will work only at speeds slow enough for truck hunting and vehicle motion to be insignificant. On the other hand, vehicle bounce is not completely undesirable if it can be measured, since

rail deflections are proportional to loading, and this allows a
modulus computation independent of the static measurement. In its
most sophisticated implementation, the signal processing system
would be required to combine the vehicle motion and track deflection measurements in an optimal manner.

7. ACKNOWLEDGEMENTS

This work was sponsored by the Canadian National Railways and
Canadian Pacific Limited.

REFERENCES

[1] KERR, A.D., "The Stress and Stability Analysis of Railroad Tracks", *ASME Paper 74-WA/APM-23, Winter Annual Meeting*, 1974, also in *Trans. ASME Journal of Appl. Mech.*

[2] HETÉNYI, M., *Beams on Elastic Foundation*, Ann Arbor: University of Michigan Press, 1958.

STRUCTURAL CONTROL, H.H.E. Leipholz (ed.)
North-Holland Publishing Company & SM Publications
© IUTAM, 1980

FLEXURE OSCILLATIONS OF BRIDGES - ESTIMATION OF THE VELOCITY
AND OF THE ACCELERATION BY USING LUENBERGER OBSERVERS

J. Auloge

Laboratoire de Mécanique, Ecole Centrale de Lyon
36, route de Dardilly, 69130 Ecully, France

INTRODUCTION

An ordinary bridge is exposed to vibrations when it is submitted to external forces; wind, earthquake, lorries or trains. The frequencies of vibrations are low so there is a problem measuring the velocity and the acceleration with usual transducers but it is possible to do it by optical sensors and to obtain the deflection versus time. The problem we are trying to solve is the following. How to obtain velocity and acceleration without derivation (discrete or continuous). It is possible to do it if we have a good mathematical model of vibration and we apply the Luenberger observer theory. We study only flexural oscillations of a one-span bridge described by a classical partial differential equation of the fourth order.

1. VIBRATION OF BRIDGES - CLASSICAL THEORY

We consider a bridge built like this:

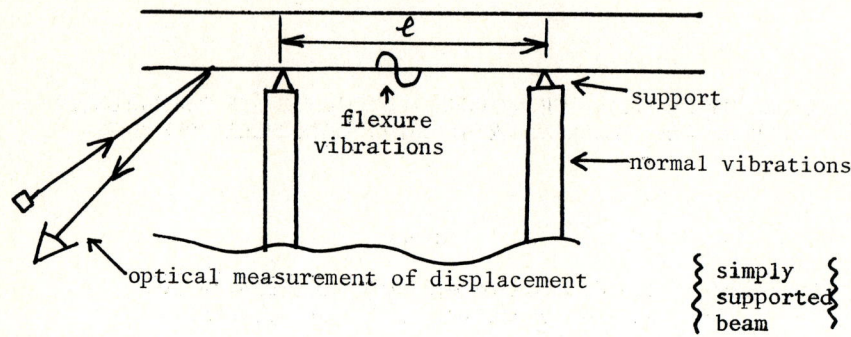

Normal vibrations or oscillations of piers are neglected. Supports may be represented by a linear differential equation in a neighbourhood of the equilibrium position.

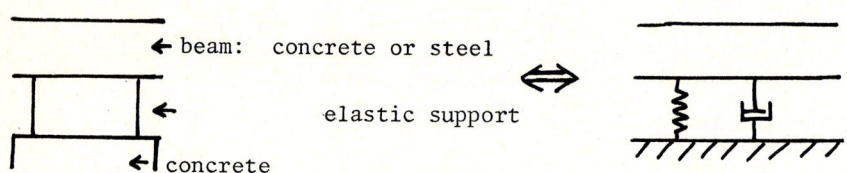

Flexural oscillations are governed by the partial differential equation:

$$\frac{\partial^2}{\partial x^2}\left(EI \frac{\partial^2 v}{\partial x^2}\right) + \rho \frac{\partial^2 v}{\partial t^2} + c \frac{\partial v}{\partial t} = F(x,t) \qquad (1)$$

where: E - Young's Modulus
         I - Inertia
         C - Damping

$F(x,t)$ - External Force Applied to the Bridge
ρ - Mass per unit of Length.

1.1 Boundary Conditions

Beam Simply Supported by Inelastic Supports

$$
\begin{aligned}
&* \ v(0,t) = 0 \\
&* \ v(\ell,t) = 0 \\
&* \ \frac{\partial^2 v}{\partial x^2}(0,t) = 0 \\
&* \ \frac{\partial^2 v}{\partial x^2}(\ell,t) = 0
\end{aligned}
\tag{2}
$$

Beam Simply Supported by Elastic Support of Stiffness K and Damping γ

$$
\begin{aligned}
&* \ v(0,t) = 0 \\
&* \ v(\ell,t) = 0 \\
&* \ T(0) = -EI\frac{\partial^3 v}{\partial x^3}(0,t) = -kv(0,t) - \gamma \cdot \frac{\partial v}{\partial t}(0,t) \\
&* \ T(\ell) = -EI\frac{\partial^3 v}{\partial x^3}(\ell,t) = -kv(\ell,t) - \gamma \cdot \frac{\partial v}{\partial t}(\ell,t) \ .
\end{aligned}
\tag{3}
$$

1.2 Analysis of Undamped Vibration

In the case of a beam simply supported by inelastic support, eigenfrequencies are given by:

$$\omega_n = n^2 \pi^2 \sqrt{\frac{EI}{\rho \ell^4}} \ , \tag{4}$$

$$\tilde{y}_n(x) = \sin\left(\frac{n\pi x}{\ell}\right) \qquad n = 1,\ldots,\infty \ . \tag{5}$$

In the case of an elastic support of stiffness K and damping γ, if we search for a solution by the method of separation of variables,

where
$$v(x,t) = y(x) \cdot \alpha(t)$$

$$y(x) = A\text{ch}\beta x + B\text{sh}\beta x + C\cos\beta x + D\sin\beta x .$$

The following conditions must be satisfied:

* $\dfrac{\partial^2}{\partial x^2}(0,t) = \dfrac{\partial^2}{\partial x^2}(\ell,t) = 0 ,$

* $-EI \dfrac{\partial^3 v}{\partial x^3}(0,t) = -kv(0,t) - \gamma \dot{v}(0,t) ,$ \hfill (6)

* $-EI \dfrac{\partial^3 v}{\partial x^3}(\ell,t) = -kv(\ell,t) - \gamma \dot{v}(\ell,t) .$

The calculation is very cumbersome. In fact, supports are inelastic, therefore:

$$\omega_n \# n^2\pi^2 \sqrt{\dfrac{EI}{\rho \ell^4}} . \qquad (7)$$

1.3 *Beam with Damping on Inelastic Supports*

The equation governing the oscillations is

$$\rho \dfrac{\partial^2 v}{\partial t^2} + EI \dfrac{\partial^4 v}{\partial x^4} + c \dfrac{\partial v}{\partial t} = 0 , \qquad (8)$$

with the boundary conditions:

* $v(0,t) = 0 ,$

* $v(\ell,t) = 0 ,$

* $\dfrac{\partial^2 v}{\partial x^2}(0,t) = 0 ,$

* $\dfrac{\partial^2 v}{\partial x^2}(\ell,t) = 0 .$

(9)

Then we search for $v(x,t) = y(x)\alpha(t)$ and obtain

$$(\rho\ddot{\alpha}+c\dot{\alpha}) \cdot y(x) + EI\,\alpha\,\frac{d^4y}{dx^4} = 0,$$

$$\rho\ddot{\alpha} + c\dot{\alpha} - \lambda\alpha = 0, \qquad (10)$$

$$\ddot{\alpha} + 2\xi_n\omega_n\dot{\alpha} + \omega_n^2\alpha = 0,$$

where

$$\lambda = -\omega_n^2\rho,$$

$$c = 2\xi_n \cdot \omega_n.$$

We obtain complex modes decreasing with time. Let

$$\frac{\omega_n^2\rho}{EI} = \Omega_n^4. \qquad (11)$$

Then we obtain the differential equation in terms of x,

$$\frac{d^4y}{dx^4} - \Omega_n^4 y = 0. \qquad (12)$$

The solution of this equation is given by:

$$y(x) = A \cdot \sin\Omega_n x + B \cdot \cos\Omega_n x + C \cdot \text{sh}\Omega_n x + D \cdot \text{ch}\Omega_n x, \qquad (13)$$

and according to the boundary conditions:

$$\Omega_n = \frac{n\pi}{\ell}, \qquad (14)$$

$$y_n(x) = \sin\frac{n\pi x}{\ell}. \qquad (15)$$

Therefore,

$$v(x,t) = \sum_{n=0}^{\infty} A_n \cdot \sin\frac{n\pi x}{\ell} \cdot e^{-\zeta\omega_n t} \cdot e^{i(\omega_n\sqrt{1-\zeta^2}\,t - \phi_n)}. \qquad (16)$$

1.4 *Oscillations Induced by an External Force and a Moving Vehicle*

An approximate equation is the P.D.E.

$$\rho \frac{\partial^2 v}{\partial t^2} + c \frac{\partial v}{\partial t} + EI \frac{\partial^4 v}{\partial x^4} = F(x,t) \ . \tag{17}$$

We search for an expansion on the modal basis,

$$v(x,t) = \sum_{\rho \in N} \alpha_\ell(t) \cdot \tilde{y}(x) \rightarrow \frac{\partial^4 v}{\partial x^4} = \sum \alpha_\ell(t) \cdot \Omega_\ell^4 \cdot \tilde{y}_\ell(x) \ ,$$

therefore, we obtain

$$\sum_{\ell \in N} (\rho \ddot{\alpha}_\ell + c \dot{\alpha}_\ell + \Omega_\ell^4 EI \alpha_\ell) \cdot \tilde{y}_\ell(x) = F(x,t) \ .$$

Then we apply the principle of orthogonality of modes:

$$\tilde{y}_\ell \perp \tilde{y}_n \quad \text{if } n \neq \ell,$$

$$\ddot{\alpha}_n + 2\xi_n \omega_n^2 \dot{\alpha}_n + \omega_n^2 \alpha_n = \frac{1}{\rho \langle \tilde{y}_n | \tilde{y}_n \rangle} \langle F(x,t) | \tilde{y}_n \rangle \tag{18}$$

with $n \in N$.

In the case of a simply supported beam

$$\langle \tilde{y}_n | \tilde{y}_n \rangle = \frac{\ell}{2} \ .$$

If we consider a lorry or a train moving at a uniform velocity:

$$F(x,t) = F_o \cdot \delta(x-vt) \quad , \quad 0 < x < \ell$$

$$\langle F(x,t) | \tilde{y}_n \rangle = F_o \cdot \sin \frac{n\pi vt}{\ell} = \psi_n(t) \ .$$

So we obtain the modal equation:

$$\ddot{\alpha}_n + 2\xi_n \cdot \omega_n \dot{\alpha}_n + \omega_n^2 \alpha_n = \frac{2F_o}{\rho \ell} \sin \frac{n\pi vt}{\ell} = \psi_n(t) \ .$$

2. ESTIMATION OF THE VELOCITY AND OF THE ACCELERATION BY USING MINIMAL ORDER LUENBERGER OBSERVERS

There is an infinity of equations of type (19). These equations are second order equations. We assume we know $\alpha(t)$ by measurements. There is a problem of locating the optical sensors. The forced solution of equation (19) is of the form:

$$v(x,t) = \sum_{n \in N} \alpha_n(t) \cdot \tilde{y}_n(x),$$

where v is the n-th solution of forced equation (19) with null conditions and is given by:

$$\alpha_n(t) = \frac{<v(x,t) \mid \tilde{y}_n(x)>}{<\tilde{y}_n(x) \mid \tilde{y}_n(x)>} = \frac{2}{\ell} <v(x,t) \mid \tilde{y}_n(x)>,$$

that is to say,

$$\alpha_n(t) = \frac{2}{\ell} \int_0^\ell v(x,t) \cdot \tilde{y}_n(x) \cdot dx, \qquad (20)$$

and the deflection $v(x,t)$ is known by measurements $\tilde{v}$. If this deflection is known with an error ε.

$$|v(x,t) - \tilde{v}(x,t)| < \frac{\varepsilon}{2}, \qquad \forall x \in [0,\ell],$$

then

$$|\alpha_n(t) - \tilde{\alpha}_n(t)| < \varepsilon,$$

so it is possible to use characteristic functions

$$X\left[\frac{m}{\ell}, \frac{m+1}{\ell}\right],$$

where

$$\ell \in [0, E(\ell) - 1].$$

If,

$$x_o \in \left[\frac{m}{\ell}, \frac{m+1}{\ell}\right[\to \tilde{v}(x,t) = v(x_o,t) \cdot X_{\left[\frac{m}{\ell}, \frac{m+1}{\ell}\right[}.$$

We search an asymptotic estimate of the modal velocity $\dot{\alpha}_n$ from the knowledge of $\alpha_n(t)$. To do that, we use the minimal order Luenberger observer theory. From equation (19), we see that $X_n \big| \substack{\alpha_n \\ \dot{\alpha}_n}$ is a state vector of the system, solution of the matricial differential equation:

$$\dot{X}_n = A_n \cdot X_n + B_n \cdot U_n, \tag{21}$$

$$\delta_n = C_n \cdot X_n, \tag{22}$$

where

$$A_n = \begin{bmatrix} 0 & +1 \\ -\omega_n^2 & -2\xi_n\omega_n \end{bmatrix} \tag{23}$$

$$B_n = \begin{bmatrix} 0 \\ 1 \end{bmatrix}, \tag{24}$$

$$C = [1,0] \text{ independent from } n. \tag{25}$$

$$U_n = \psi_n(t). \tag{26}$$

The dynamics of the two order process depends on n. System (21) is observable because $\omega_n^2 \neq 0$, i.e., mode n is observable. We search for the one dimensional vector (here a scalar) solution of differential equation: see Luenberger [2, 3, 4, 5]

$$\dot{Z}_n = S_n \cdot Z_n + G_n \cdot \tilde{X}_n + R_n \cdot U_n, \tag{27}$$

where

$$Z_n = T_n \cdot \tilde{X}_n.$$

Flexure Oscillations of Bridges 103

The observer gives an estimation of the state vector of the form:

$$\hat{X}_n = V_n Z_n + W_n \cdot \delta_n. \qquad (28)$$

It is possible to find a minimal order Luenberger observer if the following conditions are satisfied:

* matrix S_n is asymptotically stable and its dynamics depends on the mode n,
* $T_n \tilde{A}_n - S_n \cdot T_n = G_n$,
* $R_n = T_n \cdot \tilde{B}_n$, $\qquad (29)$
* the dynamics of the observer must be different from the dynamics of the oscillation, so the modes must be approximately known, i.e., the pole of the observer must be different from the vibration modes,
* $V_n T_n + W_n \cdot C_n = I_2$.

So the matrices are given by:

* $S_n = -b_n$,
* $T_n = (-b_n, 1)$,
* $G_n = (b_n(2\xi_n\omega_n - b_n) - \omega_n^2, 0) = (g_n, 0)$, $\qquad (30)$
* $h_n = b_n - 2\xi_n\omega_n$.

If these velocities are satisfied, the asymptotic estimation of the velocity is given by:

$$\hat{\dot{\alpha}}_n = Z_n + h_n \cdot \alpha_n. \qquad (31)$$

Its convergence depends on the dynamics of the observer, which is given by the poles of S_n. It is possible to choose a sequence (b_n) $n \in N$, different from the sequence of the real part of complex modes. So we define a modal observer. The modal estimation of the acceleration is given by:

$$\hat{\ddot{\alpha}}_n = \psi_n(t) - 2\xi_n \cdot \omega_n \hat{\dot{\alpha}}_n - \omega_n^2 \cdot \alpha_n \ . \tag{32}$$

So the difference between the acceleration tends towards zero.

$$|\ddot{\alpha}_n - \hat{\ddot{\alpha}}_n| = 2\xi_n \omega_n |\dot{\alpha}_n - \hat{\dot{\alpha}}_n| \to 0 \text{ if } t \to \infty .$$

2.1 Problem of Difference in Phase

If we apply a harmonic force $F(t)$, (it is the case in this problem), and null conditions, does the observer introduce a difference in phase between the velocity and its estimation? The answer is no. Here, we give a scheme of the calculation. Let,

$$\alpha_n(t) = (A_1)_n(\omega) \cdot e^{-i(\phi_1)_n(\omega)} \cdot e^{i\omega t} , \tag{33}$$

where

$$(A_1)_n = \frac{1}{(\omega_n^2 - \omega^2) + 2i\zeta_n \omega \omega_n} ,$$

$$\dot{\alpha}_n(t) = \frac{i\omega}{(\omega_n^2 - \omega^2) + 2i\zeta_n \omega \omega_n} e^{-i(\phi_1)_n(\omega)} \cdot e^{i\omega t} . \tag{34}$$

The differential equation of the observer admits a sinusoidal solution.

$$\dot{Z}_n + b_n Z_n = g_n \alpha_n + U_n , \tag{35}$$

$$Z_n(t) = \frac{g_n \cdot (A_1)_n(\omega) e^{-i(\phi_1)_n(\omega)} + 1}{b_n + i\omega} . \tag{36}$$

Therefore:

and

$$\hat{\dot{\alpha}}_n(t) = (\hat{A}_2)_n(\omega) \cdot e^{-i(\hat{\phi}_2)_n(\omega)} \cdot e^{i\omega t} = Z_n + h_n \cdot \alpha_n \tag{37}$$

$$(\hat{A}_2)_n(\omega) \cdot e^{-i(\hat{\phi}_2)_n(\omega)} = (A_2)_n(\omega) \cdot e^{-i(\phi_2)_n(\omega)} , \tag{38}$$

$$\boxed{\hat{\dot{\alpha}}(t) \to \dot{\alpha}(t) \quad \text{if} \quad t \to \infty} \ . \tag{39}$$

2.2 Convergence of the Solution of the Partial Differential Equation

We define:

$$\dot{y}(x,t) = \sum_{n \in N} \dot{\alpha}_n(t) \cdot \tilde{y}_n(x) ,$$

$$\dot{\hat{y}}(x,t) = \sum_{n \in N} \dot{\hat{\alpha}}_n(t) \cdot \tilde{y}_n(x) .$$

We have the convergence of the estimation $\hat{\alpha}_n$ towards $\dot{\alpha}_n$

$$\dot{\alpha}_n = <\dot{y}(x,t) | \tilde{y}_n(x)> = \int_0^\ell \dot{y}(x,t) \cdot \tilde{y}_n(x) \cdot dx .$$

Therefore,

$$\boxed{\begin{aligned} \lim_{t \to \infty} \dot{\hat{y}}(x,t) &= \dot{y}(x,t) \\ \lim \ddot{\hat{y}}(x,t) &= \ddot{y}(x,t) \end{aligned}} \qquad (40)$$

2.3 Location of Sensors

An interesting problem we did not solve consists in the optimal location and the number of sensors.

They must be disposed to describe the excited modes and their number must be sufficient to have a good accuracy of the second member of the modal equation.

3. CONCLUSION

If the displacement of a bridge is measured, it is possible to estimate the velocity and the acceleration of the flexure oscillations by using Luenberger observers.

REFERENCES

[1] AULOGE, J., "Luenberger Observers for Mechanical Systems: Application of Mechanical Vibrations", *Meco*, 1978, Athens.

[2] LUENBERGER, D., "Canonical Form for Linear Multivariable Systems", *IEEE Trans. on Automatic Control*, Vol. AC12, June, 1967, pp. 290-293.

[3] LUENBERGER, D., "Observing the State of a Linear System", *IEEE Trans. Military Electronics*, Vol. MIL8, April, 1966, pp. 74-80.

[4] LUENBERGER, D., "Observers for Multivariable Systems", *IEEE Trans. on Automatic Control*, Vol. AC11, April, 1966, pp. 190-197.

[5] LUENBERGER, D., "An Introduction to Observers", *IEEE Trans. on Automatic Control*, Vol. AC16, December, 1971, pp. 596-602.

[6] MEIROVITCH, L., *Elements of Vibration Analysis*, MacGraw-Hill, 1975.

[7] CLOUGH, R.W. and PENZIEN, *Dynamics of Structures*, MacGraw-Hill, 1975.

STRUCTURAL CONTROL, H.H.E. Leipholz (ed.)
North-Holland Publishing Company & SM Publications
© IUTAM, 1980

ACTIVE CONTROL OF LARGE CIVIL ENGINEERING STRUCTURES:
A NAÏVE APPROACH

M.J. Balas

Electrical and Systems Department
Rensselaer Polytechnic Institute
Troy, New York 12181, U.S.A.

1.0 INTRODUCTION

This paper summarizes certain concepts of active control of mechanically flexible structures which, we believe, have a bearing on the problems encountered in control of large civil engineering structures (LCES). However, we admit that this approach was originated with a different application in mind, namely large structures in space (LSS), such as large flexible spacecraft and satellites. Hence, we claim no deep expertise in LCES, per se, and consequently have subtitled this paper - A Naïve Approach. In addition to the many helpful conversations we have had with some of the experts in civil engineering structures, two things calm our trepidation in placing these ideas before a predominantly civil engineering audience: the basic similarity of all mechanically flexible structures, be they LSS or LCES, and the fundamental tolerance that must be inherent in any audience that has contemplated attaching wings and engines to unruly buildings and bridges to tame them down. We hope that this naïve approach will help to provide a sufficiently general framework to address the many complex issues of LCES control and will contribute to an understanding of the underlying theoretical mechanisms through which the various proposed LCES control schemes operate.

Large civil engineering structures (LCES), such as bridges and buildings, are extremely susceptible to transient loadings like strong wind gusts and earthquakes. These random loadings usually occur for short time intervals during the lifetime of the structure and may produce large deflections and accelerations. LCES must be able to withstand these forces without being "over designed".

Although passive techniques, such as dampers and tuned shock absorbers, have been used in the past (and should not be ignored in certain applications) e.g., [1], active control is a promising approach for suppression of vibrations in LCES because large amounts of damping can be generated with a minimum of control effort. Also, active controllers can be added to existing structures without substantial modification of the structure. Active control of LCES refers to the use of a feedback control system that processes information from vibration sensors (e.g., accelerometers) located at various points on the structure and generates control commands for force or torque actuators to suppress the vibration. Recent microprocessor and mini-computer technology makes it possible to design such feedback controllers effectively and cheaply.

Although LCES may be quite rigid under normal operation, during periods of high loading, they may become extremely mechanically flexible. In theory, flexible structures are distributed parameter systems (DPS), in the sense that their behaviour is best modelled by partial differential equations, i.e., a continuum model. This necessitates an active control theory that can handle the special nature of DPS, e.g., an infinite number of states. In practice, there are various ways of lumping the parameters of the structure to produce a large finite dimensional model which approximates the behaviour of the structure. The dividing line between large dimensional and infinite dimensional systems becomes fuzzy

in practical applications; the structure models will be very
large for the sake of accuracy but the active controller will be
forced to deal with a much smaller dimensional model due to on-
line computational limitations. Whether one looks at LCES as DPS
or large scale systems, the "curse of dimensionality" will always
be present in active controller design and must be carefully
taken into account.

Active control of mechanically flexible structures
appears in several related but distinct areas:

(1) distributed parameter systems (DPS) or systems governed
by partial differential equations, e.g., [2],

(2) large civil engineering structures (LCES), e.g., [3]-[7],

(3) large structures in space (LSS), e.g., [8]-[10],

(4) aeroelasticity and flutter, e.g., [11]-[14].

My own work falls mainly into (1) and (3) and serves as the basis
for this paper, e.g., [15]-[19]; proofs and other details may be
found there. The interpretation of this work in the context of
LCES is the goal of this paper.

2.0 SOME RELATIONSHIPS BETWEEN LSS AND LCES

In Figure 1, some relationships between LSS and LCES are presented;
of course, this figure represents my own preconceptions. LSS like
the Solar Power Station Satellite may very well be the size of
Manhattan Island. Such structures would be very lightweight and
have very low natural damping. They would be constructed in space
and operated at very high earth orbits under stringent requirements
for pointing accuracy, shape fidelity, and attitude orientation;
such requirements would necessitate control of rigid body and many
structural modes. Principal disturbances would be large transients.

	L S S	L C E S
SIZE	MANHATTAN	COMPARABLE — BUILDINGS IN MANHATTAN
TYPE	CONTINUUM OR LARGE DIMENSIONAL MODELS (OSCILLATORY)	
NUMBER CONTROLLED MODES	LARGE	SMALL (SINGLE)
CRITICAL RESONANT FREQUENCIES	MANY (THROUGHOUT SPECTRUM CLOSELY PACKED)	FEW (LOW FREQUENCY SEPARATED)
DAMPING	VERY LOW ($< \frac{1}{2}$ %)	MEDIUM ($\frac{1}{2}$ - 10 %)
DISTURBANCES	LARGE TRANSIENT (MANEUVRES, DOCKING, ACCIDENTS) SMALL STOCHASTIC (SOLAR PRESSURE, GRAVITY GRADIENT)	LARGE STOCHASTIC (WIND GUSTS, EARTHQUAKES) SMALL TRANSIENT (AIRPLANES)

Figure 1 - Relationships Between LSS and LCES

Correspondingly, LCES, such as bridges and buildings, may be comparable in size to small LSS and would have higher natural damping and fewer frequencies requiring control. Generally, rigid body modes would not exist unless the structure was in very deep trouble; often, only a single bending mode would need control. The critical structural frequencies are fewer and more widely separated in LCES than those in LSS. Large nonstationary stochastic disturbances, such as wind gusts and earthquakes, are the most likely disturbances on LCES and these are more likely to interact nonlinearly with the structure.

3.0 LCES DESCRIPTION

Since most LCES are continua or composed of separate interconnected continua, they are DPS and may be described best by a system of partial differential equations of the form:

$$\begin{cases} m(x) u_{tt}(x,t) + D_0 u_t(x,t) + A_0 u(x,t) = F_0(x,t) \\ u(x,0) = u^0, \quad u_t(x,0) = u_t^0, \end{cases} \quad (1)$$

where $u(x,t)$ represents the (vector of) generalized displacements (translational and rotational) of the structure Ω from its equilibrium position due to initial loadings and external torques and forces $F_0(x,t)$. The mass density is $m(x)$ and can be eliminated with a change of variables. The damping D_0 and internal restoring forces A_0 are given by appropriate differential operators (e.g., A_0 may be the Laplacian or biharmonic operator) whose domains of influence contain all functions in $H_0 = L^2(\Omega)$ which satisfy the structural boundary conditions. These operators have complete eigendata:

$$A_0 \phi_k = \omega_k^2 \phi_k, \quad (2)$$

$$D_0 \phi_k = 2\xi_k \omega_k \phi_k, \quad (3)$$

where $\phi_k(x)$ are the structure *mode shapes*, ω_k are the *mode* (or resonant) *frequencies*, and ξ_k are the *modal damping coefficients*. When the displacement is expanded in terms of these mode shapes

$$u(x,t) = \sum_{k=1}^{L} u_k(t) \phi_k(x), \quad (4)$$

the mode amplitudes $u_k(t)$ satisfy

$$\ddot{u}_k(t) + 2\xi_k \omega_k \dot{u}_k(t) + \omega_k^2 u_k(t) = F_k(t), \quad (5)$$

where $F_k(t)$ is the force $F_0(x,t)$ resolved into the kth mode. In theory, $L = \infty$ and, in practice, L is often very large.

Equation (1) may be put into so-called state variable form with the choice $v(x,t) = [u(x,t) u_t(x,t)]^T$:

$$v_t = Av + F, \qquad (6)$$

where

$$A \equiv \begin{bmatrix} 0 & I \\ -A_0 & -D_0 \end{bmatrix} \quad \text{and} \quad F \equiv \begin{bmatrix} 0 \\ F_0 \end{bmatrix};$$

this description is set in a Hilbert space H called the *full state space* [9]. The transition operator or semigroup U(t) generated by A has the property:

$$v(t) = U(t)v_0, \qquad (7)$$

when $F \equiv 0$ and $v_0 = [u^0 \ u_t^0]^T$. When the *energy norm*

$$||v(t)||_E^2 = (mu_t, u_t)_0 + (A_0 u, u)_0, \qquad (8)$$

is used, where $(\cdot,\cdot)_0$ is the H_0 inner product, the system (6) is *dissipative*:

$$||v(t)||_E \leq M e^{-\delta t} ||v_0||_E, \qquad (9)$$

where M, δ are nonnegative constants determined by the damping in the structure, e.g., the system is energy conserving when $D_0 = 0$ (and consequently $M = 1$ and $\delta = 0$). All LCES will have some damping and, hence, be dissipative; however, the stability margin δ may be quite small.

The external force distribution F_0 in (1) is given by

$$F_0(x,t) = F_D(x,t) + F_C(x,t), \qquad (10)$$

where F_D represents disturbance forces on the structure and F_C represents the control forces due to M actuators:

$$F_C(x,t) = \sum_{i=1}^{M} b_i(x) f_i(t) . \qquad (11)$$

The actuator amplitudes f_i are the controls and the influence functions b_i indicate the range of actuator influence, e.g., they are Dirac delta functions (or their derivatives) when *point* force (or torque) actuators are used. Many different actuators are available for LCES: adjustable bridge stanchions and cable stays [20], tendon control for tall buildings [21]-[22], gyroscopes, thrusters, and aerodynamic surfaces [23]-[26].

Observations are produced by P sensors:

$$y(t) = C_0 u + C_0' u_t , \qquad (12)$$

where $y_j(t) = (c_j, u)_0 + (c_j', u_t)_0$, $1 \leq j \leq P$, with c_j, c_j' the influence functions of the position and velocity sensors, respectively. These influence functions are Dirac delta functions when the sensors are point devices. Any mixture of position, velocity, or acceleration measurements of translation or rotation can be included in (12). A stochastic process representing sensor noise could be added to (12) without any serious changes in what follows.

Therefore, (6) becomes the DPS

$$\begin{cases} v_t = Av + Bf + F_E , \\ y = Cv , \quad v(o) = v_0 , \end{cases} \qquad (13)$$

where f, y are the actuator command and sensor output vectors, respectively, and F_E is the external disturbance distribution:

$$F_E \equiv \begin{bmatrix} 0 \\ F_D \end{bmatrix} , \quad B \equiv \begin{bmatrix} 0 \\ B_0 \end{bmatrix} , \quad \text{and} \quad C \equiv [C_0 \ C_0'] .$$

Note that B,C have ranks M and P, respectively. This DPS (13) must be controlled by a few actuators and sensors even though it has an infinite dimensional state. The control computer has limited capacity and must use this capacity to process the sensor outputs y to produce actuator commands f in something like real-time.

4.0 FUNDAMENTAL ISSUES OF LCES CONTROL

4.1 *Structure Modelling*

Any control system for LCES is only as good as the mathematical model of the structure. These models may be developed from simple floor-to-floor relative displacements for tall buildings, e.g., [27], to finite element modal approximation of the structure, e.g., [24]. When the model is crude, the control must necessarily be less effective. For systems where the structural parameters are poorly known or time-varying, some improvement may be achieved with adaptive controllers that start with the original structure model and "tune themselves up" to the actual structure while providing control; however, there are certain trade-offs that must be made when adaptive procedures are applied to flexible structures [28].

4.2 *Control Device Placement*

The type, number, and placement of acutators and sensors is a critical problem for LCES. These devices cannot always be freely chosen or located arbitrarily on the structure. Cost and reliability dictate that a small number of devices should be used; however, some trade-off between the number of devices used and the control computer capacity can be made, as will be indicated later.

4.3 Reduced-Order Modelling

Very large dimensional models may be used for evaluation of DPS like LCES but, in order to implement real-time controllers, substantially reduced-order models (ROM) must be used in the on-line control; the size of the ROM is usually dictated by the on-line computer capacity and its ability to perform real-time integration of systems of ordinary differential equations. Again, trade-offs are possible with the size of the ROM (computer capacity), control accuracy and speed, and number of control devices. Even when the actual mode shapes of the structure are known, e.g., [20]-[21], there will always be residual (uncontrolled) modes and the controller design will depend on the choice of critical modes to be controlled.

4.4 Multivariable Controller Design and Desired Performance

Outputs from a variety of sensors must be processed to deliver several actuator commands for effective and efficient structure control. The devices may not be collocated in some applications nor will the same number of actuators and sensors be present. The control must be designed to deal with these multivariable considerations, as well as stochastic disturbances and control force limitations; these considerations must also be included in the desired performance to be asked of the actively controlled LCES.

4.5 Controller Interaction with Residuals

Since the controller must be designed from a ROM, the interaction of the controller with the unmodelled residuals is extremely important. This interaction, if not analyzed and compensated, can negate the benefits of active structure control.

All of these issues are interconnected and the trade-offs available in them should be carefully investigated before any LCES active controller design is recommended.

5.0 IMPLEMENTABLE LCES CONTROLLERS: A FRAMEWORK FOR DESIGN AND ANALYSIS

In this section we present a framework to address the issues and trade-offs discussed in Section 4.

Any linear implementable controller for LCES will take the following form (Figure 2):

$$\begin{cases} f(t) = H_{11} y(t) + H_{12} z(t) , \\ \dot{z}(t) = H_{21} y(t) + H_{22} z(t) , \end{cases} \quad (14)$$

where dim $z = S$ which is a direct function of the on-line computer capacity and hence should be as small as possible. It is well known that digital computers must process data in discrete time; so, the proper form of (14) would be the discrete time version but it is clear that discrete time versions of the results that follow can be obtained and we will refrain from introducing this extra complication.

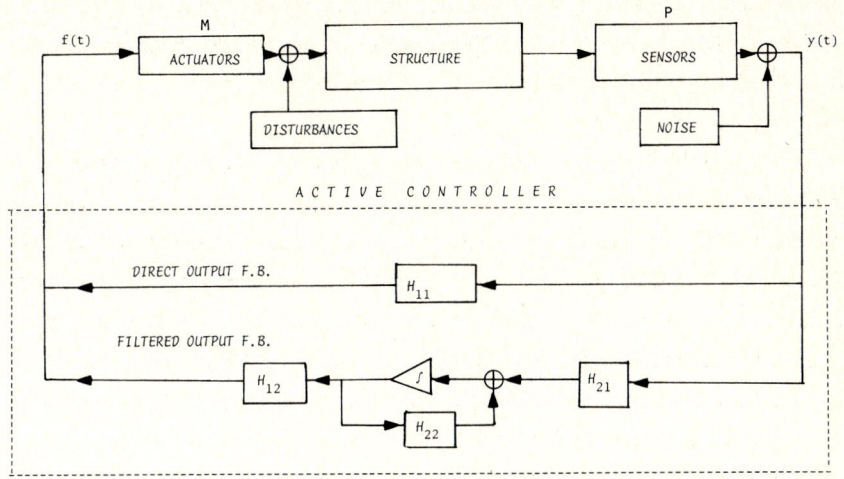

Figure 2 - *Actively Controlled Structure*

The design of the controller in (14) is based on some ROM of the DPS (13) representing the structure to be controlled. A ROM is produced by selecting an appropriate finite dimensional subspace H_N in the full state space H and projecting (13) onto this subspace; the projection is often orthogonal but not always, e.g., [29]. Let P be the projection onto H_N and Q the projection onto the residual subspace and (13) becomes

$$Pv_t = A_N Pv + A_{NR} Qv + B_N f + PF_E , \qquad (15)$$

$$Qv_t = A_{RN} Pv + A_R Qv + B_R f + QF_E , \qquad (16)$$

$$y = C_N Pv + C_R Qv , \qquad (17)$$

where $A_N = PAP$, $B_N = PB$, $A_{NR} = PAQ$, etc. The ROM is given by

$$\begin{cases} \dot{v}_N = A_N v_N + B_N f + PF_E , \\ y = C_N v_N , \end{cases} \qquad (18)$$

where $\dim v_N = \dim H_N = N$ and it is upon this ROM that the controller (14) is designed. The terms $A_{NR}Qv$ and $A_{RN}Pv$ are *model error* and $B_R f$ and $C_R Qv$ are *control* and *observation spillover*, respectively.

If the exact mode shapes $\phi_k(x)$ are known (as they are for simple structures like beams and plates), then the *modal subspace* is a logical choice for H_N; $\frac{N}{2}$ critical modes are selected for control and the ROM is obtained by orthogonal projection onto this modal subspace. In many cases for complex structures, only approximate modal data are available via finite element numerical approximation; this approximate data can be used as the basis for the subspace. When this modal data is used for the ROM, any controller design based on such an ROM is customarily called *modal control*; however, when the exact mode shapes are known as in [20]-[22] and [30], the model error is zero but the spillover is usually nonzero. The choice of critical modes in LCES may be quite simple in many cases; in [22] and [30], only the first bending mode is retained and controlled. Other ROM are possible for LCES, e.g., [5], [31].

The performance desired of the actively controlled structure must be specified in advance in order to design the controller (14). The desired performance, stated rather loosely, is to tailor the open-loop (uncontrolled) structural response to that of a more benign closed-loop (controlled) structure. Several methods (and combinations of these) exist to accomplish this task:

(a) pole placement: the control law is designed to relocate critical resonant frequencies and increase their damping, e.g., [32],

(b) stochastic optimization (LQG): optimal regulator and Kalman filter techniques are used based on a quadratic performance index and stochastic disturbance covariances, e.g., [10], [15], [20]-[21], [31]-[34]. Note that these techniques can also be used to achieve (a) indirectly.

(c) disturbance accommodation: when disturbances have known waveforms but unknown amplitudes (i.e., superposition of a few known frequencies), they can be handled with the techniques of [35]-[36].

To satisfy any of the above performance requirements for the full DPS is generally impossible with an implementable controller; consequently, the performance is desired of the ROM alone rather than the full DPS, with the consequences of this left until later analysis. We emphasize that optimal control is less important than "good" control and the above techniques for performance modification can usually be "tuned up" by some trial and error to produce good control.

The number and location of control devices is directly affected by the controllability and observability of the ROM. When the modal ROM is used and the damping is small, easily verifiable conditions for controllability and observability can be obtained in terms of the device locations and the critical mode shapes [15]. The minimum number of actuators or sensors is the maximum mode frequency multiplicity (i.e., number of times the frequency is repeated); the devices must be located away from the mode shape zeros with some additional independent control required for repeated frequencies. For the simple Euler-Bernoulli beam with pinned ends, a single actuator and position or velocity sensor are sufficient if properly located; however, for a circular pinned plate or membrane at least two actuators and sensors are required and, for a pinned rectangular plate or membrane, the minimum number of devices is a nondecreasing function of the number of controlled modes. In addition, if pole placement alone is desired, then, from [18],

$$M + P + S \geq N + 1 , \qquad (19)$$

i.e., the total number of devices plus the controller order must exceed the dimension of the ROM (for the modal case N is twice

the number of critical modes). These results are not the final
answer for controller design but they do indicate the trade-offs
available in control devices, device placement, and controller
order (on-line computer capacity).

In most previous designs for LCES, the full order state
estimator is used (S = N) and the controllability-observability
conditions dictate the minimum number of devices used. Of course,
more devices could always be used (e.g., for reliability) and,
in fact, some especially nice properties occur when the number of
actuators or sensors can be as large as N/2 and a decoupling controller can be designed [24]. At the other end of the control
spectrum is direct output feedback (DOFB) where the sensor outputs
are multiplied by gains and fedback directly to the actuators [17];
from (19), this requires at least that the total number of devices
exceed N - many devices but very little computer capacity required.
A special case of DOFB which characterizes the effect of passive
and active dampers on structures is discussed in [37].

6.0 CONTROLLER INTERACTION WITH RESIDUALS: BOUNDS ON SPILLOVER AND MODEL ERROR

All controllers for LCES will be designed from ROM but they operate
in closed-loop with the full structure - equations (15)-(17). The
controller interacts with the residuals (16) through the model
error and spillover terms. Even when model error is sufficiently
small as to be negligible, the spillover remains as in Figure 3
and can produce pole-shifting (Figure 4) which degrades the
desired system performance and, in some cases, produces instabilities
especially in the lightly damped residual modes, e.g., [15], [17],
[38].

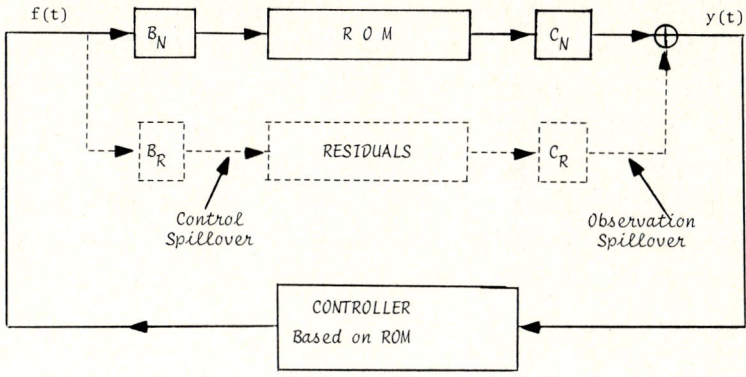

Figure 3 - Spillover

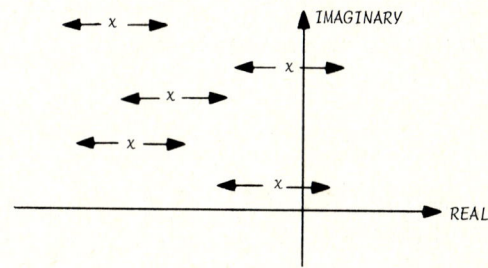

Figure 4 - Pole-Shifting due to Spillover

The question of *how much spillover and model error can be tolerated* is partially answered by obtaining energy bounds on the closed-loop system, e.g., [15]-[19]. These results take the following form:

$$||v(t)||_E \leq M(\beta)\, \bar{e}^{\delta' t} ||v_0||_E , \qquad (20)$$

when

$$\delta' \geq \delta - M(\beta)\Gamma , \qquad (21)$$

where δ and δ' are the spillover-free and spillover-present stability margins, respectively, and Γ, β are the observation and control spillover coefficients and $M(\beta)$ is an explicit function of β and the spillover-free data. Similar results can be obtained which include bounded model errors in β and Γ; crude models can increase both model error and spillover. The bounds obtained in [19] apply to control in the presence of some nonlinearity - a situation that is highly likely in LCES under high transient loading.

If (21) is not satisfied, i.e., the closed-loop stability margin is too deteriorated by spillover or model error, this warns that some compensation may be necessary to try to counteract or reduce these error terms, e.g., pre or post-filtering [15], orthogonal filtering [39], direct reduction via controller enhancement [40], or some combincation of these. After compensation has been added, the bounds (20) and (21) may be rechecked to decide if adequate error reduction has been achieved.

7.0 CONCLUSIONS

From our experience with distributed parameter systems and large structures in space, we have tried to generate a rather general framework for the design and analysis of actively controlled large civil engineering structures (LCES). Our emphasis has been on the design trade-offs and problem interactions rather than a proposed *unique* design algorithm (which we do not believe exists). We have also stressed the analysis of the controller interaction (through spillover and model error) with the unmodelled residuals and have presented energy bounds for the closed-loop system which could be used to aid the trial and error design process for practical LCES controllers.

In closing, we point out that structure control is really a combination of optimal design, e.g., [41]-[42], and active control through actuators and sensors. We hope the framework presented here can be used to give a clearer picture of the interrelationship and interaction of structural design and active controllers. We agree wholeheartedly with Professor Mroz that future LCES should be *control-configured structures*, i.e., the structural design and active control will be integrated, rather than independently optimized, to produce a new strain of civil engineering structures.

ACKNOWLEDGEMENT

We thank Professor J.T.P. Yao for his kind encouragement to prepare this paper, and to Professor H.H.E. Leipholz and the Solid Mechanics Division, University of Waterloo, for their support to attend and participate in this conference.

REFERENCES

[1] PETERSEN, N.R., "Design of Large Scale Tuned Mass Dampers", *Proc. IUTAM Symposium on Structural Control*, held at the University of Waterloo, Waterloo, Ontario, Canada, June 4-7, 1979.

[2] AZIZ, A., WINGATE, J. and BALAS, M., (Editors), *Control Theory of Systems Governed by Partial Differential Equations*, Academic Press, 1977.

[3] LEIPHOLZ, H.H.E., (Editor), *Proc. of the IUTAM Symposium on Structural Control*, held at the University of Waterloo, Waterloo, Ontario, Canada, June 4-7, 1979.

[4] YAO, J.T.P., "Concept of Structural Control", *ASCE, J. of the Structural Div.*, Vol. 98, 1972, pp. 1567-1574.

[5] YAO, J.T.P. and TANG, J-P., "Active Control of Civil Engineering Structures", *Technical Report CE-STR-73-1*, School of Civil Engineering, Purdue University, 1973.

[6] YANG, J-N. and YAO, J.T.P., "Formulation of Structural Control", *Technical Report CE-STR-74-2*, School of Civil Engineering, Purdue University, 1974.

[7]   ZUK, W., "Kinetic Structures", *ASCE, Civil Engineering*, Vol. 39, 1968, pp. 62-64.

[8]   MEIROVITCH, L., (Editor), *Proc. of the First and Second Symposia on Dynamics and Control of Large Flexible Spacecraft*", Blacksburg, Virginia, June 13-15, 1977 and June 21-23, 1979.

[9]   BALAS, M., "Some Trends in LSS Control Theory: Fondest Hopes; Wildest Dreams", *Proc. Joint Autom. Control Conf.*, Denver, Colorado, June 17-20, 1979, pp. 42-55.

[10] MEIROVITCH L. and ÖZ, H., "An Assessment of Methods for Control of LSS", *Proc. Joint Autom. Control Conf.*, Denver, Colorado, June 17-20, 1979, pp. 34-41.

[11] SENSBURG, O., BECKER, J. and HÖNLINGER, H., "Active Control of Flutter and Vibration of an Aircraft", *Proc. of the IUTAM Symposium on Structural Control*, held at the University of Waterloo, Waterloo, Ontario, Canada, June 4-7, 1979.

[12] EDWARDS, J., BREAKWELL, J.V. and BRYSON, A., "Active Flutter Control using Generalized Unsteady Aerodynamic Theory", *J. Guidance and Control*, Vol. 1, 1978, pp. 32-40.

[13] HOLMES, P. and MARSDEN, J., "Bifurcation to Divergence and Flutter in Flow-Induced Oscillations: An Infinite Dimensional Analysis", *Automatica*, Vol. 14, 1978, pp. 367-384.

[14] ROBINSON, A., "Survey of Dynamic Analysis Methods for Flight Control Design", *J. Aircraft*, Vol. 6, 1969, pp. 81-98.

[15] BALAS, M., "Feedback Control of Flexible Systems", *IEEE Trans. Autom. Control*, Vol. AC-23, 1978, pp. 673-679.

[16] BALAS, M., "Modal Control of Certain Flexible Dynamic Systems", *SIAM, J. Control*, Vol. 16, 1978, pp. 450-462.

[17] BALAS, M., "Direct Output Feedback Control of LSS", *J. Astronautical Sciences*, Vol. 27, 1979, pp. 157-180.

[18] BALAS, M., "Reduced-Order Control of LSS", presented at the 17th AIAA Aerospace Sciences Meeting, New Orleans, Louisianna, January 15-17, 1979.

[19] BALAS, M., "Control of Flexible Structures in the Presence of Certain Nonlinear Disturbances", *Proc. Joint Autom. Control Conf.*, Denver, Colorado, June 17-20, 1979, pp. 23-29.

[20] YANG, J-N. and GIANNOPOULOS, F., "Dynamic Analysis and Active Control of Two Cable-Stayed Bridge", presented at the ASCE Convention and Exposition, Boston, Massachusetts, April 2-6, 1979.

[21] YANG, J-N. and GIANNOPOULOS, F., "Active Tendon Control of Structures", *ASCE, J. of the Eng. Mech. Div.*, Vol. 104, 1978, pp. 551-568.

[22] ROORDA, J., "Tendon Control in Tall Structures", *ASCE, J. of the Structural Div.*, Vol. 101, 1975, pp. 505-521.

[23] YAO, J.T.P., "Identification and Control of Structural Damage", *Proc. of the IUTAM Symposium on Structural Control*, held at the University of Waterloo, Waterloo, Ontario, Canada, June 4-7, 1979.

[24] MEIROVITCH, L. and ÖZ, H., "Active Control of Structures by Modal Synthesis", *Proc. of the IUTAM Symposium on Structural Control*, held at the University of Waterloo, Waterloo, Ontario, Canada, June 4-7, 1979.

[25] CHANG, J. and SOONG, T.T., "The Use of Aerodynamic Appendages for Tall Building Control", *Proc. of the IUTAM Symposium on Structural Control*, held at the University of Waterloo, Waterloo, Ontario, Canada, June 4-7, 1979.

[26] KLEIN, R. and SALHI, H., "The Time-Optimal Control of Wind Induced Structural Vibrations using Active Appendages", *Proc. of the IUTAM Symposium on Structural Control*, held at the University of Waterloo, Waterloo, Ontario, Canada, June 4-7, 1979.

[27] SAE-UNG, S. and YAO, J.T.P., "Active Control of Building Structures", *ASCE, J. of the Eng. Mech. Div.*, Vol. 104, 1978, pp. 335-350.

[28] BALAS, M. and JOHNSON, C.R., "Adaptive Control of LSS", presented at Yale Workshop on Adaptive Control, New Haven, Connecticut, August 23-25, 1979.

[29] SKELTON, R. and GREGORY, C., "Measurement Feedback and Model Reduction by Modal Cost Analysis", *Proc. Joint Autom. Control Conf.*, Denver, Colorado, June 17-20, 1979, pp. 211-218.

[30] MARTIN, C. and SOONG, T.T., "Modal Control of Multi-Story Structures", *ASCE, J. of the Eng. Mech. Div.*, Vol. 102, 1976, pp. 613-623.

[31] YANG, J-N., "Application of Optimal Control Theory to Civil Engineering Structures", *ASCE, J. of the Eng. Mech. Div.*, Vol. 101, 1975, pp. 819-838.

[32] CHANG, M. and SOONG, T.T., "Optimal Control Configuration for Control of Complex Structures", *Proc. IUTAM Symposium on Structural Control*, held at the University of Waterloo, Waterloo, Ontario, Canada, June 4-7, 1979.

[33] ABDEL-ROHMAN, M. and LEIPHOLZ, H.H.E., "A General Approach to Active Structural Control", *Proc. of the IUTAM Symposium on Structural Control*, held at the University of Waterloo, Waterloo, Ontario, Canada, June 4-7, 1979.

[34] JUANG, J-N., YANG, J-N. and SAE-UNG, S., "Active Control of Large Civil Engineering Structures", *Proc. of the IUTAM Symposium on Structural Control*, held at the University of Waterloo, Waterloo, Ontario, Canada, June 4-7, 1979.

[35] JOHNSON, C.D., "Theory of Disturbance Accommodating Controllers", from *Control and Dynamic Systems: Advances in Theory and Applications*, (C.T. Leondes, Editor), Vol. 12, 1976, Academic Press, New York, New York.

[36] BALAS, M., "Disturbance Accommodating Control of Certain Distributed Parameter Systems", presented at Duke University Symposium on Policy Analysis and Information Systems", Durham, North Carolina, June, 1979.

[37] BALAS, M., "Direct Velocity Feedback Control of Large Structures in Space", *J. Guidance and Control*, Vol. 2, 1979, pp. 252-253.

[38] McLEAN, D., "Effects of Reduced Mathematical Models on Dynamic Response of Flexible Aircraft with Closed-Loop Control", *Proc. of the IUTAM Symposium on Structural Control*, held at the University of Waterloo, Waterloo, Ontario, Canada, June 4-7, 1979.

[39] SKELTON, R. and LIKINS, P., "Orthogonal Filters for Model Error Compensation in the Control of Nonrigid Spacecraft", *J. Guidance and Control*, Vol. 1, 1978, pp. 41-49.

[40] BALAS, M., "Enhanced Modal Control of Flexible Structures via Innovations Feedthrough", presented at the 2nd Symposium on Dynamics and Control of Large Flexible Spacecraft, Blacksburg, Virginia, June, 1979.

[41] MROZ, Z., "An Optimal Force Action and Reaction on Structures", *Proc. of the IUTAM Symposium on Structural Control*, held at the University of Waterloo, Waterloo, Ontario, Canada, June 4-7, 1979.

[42] CARMICHAEL, D. and CLYDE, D., "Multilevel Control Concepts in Relation to the Structural Design Problem", *Proc. of the IUTAM Symposium on Structural Control*, held at the University of Waterloo, Waterloo, Ontario, Canada, June 4-7, 1979.

STRUCTURAL CONTROL, H.H.E. Leipholz (ed.)
North-Holland Publishing Company & SM Publications
© IUTAM, 1980

OPTIMIZATION OF CONTROL DEVICES IN BASE ISOLATION
SYSTEMS FOR ASEISMIC DESIGN

M.A. Bhatti, K.S. Pister and E. Polak

Earthquake Engineering Research Center
University of California
Berkeley, California 94720, U.S.A.

1. INTRODUCTION

A number of tests of steel frames supported on a flexible, hysteretic foundation system have been conducted on the Earthquake Simulator at the Earthquake Engineering Research Center, University of California, Berkeley. These tests, along with a description of the test configuration and details of the support bearings and energy dissipation devices can be found in [1] and are also reported by Kelly in this Proceedings [2]. The tests show that for small earthquakes the structure behaves as if attached to a rigid foundation, while for strong earthquakes, the foundation isolation system (control device) yields and absorbs amounts of energy equivalent to as much as 35 percent of critical viscous damping. The question naturally arises, what is the "best" choice of control device for a particular structure and site earthquake hazard? This paper formulates the design of the control device as a constrained optimization problem utilizing nonlinear programming techniques. The format is that of a min-max problem with time-dependent constraints whose solution is obtained utilizing a method of feasible directions. An example is included to illustrate the methodology and indicate the significance of the results.

The test structure is a three-story steel frame with added mass at each floor level and whose dimensions are given in Figure 1. The structure is supported vertically by specially designed rubber bearings whose properties are specified. These bearings also provide nominal shear resistance. An energy absorbing device is linked to the base of the structure as shown. This device acts as an hysteretic passive controller, supplying a time-dependent horizontal force to the base. Selection of the mechanical properties of the controller constitutes the design problem.

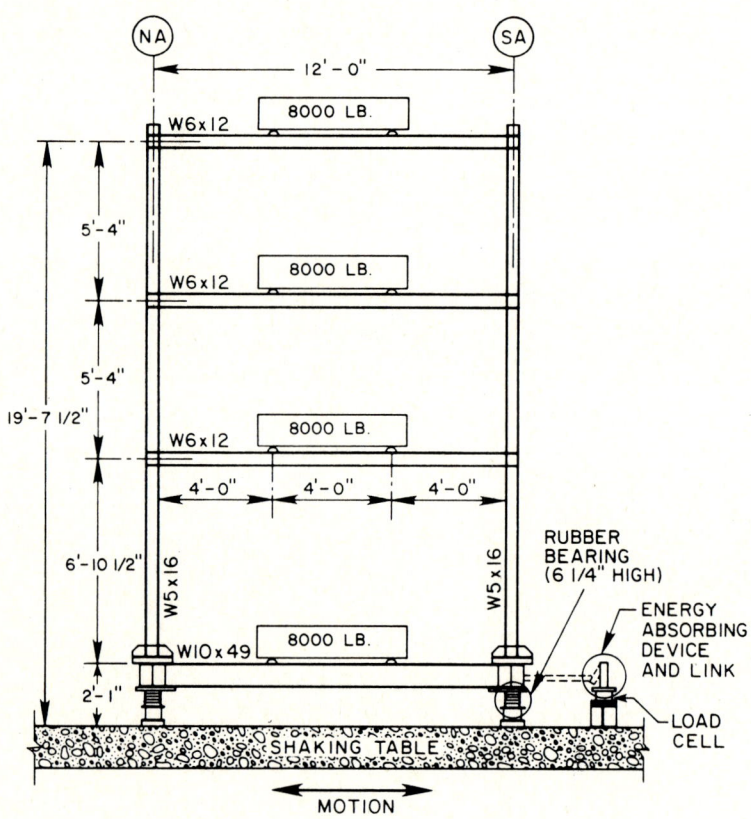

Figure 1 - Dimensions of Structure

2. CHARACTERISTICS OF THE CONTROL DEVICE

As described in [1, 2] the control device dissipates energy through hysteresis associated with cyclic inelastic torsion of a mild steel bar. The device is linked to the base floor of the test structure in such a way that a horizontal force is applied to the base, as shown in Figure 1. Based on work reported in [3], it has been found that an appropriate model for the control device can be expressed by the pair of equations:

$$\dot{F}(t) = K_0 \left[\dot{U}(t) - |\dot{U}(t)| \left(\frac{F(t)}{F_0} - S \right)^n \right], \qquad (1)$$

$$S(t) = \alpha \left[\frac{U(t)}{U_0} - \frac{F(t)}{F_0} \right], \qquad (2)$$

where

$F(t)$ = horizontal force in the device,
$U(t)$ = displacement of the device,
$\dot{U}(t)$ = velocity of the device,
K_0 = F_0/U_0,
F_0 = yield force,
U_0 = yield displacement,
α = a constant which controls the slope after yielding,

$$K_y \approx K_0 \frac{\alpha}{1+\alpha},$$

n = a material parameter, taken as an odd integer which controls the sharpness of transition from the elastic to the inelastic region. As $n \to \infty$ the model approaches a bilinear model.

A typical hysteresis loop associated with a sinusoidal displacement $U(t)$ is shown in Figure 2 for a prescribed set of model parameters F_0, U_0, α, n.

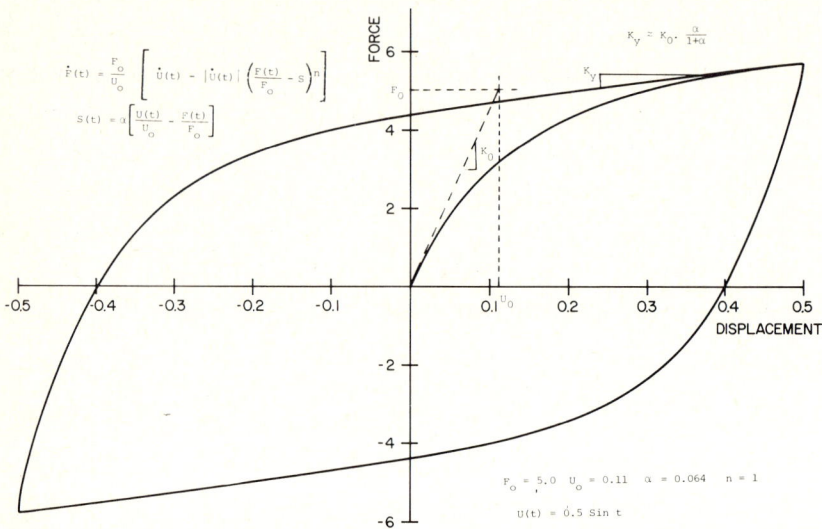

Figure 2 - *Hysteresis Loops Generated by Rate Independent Hysteresis Model under Sinusoidal Excitation*

3. EQUATIONS OF MOTION FOR THE STRUCTURE

As shown in Figure 1, the structural system consists of an assemblage of beam and column sections. Axial deformation in both beams and columns is neglected; thus, we have one lateral and two rotational degrees of freedom per story. It is assumed that story masses are lumped at the floor levels and rotational inertia is neglected. The structure is designed to remain elastic during an earthquake, so that nonlinearity in the system is confined to the control device. Rotational degrees of freedom may be eliminated by appropriate partitioning of the mass, damping and stiffness matrices associated with the discretized structural model of the frame. The resulting equations of motion can be written [1];

$$\underline{M}\,\ddot{\underline{U}} + \underline{C}\,\dot{\underline{U}} + \underline{K}^E\,\underline{U} + \underline{F} = -\underline{M}\,\underline{e}\,\ddot{U}_g(t) , \qquad (3)$$

where

$\ddot{U}_g(t)$ = ground acceleration history,
$\underline{U}^T$ = $[U_1, U_2, U_3, U_4]$, vector of lateral floor displacements. The superscript T denotes the matrix transpose,
$\underline{F}^T$ = $[0, 0, 0, F_4]$,
$\underline{e}^T$ = $[1, 1, 1, 1]$,
F_4 = force in the energy absorbing device, whose constitutive equations from (1), (2) are:

$$\dot{F}_4(t) = K_0 \left[\dot{U}_4(t) - |\dot{U}_4(t)| \left(\frac{F_4(t)}{F_0} - S \right)^n \right], \qquad (4)$$

$$S(t) = \alpha \left[\frac{U_4(t)}{U_0} - \frac{F_4(t)}{F_0} \right]. \qquad (5)$$

The stiffness, mass and damping matrices $\underline{K}^E$, $\underline{M}$, $\underline{C}$, respectively, for the structure shown in Figure 1 can be routinely calculated. Thus, the lateral stiffness matrix of the complete structure, including the stiffness of rubber bearings is (units are kips, inches)

$$\underline{K}^E = \begin{bmatrix} 46.38 & -66.64 & 23.04 & -2.78 \\ & 144.40 & -96.50 & 18.74 \\ & \text{SYMMETRIC} & 122.43 & -48.97 \\ & & & 34.21 \end{bmatrix}. \qquad (6)$$

The mass matrix of the structure corresponding to the lateral degrees of freedom is

$$\underline{M} = \begin{bmatrix} 0.02438 & & & \\ & 0.02438 & & \\ & & 0.02514 & \\ & & & 0.02832 \end{bmatrix}. \qquad (7)$$

Rayleigh damping is assumed in constructing the damping matrix

$$\underline{C} = \alpha \underline{M} + \beta \underline{K}^E .$$

The coefficients α and β are computed from

$$\begin{bmatrix} \frac{1}{\omega_1} & \omega_1 \\ \frac{1}{\omega_2} & \omega_2 \end{bmatrix} \begin{Bmatrix} \alpha \\ \beta \end{Bmatrix} = \begin{Bmatrix} \xi_1 \\ \xi_2 \end{Bmatrix} ,$$

where ω_1 and ω_2 are first and second mode frequencies, and ξ_1 and ξ_2 are the respective critical damping ratios in these modes. The damping matrix for the present structure, assuming ξ_1 = 3% and ξ_2 = 1%, is given below,

$$\underline{C} = \begin{bmatrix} .0279 & -.0332 & .0115 & -.0014 \\ & .0768 & -.0481 & .0093 \\ & & .0660 & -.0244 \\ \text{SYMMETRIC} & & & .0226 \end{bmatrix} . \qquad (8)$$

Note that in the equations of motion (3) the characteristic parameters of the device (F_0, U_0, α, n), which are the design variables for the problem, appear explicitly only in the force vector $\underline{F}$. For a prescribed set of these parameters and a ground acceleration, solutions of (3) are obtained by time discretization and step-by-step integration of (3) using Newmark's method, in which the internal forces in the device are computed using a fourth-order Runge-Kutta algorithm. Details may be found in [1].

The calculated structural response of the structure for a prescribed ground motion and assumed design variables constitutes a simulation of the performance of the structure. We are now in a position to define a design problem in which certain measures of performance are to be constrained and optimized.

4. AN OPTIMAL DESIGN PROBLEM

As an example we will examine the following design problem. For brevity a discussion of the design algorithm is omitted; details may be found in [1]. Referring to Figure 1, we wish to minimize the pseudo-shear in the structure subject to the condition that the maximum displacement of the bottom floor remains less than a certain prescribed value. Accordingly, we have

$$\min_{\underline{z}} \left[\max_{t \in T} \left\{ \left(K_1(U_1(\underline{z},t) - U_2(\underline{z},t)) \right)^2 + \left(K_2(U_2(\underline{z},t) - U_3(\underline{z},t)) \right)^2 + \left(K_3(U_3(\underline{z},t) - U_4(\underline{z},t)) \right)^2 \right\} \right] , \quad (9)$$

subject to:

$$\max_{t \in T} (U_4(\underline{z},t))^2 \leq \delta^2 ,$$

$$F_0, U_0, \quad \alpha > 0 \quad (10)$$

$T = [t_0, t_f]$, duration of ground shaking,

$\underline{z}^T = [z^1, z^2, z^3] = [F_0, U_0, \alpha]$, design vector,

K_1, K_2, K_3 = the story stiffnesses (top down), from (6),

U_1, U_2, U_3, U_4 = displacement response histories at the floor levels (top down).

δ = prescribed limit on the bottom floor displacement, selected to be 4 inches in the present problem.

The following comments are in order. The objective function appearing in (9), called a "pseudo-shear", is the sum of the squares of the individual story shear forces; clearly, other measures of performance could have been chosen. The choice here was dictated by an *a posteriori* examination of the story shear forces resulting from the solution of the example. Further, the design vector has three components F_0, U_0, α. The value of the material parameter "n" was set equal to unity.

The problem stated in (9) and (10) can be transformed to an equivalent canonical form of nonlinear programming problem for which an algorithm and program are available [1, 4]. The equivalent nonlinear programming problem is,

$$\min_{\underline{z}} z^4 \,, \tag{11}$$

subject to:

$$\max_{t \in T} \left[\left\{ K_1 \bigl(U_1(\underline{z},t) - U_2(\underline{z},t) \bigr) \right\}^2 + \left\{ K_2 \bigl(U_2(\underline{z},t) - U_3(\underline{z},t) \bigr) \right\}^2 \right.$$
$$\left. + \left\{ K_3 \bigl(U_3(\underline{z},t) - U_4(\underline{z},t) \bigr) \right\}^2 \right] \leq z^4 \,, \tag{12}$$

$$\max_{t \in T} [U_4(\underline{z},t)]^2 \leq \delta^2$$

$$F_0, U_0, \alpha > 0 \,.$$

It can be seen that the dummy cost parameter z^4 represents an upper bound on the pseudo-shear.

The El Centro 1940 NS ground motion with modified time scale and amplitude as used in the experimental investigations [1, 2] was used for this programme. Initial values of the design parameters were

$$F_0 = 5.0, \ U_0 = 0.11, \ \alpha = 0.064, \ n = 1.$$

Note that this represents a feasible design; i.e., constraints are met. The value of the dummy cost associated with these parameters was 35.0. Optimal values of the design parameters were found to be:

$$F_0 = 4.28, \quad U_0 = 1.75, \quad \alpha = .00577$$

and the dummy cost was 9.15. Reduction in the value of the cost parameter versus design iteration number is shown in Figure 3.

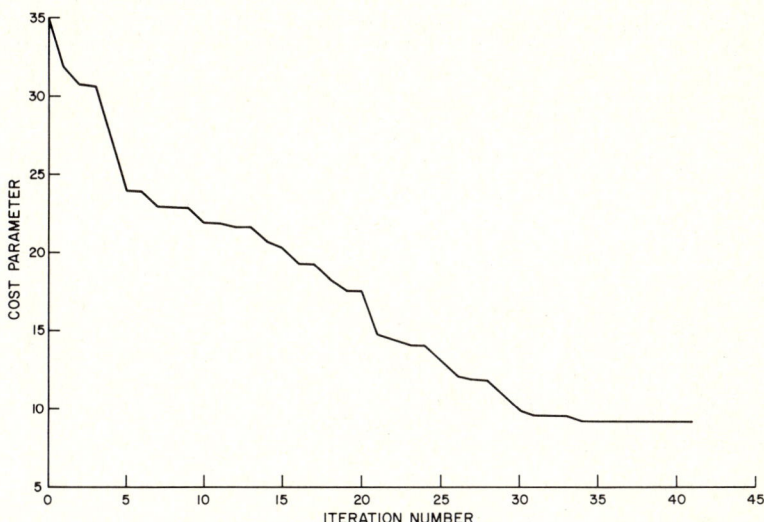

Figure 3 - Cost Parameter and Design Iterations

Figures 4, 5, 6 and 7 show the story shear forces and the bottom floor displacement for both initial and optimal device parameters. These plots clearly demonstrate the effectiveness of the optimal isolation device, which reduces the story shears by as much as a factor of two. Although the bottom floor displacement is increased, the maximum remains within the allowable limit of 4 inches.

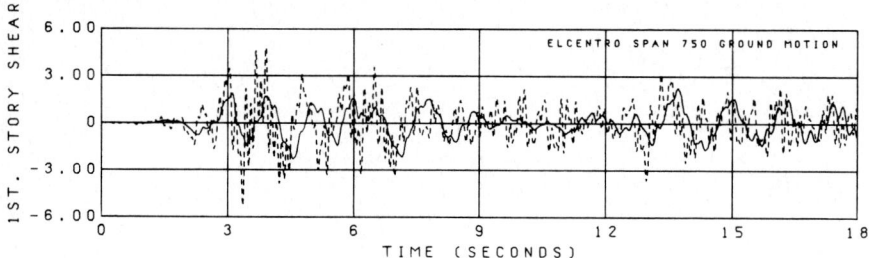

Figure 4 - First Story Shear Time History
——— *Optimal Design* ---- *Initial Design*

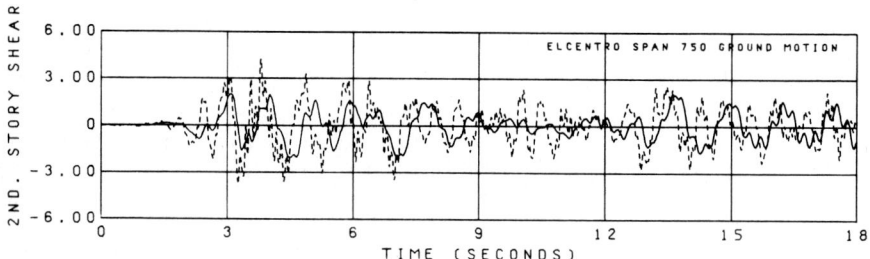

Figure 5 - Second Story Shear Time History
——— *Optimal Design* ---- *Initial Design*

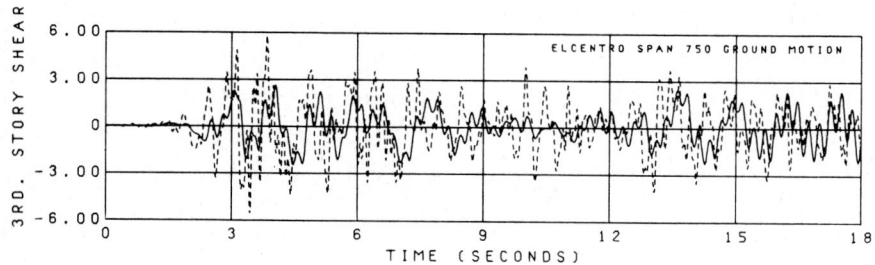

Figure 6 - Third Story Shear Time History
——— *Optimal Design* ---- *Initial Design*

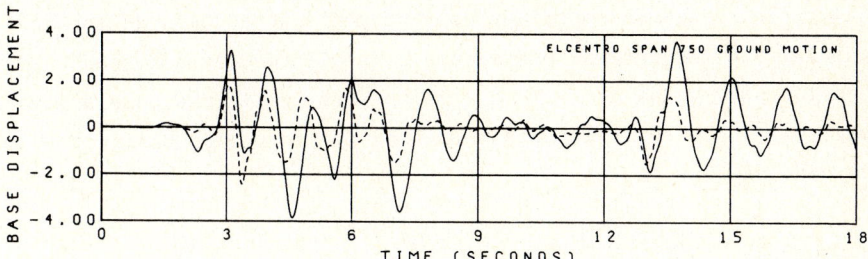

Figure 7 - Base Displacement Time History
——— Optimal Design ---- Initial Design

As is clear from the preceding development, only one earthquake ground motion has been used in the optimal design process. Since earthquakes are random in nature, it is unlikely that the same earthquake ground motion will be repeated at some future time. Therefore, it becomes necessary to examine the effectiveness of the optimal design process for different earthquakes. To get meaningfull results, these additional earthquakes should have characteristics similar to the one used in the design process. This requirement prohibits the use of actual past earthquake records, inasmuch as for a given site there is typically an insufficient number of past records. An alternative is the use of artificially generated earthquakes of the same class. For the present study a family of five earthquakes having characteristics similar to the El Centro 1940 NS earthquake was generated using the programme PSEQGN developmentd by P. Ruiz and J. Penzien [5] and later modified by M. Murakami [6]. The earthquake accelerograms were generated by passing nonstationary shot noise through two second-order linear filters and applying a base line correction. Each accelerogram was of thirty seconds duration with four seconds of parabolic built up, eleven seconds of constant intensity followed by fifteen seconds of exponential decay. The maximum acceleration in each record was about 0.30g. The structure was analyzed twice, with initial and optimal parameters, subjected to these five earthquakes. Story shears and bottom floor displacements were compared for the initial and the optimal parameters. In all the cases response was considerably reduced in the optimal system, while the base displacement was increased, although remaining within the specified constraint of four inches.

ACKNOWLEDGEMENTS

This research was conducted under National Science Foundation Grant ENV76-04264.

REFERENCES

[1] BHATTI, M.A., PISTER, K.S. and POLAK, E., "Optimal Design of an Earthquake Isolation System", *Report No. UCB/EERC-78/22*, Earthquake Engineering Research Center, University of California, Berkeley, California, October, 1978.

[2] KELLY, J.M., "Control Devices in Base Isolation Systems for Aseismic Design", *Proc. IUTAM Symposium on Structural Control*, held at the University of Waterloo, June 4-7, 1979, Waterloo, Ontario, Canada.

[3] ÖZDEMIR, H., "Nonlinear Transient Dynamic Analysis of Yielding Structures", *Ph. D. Dissertation*, Division of Structural Engineering and Structural Mechanics, Department of Civil Engineering, University of California, Berkeley, California, 1976.

[4] GONZAGA, G., POLAK, E. and TRAHAN, R., "An Improved Algorithm for Optimization Problems with Functional Inequality Constraints", *Memorandum No. UCB/ERL M78/56*, Electronics Research Laboratory, University of California, Berkeley, California, September, 1977.

[5] RUIZ, P. and PENZIEN, J., *PSEQGN - Artificial Generation of Earthquake Accelerograms*, A Computer Program distributed by NISEE/Computer Applications, University of California, Berkeley, California, March, 1969.

[6] MURAKAMI, M. and PENZIEN, J., "Nonlinear Response Spectra for Probabilistic Seismic Design and Damage Assessment of Reinforced Concrete Structures", *Report No. UCB/EERC 75-38*, Earthquake Engineering Research Center, University of California Berkeley, California, November, 1975.

STRUCTURAL CONTROL, H.H.E. Leipholz (ed.)
North-Holland Publishing Company & SM Publication
© IUTAM, 1980

OPTIMALLY STABLE STRUCTURES SUBJECTED TO
FOLLOWER FORCES

R. Bogacz*

Institute of Fundamental Technological Research
Polish Academy of Sciences
Warsaw, Poland

O. Mahrenholtz

Institut für Mechanik
Universität Hannover
Hannover, West Germany

1. INTRODUCTION

The study of the behaviour of structures under follower forces is an important problem in engineering practice. In the past, several research workers have investigated this problem from the point of view of stability. Another aspect which is equally important is the optimal design of the columns. This also has been studied by several authors. Niordson [1] was probably the first who considered the optimal design of vibrating beams.

Prager and Taylor [2] have developed a unified theory of optimal design under stability and frequence constraints. Some further generalizations have been presented by Vepa and Roorda [3]. The optimal shape of cantilevers under follower forces has been investigated by Vepa [4], Claudon [5] and Plaut [6]. It may be mentioned here that the optimal forms of stepped columns and beams have always found great application. With this in view Mroz with

---
*Fellow of the Alexander von Humboldt Foundation 1978/79, visiting the University of Hannover.

co-authors [7, 8, 9] have considered the optimal segmentation of beams with discrete variations in the sectional properties. Some optimal segmentation problems for columns have been studied previously by the authors [10]. The present paper complements the previous research devoted to the optimal segmentation problem.

The structure considered consists of a stepped column with segments of lengths $L_1, \ldots, L_n$ and corresponding cross-sections $A_1, \ldots, A_n$.

The structure is very general in that it is assumed to be supported at p points at distances $r_1, \ldots, r_p$ from the fixed end on supports of dynamical stiffness $\kappa_1, \ldots, \kappa_p$ respectively. The problem is to find the optimal cross-sections or parameters of interacting systems so that the total cost, including that of the connections $\delta_1, \ldots, \delta_{n-1}$, is a minimum:

$$J = \phi(A_1,\ldots,A_n, L_1,\ldots,L_n, \kappa_1,\ldots,\kappa_n, r_1,\ldots,r_p, \delta_1,\ldots,\delta_{n-1}). \tag{1}$$

It is required that the optimal structure by definition must fulfill a further stability constraint condition [5, 11]

$$P_{cr} \geq P, \tag{2}$$

here P is the given value of the follower force.

2. ANALYSIS

2.1 *Columns with Concentrated Load*

The simplest form of equation of motion for columns with concentrated axial load is

$$\frac{\partial^2}{\partial x^2}\left(EI \frac{\partial^2 y}{\partial x^2}\right) + P \frac{\partial^2 y}{\partial x^2} + \rho A \frac{\partial^2 y}{\partial t^2} = 0. \tag{3}$$

The boundary conditions for the case of Beck's column are:

for $x = 0$: $\quad y = 0 \quad ; \quad \dfrac{\partial y}{\partial x} = 0$

for $x = L$: $\quad \dfrac{\partial^2 y}{\partial x^2} = 0 \quad ; \quad \dfrac{\partial}{\partial x}\left[EI \dfrac{\partial^2 y}{\partial x^2}\right] = 0 .$ (4)

The corresponding conditions for Reut's column with y^* as the displacement are:

for $x = 0$: $\quad y^* = 0 \quad ; \quad \dfrac{\partial y^*}{\partial x} = 0$

for $x = L$: $\quad EI \dfrac{\partial^2 y^*}{\partial x^2} + Py^* = 0 \quad ; \quad \dfrac{\partial}{\partial x}\left[EI \dfrac{\partial^2 y^*}{\partial x^2} + Py^*\right] = 0 .$ (5)

It is seen that neither of the above two problems is selfadjoint. However, the two are mutually adjoint.

Only for constant ρA and EI exact closed form solutions can be found. When these parameters are varying, in general only approximate solutions are possible. Since the problem is not selfadjoint, the Leipholz generalization of Hamilton's principle, [12] can be used in the solution process. Using this concept for Beck's column the Leipholz' functional takes the form:

$$H^* = - \int_{t_1}^{t_2} \int_0^L \left[\ddot{y}(x,t) + \dfrac{EI}{\rho} y^{IV}(x,t) + \dfrac{P}{\rho} y''(x,t)\right] y''[(L-x),t] \, dx \, dt .$$ (6)

For the actual motion we have

$$\delta H^* = 0 .$$ (7)

In our case, the solution is based on a harmonic motion:

$$y(x,t) = Y(x)e^{i\omega t} .$$ (8)

By virtue of (6), (7) and (8) we have

$$\int_0^L \left[-\omega^2 Y(x) + \frac{EI}{\rho} Y^{IV}(x) + \frac{P}{\rho} y''(x) \right] \delta Y''(L-x) dx = 0 . \qquad (9)$$

If $Y_1(x)$ and $Y_2(x)$ are two admissible functions satisfying the boundary conditions (4), then according to [12] the relation

$$\det(b_{ij}) = 0 \quad , \quad (i,j = 1,2) , \qquad (10)$$

where

$$b_{ij} = \int_0^L \left[-\omega^2 Y_i(x) + \frac{EI}{\rho} Y_i^{IV}(x) + \frac{P}{\rho} Y_i'' \right] Y_j(L-x) dx , \quad (11)$$

can be used to determine the characteristic curves. In the case of a concentrated load it is also possible to use the concept of the Leipholz' adjoint variational principle [5, 10], which leads to a relation between P and ω of the form

$$\left| c_{ij} + P d_{ij} - \omega^2 k_{ij} \right| = 0 . \qquad (12)$$

The matrices (c_{ij}) and (d_{ij}) and (k_{ij}) are:

$$c_{ij} = \int_0^L EI\, f_i'' f_j'' \, dx \; ; \quad d_{ij} = \int_0^L f_i'' f_j \, dx \; ; \quad k_{ij} = \int_0^L \rho A\, f_i f_j \, dx$$

where f_i are orthonormal polynomials.

The plot of the first values of ω as functions of P has one of the following characteristic forms (Figure 1). The procedure for satisfying the condition (2) is as follows. After starting from an assumed form of the structure, for a given value of P and the number N of considered modes, N values of ω are to be determined. If N real values of ω exist, the assumed structure is stable; otherwise, we have only N - 2K (K = 1, 2, ...) real values.

The structural parameters have to be altered, until the optimal form is found.

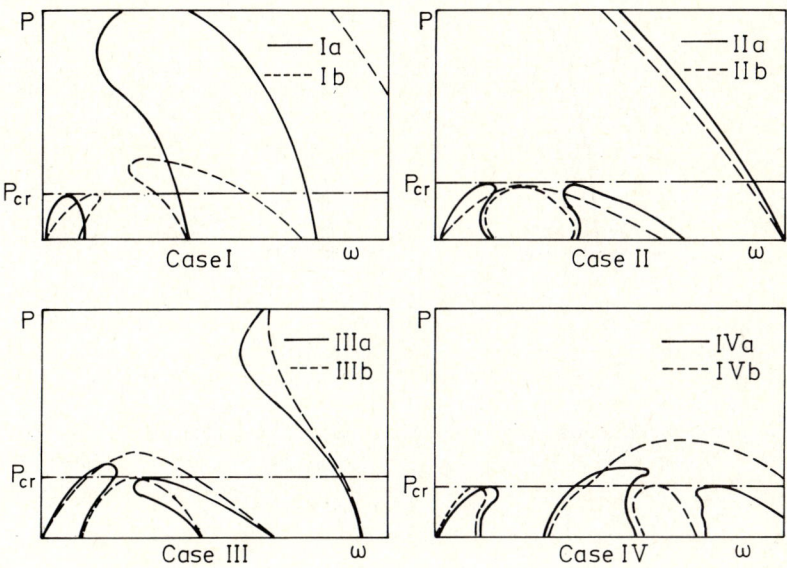

Figure 1 - *Characteristic Forms for P as a Function of ω*

2.2 Columns with Distributed Load

The other case of a simple structure, for which we look for the optimal form is the so-called Leipholz' column and adjoint system (Figure 2). This type of column is a simplification of such a system considered by Mahrenholtz.

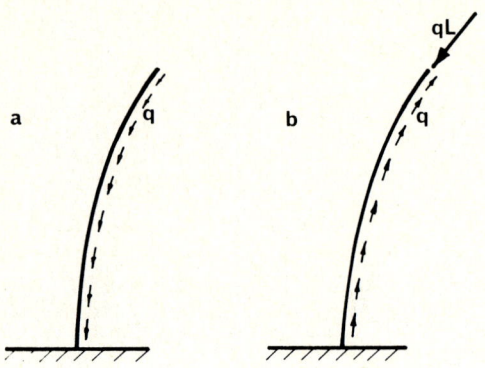

Figure 2 - (a) Leipholz Column, (b) Adjoint System

Now, the equation of motion takes the form

$$\frac{\partial^2}{\partial x^2}\left(EI \frac{\partial^2 y}{\partial x^2}\right) + P\left(1 - \frac{x}{L}\right)\frac{\partial^2 y}{\partial x^2} + \rho A \frac{\partial^2 y}{\partial t^2} = 0 , \qquad (13)$$

where $P = q_o L$.

The boundary condition is the same as in (4). The generalized adjoint system shown in Figure 2 is analytically described by boundary condition of form (4) and the following equation of motion

$$\frac{\partial^2}{\partial x^2}\left(EI \frac{\partial^2 y^*}{\partial x^2}\right) + P\frac{x}{L}\frac{\partial^2 y^*}{\partial x^2} + \rho A \frac{\partial^2 y^*}{\partial t^2} = 0 . \qquad (14)$$

The functional can now be described by

$$H^* = \int_o^L \left[EI\, Y_1^{IV}(x) + P\left(1 - \frac{x}{L}\right)Y_1''(x) - \rho A \omega^2 Y_1(x)\right] Y_2''(L-x)\, dx , \qquad (15)$$

where Y_1 and Y_2 are approximations of $y = Y\exp(i\omega t)$ in (12) and $y^* = Y^*\exp(i\omega t)$ in (13), respectively.

All the equations and procedures described before are valid except for the definition of the matrix (b_{ij}), which now has the elements

$$b_{ij} = \int_o^L \left[EI\, f_i^{IV}(x) + P\left(1 - \frac{x}{L}\right) f_i''(x) - \rho A \omega^2 f_i(x) \right] f_j''(L-x)\,dx , \quad (16)$$

where

$$Y = \Sigma\, a_k f_k(x) ; \quad Y^* = \Sigma\, a_k^* f_k(x) ,$$

and $f_k(x)$ are coordinate functions.

2.3 Columns with Concentrated Load Interacting with Other Systems

For the case, when the construction is composed of Beck's column with piecewise constant thickness and elastic or dynamical supports (Figure 3), where stiffnesses and location of points of interaction are to be determined, a solution can be found by method of transfer matrices. The equation of motion for the column including the effect of shear deflection and rotary inertia (Figure 4) has the form:

$$y^{IV} + \frac{\rho A \omega^2}{EI}\left(\frac{EI}{GA_o} + I_o^2\right) + \frac{P}{EI} y'' - \frac{\rho A \omega^2}{EI}\left(1 - \frac{\rho A I \omega^2}{GA_o^o}\right) y = 0 , \quad (17)$$

with boundary conditions

$$y(o) = 0 , \quad \phi(o) = 0 , \quad M(L) = 0 , \quad Q(L) = 0 , \quad (18)$$

where

$$y = \frac{1}{\rho A \omega^2} \frac{dQ}{dx} ; \quad \phi = \frac{1}{\rho A I_o \omega^2}\left(Q - \frac{dM}{dx}\right) , \quad M = EI \frac{d\phi}{dx} ,$$

$$Q = GA_o \left(\frac{dy}{dx} + \phi\right) .$$

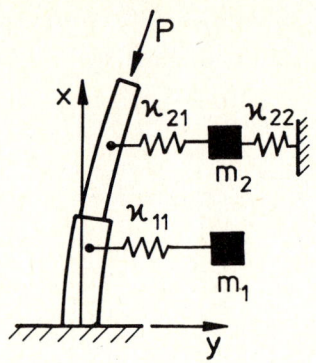

Figure 3 - Columns with Concentrated Load Interactions with Other Systems

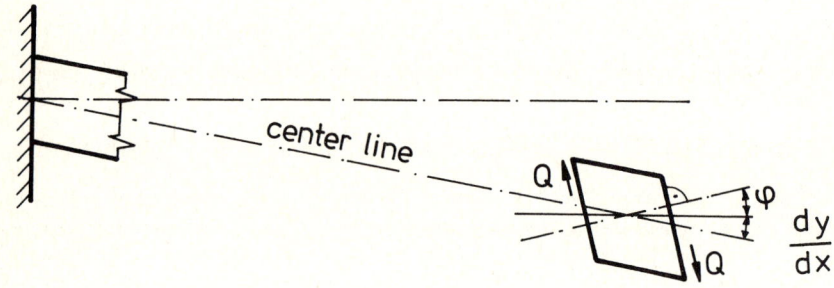

Figure 4 - Column Including Effect of Shear Deflection and Rotary Inertia

The exact solution of equation (17) when EI is constant and the motion is harmonic, is given by

$$y(x) = A_1 \text{sh}\lambda_1 x + A_2 \text{ch}\lambda_1 x + A_3 \sin\lambda_2 x + A_2 \cos\lambda_2 x , \quad (19)$$

where $\lambda_1 = k_{1/L}$,

$$k_{1/2} = \lambda_{1/2} L = \left\{ \sqrt{\frac{\rho A \omega^2 L^4}{EI} + \frac{1}{4}\left[\frac{PL^2}{EI} + \rho^2 A^2 \omega^4 L^4 \left(\frac{1}{GA_0} - \frac{I_0}{EI}\right)\right]^2} \right. $$
$$\left. \mp \frac{1}{2} \rho A \omega^2 L^2 \left(\frac{1}{GA_0} + \frac{I_0}{EI}\right) + \frac{PL^2}{EI} \right\}^{1/2}. \tag{20}$$

Since all the dependent variables Y, φ, M, Q have a similar constitutive form, we can state

$$\underline{G} = [y \; \phi \; M \; Q]^T. \tag{21}$$

Looking for a transfer matrix $\underline{T}_i$, which connects the state vector $\underline{G}_i^0 = \underline{G}(x_i = 0)$, (Figure 5) with $\underline{G}_{i+1}^0 = \underline{G}(x_{i+1} = 0)$ by the equation

$$\underline{G}_{i+1}^0 = \underline{T}_i \; \underline{G}_i^0, \tag{22}$$

we obtain for a segment of constant thickness the transfer matrix:

$$\underline{T}_i = \begin{bmatrix} t_0 - \frac{\Omega L^2 t_2}{GA_0} & L(t_1 - C_3 t_3) & \frac{Lt_2}{EI} & -\frac{Lt_1}{GA_0} + \frac{L^3 t_3}{EI} + \frac{t_3}{GA_0 L^2} \\ -\frac{\Omega L^4 t_3}{EI} & t_0 - \left(\frac{\Omega}{GA_0} + \frac{P}{EI}\right) L^2 t_2 & \frac{Lt_1}{EI} - \frac{L^2}{EI}\left(\frac{\Omega}{GA_0} + \frac{P}{EI}\right) & \frac{L^2 t_2}{EI} \\ -\Omega L^2 t_2 & (P + \Omega I_0^2) L t_1 + C_2 t_3 & t_0 - \frac{\Omega I_0^2 + P}{EI} L^2 t_2 & Lt_1 - C_3 t_3 \\ -\Omega L\left(t_1 - \frac{\Omega L}{GA} t_3\right) & \Omega L^2 t_2 & \frac{\Omega L^2 t_3}{EI} & t_0 - \frac{\Omega L^2 t_2}{GA_0} \end{bmatrix}$$

where

$$\Omega = \rho A \omega^2,$$
$$C_1 = \Omega L^4 \left(\frac{1}{EI} + \frac{\Omega}{GA_0}\right), \quad C_2 = \frac{\Omega L^4}{EI} + \left(\frac{\Omega I_0^2 + P}{EI}\right)^2 L^4,$$
$$C_3 = \Omega L^2 \left(\frac{1}{GA_0} + \frac{I_0^2}{EI}\right),$$
$$t_i = \frac{c_i}{\lambda_1^2 + \lambda_2^2} \quad (i = 0, 1, 2, 3),$$

$$c_0 = \lambda_2^2 \cosh\lambda_1 + \lambda_1^2 \cos\lambda_2, \quad c_2 = \cosh\lambda_1 - \cos\lambda_2,$$

$$c_3 = \cosh\lambda_1 - \cos\lambda_2, \quad c_3 = \frac{1}{\lambda_1}\sinh\lambda_1 - \frac{1}{\lambda_2}\sin\lambda_2.$$

$A_0 = A\beta^{-1}$ where β is a form factor depending on the shape of the cross section.

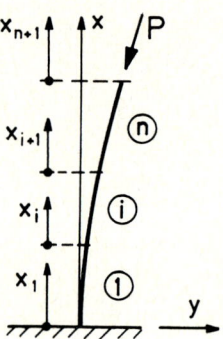

Figure 5 - Co-ordinate System

For an interacting discrete system (with spring stiffnesses κ_{ij} and concentrated masses m_i) the transfer matrix $\underline{T}_i$ takes the form:

$$\underline{T}_i = \begin{bmatrix} 1 & 0 & 0 & 0 \\ 0 & 1 & 0 & 0 \\ 0 & 0 & 1 & 0 \\ \eta_i & 0 & 0 & 1 \end{bmatrix} \qquad (23)$$

where:

$$\eta_i = \frac{\kappa_{i_1}(\kappa_{i_2} - m_i\omega^2)}{\kappa_{i_1} + \kappa_{i_2} - m_i\omega^2}. \qquad (24)$$

Optimally Stable Structures   149

The transfer matrix of the complete structure is given by

$$\underline{T} = \underline{T}_n \underline{T}_{n-1}, \ldots, \underline{T}_2 \underline{T}_1 . \tag{25}$$

We obtain the characteristic equation by fulfilling the boundary conditions (18), i.e.,

$$\underline{G}_1^0 = [0 \quad 0 \quad M_1^0 \quad Q_1^0]^T , \tag{26}$$

$$\underline{G}_{n+1}^0 = [y_{n+1}^0 \quad \phi_{n+1}^0 \quad 0 \quad 0]^T , \tag{27}$$

which yields

$$\begin{vmatrix} t_{33} & t_{34} \\ t_{43} & t_{44} \end{vmatrix} = 0 . \tag{28}$$

3. NUMERICAL RESULTS

3.1 *Columns with Concentrated Load*

As an illustrative example we first look for the optimal forms of the Beck's and the Reut's columns, for the following functional of total cost

$$J = r \sum_{j=1}^{m} A_j L_j + (m-1)\delta . \tag{29}$$

Here, r is the cost of the unit volume, m the number of segments and δ the cost of a connection between two segments. The bending stiffness EI is taken proportional to $(\rho A)^2$. For different values of m the maximal value A of A_j is taken to be constant. In Figure 6, the variation of $(\rho A)^2$ along the column, for columns, which have the minimum volume v_i, and m = 1, 2 and 3 is shown.

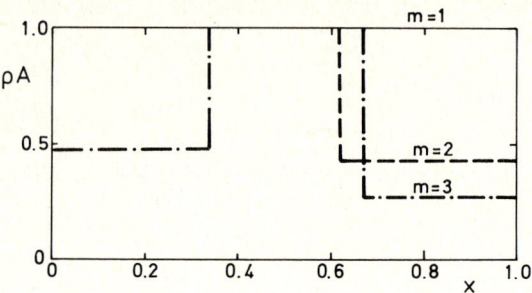

Figure 6 - Variation of the Cross-Section for the Case of Concentrated Load

For the column with m = 2, Figure 7 shows the critical force P_{cr} as a function of the volume $v = (1-A_2/A)$. L_2/L cut off from the column with m = 1. The parameter on the curves is the ratio L_2/L. In these calculations it is taken $GA_0 \to \infty$ and $I_0 = 0$. For the given value of $P_{cr} = 20.05$ EI/L, the minimum volume case is reached for $L_1/L = 0.62$. The maximum critical force case for $v < v_1$ and constraint $A_j < A$ is reached for $L_1/L = 0.70$.

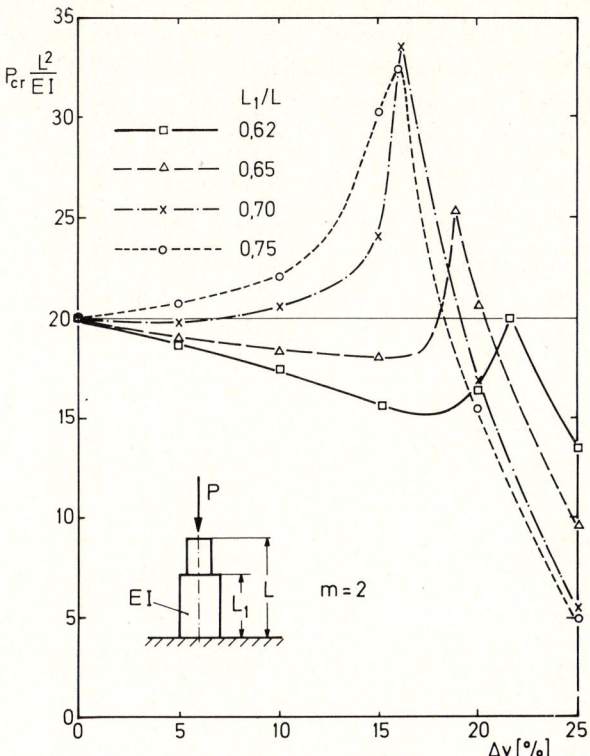

Figure 7 - *Critical Load as a Function of the Cut off Volume*

The dependence $P(\omega)$ as shown in Figure 2, Case I is characteristic for such column shapes, for which the P_{cr} versus Δv curve would lie to the left of the peak shown in Figure 7. Similarly, the Case III is the typical variation of $P(\omega)$ for the ration right of the peak. Case II and IV are characteristic for columns which in Figure 7 would lie at the peak itself, and the critical force or volume of columns for such a configuration of the characteristic curves are extremized.

The maximum of P_{cr} as a function of L_1/L is shown in Figure 8. For every value $L_1/L \leq 0.62$ the maximum of critical load is equal to that of a column with constant thickness ($\Delta v = 0$), that is $P_{cr} = 20.05 \ EI/L^2$. For $L_1/L > 0.62$ forms of column with $\Delta v > 0$ exist, which have a maximum of critical load greater than the above value. The same kind of analysis can be done for columns with $m > 2$.

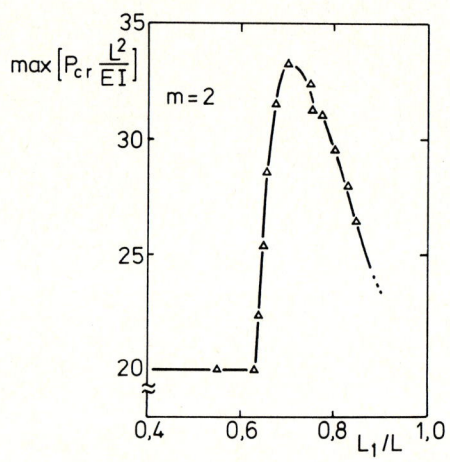

Figure 8 - Maximum of Critical Load as a Function of the Length of the First Segment

The result of this analysis shows the total cost J (equation (29)) as a function of the connection cost δ for the first four values m, where m is the number of segments. For $\delta > 0.22 \ rAL$ the optimal form is the one with $m = 1$; similarly for $0.19 \ rAL < \delta < 0.22 \ rAL$ and for $m = 2$, for $0.05 \ rAL < \delta < 0.19 \ rAL$ the optimal form corresponds to the case of $m = 2$ and $m = 3$ respectively.

The described procedure is an optimal solution for a given value of the critical load and minimum of the total cost with the constraint $A_j < A$. The inverse problem - maximum of critical load by constant volume [6], as well as by constant total cost - leads to another result. This fact can also be easily seen from Figure 7.

3.2 Columns with Distributed Load

The calculation for Leipholz' column (Figure 2) is done by using the same procedure. The optimal forms for m = 2 are shown in Figure 9. The column of form "a" has the same critical load as for m = 1, but the volume is 25% smaller. The column "b" has a volume of about 20% smaller than for m = 1, nevertheless, the critical load is about 18% greater.

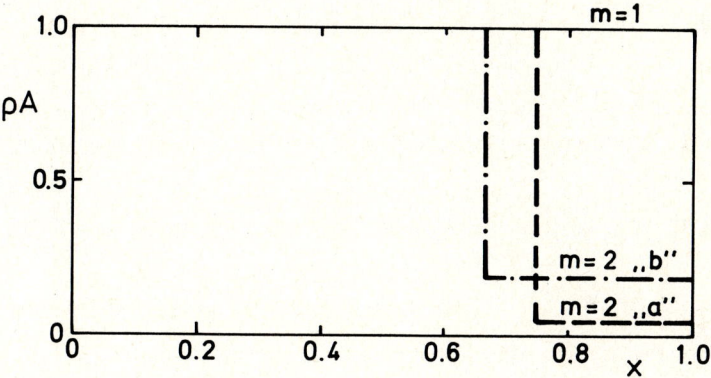

Figure 9 - Variation of the Cross-Section for the Case of Distributed Load

The corresponding curves for critical load versus the cutoff volume for m = 2 are similar as in [10]. The same kind of analysis can be done for columns with m > 2.

3.3 Columns with Concentrated Load and Elastic Supports

The interaction with other systems shown in Figure 3 may qualitatively change the configuration of characteristic curves in the (P,ω) plane. Also maximizing of the smallest value of P_{cr} for which the necessary condition [4, 5, 6] is

$$\frac{\partial P}{\partial \omega} = 0 , \qquad (30)$$

usually leads to the optimal shape of the structure. This can be easily seen from the following example. Let us look for the optimal stiffness of a spring which is located on the end of the column. The maximalizing of the smallest value of P_{cr} leads to the solution $\kappa \to \infty$, i.e., a rigid support. In this case the optimal solution can be found from the conditions for the first eigenvalue namely

$$\left.\frac{\partial \omega}{\partial P}\right|_{\omega=0} = 0 . \qquad (31)$$

Fulfilling these conditions we obtain the $\kappa_{opt} = 345$ KG cm^{-1} (Figure 10). The configuration of the characteristic curves is also shown in Figure 10. For $\kappa > 345$ the unstable solution has a divergent form. The dependence of the critical force on the elastic support location for $\kappa = 3450$ is shown in Figure 11.

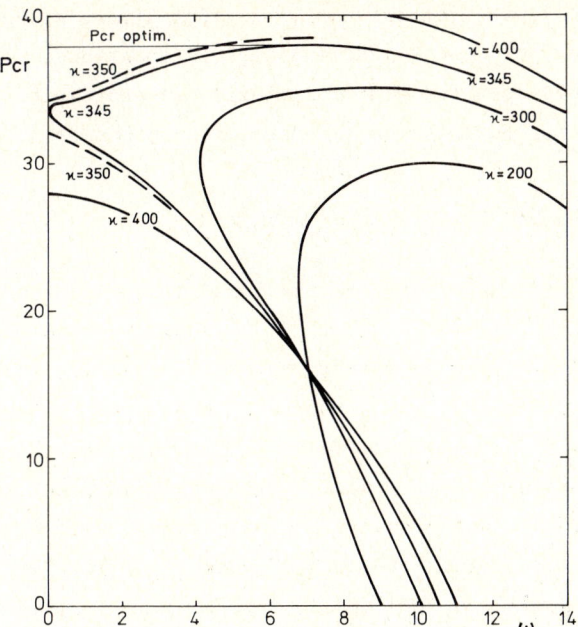

Figure 10 - Shape of the Characteristic Curves for Various Spring Stiffnesses

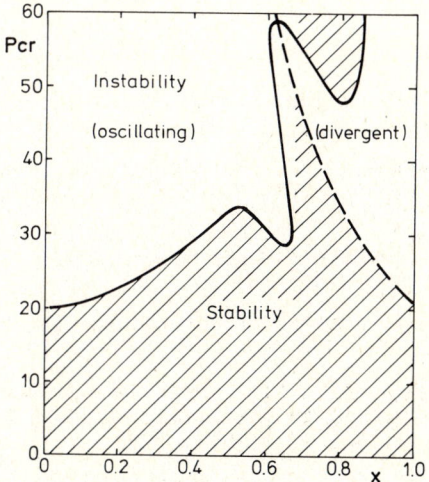

Figure 11 - Dependence of Critical Load on the Elastic Support Location for $\kappa = 3450$

It follows from this analysis that divergent instability is possible for the case of various support locations. The optimal solution depends upon the cost functional (1) and can be obtained used the same procedure as in the case without supports.

4. CONCLUDING REMARKS

The present work complements the previous research on optimal structures under follower forces. It is shown that multimodal optimal solutions are essential in optimal design with stability constraints for structures composed of elements with piece-wise constant cross-section and stiffness. This is especially important for the case of columns interacting with other systems. These results are useful in the analysis of active control of various structures, for example: steering of rockets, where the study of the choice of auxiliary jet locations is similar. It is also useful in finding the optimal power of auxiliary jets in correcting course deviations.

REFERENCES

[1] NIORDSON, F.I., "The Optimal Design of Vibrating Beam", *Quart. of Appl. Math.*, Vol. 23, 1965, pp. 47-53.

[2] PRAGER, W. and TAYLOR, J.E., "Problems of Optimum Structural Design", *J. Appl. Mech.*, Vol. 35, 1968, pp. 102-106.

[3] VEPA, K. and ROORDA, J., "Optimality Criteria for Stability Problems using Pontryagin's Maximum Principle", *Proc. of Third CANCAM*, Calgary, 1971, p. 365.

[4] VEPA, K., "Optimality Stable Structural Forms", University of Waterloo, 1972.

[5] CLAUDON, J.L., "Characteristic Curves and Optimum Design of Two Structures subjected to Circulatory Loads", *J. de Méchanique*, Vol. 14, No. 3, 1975.

[6] PLAUT, R.H., *Optimal Design for Stability under Dissipative, Gyroscopic of Circulatory Loads*, Springer-Verlag, Warsaw, 1975.

[7] MASUR, E.F., "Optimal Placement of Available Sections in Structural Eigenvalue Problems", *J. Optimization Theory Appl.*, Vol. 15, 1975, pp. 68-84.

[8] MRÓZ, Z. and ROZWANY, G.I.N., "Optimal Design of Elastic Beams with Variable Support Conditions", *J. of Optimiation Theory, Appl.*, Vol. 15, No. 1, 1975.

[9] SZELAG, D. and MRÓZ, Z., "Optimal Design of Elastic Beams with Unspecified Support Condition", *ZAMM*, 1978, (in print).

[10] BOGACZ, R., IRRETIER, H. and MAHRENHOLTZ, O., "Optimal Design of Structures under Non-Conservative Forces with Stability Constrain", *Proc. of EUROMECH 112 Colloquium on Bracketing of Eigenfrequencies of Continuous Structures*, Hungary, February, 1979, (in print).

[11] MAHRENHOLTZ, O., *"Das Stabilitätsverhalten des durchströmten, frei hängenden Rohres"*, Pflüger-Festschrift, Hannover, 1977.

[12] LEIPHOLZ, H.H.E., "On an Extension of Hamilton's Variational Principle to Nonconservative Systems which are Conservative in a Higher Sense", *Ing.-Archiv*, Vol. 47, 1978, pp. 257-266.

[13] LEIPHOLZ, H.H.E., "On a Variational Principle for the Clamped Free Rod Subjected to Tangential Follower Forces", *Mech. Res. Comm.*, Vol. 5, No. 6, 1978, pp. 335-359.

SUR LES SOLUTIONS DE PROBLEMES ASSOCIES
EN OPTIMISATION DES STRUCTURES MECANIQUES

P. Brousse

Institut de Mécanique Théorique et Appliquée
Université Pierre et Marie Curie (Paris)

1. INTRODUCTION

Dans cet article, nous étudions les couples de problèmes d'optimisation tels que l'on passe de chacun des problèmes du couple à l'autre en échangeant la fonction objectif et une fonction contrainte inégalité.

    Dans la littérature, on trouve des exemples particuliers de problèmes associés (également dits problèmes duaux) portant sur certaines structures mécaniques. C'est ainsi que l'on a cherché, par exemple, à minimiser la masse d'une barre, d'une plaque, d'une coque, etc., lorsqu'une de ses fréquences de vibration est donnée ou minorée et à maximiser cette fréquence lorsque la masse est donnée ou majorée. Dans cet exemple, et dans les autres exemples que nous connaissons, les auteurs montrent, qu'en associant convenablement les constantes données, les solutions de l'un des problèmes envisagés sont solutions du problème associé et même que les deux problèmes associés ont les mêmes solutions [1-6]. La plupart des démonstrations sont particulières à chaque cas traité; elles utilisent des propriétés d'extrema, de contrôle optimal, de pseudo-convexité, etc.. Citons une méthode basée uniquement sur des propriétés extrémales [7] grâce à laquelle on a montré qu'une condition déjà connue pour la minimisation de la masse d'un milieu continu à frontière libre lorsqu'une certaine intégrale étendue à ce milieu est donnée, est suffisante à la fois pour ce problème et pour le problème associé [8].

Ici, nous formons et nous démontrons des théorèmes *généraux* d'optimisation valables pour *tous* les problèmes associés revêtant une forme très générale; nous n'avons besoin d'aucune propriété de continuité, de différentiabilité ou de convexité. Nous montrons comment ces théorèmes peuvent servir à la résolution d'un problème ayant la forme envisagée lorsqu'on connaît les solutions du problème associé. Nous indiquons des situations générales où les hypothèses des théorèmes sont automatiquement réalisées; ici nous avons besoin de propriétés de croissance et de continuité. Enfin, nous donnons plusieurs applications à des problèmes d'optimisation de structures mécaniques.

2. NOTATIONS ET THEOREME 1

On se donne

- une partie S d'un espace ε de dimension finie ou infinie; le point générique de ε sera désigné par a,
- deux fonctions numériques f et g définies sur S et deux nombres constants m_o et p_o.

On désigne par $G(p_o)$ (resp. $F(m_o)$) la partie de S où l'on a $g(a) \geq p_o$ (resp. $f(a) \leq m_o$). *On fait l'hypothèse que* $G(p_o)$ *n'est pas vide*; l'infimum de la fonction f sur $G(p_o)$ sera désigné par $f^*_{p_o}$.

On considère alors les deux problèmes suivants:

- $P(p_o)$: minimiser f sur $G(p_o)$
- $Q(m_o)$: maximiser g sur $F(m_o)$.

L'ensemble des solutions du problème $P(p_o)$ (resp. $Q(m_o)$)

sera noté $X(p_o)$ (resp. $Y(m_o)$).

Enoncé du théorème 1. *Si $P(p_o)$ a au moins une solution, alors le problème $Q(f^*_{p_o})$ équivaut au problème: maximiser g sur $X(p_o)$, et toute solution de ces deux problèmes vérifie $g(a) \geq p_o$. Si $P(p_o)$ n'a pas de solution, alors $Q(f^*_{p_o})$ n'a aucune solution vérifiant $g(a) \geq p_o$.*

Démonstration. Rappelons d'abord que le problème $Q(f^*_{p_o})$ consiste à maximiser la fonction g sur $F(f^*_{p_o})$ c'est-à-dire sur la partie de S où l'on a $f(a) \leq f^*_{p_o}$.

Dans la première partie du théorème on suppose que $X(p_o)$ n'est pas vide. Envisageons successivement les parties de $F(f^*_{p_o})$ qui sont dans $G(p_o)$ et dans le complément de $G(p_o)$ par rapport à S. Dans $G(p_o)$, les seuls points qui appartiennent à $F(f^*_{p_o})$ sont ceux de $X(p_o)$, cela d'après la définition des solutions de $P(p_o)$; sur $X(p_o)$ on a $g(a) \geq p_o$. Dans le complément de $G(p_o)$ par rapport à S on a $g(a) < p_o$, et les points de $F(f^*_{p_o})$ qui sont dans ce complément vérifient en particulier cette inégalité. Le problème: maximiser g sur $F(f^*_{p_o})$, équivaut donc au problème: maximiser g sur $X(p_o)$, et toute solution éventuelle de ces problèmes vérifie $g(a) \geq p_o$.

Dans la deuxième partie du théorème on suppose que $X(p_o)$ est vide. Si $Q(f^*_{p_o})$ n'a pas de solution, la conclusion est évidemment exacte. Si $Q(f^*_{p_o})$ a au moins une solution, aucune des solutions de ce problème n'est dans $G(p_o)$ puisque dans $G(p_o)$ on a $f(a) > f^*_{p_o}$. Donc toutes les solutions de $Q(f^*_{p_o})$ vérifient $g(a) < p_o$.

Conséquences.

- *Toute solution a^* de $Q(f^*_{p_o})$ telle que l'on ait $g(a^*) \geq p_o$ est solution de $P(p_o)$.*

En effet, le problème $P(p_o)$ a au moins une solution d'après la deuxième partie du théorème 1, et d'après la première partie, le point a* appartient à $X(p_o)$, c'est-à-dire est solution de $P(p_o)$.

- *Si le problème $P(p_o)$ a une solution unique a*, c'est l'unique solution de $Q(f^*_{p_o})$.*

En effet, $X(p_o)$ se réduit au point a*, et ce point est évidemment l'unique solution du problème: maximiser g sur $X(p_o)$.

3. THEOREME 2

Dans la très grande majorité des problèmes d'optimisation dont la formulation est celle de $P(p_o)$, toutes les solutions se trouvent sur la partie $g(a) = p_o$ de la frontière. Donc, en supposant a priori qu'il en est ainsi, on fait une hypothèse qui est, pratiquement, peu restrictive. Cette remarque explique l'intérêt du théorème 2.

Nous gardons les notations données au début de la section 2.

Enoncé du théorème 2. Si le problème $P(p_o)$ a au moins une solution et si toutes les solutions de ce problème vérifient $g(a) = p_o$, alors le problème $Q(f^*_{p_o})$ a les mêmes solutions que $P(p_o)$. Autrement dit on a $X(p_o) = Y(f^*_{p_o})$.

Démonstration. Ici, tout point a de $X(p_o)$ vérifie $g(a) = p_o$. Donc $X(p_o)$ est l'ensemble des solutions du problème: maximiser g sur $X(p_o)$. Le théorème 1 entraîne alors la conclusion du théorème 2.

Commentaires

- La conclusion du théorème 1 reste valable lorsque, dans les énoncés on remplace simultanément l'inégalité « $g(a) \geq p_o$ » et l'expression « maximiser g » par « $g(a) \leq p_o$ » et « minimiser g ».

- On forme un théorème analogue au théorème 2 en intervertissant les deux problèmes $P(p_o)$ et $Q(m_o)$. En supposant que $Q(m_o)$ n'est pas vide, on a $Y(m_o) = X(g^*_{m_o})$ où $g^*_{m_o}$ est le supremum de g sur $F(m_o)$.

4. APPLICATION A LA RESOLUTION DE $Q(m_o)$ LORSQU'ON CONNAIT LES SOLUTIONS DE $P(p_o)$

D'après les commentaires précédents, les deux problèmes $P(p_o)$ et $Q(m_o)$ jouent le même rôle. Nous supposons, par exemple, que nous savons résoudre $P(p_o)$ et nous allons indiquer comment on peut alors obtenir les solutions de $Q(m_o)$.

On a d'abord la réponse suivante résultant du théorème 2 : si les hypothèses de ce théorème 2 sont vérifiées, les solutions de $Q(m_o)$ pour $m_o = f^*_{p_o}$ sont celles de $P(p_o)$.

Mais, très souvent, on peut obtenir les solutions de $Q(m_o)$, non seulement pour une valeur particulière de m_o, mais encore pour autant de valeurs que l'on veut, prises dans certains ensembles.

Supposons en effet que les hypothèses du théorème 2 (existence d'au moins une solution de $P(p_o)$, localisation des solutions de ce problème sur $g(a) = p_o$) soient vraies pour toutes les valeurs p_o d'un intervalle donné J ($\underline{p}_o$, $\overline{p}_o$). Soient deux valeurs arbitraires p_o et p'_o de cet intervalle, telles que l'on ait $p_o < p'_o$. Alors, la partie $G(p'_o)$ de S est incluse dans $G(p_o)$ et ne contient aucun point vérifiant $g(a) = p_o$ et en particulier aucun point de $X(p_o)$. Donc, sur $X(p'_o)$ on a $f(a) > f^*_{p_o}$ d'après la définition des solutions de $P(p_o)$. Ainsi, l'inégalité $p_o < p'_o$ entraîne $f^*_{p_o} < f^*_{p'_o}$. Autrement dit, $f^*_{p_o}$ est une fonction strictement croissante de p_o.

La fonction f^*, qui a tout p_o de J fait correspondre $f^*_{p_o}$, admet donc une fonction réciproque définie sur l'ensemble I des $f^*_{p_o}$. A tout nombre m_o de I, cette fonction réciproque fait correspondre un nombre que nous notons $p_o(m_o)$.

Désignons maintenant par $a^*(p_o)$ les solutions du problème $P(p_o)$. Alors pour toutes les valeurs m_o de l'ensemble I, le maximum de g sur $F(m_o)$ est $p_o(m_o)$ et les solutions du problème $Q(m_o)$ sont les $a^*(p_o(m_o))$ obtenues en remplaçant p_o par $p_o(m_o)$ dans les $a^*(p_o)$.

Dans beaucoup d'applications pratiques, la fonction f* est continue. Alors I est un intervalle, l'intervalle $(f^*_{\underline{p}_o}, f^{\underline{*}}_{p_o})$. Les situations que l'on rencontre admettent alors les deux situations extrêmes suivantes.

- On peut exprimer le nombre $p_o(m_o)$ et les solutions $a^*(p_o)$ sous forme littérale en fonction respectivement de m_o et de p_o. Alors, pour tout m_o de l'intervalle I, le maximum $p_o(m_o)$ de g et les solutions de $Q(m_o)$ s'obtiennent également sous forme littérale par substitution. Nous donnerons un exemple dans la section 6.

- On sait résoudre numériquement $P(p_o)$ pour des valeurs numériques convenables de p_o dans J. Alors des interpolations numériques peuvent permettre de trouver $p_o(m_o)$ pour les valeurs m_o de I, et par suite de résoudre numériquement $Q(m_o)$ dans cet intervalle.

5. REMARQUES SUR CERTAINES SITUATIONS AUXQUELLES S'APPLIQUENT LES THEOREMES PRECEDENTS

Nous allons maintenant indiquer une extension de la notion de problèmes associés, puis présenter un cas, fréquemment rencontré en optimisation des structures mécaniques, et pour lequel, ou bien on connaît l'unique solution du problème $Q(f^*_{p_o})$, ou bien les hypothèses du théorème 2 sont automatiquement réalisées.

Une extension des theoremes 1 et 2. La définition des problèmes associés et les théorèmes 1 et 2 supposent que, sur S, il n'y ait qu'une contrainte inégalité: $g(a) \geq p_o$. Supposons maintenant qu'il y ait plusieurs contraintes inégalités. On peut toujours mettre ces contraintes sous la forme $g_j(a) \geq p_o$, $j = 1,\ldots\overline{j}$, où les fonctions g_j sont définies sur S. Nous introduisons l'enveloppe inférieure des g_j, enveloppe que nous désignons par g. Alors les $\overline{j}$ contraintes $g_j(a) \geq p_o$ sont équivalentes à l'unique contrainte $g(a) \geq p_o$. Nous sommes ainsi conduits à définir le problème associé de:

par
: minimiser f sur l'intersection de S et des $g_j(a) \geq p_o$
: maximiser sur $F(m_o)$ la plus petite des fonctions g_j.

Les théorèmes 1 et 2 s'appliquent alors avec la définition précédente de la fonction g comme enveloppe.

Un cas général rencontré en optimisation des structures mécaniques. Nous envisageons le cas où la région S de l'espace ε est définie par par une minoration et une majoration de a : $\underline{a} \leq a \leq \overline{a}$, où $\underline{a}$ et $\overline{a}$ sont donnés. Si l'espace ε est de dimension finie, $\underline{a}$ et $\overline{a}$ sont des matrices-colonnes données, et la double inégalité précédente signifie que chaque terme de la matrice-colonne inconnue a est comprise entre (au sens large) les termes de même rang de $\underline{a}$ et de $\overline{a}$. Si ε, de dimension infinie, est un espace de fonctions définies sur une partie connexe Δ de R, R^2, R^3, à valeurs dans R (éventuellement R^p), alors $\underline{a}$ et $\overline{a}$ sont des fonctions définies sur Δ et la double inégalité signifie, que pour tout x de Δ, l'on a $\underline{a}(x) \leq a(x) \leq \overline{a}(x)$. Nous supposons en outre que la fonction f est strictement croissante (comme il arrive lorsque cette fonction est la masse ou le volume) et que la fonction g est continue.

Alors, si $\underline{a}$ est dans $G(p_o)$, c'est l'unique solution du problème $P(p_o)$, et donc *l'unique solution du problème $Q(f^*_{p_o})$* d'après la deuxième conséquence du théorème 1. Si $\underline{a}$ n'est pas dans $G(p_o)$ et si le problème $P(p_o)$ a au moins une solution (cette solution existe toujours si ε est de dimension finie), nous allons montrer que *toutes les solutions de ce problème sont sur $g(a) = p_o$*. En effet, si une solution a* vérifiait $g(a^*) > p_o$, on pourrait trouver a', situé dans $G(p_o)$, et tel que l'on ait a' < a* (c'est-à-dire a' $\leq$ a* et a' ≠ a*); on aurait alors f(a') < f(a*) et a* ne serait pas solution du problème $P(p_o)$.

6. PREMIER EXEMPLE: MINIMISATION DE LA MASSE D'UNE COLONNE ELASTIQUE ET MAXIMISATION DE LA CHARGE SUPPORTEE PAR CETTE COLONNE

La colonne, simplement supportée aux deux extrémités, est soumise à une charge axiale compressive P appliquée à une extrémité. Elle est cylindrique, en forme de tube; le rayon intérieur et le rayon extérieur sont respectivement notés r_1 et r_2; sa hauteur h est donnée. Elle est faite d'un matériau homogène, de masse volumique ρ et de module d'Young E. On néglige son poids devant P.

On impose à r_2 d'être au plus égal à une longueur donnée c. On suppose qu'il n'y a que deux défaillances possibles: le dépassement d'une contrainte de compression σ_o donnée, l'apparition du flambement d'Euler.

On envisage les deux problème suivants:

- la charge que peut supporter la colonne étant supérieure ou égale à une charge p_o donnée, choisir r_1 et r_2 de façon que le masse M de la colonne soit minimale;

- la masse M étant inférieure ou égale à une masse donnée M_o, choisir r_1 et r_2 de façon que la charge que peut supporter la colonne soit maximale.

En posant

$$P_s = \pi c^2 \sigma_o \;,\; P_b = \frac{\pi^3 c^4 E}{4h^2} \;,\; m_o = \frac{M_o}{\pi c^2 h \rho} \;,$$

$$m = \frac{M}{M_o} m_o \;,\; a_1 = \frac{r_1^2}{c^2} \;,\; a_2 = \frac{r_2^2}{c^2} \;,$$

les deux problèmes s'énoncent comme suit

$\mathcal{P}(p_o)$: minimiser $m = a_2 - a_1$

sur $G(p_o) = \{a_1, a_2 \mid 0 \leq a_1,\; a_2 \leq 1,\; P_s(a_2 - a_1) \geq p_o$

$P_b(a_2^2 - a_1^2) \geq p_o\}$

$Q(m_o)$: maximiser le plus petit des deux nombres

$$P_s(a_2-a_1) \ , \ P_b(a_2^2-a_1^2)$$

sur $F(m_o) = \{a_1, a_2 \mid 0 \leq a_1, \ a_2 \leq 1, \ a_2-a_1 \leq m_o\}$.

Le second problème se résoud de façon immédiate. Le tableau suivant donne ses solutions a_1^* et a_2^* ainsi que la charge maximale suivant les valeurs de m_o.

- $m_o > 1$: pas de solution

- $m_o \geq 2 - P_s/P_b$: $a_2^* = 1$, $a_1^* = 1-m_o$; $P_b m_o(2-m_o)$

- $m_o \leq \inf(P_s/P_b, 2-P_s/P_b)$: $\begin{cases} \frac{1}{2}(P_s/P_b + m_o) \leq a_2^* \leq 1 \\ a_1^* = a_2^* - m_o \ ; \ P_s m_o \end{cases}$

- $m_o \geq P_s/P_b$: $m_o \leq a_2^* \leq 1$, $a_1^* = a_2^* - m_o$; $P_s m_o$.

Les charges maximales définissent des fonctions p_o de m_o. Chacune de ces fonctions admet dans son intervalle de définition une fonction réciproque très facile à exprimer. Par substitution, ainsi qu'on l'a expliqué dans la section 4, on obtient directement les solutions du premier problème suivant les valeurs de p_o, ainsi que le minimum de m :

- $p_o > P_s$ ou $p_o > P_b$: pas de solution

- $p_o \geq 2 P_s - P_s^2/P_b$: $a_2^* = 1$, $a_1^* = \sqrt{1-p_o/P_b}$; $1 - \sqrt{1-p_o/P_b}$

- $p_o \leq \inf(2 P_s - P_s^2/P_b, \ P_s^2/P_b)$: $\begin{cases} \frac{1}{2}(P_s/P_b + p_o/P_s) \leq a_2^* \leq 1 \\ a_1^* = a_2^* - p_o/P_s \ ; \ p_o/P_s \end{cases}$

- $p_o \geq P_s^2/P_b$: $p_o/P_s \leq a_2^* \leq 1$, $a_1^* = a_2^* - p_o/P_s$; p_o/P_s .

7. DEUXIEME EXEMPLE: MINIMISATION DE LA MASSE ET MAXIMISATION DU COEFFICIENT DE SECURITE D'UN SYSTEME PLAN DE BARRES PARFAITEMENT PLASTIQUES

La variable est ici la matrice-colonne des moments fléchissants limites des barres. Elle est encore désignée par a. On admet l'hypothèse linéaire: la masse totale m des barres est proportionnelle à $L^T a$, où L est la matrice-colonne des longueurs des barres. Afin de respecter cette hypothèse de linéarité, et aussi pour des raisons d'ordre technologique, on impose à la matrice a d'être comprise entre (au sens large) deux matrices strictement positives fixes $\underline{a}$ et $\overline{a}$. Enfin, nous admettons l'hypothèse de la ruine par formation de rotules plastiques.

Nous avons montré [9] que la condition de stabilité est

$$a^T |D^T \gamma^\alpha| \geq p \phi^T C^T \gamma^\alpha, \quad \alpha = 1, \ldots \overline{\alpha},$$

où T est le symbole de la transposition, où p est le coefficient de sécurité, où ϕ est la matrice-colonne fixe des charges, où les matrices rectangulaires D et C ne dépendent que de la géométrie de la structure et où les γ^α sont les matrices-colonnes formées par certaines valeurs convenablement choisies de paramètres indépendants qui engendrent les mécanismes cinématiquement admissibles. Pour toutes les matrices-colonnes γ^α, le produit $\phi^T C^T \gamma^\alpha$ est strictement positif.

On envisage alors les deux problèmes associés suivants: minimiser la masse m lorsque le coefficient de sécurité est supérieur ou égal à un nombre donné p_o; maximiser le coefficient de sécurité p lorsque la masse est inférieure ou égale à une masse donnée m_o. Ces deux problèmes admettent respectivement les énoncés suivants:

$P(p_o)$ $\begin{cases} \text{minimiser } m = \mu L^T a \ (\mu, \text{ const.} > 0) \text{ sous les} \\ \text{contraintes: } \underline{a} \leq a \leq \overline{a}, \ a^T|D^T\gamma^\alpha| \geq p_o \ \phi^T C^T \gamma^\alpha \\ \text{pour tous les} \end{cases}$

$Q(m_o)$ $\begin{cases} \text{maximiser (lorsque } \alpha \text{ varie de 1 à } \overline{\alpha}) \text{ le plus petit} \\ \text{des quotients } a^T|D^T\gamma^\alpha|/\phi^T C^T \gamma^\alpha \text{ sous les} \\ \text{contraintes : } \underline{a} \leq a \leq \overline{a}, \ \mu L^T a \leq m_o \end{cases}$

Supposons que la région admissible du premier problème ne soit pas vide. Alors ce problème a au moins une solution. D'après la section 5 et le théorème 2 on a le résultat suivant: les solutions du second problème (maximisation du coefficient de sécurité) sont identiques à celles du premier (minimisation de la masse) lorsque m_o est le minimum de la masse dans le premier problème.

8. TROISIEME EXEMPLE: ELEMENTS FINIS

Envisageons le cas d'éléments de barres, de membranes, de coques, où les variables sont les aires des sections droites ou les épaisseurs (supposées constantes) des éléments. La matrice-colonne de ces variables est désignée par $a = \{a_i\}$. Des minorations et des majorations sont imposées aux a_i : $\underline{a} \leq a \leq \overline{a}$. Eventuellement il y a d'autres contraintes. D'une façon analogue à ce qui a été fait dans les sections 6 et 7 on peut appliquer les résultats des sections initiales à des problèmes tels que les suivants: 1) minimiser la masse lorsque la rigidité locale ou globale est supérieure ou égale à une rigidité locale ou globale donnée, ou lorsque la fréquence fondamentale est supérieure ou égale a une fréquence donnée; 2) maximiser la rigidité locale ou globale ou maximiser la fréquence fondamentale lorsque la masse est inférieure ou égale à une masse donnée.

9. CONCLUSION

Les théorèmes 1 et 2 ainsi que les remarques des sections 4 et 5 s'appliquent lorsque l'espace ε est de dimension infinie. Dans des articles ultérieurs, nous donnerons des exemples de problèmes d'optimisation associés relatifs à des structures correspondant à ce cas général, où les variables d'optimisation sont des fonctions.

REFERENCES

[1] TAYLOR, J.E., "Minimum mass bar for axial vibration at specified natural frequency,"*AIAA Journal*, 5 (10), 1967, pp. 1911-1913.

[2] BRACH, R.M., "On optimal design of vibrating structures," *J. Optimization Theory and Applications*, 11 (6), 1973, pp. 662-667.

[3] SIPPEL, D.L. and WARNER, W.H., "Minimum mass design of multi-element structures under a frequency constraint," *AIAA Journa* 11 (4), 1973, pp. 483-489.

[4] CARDOU, A. and WARNER, W.H., "Minimum mass design of sandwich structures with frequency and section constraints," *J. Optimization Theory and Applications*, 14(6),1974,pp.633-647

[5] JOURON, C., "Analyse théorique et numérique de quelques problèmes d'optimisation intervenant en théorie des structures," Thèse, Université de Paris XI, 1976.

[6] VAVRICK, D.J. and WARNER, W.H., "Duality among optimal design problems for torsional vibration," *J. Struct. Mech.*, 6 (2), 1978, pp. 233-246.

[7] PRAGER, W., "Optimality criteria in structural design," *Proceed. Nat. Acad. Sc.*, 61 (3), 1968, pp. 794-796.

[8] BROUSSE, P., "Les méthodes d'optimisation dans la construction," Collège international des sciences de la construction, *Séminaire tenu à Saint-Rémy-lès-Chevreuse* (France) 6-9 nov., 1973, pp. 17-38.

[9] BROUSSE, P., "Reduction of kinematic inequality and optimization for perfectly plastic frames," *J. Struct. Mech.*, (in press).

MULTILEVEL CONTROL CONCEPTS IN RELATION
TO THE STRUCTURAL DESIGN PROBLEM

D.G. Carmichael and D.H. Clyde
University of Western Australia
Nedlands, Western Australia

1. INTRODUCTION

The concept of a system provides a representation of behaviour through an assemblage of interacting subsystems [1,2]. The idea of subsystems is fundamental yet their choice is not unique with the decomposition of the system (system model) to subsystems (subsystem models) following four main formats (with a certain degree of overlapping) [3]. Emphasis in the present work is on an hierarchical decomposition of the system where the conventional relationships of structural mechanics, namely equilibrium, compatibility and constitution, assume a well defined role [4]. Such an hierarchical representation for structures is conceptually useful not only for viewing the equations of structural mechanics but also for the associated optimization problem, gaining particular distinction for the treatment of nonlinearities and the reliability problem [5,6].

Generally, however, decompositions may be broadly viewed as vertical or horizontal [7]. Vertical decomposition relates to the uncoupling of lower level subsystems from the overall system and the recognition of interaction between subsystems. The subsystem description will in general be of a different form and in different units to the system model. Horizontal decomposition relates to a system model being decomposed into its constituent parts at the same level by the removal of interaction terms; that is the uncoupling is on the one level. Typically the constituent part description is of the same form as the overall system model.

As multilevel systems theory treats the decomposition of systems into subsystems with interaction, so multilevel optimization theory looks at the decomposition of a single level design problem into subproblems with interaction (while preserving the model, constraint and criterion structures). Coordination schemes are devised whereby the solutions to the subproblems result in the solution of the original problem [2].

Much emphasis has been given to multilevel optimization theory under the argument that it is a theory dealing with complexity and large scale systems. The approach however is applicable to any optimization problem as it is a fundamental approach to problem formulation and an effective solution method. The effectiveness is more pronounced with large scale systems. An underlying principle of multilevel optimization is that with the current state of the art of optimization, several small problems with low order systems are generally easier to solve than one large problem with a high order system. Subsystem interaction may be weak or strong.

The solution of the subproblems directly does not constitute the optimum solution to the overall problem because of the coupling between subsystems. The coupling 'constraints' are typically dealt with in an iterative fashion. The supremal (upper level, second level) unit of the problem coordinates the infimal (lower level, first level) units (Figure 1) such that the resulting solution is equivalent to the original problem solution. Figure 1 implies a two level formulation; typically multilevel optimization theory has been developed for two level formulations but equivalent arguments extend to many levels. Various optimization techniques may be used to solve portions of the total problem providing versatility to the approach.

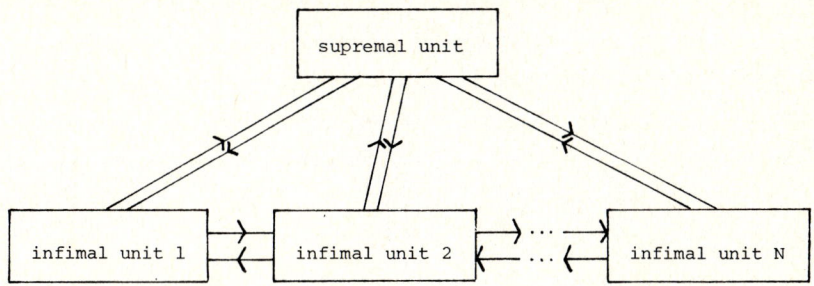

Figure 1 - Problem Decomposition and Interaction

For an hierarchical decomposition, the design problem may be carried out in the control space at each level. Similarly the reliability verification problem may be carried out in the state space at each level. It is argued that a rational framework for a code of practice should acknowledge the fundamental hierarchical nature of structures, where the ideas of reliability, failure, safety and limit states assume elementary geometric interpretations.

Ideas related to multilevel optimization have been used for structural design for example by Vanderplaats and Moses [8] using cross section and node coordinate levels. Kirsch et al. [9] used a partitioning idea to create substructures while Kirsch [10] and Kirsch and Moses [11] used model and goal coordination ideas for mathematical programming problems. Baker et al. [12] designed portal frame structures using management decisions (floor area, openings, building size,...) at the upper level and structural designer's decisions (member sizes, cladding sizes, member spacings,...) at the lower level; the lower level problem was solved using dynamic programming. See also Arora and Govril [13].

2. AN HIERARCHICAL MULTILEVEL SYSTEM REPRESENTATION FOR STRUCTURES

A structure may be viewed as an hierarchical multilevel system [4]. This is in fact implicit in the organization of the models used in modern matrix methods of structural analysis and these methods exploit the hierarchical property. The decomposed subsystems correspond to the structure, member, element and material levels (Figure 2). Alternative subsystems may be defined such as are employed in substructure analysis techniques, finite element

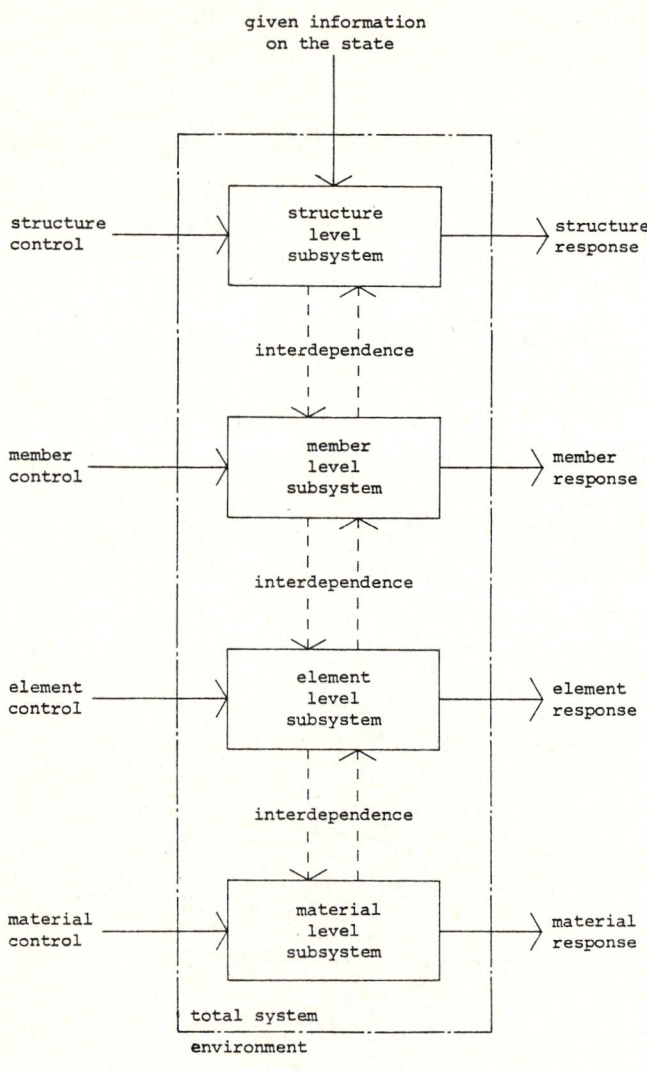

Figure 2

analysis techniques, segmental analysis for reinforced concrete, layering of cross-sections and others. The equivalent model description (input-output rule) of subsystem properties is the constitutive relationship. When combined on the next higher level with subsystem interaction (namely compatibility and equilibrium relationships), the three sets of relationships are then sufficient to define the behaviour on the next higher level (Figure 3).

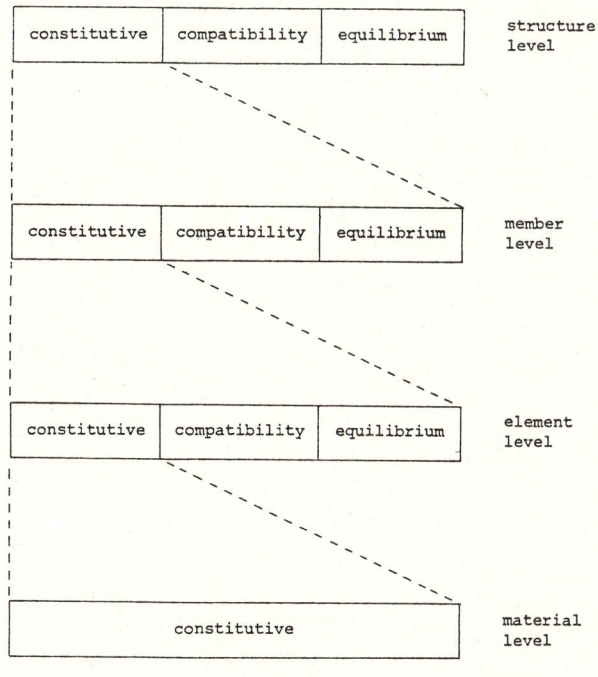

Figure 3

With reference to Figure 2, subsystem boxes indicate the constitutive relationship at the given level. Subsystem controls relate to the properties concerning components of the subsystem and the distribution of these components over space and time. The state refers to the system internal behaviour, and the response to external behaviour. The given information on the state derives from an interaction with the environment (as is the response an interaction with the environment) whereas the control derives from the designer. An illustration of these ideas is given in Figure 4 for the case of structure nodal loading. Related schemes may be worked out for other known information on the system state; for example, for the case of imposed structure nodal displacements, an 'inverted' scheme applies with the controls now becoming flexibilities in place of stiffnesses. (The use of directed paths is

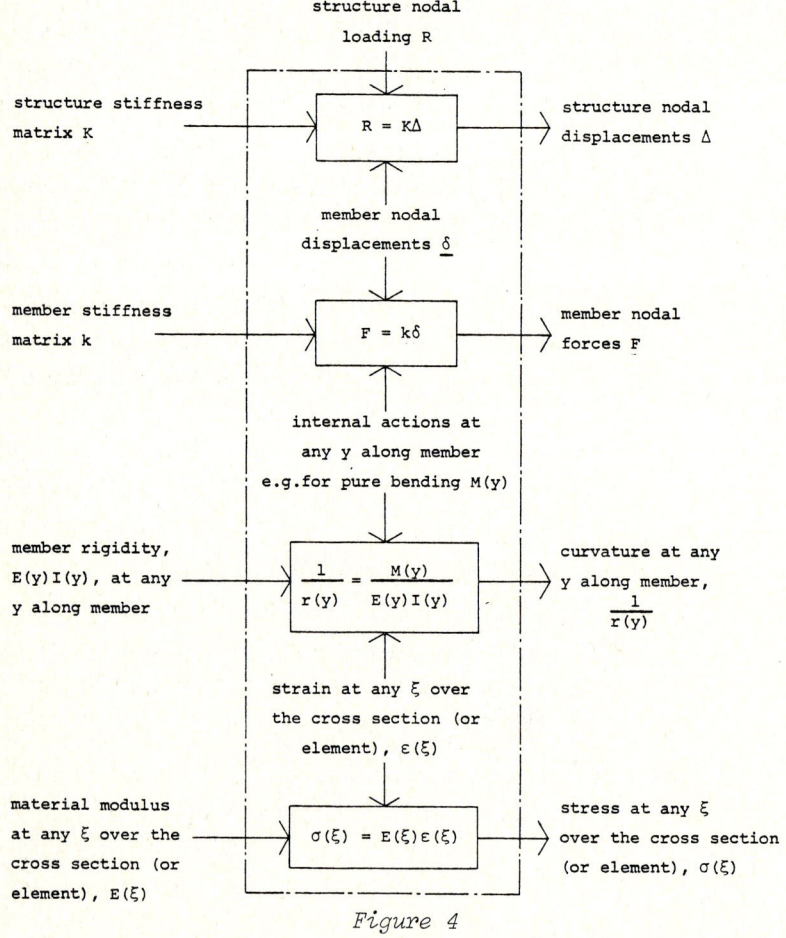

Figure 4

purely schematic and implies dependence relationships and not flows of the entities).

The equivalent of Figure 4 for the models based on the theory of plasticity cannot yet be as definitively presented because the body of theory on the models themselves has not been subjected to the same level of intensive development. The member level is no longer appropriate because the models in ascending hierarchical level progress from yield of the material to interaction diagram of the cross-section directly to structural collapse load. The linear constitutive laws in the boxes at each level of Figure 4 would be replaced by the yield criterion (an inequality) at that level and the flow rule. The step directly from section to structure is reflected equally in plate theory in yield line theory for instance. Plasticity theory emphasizes the dimensions of the model at each level and is an aid to clarity in strength design. The consequences of stepping from the material level to the element level in reinforced concrete changes the comparison space from one-dimensional to six-dimensional.

3. ALTERNATIVE FORMS OF MODEL DECOMPOSITION

Decomposition according to hierarchy frequently implies an order of magnitude change and hence a change in units on going from the system to the subsystem. Three other forms of decomposition may be recognised although overlapping occurs between the four forms in some cases [3].

(a) Decomposition according to the construction of the model. This essentially implies a partitioning of the component equations of the model such as for example between masses in a lumped parameter structural dynamics representation or according to nodes in a framed structure.

(b) Decomposition according to the nature of the control. This implies a different model formulation for each subsystem as for example in a slab-beam-column system where the subsystems are the slabs, beams and columns.

(c) Decomposition of the independent variable (time and/or space) interval. Such decompositions allow for the ready treatment of discontinuities in the state or model; the junction of the subsystems is typically taken at the point of discontinuity. The subsystems in effect are subintervals.

The above decomposition types may be classified as *vertical* or between levels (the hierarchical form) and *horizontal* or within a level (forms (a), (b) and (c)) [7]. The arguments in the following for all decompositions are developed for two level formulations but equivalent arguments extend to formulations of many levels. Obviously vertical and horizontal decompositions may be mixed when extending beyond a two level formulation. When model decomposition is used in conjunction with optimization, similar optimization strategies can be used to handle both vertically and horizontally decomposed models. The form of the decomposition adopted is at the discretion of the designer. There are no fixed rules to guide the designer in this respect. A similar flexibility exists with the choice of multilevel optimization techniques.

Forms (a) and (c) permit a general mathematical treatment while the hierarchical form and (b) vary from structure to structure, though obviously for particular structure types (for example trusses, rigid frames,...) general formulations could be postulated. Consider (a) in detail further, for example for algebraic system models. Form (a) for ordinary differential equation system models, and form (c) are given in Appendices A and B, respectively.

Consider a system, comprising N subsystems or constituent parts, described by

$$F(x,u) = 0 \qquad (1)$$

where $x = (x^1, \ldots, x^N)^T$ is a state vector, $u = (u^1, \ldots, u^N)^T$ is a control vector and $F = (F^1, \ldots, F^N)^T$ is a general nonlinear vector function of the arguments shown, and F^i, x^i and u^i may refer to a vector of quantities for the i'th subsystem. The superscript denotes the relevant subsystem.

The subsystem models are

$$F^i(x^i, u^i, \pi^i) = 0 \quad , \tag{2}$$

where π^i are the interconnection variables accounting for the interaction of subsystem i with the other (N-1) subsystems. The subsystem interaction or coupling is expressed by

$$\pi^i = \gamma^i(x^j, u^j) \quad j = 1,\ldots,N \; ; \; j \neq i \quad , \tag{3}$$

where γ^i is a general function of the arguments shown.

Example: Consider the truss shown in Figure 5. Notation follows [14]. The structural stiffness relationship is

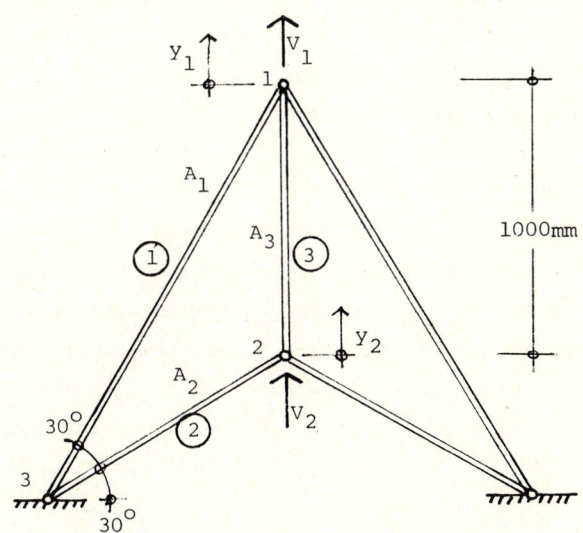

Figure 5

$$\begin{bmatrix} V_1/2 \\ V_2/2 \end{bmatrix} = \begin{bmatrix} (\sqrt{3}A_1 + 2A_3)E/4\ell & -EA_3/2\ell \\ -EA_3/2\ell & (A_2 + 2A_3)E/4\ell \end{bmatrix} \begin{bmatrix} y_1 \\ y_2 \end{bmatrix}.$$

Set $\quad \pi_1^1 \triangleq A_3 \quad\quad \pi_2^1 \triangleq y_2 E/2\ell$

$\quad\quad\quad \pi_1^2 \triangleq A_3 \quad\quad \pi_2^2 \triangleq y_1 E/2\ell \quad$.

The system model then becomes

$$V_1/2 = (\sqrt{3}A_1 + 2\pi_1^1)Ey_1/4\ell - \pi_1^1 \pi_2^1$$

$$V_2/2 = -\pi_1^2 \pi_2^2 + (A_2 + 2\pi_1^2)Ey_2/4\ell \quad,$$

which are two uncoupled equations.

4. MULTILEVEL OPTIMIZATION

Decomposition implies breaking the system into subsystems with interaction and breaking the problem constraints and criterion into constraints and criteria associated with the subproblems. (In certain circumstances it is not required that all the constraints be transferred to the lower level). Decoupling of the design problem components may be carried out by the introduction of interaction variables such that there results independent optimization problems at the lower level. There is a two way exchange of information between the optimization problems at the lower level and the overseeing problem at the upper level; the upper level coordinates the solution process by firstly relaxing the interaction (interpreted in terms of coordination variables) and then gradually enforcing the interaction, the whole process typically being done in an iterative fashion.

Coordination, then, implies that the subproblem solutions are manipulated in order that they end up equivalent to the solution of the original problem. To bring about this result, in general, modifications are required to both the system model and the problem criterion. With modifications to the system model alone, the supremal is unable to coordinate the infimal results, while with modifications to the problem criterion alone, this implies the subsystems are already decoupled, and the subproblems may be solved independently. The particular coordination methodology used depends on the usage and adjustment of the coordination variables and can follow various lines with the most common categorization being according to interaction balance methods and interaction prediction methods though other methods are used. See [15] and [16] for a discussion.

The form of decomposition chosen directly reflects upon the coordination and the success of coordination that follows. Note that in decomposing a problem, interaction variables and interaction relationships are introduced, while in coordinating the subproblems coordination variables are employed. Coordination variables may or may not be interaction variables, depending on the coordination scheme adopted.

Irrespective of the particular coordination scheme used, there are several common steps involved in coordination: Consider in relation to a two level formulation,

(a) The choice of the coordination variables. The coordination variables may be chosen from actual variables, Lagrange multipliers (or penalty weights) or any combination of these. They may be chosen as best suits the particular problem. They do not necessarily have to be the representation of subsystem interaction.

(b) At the first level these coordination variables are taken as fixed, giving independent first level problems.

(c) The first level subproblems are solved returning to the second level information necessary to update the coordination variables' values.

(d) At the second level the optimal values of these coordination variables are determined through some iterative algorithm. In essence the upper level coordinates the activities of the lower level.

The following presentation of multilevel optimization ideas is developed through existing control formulations for mathematical programming problems (and optimal control problems - Appendix C) to the design problem associated with an hierarchical decomposition of the system model. The first two permit general presentations while the third is particular to the type of structure under consideration.

Consider a general mathematical programming problem and using the decomposition (1)-(3) [17]. The problem may be written as

$$\min J = G(x,u)$$
$$\text{subject to } F(x,u) = 0 \quad (4)$$
$$H(x,u) \leq 0 \ .$$

Modification of the model involves the introduction of the interaction variables π in place of state variables, control variables or general functions $\gamma(x,u)$ of state and control variables in such a way that G is separable and F and H contain no common x, u or π variables between subsystems. That is

$$G(x,u,\pi) = \sum_{i=1}^{N} G^i(x^i,u^i,\pi^i)$$

$$F^i(x^i,u^i,\pi^i) = 0 \quad (5)$$
$$H^i(x^i,u^i,\pi^i) \leq 0 \qquad i = 1,\ldots,N \ .$$

Note that the choice of π is essentially arbitrary.

Modification of the criterion involves adjoining the interconnection relation (3) through the use of Lagrange multipliers $\beta = (\beta^1,\ldots,\beta^N)^T$ to give

$$\min J' = \sum_{i=1}^{N} \{G^i(x^i, u^i, \pi^i) + \beta^{iT}[\gamma^i(x^j, u^j) - \pi^i]\}. \quad (6)$$

It is noted that the last term contains superscripts in j and some manipulation will be required to make the total criterion separable. The specific form of the criteria for the subproblems depends on the specific coordination method used.

The coordination approach adopted determines the coordination variables to be chosen. Several approaches to coordination are available and are termed interaction balance methods (also known as 'nonfeasible' methods), interaction prediction methods and 'feasible' methods. (The terminology 'feasible' derives from the result that the coupling constraint (3) is always satisfied and hence all solutions during the iteration process are feasible; while the coupling constraint is violated at each stage of the iterations short of the optimum in 'nonfeasible' methods. The terminology 'feasible'/'nonfeasible' is not preferred here; according to this classification the interaction prediction methods belong to the 'nonfeasible' category).

Interaction balance methods, which belong to the goal (criterion) coordination group, select the value for β^i at the upper level and these values are fed to the lower level. The interaction prediction methods select both π and β at the second level. For both, the associated decomposition puts the $\beta^{iT}\pi^i$ term in the i'th subproblem criterion and all of the $\beta^{iT}\gamma^i(x^j, u^j)$ terms associated with the j'th variables in the j'th subproblem criterion. Feasible methods, which belong to the model coordination group, select values for π^i at the upper level and these values are fed to the lower level. The associated decomposition puts all of the $\beta^{iT}[\gamma^i(x^j, u^j) - \pi^i]$ terms associated with the j'th variables in the j'th subproblem criterion. See Bauman [7].

Some means or algorithm is required for adjusting the values of π and β between iterations. Generally the algorithms are iterative in nature with the computations alternating between levels. Most favoured are the gradient and related schemes. The choice of the starting values for β and π as well as the step sizes in the iteration are at the discretion of the designer [17].

Example: Consider the design of the truss in Figure 5 (with $A_3 = A_2$) according to an optimality criterion.

$$(\min)\ J = 1732\ A_1 + 1500\ A_2\ ,$$

and with constraints

$$A_1 \geq 0 \qquad A_2 \geq 0$$

$$y_1 \leq 5 \qquad y_2 \leq 5$$

$$\begin{bmatrix} 50 \\ 25 \end{bmatrix} = \begin{bmatrix} (\sqrt{3}A_1 + 2A_2)E/4000 & -EA_2/2000 \\ -EA_2/2000 & 3EA_2/4000 \end{bmatrix} \begin{bmatrix} y_1 \\ y_2 \end{bmatrix} .$$

Eliminating y_1 and y_2 through the use of the stiffness equations gives the constraints (for $E = 207\ kN/mm^2$)

$$5.196\ A_1 + 2A_2 \geq 773$$

$$5.196\ A_1 + 2A_2 - 167.43\ A_1/A_2 \geq 580\ .$$

In subsystem one, let $\pi_1 \triangleq A_2$ and in subsystem two, let $\pi_2 \triangleq A_1$. The modified criterion becomes

$$J' = 1732\ A_1 + 1500\ A_2 + \beta_1(A_2 - \pi_1) + \beta_2(A_1 - \pi_2).$$

Consider the interaction balance method as an example. The subproblems are

Subproblem one:

$$\min J^1 = 1732 A_1 - \beta_1 \pi_1 + \beta_2 A_1$$
$$\text{subject to } 5.196 A_1 + 2\pi_1 \geq 773 \quad A_1 \geq 0 \ .$$

Subproblem two:

$$\min J^2 = 1500 A_2 + \beta_1 A_2 - \beta_2 \pi_2$$
$$\text{subject to } 5.196 \pi_2 + 2A_2 - 167.43 \pi_2/A_2 \geq 580 \quad A_2 \geq 0.$$

The interaction balance method fixes the values of β_1 and β_2 at the second level. The first subproblem is then a linear programming problem while the second subproblem is a nonlinear programming problem, the solutions of which return optimal values for A_1, π_1, A_2 and π_2 to the second level. At the second level β_1 and β_2 may be updated in such a fashion that A_1 approaches π_2 and A_2 approaches π_1 in value. The process is repeated many times leading to the optimal values of $\hat{A}_1 = 111.4$ mm^2 and $\hat{A}_2 = 96.3$ mm^2. Backsubstituting into the original constitutive relationship for the structure gives $\hat{y}_1 = 5$ mm and $\hat{y}_2 = 5$ mm.

(Large scale linear and nonlinear programming is discussed for example by Lasdon [18] where decomposition and solution procedures for specialized problem structures are given).

Consider now optimization associated with an hierarchical decomposition. Unlike the other decompositions described, decomposition based on hierarchy follows no set mathematical format but instead varies with each problem. The approach permits the effective and direct treatment of nonlinearities and constraints at the lower levels, an example of the latter being stress constraints as part of an overall structure level problem.

Example: Consider the truss design problem again, but now allowing for the separate treatment of member 3. The design problem becomes allowing now for stress constraints

$$\min J = 1732 A_1 + 1000 A_2 + 500 A_3$$

subject to $\quad A_1 \geq 0 \qquad A_2 \geq 0 \qquad A_3 \geq 0$

$\qquad\qquad\quad y_1 \leq 5 \qquad y_2 \leq 5$

$$\begin{bmatrix} E/2000 & 0 \\ 0 & E/2000 \\ E/1000 & -E/1000 \end{bmatrix} \begin{bmatrix} y_1 \\ y_2 \end{bmatrix} \leq \begin{bmatrix} 0.16 \\ 0.16 \\ 0.16 \end{bmatrix}$$

$$\begin{bmatrix} 50 \\ 25 \end{bmatrix} = \begin{bmatrix} (\sqrt{3}A_1 + 2A_3)E/4000 & -EA_3/2000 \\ -EA_3/2000 & (A_2 + 2A_3)E/4000 \end{bmatrix} \begin{bmatrix} y_1 \\ y_2 \end{bmatrix}.$$

The last set of inequalities are the stress constraints and the whole formulation follows [14].

Choosing members 1, 2 and 3 as the subsystems, then their relevant constitutive relationships and interaction (compatibility and equilibrium) are respectively,

$$\begin{bmatrix} p_1 \\ p_2 \\ p_3 \end{bmatrix} = \begin{bmatrix} EA_1/1732 & & \\ & EA_2/1000 & \\ & & EA_3/2000 \end{bmatrix} \begin{bmatrix} u_1 \\ u_2 \\ u_3 \end{bmatrix} \quad \text{or } F = k\delta$$

$$\begin{bmatrix} u_1 \\ u_2 \\ u_3 \end{bmatrix} = \begin{bmatrix} 0.5\sqrt{3} & 0 \\ 0 & 0.5 \\ 1 & -1 \end{bmatrix} \begin{bmatrix} y_1 \\ y_2 \end{bmatrix} \quad \text{or } \delta = A_* \Delta$$

$$\begin{bmatrix} 50 \\ 25 \end{bmatrix} = \begin{bmatrix} 0.5\sqrt{3} & 0 & 1 \\ 0 & 0.5 & -1 \end{bmatrix} \begin{bmatrix} p_1 \\ p_2 \\ p_3 \end{bmatrix} \quad \text{or } R = A_*^T F$$

where p_i, u_i and y_i are the internal member forces, member displacements and nodal displacements respectively.

The modification of the criterion involves adjoining the interaction relations to give

$$J' = A^T \ell + \beta_c^T [A_*\Delta - \delta] + \beta_e^T [A_*^T F - R] \quad ,$$

where A is a vector of member cross sectional areas (allowing for symmetry with member 3), ℓ is a vector of member lengths and β_c and β_e are vector Lagrange multipliers.

Considering the interaction balance method or the interaction prediction method, the subproblem associated with member 1 for example becomes

$$\min J^1 = 1732 A_1 + \beta_{c1} [0.5\sqrt{3} y_1 - u_1] + \beta_{c3}[y_1]$$

$$+ \beta_{e1} [50 - 0.5\sqrt{3} p_1]$$

subject to $\quad p_1 = \dfrac{EA_1}{1732} u_1$

$\dfrac{Eu_1}{1732} \leq 0.16$

$y_1 \leq 5 \qquad A_1 \geq 0 \quad .$

Depending on which coordination method is used then some or all of u_i, p_i, β_c and β_e are chosen at the second level.

Appendix D gives notes on the hierarchical treatment for problems containing differential equation models.

5. RELIABILITY AND CODES OF PRACTICE

Present day codes with partial safety factor formats and emphasizing reliability theory and limit states philosophy, may be discussed within an hierarchical, multilevel representation of structures using the state space at each level for the reliability verification problem [6].

A safety or serviceability verification may be carried out at any level on that level's characteristic behaviour (state or load effect). Associated with this state is a resistance (used to denote 'capacity') and has a one-to-one correspondence. Failure corresponds to the probability of the state exceeding the resistance to a predefined level. Reliability may then be defined as the probability that the state does not exceed the resistance. Serviceability in general relates to the system response, and safety to the system state. In general the response $z = z(x)$, and hence a discussion on behaviour constraints can be restricted to a state space only.

At any level the state in general may be represented by an n-tuple of variables $\{x_i; i = 1,\ldots,n\}$, with the number of state boundary conditions the same. Changes in the state depend directly on changes in the boundary conditions (including loadings). It is convenient to represent the state at any time of any location in the system by some point in an n-dimensional state space. (An alternative (n+4)-dimensional space including spatial and temporal coordinates may also be used). A locus of these points may be produced by changing the state boundary conditions which themselves define lower dimensional spaces. It is assumed here that a set of controls $\{u_j; u = 1,\ldots,r\}$ exists and only the boundary controls are varied. Various sets of controls obviously define different points in the state space for given boundary conditions. The system as designed may be represented by a point in control space.

For any given state at any given level, let the corresponding resistance of the structure be characterized by elements $r \in R$. The permissible region of states corresponds to a region $S \subset R$ and the limit states are represented by the limit surface ∂S. ∂S may be deterministic or stochastic and defines the boundary of the 'limit' domain. Failure of the system corresponds to the intersection of the state with this limit surface. Reliability may then be defined as the probability that the state remains within the permissible region S. The shortest distance from the boundary of ∂S represents the safety margin of the system. Where non-linearity of ductility exists the reliability assessment is transferred to successively higher levels [5].

6. CONCLUSION

It is seen that a multilevel representation and in particular an hierarchical version, is fundamental to the equations and principles of structural mechanics, with control systems theory being a natural vehicle for descriptive purposes. The multilevel representation is conceptually useful for not only the modelling aspects but also the associated optimization and reliability verification problems.

Several system decompositions and associated design problem decompositions were put forward relating to both vertical and horizontal formats. The treatment of the associated interaction between subproblems provides the distinguishing features of the presently available multilevel optimization techniques. Coordination schemes ensure that the solutions of the subproblems result in the solution to the original single level problem.

As the use of a control space at each level of the structure hierarchy generalises the design problem discussion, so the use of a state space simplifies the reliability and associated problems discussion. Again the hierarchical representation proves fundamental.

REFERENCES

[1] HALL, A.D., *A Methodology for Systems Engineering*, Van Nostrand, 1962.

[2] MESAROVIC, M.D., MACKO, D. and TAKAHARA, Y., *Theory of Hierarchical, Multilevel Systems*, Academic Press, 1970.

[3] SCHOEFFLER, J.D., "On-Line Multilevel Systems," *Optimization Methods for Large Scale Systems*, (ed. D.A. Wismer), McGraw Hill, 1971, pp. 291-330.

[4] CARMICHAEL, D.G. and CLYDE, D.H., "A Basis for the Use of Control Systems Theory in the Mathematical Modelling and Design of Structures," Fifth Australas. Conf. Mech. Struct. Mat., Melbourne, 1975, pp. 47-62.

[5] CLYDE, D.H., "The Safety Verification Problem in Structures," Fifth Australas. Conf. Mech. Struct. Mat., Melbourne, 1975, pp. 103-112.

[6] CARMICHAEL, D.G. and CLYDE, D.H., "A Rational Approach to Design Code Formats," I.E. Aust. Metal Structures Conf., Adelaide, 1976, pp. 104-108.

[7] MESAROVIC, M.D., "On Vertical Decomposition and Modelling of Large Scale Systems," *Decomposition of Large-Scale Problems*, (ed. D.M. Himmelblau), North-Holland, 1973, pp. 323-340.

[8] VANDERPLAATS, G.N. and MOSES, F., "Automated Design of Trusses for Optimum Geometry," *ASCE J. Struct. Div.*, 98, ST3, 1972, pp. 671-690.

[9] KIRSCH, U., REISS, M. and SHAMIR, U., "Optimum Design by Partitioning into Substructures," *ASCE J. Struct. Div.*, 98, ST1, 1972, pp. 249-267.

[10] KIRSCH, U., "Multilevel Approach to Optimum Structural Design," *ASCE J. Struct. Div.*, 101, ST4, 1975, pp. 957-974.

[11] KIRSCH, U. and MOSES, F., "Decomposition in Optimum Structural Design," *ASCE J. Struct. Div.*, 105, ST1, 1979, pp. 85-100.

[12] BAKER, K.N., CARMICHAEL, D.G. and VAN DER MEER, A.T., "A Multilevel Staged Approach to the Optimal Design of Industrial Portal Frame Structures," I.E. Aust. Metal Structures Conf., Perth, 1978.

[13] ARORA, J.S. and GOVRIL, A.K., "An Efficient Method for Optimal Structural Design by Substructuring," *Comp. and Struct.*, 7, 1977, 507-515.

[14] MAJID, K.I., *Optimum Design of Structures*, Butterworths, 1974.

[15] MAHMOUD, M.S., "Multilevel Systems Control and Applications: A Survey," *IEEE Trans. Syst. Man Cybern.*, SMC-7, 3, 1977, pp. 125-143.

[16] SMITH, N.J. and SAGE, A.P., "An Introduction to Hierarchical Systems Theory," *Comput. and Elect. Eng.*, 1, 55-71, 1973.

[17] BAUMAN, E.J., "Multilevel Optimization Techniques with Application to Trajectory Decomposition," *Advances in Control Systems: Theory and Applications*, (ed. C.T. Leondes), Vol. 6, 1968, pp. 159-220.

[18] LASDON, L.S., *Optimization Theory for Large Systems*, Macmillan, 1970.

[19] GUINZY, N.J. and SAGE, A.P., "System Identification in Large Scale Systems with Hierarchical Structures," *Comput. and Elec. Eng.*, 1, 1973, pp. 23-42.

[20] CARMICHAEL, D.G., "Singular Optimal Control Problems in the Design of Vibrating Structures," *Jnl. Sound Vibrn.*, 53, 2, 1977, pp. 245-253.

[21] CARMICHAEL, D.G., "On a Minimum Weight Disk Design Problem," *Trans. ASME, Jnl. App. Mech.*, Ser. E, 44, 3, 1977, pp. 506-507.

[22] CARMICHAEL, D.G., "Optimal Control in the Design of Material Continua," *Archives of Mechanics (Arch. Mech. Stos.)*, 30, 6, 1978, pp. 743-755.

[23] CARMICHAEL, D.G. and GOH, B.S., "Optimal Vibrating Plates and a Distributed Parameter Singular Control Problem," *Int. J. Control*, 26, 1, 1977, pp. 19-31.

[24] FRY, C.M. and SAGE, A.P., "On Hierarchical Estimation and System Identification Using the MAP Criterion," *Comput. and Elec. Eng.*, 1, 1973, pp. 361-389.

[25] WISMER, D.A., "Distributed Multilevel Systems," in *Optimization Methods for Large-Scale Systems with Applications*, (ed. D.A. Wismer), McGraw-Hill, 1971, pp. 223-273.

APPENDIX A. DECOMPOSITION OF DIFFERENTIAL EQUATION MODELS ACCORDING TO THE CONSTRUCTION OF THE MODEL

Consider an n'th order system

$$\frac{dx(y)}{dy} = f[x(y), u(y), y] \quad y \in [y^L, y^R] \tag{A1}$$

where x is an n dimensional state vector, u is an r dimensional control vector, f is a general nonlinear n vector function of the arguments shown and y is the independent variable (space or time).

Decompose the system into N subsystems or constituent parts, each of dimension n^i such that

$$\sum_{i=1}^{N} n^i = n \qquad (A2)$$

$$\frac{dx^i(y)}{dy} = f^i[x^i(y), \pi^i(y), u^i(y), y] \ . \qquad (A3)$$

Each subsystem is related to the other subsystems only through the interconnection,

$$\pi^i(y) = \gamma^i(x^j, u^j) \qquad j = 1,\ldots,N \ ; \ j \neq i \qquad (A4)$$

where x^i, u^i, π^i, f^i and γ^i may all be vector quantities.

Example: Consider the beam equation

$$\frac{d^2}{dy^2}\left[EI \frac{d^2w}{dy^2}\right] = q \ .$$

Notation follows [4]. In state equation form

$$\frac{dx_1}{dy} = x_2$$

$$\frac{dx_2}{dy} = x_3 / u$$

$$\frac{dx_3}{dy} = x_4$$

$$\frac{dx_4}{dy} = q$$

where the state variables x_1, x_2, x_3 and x_4 are deflection (w), slope, bending moment and shearing force respectively. The control variable u is the rigidity EI.

Let $\pi^1 \triangleq x_3$ giving two subsystems

$$\frac{dx_1}{dy} = x_2 \qquad \frac{dx_3}{dy} = x_4$$

$$\frac{dx_2}{dy} = \frac{\pi^1}{u} \qquad \frac{dx_4}{dy} = q \quad .$$

The subsystems are the compatibility and equilibrium equations and there is an important class of problem, i.e. the statically determinate case, for which they are uncoupled.

A similar approach may be adopted for difference equation and partial differential equation models.

APPENDIX B: DECOMPOSITION OF THE INDEPENDENT VARIABLE INTERVAL

Consider for the same n'th order system (equation (A1)) with given state boundary conditions at y^L and Y^R, that at some known conditions of the form

$$\psi[x(y'),y'] = 0 \tag{B1}$$

there exists a discontinuity in the model or in the state. Under such conditions it will be found convenient to decompose the interval $[y^L, y^R]$ into two subintervals from y^L to y' and from y' to y^R. For the first subinterval

$$\frac{dx^1}{dy} = f^1[x^1, u^1, y] \tag{B2a}$$

and for the second subinterval

$$\frac{dx^2}{dy} = f^2[x^2, u^2, y] \qquad (B2b)$$

with the relation between x^1 and x^2 and between y' and y'^+ given by

$$y' - y'^+ = 0$$

$$h[x^1(y'), y'] - x^2(y'^+) = 0 . \qquad (B3)$$

Example: Consider a symmetrical two span continuous beam. At the centre support at y'

$$x_1(y') = 0$$
$$x_2(y') = 0$$

which define (B2) and

$$x_1^1(y') - x_1^2(y'^+) = 0$$
$$-x_2^1(y') - x_2^2(y'^+) = 0$$
$$x_3^1(y') - x_3^2(y'^+) = 0$$
$$-x_4^1(y') - x_4^2(y'^+) = 0$$

which define (B3).

APPENDIX C: MULTILEVEL OPTIMIZATION FOR THE OPTIMAL CONTROL PROBLEM WITH DECOMPOSITION ACCORDING TO THE CONSTRUCTION OF THE MODEL AND OF THE INDEPENDENT VARIABLE INTERVAL

Consider now a general optimal control problem using the decomposition (A1)-(A4) [16,17,19]. The problem may be written as

$$\min J = g[x(y), y]_{y=y^L}^{y=y^R} + \int_{y^L}^{y^R} G[x(y), u(y), y] dy$$

$$h[x(y), u(y), y] \leq 0$$

$$\frac{dx(y)}{dy} = f[x(y), u(y), y] \quad y \in [y^L, y^R] \tag{C1}$$

together with state boundary conditions at y^L and y^R.

Modification of the criterion firstly involves adjoining the interaction relationship (A4) to the original optimality criterion via Lagrange multipliers $\beta(y) = (\beta^1, \ldots, \beta^N)$. The new optimality criterion becomes

$$J' = g[x(y), y]\Big|_{y=y^L}^{y=y^R} + \int_{y^L}^{y^R} \{G[x(y), u(y), y]$$

$$+ \sum_{i=1}^{N} \beta^{iT}(y) [\pi^i(y) - \gamma^i(x^j, u^j, y)]\} dy \quad . \tag{C2}$$

For a decomposition associated with the interaction balance and interaction prediction methods and for g and G additively separable [16,19],

$$J^i = g^i[x^i, y]\Big|_{y=y^L}^{y=y^R} + \int_{y^L}^{y^R} \{G^i(x^i, u^i, y) + \beta^{iT}\pi^i$$

$$- \sum_{j \neq i}^{N} \beta^{jT} \gamma^{ji}(x^i, u^i, y)\} dy \tag{C3}$$

where

$$\gamma^i(x^j, u^j, y) = \sum_{j \neq i}^{N} \gamma^{ij}(x^j, u^j, y) \quad . \tag{C4}$$

A separable form of the constraints is also required. Note that the assumption of the criterion and the constraints being separable is not restrictive as the interaction variables π may be freely chosen to give this desired effect when creating the subsystems.

For this problem, forms of coordination other than the interaction balance, interaction prediction and 'feasible' methods are available [15]. In particular costate coordination or control coordination may be used. The iterations on the coordination variables at the second level may be done by gradient or related or other methods [16,17,19]. Where π appears as a control type variable in the subproblems (interaction balance methods), it occurs linearly and the possibility of singular optimal control [20-23] should be checked [16]. Alternatively the interaction relation may be expressed in a quadratic form or else a feasible-type penalty function approach may be used.

Example: Consider the design of a beam according to an optimality criterion

$$(\min) \; J = \int_0^L EI \; dy$$

and constraints related to deflections.

Choosing subsystems of equilibrium-type relationships and compatibility-type relationships gives the modified criterion of

$$J' = \int_0^L \{u + \beta^1(\pi^1 - x_3)\} \; dy \; .$$

Using the interaction balance method or the interaction prediction method the subproblems become,

Subproblem one:

$$\frac{dx_1}{dy} = x_2$$

$$\frac{dx_2}{dy} = \frac{\pi^1}{u}$$

$$(\min) \ J^1 = \int_0^L \{u + \beta^1 \pi^1\} \ dy$$

constraints related to x_1.

Subproblem two:

$$\frac{dx_3}{dy} = x_4$$

$$\frac{dx_4}{dy} = q$$

$$(\min) \ J^2 = \int_0^L - \beta^1 x_3 \ dy \ .$$

For the design problem associated with the decomposition of the independent variable interval y, Bauman [17] gives the necessary conditions for optimality for the subproblems and for example for a two subproblem feasible method, x^2 and y'^+ are updated at the second level.

APPENDIX D: MULTILEVEL OPTIMIZATION WITH HIERARCHICAL DECOMPOSITION OF DIFFERENTIAL EQUATION MODELS

Example: Consider the beam design problem again. The subsystem at the element level is the moment-curvature relationship

$$\frac{M}{EI} = \frac{1}{r}$$

with subsystem interaction expressed by equilibrium and compatibility relationships which are both second order differential equations. This results in an infinite number of subproblems. Discretization of the beam length may be used to produce a formulation with a finite number of subproblems, here differential equations become difference equations and the criterion integral becomes a summation and the solution methodology follows that for discrete control problems (see [24]).

Distributed parameter control problems may be handled similarly. See for example [25].

STRUCTURAL CONTROL, H.H.E. Leipholz (ed.)
North-Holland Publishing Company & SM Publications
© IUTAM , 1980

THE USE OF AERODYNAMIC APPENDAGES FOR
TALL BUILDING CONTROL

J.C.H. Chang and T.T. Soong

Department of Civil Engineering
State University of New York at Buffalo
Buffalo, New York, U.S.A.

1. INTRODUCTION

The use of aerodynamic surfaces or appendages in reducing building sways was first proposed by Klein et al [1]. Its feasibility was demonstrated for one type of structure on which a small appendage was used as a control device for minimizing building acceleration. The control scheme used is a simple on-off type in which the appendage is fully extended when the structural velocity at controller location is in the direction of the wind and is fully closed when the structural velocity is in the opposite direction. One of the attractive features of this control scheme is that it utilizes energy contained in the mean component of the excitation, thus requiring practically no additional power to operate the controller.

Some possible configurations of such appendages are shown in Figure 1. A single appendage situated at the top of a structure is shown in Figure 1(a) which is raised and lowered by an actuator moving in the horizontal direction. The appendage shown in Figure 1(b) operated very much like a venetian blind. A series of appendages can also be designed as shown in Figure 1(c) whose independent movements can provide both translational and rotational control.

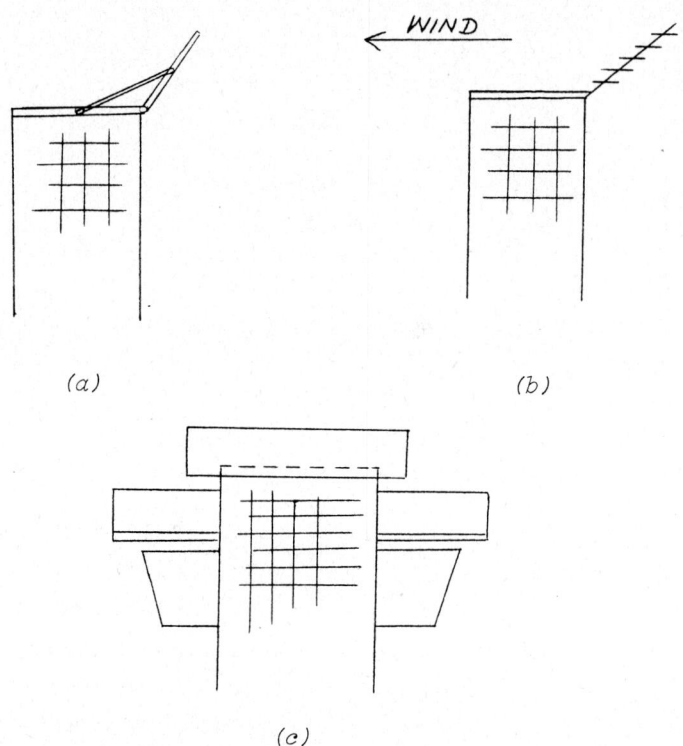

Figure 1 - Possible Configurations of Appendages
    (a) Side View
    (b) Side View
    (c) Front View

In this paper, we consider this problem from the point of view of optimum control theory. An appendage is situated at the top of an idealized cantilever tube structure and it operates as a feedback controller in minimizing wind-induced building motion in the wind direction. Results are presented for on-off as well as continuous control schemes and they are compared with ad hoc procedures such as the one proposed by Klein et al [1].

2. PROBLEM FORMULATION

Consider tall buildings modeled by a cantilever tube-type structure with the equation of motion

$$\frac{\partial^2}{\partial z^2}\left[EI(z)\frac{\partial^2 v(z,t)}{\partial z^2}\right] + c(z)\frac{\partial v(z,t)}{\partial t} + m(z)\frac{\partial^2 v(z,t)}{\partial t^2} =$$

$$= F_1(z,t) + F_1(z,t) , \qquad (1)$$

where $F_1(z,t)$ represents the wind load on the structure which is assumed to take the form

$$F_1(z,t) = \frac{A_s}{2}\frac{\rho}{\ell} C_D(z,t) V^2(z,t) , \qquad (2)$$

and $F_2(z,t)$ denotes the wind load on the appendage with

$$F_2(z,t) = \frac{A_p}{2} \rho C_D'(\ell,t) V^2(\ell,t) \delta(z-\ell) u(t) . \qquad (3)$$

In the above, the area of the appendage, A_p, ranges from one to five percent of A_s, the windward area of structure, and $u(t)$ denotes appendage deployment factor; it takes the value of one when appendage is fully deployed and zero when fully closed.

The wind velocity $V(z,t)$ can be written as [2]

$$V(z,t) = \bar{V}(z) + v(t) , \qquad (4)$$

where $\bar{V}(z)$ is its mean steady component which is a function of the height and $v(t)$ is the random fluctuating part. The member functions of $v(t)$ can be generated from

$$v(t) = \sqrt{2} \sum_{i=1}^{n} [2S_v(\omega_i)\Delta\omega]^{1/2} \cos(\omega_i t + \Phi_i) , \qquad (5)$$

where the random phase angles Φ_i are dependent and uniformly distributed in the interval $(0, 2\pi)$, $\Delta\omega = \omega_u/n$ where ω_u is the upper cutoff frequency, $\omega_i = [i - (1/2)]\Delta\omega$, and $S_v(\omega)$ is the spectral density of $v(t)$. For the purpose of this study, the spectral density is assumed to take the form [3]

$$S_v(\omega) = 2K\phi^2|\omega| \bigg/ \pi^2 \left\{1 + \left(\frac{\phi\omega}{\pi V_z}\right)^2\right\}^{4/3}, \qquad (6)$$

where V_z = mean wind speed at $z = 33$ ft, ϕ = the scale of turbulence ($\phi \cong 2,000$ ft), and K = the surface drag coefficient.

Let $\Phi(z)$ be the shape function of the structure and let

$$\begin{aligned}
m^* &= \int_0^\ell m(z)\Phi^2(z)\,dz, \\
k^* &= \int_0^\ell EI(z)(\Phi''(z))^2 dz, \\
c^* &= \int_0^\ell c(z)\Phi^2(z)\,dz, \qquad (7) \\
F_1^* &= \int_0^\ell F_1(z,t)\Phi(z)\,dz, \\
F_2^* &= \int_0^\ell F_2(z,t)\Phi(z)dz.
\end{aligned}$$

Then equation (1) becomes

$$m^*\ddot{y}(t) + c^*\dot{y}(t) + k^*y(t) = F_1^*(t) + F_2^*(t). \qquad (8)$$

Equation (8) gives an approximation of the actual response of the structure at some reference point on the structure. In this case, $y(t)$ is used to approximate the structural displacement at the controller location, i.e., at the top of the structure. The accuracy with which the shape function is defined determines whether equation (8) gives a good approximation of this displacement.

Aerodynamic Appendages

In vector-matrix form, equation (8) can be written as

$$\dot{x} = Ax + Bu_1(t)u(t) + w, \quad (9)$$

where

$$x = \begin{bmatrix} y(t) \\ \dot{y}(t) \end{bmatrix} = \begin{bmatrix} x_1 \\ x_2 \end{bmatrix}, \quad (10)$$

$$A = \begin{bmatrix} 0 & 1 \\ \dfrac{-k^*}{m^*} & \dfrac{-c^*}{m^*} \end{bmatrix}, \quad (11)$$

$$B = \begin{bmatrix} 0 \\ \dfrac{1}{m^*} \end{bmatrix}, \quad (12)$$

$$w = \begin{bmatrix} 0 \\ \dfrac{F_1^*(t)}{m^*} \end{bmatrix}, \quad (13)$$

$$F_2^*(t) = u_1(t)u(t), \quad (14)$$

and

$$0 \leq u(t) \leq 1. \quad (15)$$

3. OPTIMUM CONTROL SCHEMES

Our objective is to devise optimum linear feedback control laws for $u(t)$ using several objective functions. Let us consider the following cases.

(a) Minimization of the cost function

$$J = \frac{1}{2} \int_0^{t_f} x^T Q x \, dt, \quad t_f \to \infty, \quad (16)$$

where Q is a constant positive semidefinite weighting matrix.

Then, the control is determined by [4]

$$u(t) = \begin{cases} 1, & (Px)^T B < 0 \\ 0, & (Px)^T B \geq 0 \end{cases}, \qquad (17)$$

where P is solved from the Ricatti equation

$$PA + A^T P + I = 0. \qquad (18)$$

(b) Consider now the minimization of the cost function

$$J = \frac{1}{2} \int_0^{t_f} (x^T Q x + u_2^2(t)) dt, \qquad (19)$$

where

$$u_2(t) = F_2^*(t). \qquad (20)$$

Then, the control is determined by [4]

$$u(t) = \begin{cases} 1, & r(t) > 1 \\ r(t) = -B^T Px/u_1(t), & 0 < r(t) \leq 1 \\ 0, & r(t) \leq 0 \end{cases}, \qquad (21)$$

and P is found from the Ricatti equation,

$$Q + PA + A^T P - PBB^T P = 0. \qquad (22)$$

(c) The feedback control algorithm employed is a simple on-off control as a function of the structure velocity at the controller location, i.e.,

$$u(t) = \begin{cases} 1, & \dot{y}(t) \leq 0 \\ 0, & \dot{y}(t) > 0 \end{cases} \qquad (23)$$

(d) The control is determined by

$$u(t) = \begin{cases} 1, & r(t) > 0 \\ 0, & r(t) \leq 0 \end{cases}, \qquad (24)$$

which is an on-off control scheme based on (b).

4. NUMERICAL RESULTS AND CONCLUDING REMARKS

In what follows, numerical results are presented to study the effectiveness of the control schemes proposed above. The parameter values given below are chosen to approximate those associated with a typical tall structure.

ℓ = 1000 ft. (3.05 x 10^2 m)

ρ = 0.0024 slugs/cu.ft. (1.99 kg/m³)

$C_D = C_D'$ = 1.2

A_p/A_s = 0.04

$m(z)$ = 2370 slug/ft. (0.02 kg/m)

$[EI/m(z)]^{1/2}$ = 1.45 x 10^5 ft²/sec (1.35 x 10^5 m²/sec)

A_s/ℓ = 210 ft (64 m)

$$Q = \begin{bmatrix} k^* & 0 \\ 0 & 0 \end{bmatrix}$$

Assuming that the shape function is

$$\Phi = (1 - \cos\frac{\pi z}{2\ell}), \qquad (25)$$

then

$$m^* = 0.2267\ m(z)\ \ell/32.2\ ,$$
$$k^* = 12.176\ EI/\ell^3\ ,$$
$$c^* = 0.22676 c\ell\ .$$

Additional parameter values used for simulating the wind velocity are:

$$n = 25\ ,$$
$$k = 0.04\ ,$$
$$\omega_u = 40\ rad/sec\ ,$$
$$V_z = 150\ ft/sec\ ,$$
$$\alpha = 0.5\ ,$$
$$\phi = 2000\ ft\ .$$

Figures 2 and 3 show the response of the building dynamics by using control algorithms (a) and (b), respectively. It is interesting to note that the results for cases (a) and (c) are nearly identical. Furthermore, cases (b) and (c) produce almost identical structural response. The efficiency of each of the control schemes in terms of displacement and acceleration reductions and the motion of the appendage can be assessed by evaluating the integrals

$$\left.\begin{aligned} J_1 &= \frac{1}{2} \int_0^T y^2(t)\ dt\ , \\ J_2 &= \frac{1}{2} \int_0^T u^2(t)\ dt\ , \\ J_3 &= \frac{1}{2} \int_0^T \ddot{y}^2(t)\ dt\ . \end{aligned}\right\} \quad (26)$$

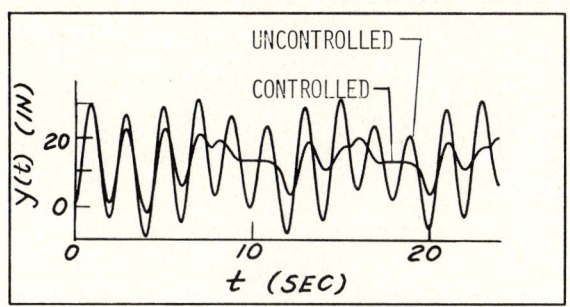

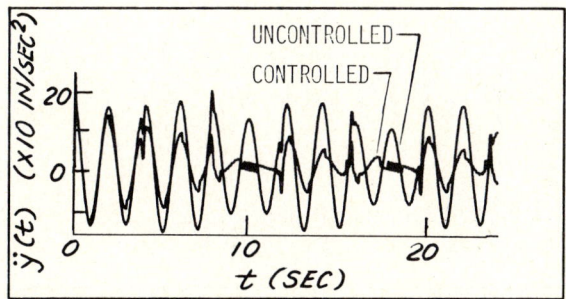

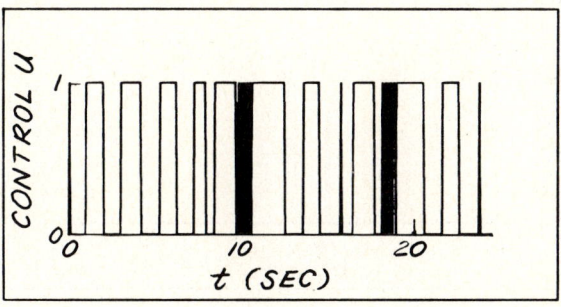

Figure 2 - Structural Response [Case (a)]

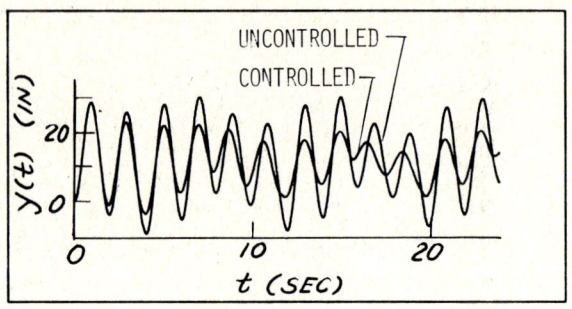

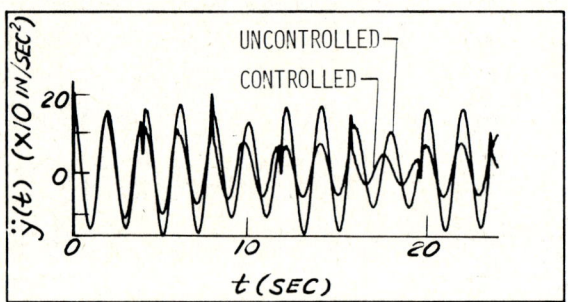

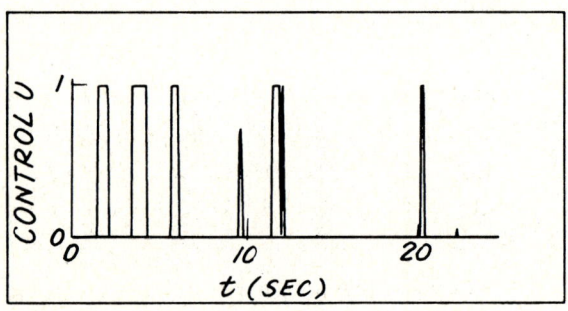

Figure 3 - Structural Response [Case (b)]

Their values with T = 36 for each control scheme are given in Table 1. While all control schemes appear to have similar effectiveness in displacement and acceleration reductions, the values of J_2 show that schemes (b) and (d) require much less movements on the part of the appendage. This can be a significant advantage in control implementation.

Table 1 - Values of J_1, J_2, and J_3

Cases	J_1	J_2	J_3
(a)	3.07×10^3	9.90	3.36×10^4
(b)	2.84×10^3	1.48	5.06×10^4
(c)	3.07×10^3	9.85	3.35×10^4
(d)	2.81×10^3	1.58	4.70×10^4
Without Control	3.94×10^3	-	1.60×10^5

These numerical results using simple examples show that significant reduction in building displacement and acceleration can be materialized by means of small appendages. This, coupled with energy savings, appears to offer a viable mechanism for control of tall structures.

ACKNOWLEDGEMENT

This work was partially supported by the National Science Foundation under Grant No. ENG 7682226.

NOTATION

The following symbols are used in this paper:
- A — uncontrolled system matrix
- A_s — cross-sectional area of building
- A_p — cross-sectional area of appendage
- B — control coefficient matrix
- C_D — drag coefficient of main structure
- C_D' — drag coefficient of appendage

EI	−	elastic properties of structure
F_1	−	force per unit length on structure
F_2	−	control force on structure
F_1^*, F_2^*	−	generalized forces associated with F_1 and F_2
k^*	−	generalized spring constant
ℓ	−	structure height
$m(x)$	−	mass per unit length
m^*	−	generalized mass
Q	−	a constant positive semidefinite matrix
$V(z,t)$	−	instantaneous wind velocity
$x(t)$	−	state vector
$y(t)$	−	displacement at the top
ρ	−	density of air

REFERENCES

[1] KLEIN, R.E., CUSANO, C. and SLUKEL, J.J., "Investigation of a Method to Stabilize Wind Induced Oscillations in Large Structures", presented at the *1972 ASME Winter Annual Meeting*, Paper No. 72-WA/AUT-11, New York, New York, November, 1972.

[2] VAICAITIS, R., SHINOZUKA, M. and TAKENO, M., "Parametric Study of Wind Loading on Structures", *Trans. ASCE, J. of the Structural Division*, Vol. 99, 1973, pp. 453-468.

[3] DAVENPORT, A.G., "The Application of Statistical Concepts to Wind Loading of Structures", *Proc. Inst. Civil Engineers*, Vol. 19, 1961, pp. 449-472.

[4] SAGE, A.P. and WHITE, C.C., III, *Optimum Systems Control*, Prentice-Hall Inc., Englewood Cliffs, New Jersey, 1977, pp. 53-86.

STRUCTURAL CONTROL, H.H.E. Leipholz (ed.)
North-Holland Publishing Company & SM Publications
© IUTAM, 1980

PRINCIPLE TYPES OF OPTIMIZATION PROBLEMS IN THE MECHANICS OF DEFORMED SOLIDS AND THEIR MATHEMATICAL MODELS

A. A. Čyras

Vilnius Civil Engineering Institute
Vilnius, Lithuanian S.S.R., U.S.S.R.

INTRODUCTION

A conventional boundary value problem concerned with the mechanics of a continuum consists of determining the stress and strain field of the solid which is loaded by external forces, while the distribution of physical parameters of the solid and its configuration and boundary conditions are being described. Such problems are referred to as one-parameter problems, because the loading and the parameters of the solid are prescribed up to one parameter. Apart from the one-parameter problems, there are also the so-called optimization problems. In this type of problem, the stress and strain fields of the characteristics of the solid are completely defined, however a certain prescribed optimality criterion of these variables must be given instead.

The aim of the paper is to offer the classification of optimization problems, to suggest the principle ways of derivation of the mathematical models and lastly, to present the structure of these problems for certain properties of the material of the deformed solid. These statements are illustrated by a sample solution for an elastic-perfectly plastic solid.

1. PRINCIPLE TYPES OF OPTIMIZATION PROBLEMS

To determine the principle types of optimization problems, it is necessary to consider the formulation of the boundary value problem in mechanics of deformed solids as well as to define the values it is based upon. The principle elements of the mechanical problems are the external forces, the body and its material. These elements are specified by certain characteristics:

(1) *The external actions* (loading and displacement) are specified by their value, direction, position and manner of application.

(2) *The body* (a mechanical model of the actual structure) is defined by the distribution of the physical constants as well as by the shape (configuration) and kinematic boundary conditions.

(3) *The material of the body* (material behaviour under deformation) is described by the appropriate constitutive law.

Thus, there may be two principle ways of optimization of the prescribed physical model of the body (elastic, plastic, locking, etc.) - with respect to the external forces and with respect to the solid. In the optimization problem only one of the above characteristics may be used together with the values defining the stress and strain field of the solid by variable parameters; other characteristics must be prescribed.

In the optimization of the external forces, the optimization may be performed with respect to the magnitude of the applied load, while the distribution of the external and load and its direction on the prescribed surface of the body are assumed to be known. On the other hand, it is possible to find an optimum direction of the loading, while the distribution and the location of loading are prescribed. The last type of problems may concern the determination of the position of load, with its direction and magnitude being prescribed on a given surface of the body. The characteristics of the body in these problems must be completely defined.

If the body is to be optimized, then three types of optimization problems are possible as well. Firstly, the distribution of the physical constants may be optimized and the configuration of the solid and the kinematic boundary condition are prescribed. Secondly, the optimum configuration of the body may be found, if the distribution of the physical constants is prescribed and the kinematic conditions are assumed to be known. The third type of optimization problems may consist in the determination of the most appropriate location of support on a given surface. It should be noted, that in these three cases the external loads are to be completely defined.

2. OPTIMALITY CRITERIA AND THEIR MODIFICATIONS FOR DIFFERENT TYPES OF OPTIMIZATION PROBLEMS

Consider an isotropic solid of volume V with respect to Cartesian coordinates $\underline{X} \equiv (X_1, X_2, X_3)$. The surface S of the body is divided into two parts: S_p is the part subjected to loading, S_u is that part where the displacements are completely prescribed. Henceforth, the displacements on S_u are assumed to be zero. Thus, here we shall deal with unknown external reactions defined by the vector field $\underline{R} \equiv (R_1, R_2, R_3)^T$.

Now, the determination of analytical expressions defining the optimality criteria in the principle types of optimization problems will be considered.

2.1 *Optimization Problem of External Load*

In this problem, the main object to be varied is the vector field of the external load. Therefore, the optimality criterion in this case is a scalar function of the argument $\underline{P}$. A problem of this type logically implies that the maximum value of this scalar function is to be determined, i.e.,

$$\int_{S_p} \psi(\underline{P})ds \to \max . \tag{1}$$

The function ψ may be expressed in different ways. In load optimization problems, a linear form of the optimality criterion is most suitable:

$$\int_{S_p} \underline{T}^T \cdot \underline{P}\, ds \to \max , \tag{2}$$

where $\underline{T} \equiv (T_1, T_2, T_3)^T$ is usually referred to as the vector field of the weight multipliers of the optimality criterion. This expression shows in relative terms the effect of the component forces at a given point on the optimality of the external load distribution.

This linear form of the optimality criterion of the external load will be used in this paper. Now, the expressions for the optimality criterion in different formulations of the problems will be dealt with.

2.1.1 Problem of Optimum Distribution of External Load

In this problem, the position of the external load and its direction are assumed to be defined. It is necessary to determine the law of the distribution of its magnitude corresponding to the optimality criterion (2). In this case the external load may be determined to within the scalar field P, with its direction, which is defined by the unit vector field $\underline{e} \equiv (e_1, e_2, e_3)^T$, being prescribed, i.e.,

$$\underline{P} = P \cdot \underline{e} . \tag{3}$$

Then, the optimality criterion for the present problem will be:

$$\int_{S_p} \underline{T}^T \cdot P\underline{e}ds \to \max . \qquad (4)$$

Here the value to be searched for is the scalar field P, subject to the sign constraint $P \geq 0$. This constraint is included in the conditions of the problem and it shows that the value of the multiplier is positive at every point on the surface S_p and the prescribed direction $\underline{e}$.

2.1.2 Problem of Determination of Optimum Direction of External Load

In this problem, the law of distribution of loading (the scalar field P) is assumed to be prescribed involving the general multiplier P°. It is necessary to determine the unit vector field $\underline{e}$ defining the direction of the load on the surface S_p which corresponds to the optimality criterion (2). Hence, the optimality criterion for the present problem will be:

$$P° \int_{S_p} \underline{T}^T \cdot \eta\underline{e}ds \to \max , \qquad (5)$$

where η is the scalar field defining the distribution of loading on S_p.

The values to be searched for are the multiplier P° and the unit vector field $\underline{e}$, which is subject to the following additional constraints resulting from the definition of a unit vector:

$$\underline{e}^T \cdot \underline{e} = 1 . \qquad (6)$$

It should be noted, that next to the load optimization problems using criterion (5), there are also other problems which necessitate the selection of the loading direction at which a maximum or a minimum value of the limit loading parameter is obtained.

Here all directions are the same and the optimality criterion is either $\underline{P}^\circ \to \max$ or $\underline{c}^\circ \to \min$, where $\underline{c}^\circ$ is the physical parameter of the solid. In such a case, the variable $\underline{e}$ comes into play only in the constraints of the problem.

In some cases depending on the actual conditions of the problem it is necessary to reduce the range of change of the loading direction. Then some additional constraints of the following type are introduced:

$$a_i \leq e_i \leq b_i \,, \tag{7}$$

where a_i and b_i are the prescribed constraints on the direction of the unit vector, provided that $-1 \leq a_i \leq 1$ and $-1 \leq b_i \leq 1$.

2.1.3 Problem of Determination of Optimum Position of the External Load

Let the vector of the load $\underline{P}$ be prescribed up to the multiplier P°, the load being distributed on the area S^* of the free surface $S(S^* \subset S_p)$ of the body. The quantity of various positions of the area S^* on the surface S_p is assumed to be finite. Let any position of this area be denoted by the index ξ, and the set of the indicies ξ - by the sympol J. Then quantity of the areas S^* on the surface S_p will be equal to $\{J\}$. It is necessary to find the position of the area S^*_ξ that satisfied the optimality criterion (2).

Hence, the load at the ξ-th position may be expressed by:

$$\underline{P} = P^\circ \eta \underline{e}^\xi \,, \tag{8}$$

where η is the scalar field determining the value of the load.

Thus, expression (2) for the optimality criterion may be written as:

$$p° \int_{S_p} \underline{T}^T \cdot \underline{\eta e}^\xi ds \to \max . \tag{9}$$

The subintegral expression in (9) is known for every value of ξ, and therefore, $\{J\}$ number of one-parametric problems are obtained. By solving these problems and comparing the results, that particular position of the load is chosen, which corresponds to the maximum value of expression (9). Note that the solution of the problem obtained in this way refers to the monotonically increasing load, i.e., the load is applied on the free surface in one position only. If the load is permitted to take the position of ξ in any succession, then a movable load is dealt with. In this case, in the conditions of problems concerning dissipative structures the accumulation of plastic strains must be taken into consideration.

Thus, three modification of the objective function for the optimization problems of the external load, corresponding to the linear form of the optimality criterion, have been obtained. Similarly, the expressions related to any other functional form of the optimality criterion may be derived, but then the problem becomes more complicated to solve.

2.2 Optimization Problem of the Body

In problems of this type, the external forces are assumed to be completely prescribed, and three parameters are to be varied, namely the configuration of the body, the distribution of the physical parameters and the support conditions.

Let the physical parameters of the solid (elasticity modulus, plasticity constant, locking constant, etc.) be defined by the scalar field $C \equiv C(X)$ which is positive. If its value is zero, then the volume of the solid at these points is understood to be zero. In this case, the scalar field of the physical constants C is considered to be the parameter determining the optimality of the body. Hence, the optimality criterion in general will be the scalar function of the argument C. The problem of this type logically implies that the minimum value of this scalar function in the volume is to be found, i.e.,

$$\int_V \phi(c)dv \rightarrow \min . \tag{10}$$

The form of the function ϕ may be different. Frequently, in the optimization problem of solids its linear form is used:

$$\int_V \Lambda C dv \rightarrow \min , \tag{11}$$

where $\Lambda \equiv \Lambda(X)$ is the scalar field of the weight multipliers of the optimality criterion. The value of Λ at every point shows the relation of the optimality of the body in general to the value of the constant C at the point being considered. By properly chosing the weight multipliers and the expressions for the constant C, optimization problems of the body having a different physical meaning may be obtained (e.g., minimum cost, minimum weight, etc.). Henceforth, a linear form of the optimality criterion will be used.

2.2.1 Problem of Determination of Optimum Distribution of the Physical Constant of the Material

In this problem, the volume of the body, condition and the applied load are assumed to be prescribed. It is necessary to find the law of distribution of the physical constant C of the solid corresponding to the optimality criterion:

$$\int_V \Lambda C dv \rightarrow \min . \tag{12}$$

Principle Types of Optimization Problems 219

The scalar field C which is the value to be searched for is subject to the logical constraint $C \geq 0$, which is included in the conditions of the problem. The constraints of the scalar field C may be different. The upper and lower bounds of the scalar field are prescribed in relation to the statement of the problem. It should be noted that any constraint always increases the value of the cost function.

Since the value of the physical constant in some way may be related to the cost of the material, such optimization problems are sometimes referred to as those of the minimum cost of the solid.

2.2.2 Problem of Determination of Optimum Configuration of the Body

Let the distribution of the physical constant of the material in the volume V be prescribed by:

$$C = C^\circ \gamma \rho , \qquad (13)$$

where C° is the value to be searched for; γ is the prescribed scalar field determining the distribution law of the material constant; ρ is a scalar field of Boolean (zero-one) variables $\rho_i = (0,1)$.

Now, expression (11) for the optimality criterion in the problem of the optimal configuration of the solid will take the following form:

$$C^\circ \int_V \Lambda \gamma \rho dv \to \min . \qquad (14)$$

The value of zero for ρ_i means that at these points the value of the physical constant of the material is zero, and it follows that there is no material there. It should be especially noted, that the boundary condition at the points of application of the load and on S_u must be satisfied completely. Hence, the values of ρ_i at these points cannot be equal to zero. Thus, the material can disappear only on that part of the surface S_p which is not subject to loading.

The problem of the optimal configuration of the solid possessing similar properties at all points is concerned with a minimum volume (weight). This is a common problem in the optimal design, nevertheless. G. Maier [16] was the first to formulate and solve it by applying Boolean variables to discrete systems.

The optimality criterion defined by (14) presents a more general form, as it allows the optimal value of the material to be considered at every design point.

2.2.3 Problem of Determination of the Optimum Position of Support of the Body

Let there be a finite number of areas S* on the surface S_u, where the body may be supported according to the boundary conditions which are assumed to be known. Let any position of the area S* be denoted by the index; and the set of all indices ξ - by the symbol J. It is necessary to find the position of the area S^*_ξ satisfying the optimality criterion defined by (11). The load and the configuration of the solid are completely prescribed, the plasticity constant is prescribed up to the parameter C°, i.e., the value of C is equal to $C^°_\gamma$, where γ is the scalar field of the distribution of the material constant. Thus, optimality criterion (11) will be:

$$C^°_\xi \int_v \Lambda\gamma dv \to \min_{\xi \in J} . \qquad (15)$$

Since the subintegral value in expression (15) is known, by fixing

$$\int_v \Lambda\gamma dv \equiv 1 ,$$

a one-parameter problem can be obtained for every position of support, where the following expression is determined:

$$C_\xi^o \to \min . \tag{16}$$

The minimum value of C_ξ^o in the set $\{J\}$ yields the solution of the present optimization problem.

3. CONDITIONS OF OPTIMIZATION PROBLEMS

The above modifications of the optimality criterion for optimization problems of deformed solids and relevant specific conditions may be applied to different physical models of the solid (elastic, plastic, locking, etc.). The procedure in deriving mathematical optimization models is to choose conditions corresponding to the physical model of the solid, to the type of loading and design conditions. It is apparent that there will be a variety of conditions, for example, linear elastic and elastic-plastic solids, strength and stability problems, monotonically increasing and cyclic loading, etc. General principles applied to the conditions of the optimization problems of strength will now be considered.

Determination of the parameters of either the load or the solid must correspond to the actual stress and strain field of the solid. The stress and strain field of the solid is determined by dual extremum energy principles or by the relevant generalized Lagrange problem including the physical law and static and geometric conditions. In deriving the mathematical models of optimization problems, the application of conventional extremum energy principles is preferable, because in some cases the optimality criterion of the problem and the expression of the energy value differs only by a certain multiplier. Hence, the final optimization problem includes the optimality criterion and the conditions of the extremum principle. This statement is valid for all types of loading applied to elastic-perfectly plastic or perfectly locking solids as well as for the optimization problems of linear elastic hinged truss systems, with the stresses being prescribed. For the physical models, where the relation between stresses and strains is not represented by a straight line parallel to either of the coordinate axes, the conditions determining the actual stress and strain field will be those of the generalized Lagrange problem or various modifications of them.

The conditions of the problems may include the constraints on mechanical quantities (stresses, strains and displacements). The conditions of this type are formulated in either linear or nonlinear inequalities.

Note that construction limitations related to load parameters or to actual structures and used in practical optimal design problems are not dealt with.

Now, to illustrate the above statements, an elastic-perfectly plastic solid under three types of loading will be considered.

3.1 *Monotonically Increasing Loading*

For this type of loading, the actual stress field, corresponding to plastic failure, is determined by the following extremum energy theorems:

Of all statically admissible stress fields at simple plastic failure that one is actual, which corresponds to the maximum value of external power (to the minimum value of plastic dissipation).

The mathematical models related to these theorems are as follows:

$$\int_{S_p} \underline{\dot{u}}^T \cdot \underline{P} ds \to \max \text{ (a)} \quad \bigg|\bigg| \quad \int_V \lambda C dv \to \min \text{ (}\delta\text{)} \qquad (16)$$

$$\left. \begin{array}{l} f(\underline{\sigma}) \leq c \\ A(\underline{\sigma}) = 0 \end{array} \right\} \text{ in } V,$$

$$N(\underline{\sigma}) = \underline{P} \text{ on } S_p,$$

$$N(\underline{\sigma}) = \underline{R} \text{ on } S_u.$$

By comparing the cost functions in problems (16a) and (16b) with the expressions for the optimality criterion in external load optimization problem (2) and in optimization problem of solids (11), it is seen that these differ by a constant multiplier. On the other hand, since the displacement velocities $\underline{\dot{u}}$ and the multipliers λ are not included in conditions (16), the extremum energy theorems are satisfied at any fixed value of the fields $\underline{\dot{u}}$ and λ, including $\underline{\dot{u}} \equiv \underline{T}$ and $\lambda \equiv \Lambda$. Thus, for the optimization problems of rigid-plastic solids under monotonically increasing loading, the conditions of structural design will be as follows:

$$\left. \begin{array}{l} f(\underline{\sigma}) \leq c \\ A(\underline{\sigma}) = 0 \end{array} \right\} \text{ in } v, \qquad (17)$$

$$N(\underline{\sigma}) = \underline{P} \text{ on } S_p,$$

$$N(\underline{\sigma}) = \underline{R} \text{ on } S_u.$$

These conditions are necessary. Depending on the type of optimization problems presented in Section 2, the expressions for the vector field $\underline{P}$ or for the scalar field C are transformed and specific conditions for these values are introduced.

3.2 *Cyclic Loading*

This type of loading is represented by a set of forces. Every force or groups of forces can vary independently or within a prescribed range. The upper bound is defined by the vector field $\underline{P}^+$, the lower bound by $\underline{P}^-$. If loading is cyclic, every combination of loads does not necessarily lead to plastic failure, nevertheless, a certain combination of these can result in cycles of plastic strains causing cyclic plastic failure. The actual stress field corresponding to this state is determined by the following extremum energy theorem:

Of all statically admissible fields of residual stresses at cyclic plastic falure, that one is actual, which corresponds to the maximum value of external power (to the minimum value of plastic dissipation in a cycle).

The extremum problems corresponding to these theorems are as follows:

$$\left\{ \int_{S_p} (\dot{\underline{u}}^+)^T \cdot \underline{P}^+ ds + \int_{S_p} (\dot{\underline{u}}^-)^T \cdot \underline{P}^- ds \right\} \to \max(a) \quad \left\| \quad \int_v (\lambda^+ + \lambda^-) C dv \to \min \right. \tag{18}$$

$$\left. \begin{array}{l} f(\underline{\sigma}^+ + \underline{\rho}) \leq c \\ f(\underline{\sigma}^- + \underline{\rho}) \leq c \\ A\underline{\rho} = 0 \end{array} \right\} \text{ in } v \ ,$$

$$N\underline{\rho} = 0 \qquad \text{on } S_p \ ,$$
$$N\underline{\rho} = \underline{r} \qquad \text{on } S_u \ .$$

The optimality criterion of the load optimization problem, that permits both the upper and lower bounds to be optimized for this type of loading, is as follows:

$$\left\{ \int_{S_p} (\underline{T}^+)^T \cdot \underline{P}^+ ds + \int_{S_p} (\underline{T}^-)^T \cdot \underline{P}^- ds \right\} \to \max \ . \tag{19}$$

In case of the optimization problem of the solid, expression (11) is also used, i.e.,

$$\int_v \Lambda C dv \to \min \ .$$

Similarly to the case of monotonically increasing loading, by comparing the cost functions of problems (18a) and (18b), with respect to the optimality criteria and by fixing the fields of displacement velocities $\underline{u}^+ \equiv \underline{T}^+$, $\underline{\dot{u}}^- \equiv \underline{T}^-$ and the fields of the multipliers $\lambda^+ + \lambda^- \equiv \Lambda$, the conditions necessary for the optimization problem of the elastic-plastic solid under cyclic loading are obtained:

$$\left. \begin{array}{l} f(\underline{\sigma}^+ + \underline{\rho}) \leq c \\ f(\underline{\sigma}^- - \underline{\rho}) \leq c \\ A\underline{\rho} = 0 \\ N\underline{\rho} = 0 \quad \text{on } S_p, \\ N\underline{\rho} = \underline{r} \quad \text{on } S_u. \end{array} \right\} \quad \text{in v}, \qquad (20)$$

Note, that in the load optimization problem the values $\underline{P}^+$ and $\underline{P}^-$ which are searched for are included in condition (20) by way of extremum "elastic" stresses expressed by:

$$\left. \begin{array}{l} \underline{\sigma}^+ = \int\limits_{S_p} (\omega^+ \underline{P}^+ - \omega^- \underline{P}^-) ds \\ \underline{\sigma}^- = \int\limits_{S_p} (\omega^- \underline{P}^+ - \omega^+ \underline{P}^-) ds \end{array} \right\} \qquad (21)$$

where ω^+ and ω^- are the influence operators of an "elastic" design which are derived from various combinations of the vectors ω_j of the operator ω. It should be kept in mind that:

$$\omega_j^- = 0 \quad, \quad \text{subject to } \omega_j^+ = \omega_j$$

and

$$\omega_j^+ = 0 \quad, \quad \text{subject to } \omega^- = \omega_j \; .$$

Thus, equations (21) allow the surface of the extremum "elastic" stresses to be obtained, which is always symmetric to its centre.

3.3 Movable Load

The movable load may be visualized as a set of forces determined by the vector field $\underline{P}$, which is distributed on the movable area S^* of the free surface $S(S^* \subset S_p)$ of the solid. The number of various positions of the area S^* on S_p is assumed to be finite. The load applied to the ξ-th area can cause plastic strains in the solid, and these will lead to redistribution of the strains due to residual strains, when the position of the load $\underline{P}^\xi$ is changed. Thus, this is the case which was dealt with when static cyclic loading was considered.

At certain combinations of positions of the load, cycles of plastic strains may occur, and these, when repeated, may lead to cyclic plastic failure of the solid. Here, a cycle is understood to be a period of time during which the area S^* takes all possible positions on the surface S_p in any sequence. Because similar phenomena in the solid arise from the application both of movable and cyclic loading, the actual stress and strain field of the solid is determined by the same extremum theorems as in the preceding section. Hence, it follows:

$$\sum_{(\xi)} \int_{S^*_\xi} (\underline{\dot{u}}^\xi)^T \cdot \underline{P}^\xi ds \to \max \quad (a) \quad \Big|\Big| \quad \sum_{(\xi)} \int_v \lambda^\xi C dv \to \min$$

$$\left. \begin{array}{l} f(\underline{\sigma}^{\circ\xi} + \underline{\rho}) \leq c, \quad \xi \in J \\ A\underline{\rho} = 0 \\ N\underline{\rho} = 0 \quad \quad \text{on } S_p, \\ N\underline{\rho} = \underline{r} \quad \quad \text{on } S_u. \end{array} \right\} \quad \text{in } v, \quad (22)$$

Here, "elastic" strains caused by the load, applied on the ξth area, are derived from the expression:

$$\underline{\sigma}^{\circ\xi} = \int_{S^*} \omega \underline{P}^\xi ds. \quad (23)$$

The use of extremum theorems (22a) and (22b) in the derivation of the conditions for various types of optimization problems is similar to the procedures presented above. By comparing the cost functions and the optimality criteria, it is observed that for the movable load the following conditions are necessary:

$$\left.\begin{array}{l} f(\underline{\sigma}^{o\xi}+\underline{\rho}) \leq c, \quad \xi \in J \\ A\underline{\rho} = 0 \\ N\underline{\rho} = 0 \quad \text{on } S_p, \\ N\underline{\rho} = \underline{r} \quad \text{on } S_u. \end{array}\right\} \text{in } v \quad (24)$$

Besides, it should be noted that these conditions define the entire process of loading. Hence, the optimization problem of the position of application of the external load, in which the load is monotonically increasing, because it is possible to load only one area on the surface S_p, will not be valid.

4. FORMULATION OF MATHEMATICAL MODELS FOR THE OPTIMIZATION PROBLEMS

The mathematical models of the optimization problems include the relevant optimality criterion and the conditions which may be classified as necessary for defining the actual stress field of the solid and others for defining the type of the problem. Thus, the problem formulated in this way is an extremum problem in functional spaces. If the cost function and the conditions of the problem satisfy the convexity conditions, then by way of formal consideration, we can derive a problem dual to this which has a certain physical meaning. Two fundamental duality theorems are valid for the dual pair of problems; the additional orthogonality condition and the values of the cost functions for the optimal solutions of dual problems are equal. These theorems are also the physical meaning of some conditions of dual problems, which are obtained by way of formal mathematical consideration, and show that optimization problems in the mechanics of solids have a mechanical rather than an economical basis. The fact is that the optimality of the load or of the solid is achieved by satisfying the corresponding constraints with respect to the stress and strain fields. This statement is proven by the analysis of the mathematical models of the optimization problems of an elastic-perfectly plastic solid under monotonically increasing loading.

Now, the dual formulation of the mathematical models of the optimization problems for a rigid-plastic solid under monotonically increasing loading will be considered.

Let us consider some mathematical models of the optimization problems in the dual formulation for a rigid-plastic solid under monotonically increasing loading. The primary problem will be the relevant optimization problem in the static formulation, its dual one - in the kinematic formulation. This is accounted for by both the main variables and the relations included in the problems. The static formulation includes dynamic variables (forces, attraction) and equilibrium equations, the kinematic formulation includes kinematic variables (displacements, strains) and geometrical equations. In the conditions of the problems, the values to be searched for will be written in the left-hand part of the relations, the values prescribed will be written in the right-hand part.

Optimum Distribution of External Load

Static Formulation

$$\int_{S_T} \underline{T}^T \cdot \underline{P} \underline{e} ds \to \max_{P}$$

$$\left. \begin{array}{l} f(\underline{\sigma}) \leq C \\ A\underline{\sigma} = 0 \end{array} \right\} \text{in } v,$$

$$\left. \begin{array}{l} N\underline{\sigma} - P\underline{e} = 0 \\ P \geq 0 \end{array} \right\} \text{on } S_p,$$

$$N\underline{\sigma} - R = 0 \quad \text{on } S_u.$$

Principle Types of Optimization Problems

Kinematic Formulation

$$\left\{ \int_v \lambda \left[\frac{\partial f(\underline{\sigma})}{\partial \underline{\sigma}}\right]^T \cdot \underline{\sigma} dv + \int_v \lambda [C-f(\underline{\sigma})] dv \right\} \to \min$$

$$\left. \begin{array}{l} A^T \underline{\dot{u}} - \lambda \dfrac{\partial f(\underline{\sigma})}{\partial \underline{\sigma}} = 0 \\ \lambda \geq 0 \end{array} \right\} \text{ in } v,$$

$$\begin{array}{ll} \underline{\dot{u}} \geq \underline{T} & \text{on } S_p, \\ \underline{\dot{u}} = 0 & \text{on } S_u. \end{array}$$

Direction of the Maximum Limit Load

Static Formulation

$$P^U \to \max$$

$$\left. \begin{array}{l} f(\underline{\sigma}) \leq C \\ A\underline{\sigma} = 0 \end{array} \right\} \text{ in } v,$$

$$\left. \begin{array}{l} N\underline{\sigma} - P^\circ \eta \underline{e} = 0 \\ \underline{e}^T \cdot \underline{e} = 1 \\ P^\circ \geq 0 \end{array} \right\} \text{ on } S_p,$$

$$N\underline{\sigma} - \underline{R} = 0 \quad \text{on } S_u.$$

Kinematic Formulation

$$\left\{ \int_v \lambda \left[\frac{\partial f(\underline{\sigma})}{\partial \underline{\sigma}}\right]^T \cdot \underline{\sigma} dv + \int_v \lambda [C-f(\underline{\sigma})] dv + \int_{S_p} \mu(\underline{e}^T \cdot \underline{e} - 1) ds \right\} \to \min$$

$$\left. \begin{array}{l} A^T \underline{\dot{u}} - \lambda \dfrac{\partial f(\underline{\sigma})}{\partial \underline{\sigma}} = 0 \\ \lambda \geq 0 \end{array} \right\} \text{ in } v,$$

$$\left. \begin{array}{l} \int_{S_p} \underline{\dot{u}}^T \cdot \eta \underline{e} ds \geq 1 \\ 2\mu \underline{e} - P^\circ \eta \underline{\dot{u}} = 0 \end{array} \right\} \text{ on } S_p,$$

$$\underline{\dot{u}} = 0 \quad \text{on } S_u.$$

Optimum Position of the External Load

Static Formulation

$$P° \int_{S^*} \underline{T}^T \cdot \eta \cdot \underline{e}^\xi ds \to \max_{\xi \in J}$$

$$\left.\begin{array}{l} f(\underline{\sigma}) \leq c \\ A\underline{\sigma} = 0 \end{array}\right\} \text{ in } v,$$

$$\left.\begin{array}{l} N\underline{\sigma} - P°\eta\underline{e} = 0 \\ P° \geq 0 \end{array}\right\} \text{ on } S^*,$$

$$N\underline{\sigma} - \underline{R} = 0 \quad \text{on } S_u.$$

Kinematic Formulation

$$\left\{ \int_v \lambda \left[\frac{\partial f(\underline{\sigma})}{\partial \underline{\sigma}}\right]^T \cdot \underline{\sigma} dv + \int_v \lambda \left[c - f(\underline{\sigma})\right] dv \right\} \to \min$$

$$\left.\begin{array}{l} A^T \cdot \underline{\dot{u}} - \lambda \dfrac{\partial f(\underline{\sigma})}{\partial \underline{\sigma}} = 0 \\ \lambda \geq 0 \end{array}\right\} \text{ in } v,$$

$$\int_{S^*} \underline{\dot{u}}^T \cdot \eta \cdot \underline{e}^\xi ds \geq \int_{S^*} \underline{T}^T \cdot \eta \cdot \underline{e}^\xi ds \quad \text{on } S_p,$$

$$\underline{\dot{u}} = 0 \quad \text{on } S_u.$$

Optimum Distribution of the Physical Constant

Static Formulation

$$\int_v \Lambda C \, dv \to \min$$

$$\left.\begin{array}{l} C - f(\underline{\sigma}) \geq 0 \\ C \geq 0 \\ A\underline{\sigma} = 0 \end{array}\right\} \text{ in } v,$$

$$N\underline{\sigma} = \underline{P} \quad \text{on } S_p,$$

$$N\underline{\sigma} - \underline{R} = 0 \quad \text{on } S_u.$$

Kinematic Formulation

$$\left\{ \int_{S_p} \underline{\dot{u}}^T \cdot \underline{P} ds + \left[\int_v \lambda f(\underline{\sigma}) dv - \int_v \left[\lambda \frac{\partial f(\underline{\sigma})}{\partial \underline{\sigma}} \right]^T \cdot \underline{\sigma} dv \right] \right\} \to \min$$

$$\left. \begin{array}{l} \lambda \dfrac{\partial f(\underline{\sigma})}{\partial \underline{\sigma}} - A^T \cdot \underline{\dot{u}} = 0 \\[4pt] \lambda \geq 0 , \\ \lambda \leq \Lambda \end{array} \right\} \quad \text{in } v ,$$

$$\underline{\dot{u}} = 0 \qquad \text{on } S_u .$$

Optimum Configuration of the Body

Static Formulation

$$\int_v \Lambda C \zeta dv \to \min$$

$$\left. \begin{array}{l} C\zeta - f(\underline{\sigma}) \geq 0 \\ \zeta = (0,1) \\ A\underline{\sigma} = 0 \end{array} \right\} \quad \text{in } v ,$$

$$N\underline{\sigma} = \underline{P} \qquad \text{on } S_p ,$$
$$N\underline{\sigma} - \underline{R} = 0 \qquad \text{on } S_u .$$

Kinematic Formulation

$$\left\{ \int_{S_p} \underline{\dot{u}}^T \cdot \underline{P} ds + \int_v \lambda [C\zeta - f(\underline{\sigma})] dv - \int_v \left[\lambda \frac{\partial f(\underline{\sigma})}{\partial \underline{\sigma}} \right]^T \cdot \underline{\sigma} dv + \int_v \Lambda C \zeta dv \right\} \to \max$$

$$\left. \begin{array}{l} \lambda \dfrac{\partial f(\underline{\sigma})}{\partial \underline{\sigma}} - A^T \underline{\dot{u}} = 0 \\[4pt] \zeta' = (0,1) , \quad \lambda \geq 0 \end{array} \right\} \quad \text{in } v ,$$

$$\underline{\dot{u}} = 0 \qquad \text{on } S_u .$$

Optimum Position of Support of the Body

Static Formulation

$$C_\xi^o \to \min$$
$$\xi \in J$$

$$\left.\begin{array}{l} C_\xi^o \gamma - f(\underline{\sigma}) \geq 0 \\ A\underline{\sigma} = 0 \\ C_\xi^o \geq 0 \end{array}\right\} \text{ in } v,$$

$$N\underline{\sigma} = \underline{P} \qquad \text{on } S_p,$$
$$N\underline{\sigma} - \underline{R}_\xi = 0 \qquad \text{on } S_\xi^*.$$

Kinematic Formulation

$$\left\{ \int_{S_p} \underline{\dot{u}}^T \cdot \underline{P} ds + \int_v \lambda f(\underline{\sigma}) dv - \int_v \lambda \left[\frac{\partial f(\underline{\sigma})}{\partial \underline{\sigma}}\right]^T \cdot \underline{\sigma} dv \right\} \to \min$$

$$\left.\begin{array}{l} \lambda \dfrac{\partial f(\underline{\sigma})}{\partial \underline{\sigma}} - A^T \underline{\dot{u}} = 0 \\[4pt] \lambda \geq 0 \\[4pt] \int_v \lambda \gamma dv \leq 1 \end{array}\right\} \text{ in } v,$$

$$\underline{\dot{u}}^\xi = 0, \quad \xi \in J, \quad \text{on } S_\xi^*.$$

NOTATION

v	-	volume of the body
S	-	surface of the body
S_p	-	part of S; surface subjected to external loading
S_u	-	part of S; surface with the displacements prescribed completely
S*	-	movable sphere on the surface S_p
x	-	coordinate of a particle of a body
ξ	-	coordinate of a movable sphere on the surface S_p

Principle Types of Optimization Problems

C	−	plasticity constant of a material
C^o	−	parameter of plasticity constant
γ	−	distribution of a plasticity constant
f	−	yield function
$\underline{P}$	−	loading on S_p
$\underline{P}^+$	−	upper bound of loading
$\underline{P}^-$	−	lower bound of loading
$\underline{P}^o$	−	loading parameter
η	−	distribution of external actions on S_p
$\underline{e}$	−	direction of external loading
$\underline{R}$	−	reactions on supported boundary on S_u
$\underline{r}$	−	residual reaction on S_u
$\underline{\sigma}$	−	stresses
$\underline{\sigma}^o$	−	"elastic" stresses
$\underline{\sigma}^+$	−	upper bound of extreme "elastic" stresses
$\underline{\sigma}^-$	−	lower bound of extreme "elastic" stresses
$\underline{\rho}$	−	residual stresses
$\underline{u}$	−	displacements
$\underline{W}$	−	residual displacements
λ	−	flow law multiplier
$\underline{u}^+$	−	upper bound of $\underline{u}$
$\underline{u}^-$	−	lower bound of $\underline{u}$
A	−	differential operator of static equilibrium in v
N	−	algebraic operator of static equilibrium on S_p
ω	−	influence matrix of "elastic" solution
ψ	−	function of optimality criterion for external loading
$\underline{T}$	−	weight multiplier of optimality criterion for external loading
ϕ	−	function of optimality criterion for a body
Λ	−	weight multiplier of optimality criterion for a body

Vector fields are denoted with an underscore, and scalar fields are without the underscore. To denote velocities of the corresponding values the letters are marked with dots. The upper index "T" denotes transposition.

REFERENCES

[1] BORKAUSKAS, A. and ČYRAS, A., "On Duality in Limit Analysis and Design of Plates", *Bull. de l'Academie Polonaise des Sciences*, Ser. sc. Techn., Warszawa, Vol. XVI, No. 6, 1968.

[2] BORKAUSKAS, A. and ČYRAS, A., "Duality in Optimization of Rigid-Plastic Solids", (in Russian), *Stroit. Mekh. i Ras. Sooruzkenii*, No. 4, 1969.

[3] COLLINS, J.F., "An Optimum Loading Criterion for Rigid-Plastic Materials", *J. Mech. Phys. Solids*, Vol. 16, No. 2, 1968.

[4] COHN, M.Z. and PARIMI, S.R., "Optimal Design of Plastic Structures for Fixed and Shakedown Loading", *J. Appl. Mech., A.S.M.E.*, Vol. 40, 1973.

[5] ČYRAS, A., "Design of Elastic-Plastic Redundant Beams by Linear Programming Technique", (in Lithuanian), *Liet. TSR Aukstuju mokyklu darbai, stratyba ir Architektura*, Vol. III, No. 2, 1963.

[6] ČYRAS, A., "Correlations of Fundamental Mathematical Analogues in Design and Analysis of Elastic-Plastic Structures", (in Russian), *Lit. Mekh. Sbornik.*, Vol. 2, No. 1, 1968.

[7] ČYRAS, A., "Duality in Structural Mechanics, Elastic and Plastic Theory Problems", (in Russian), *Lit. Mekh. Sbornik*, Vol. 3, No. 2, 1968 and Vol. 4, No. 1, 1969.

[8] ČYRAS, A., "Linear Programming and Analysis of Elastic-Plastic Structures", (in Russian), *Stroiizdat*, Leningrad, 1969.

[9] ČYRAS, A. and ATKOČIUNAS, J., "Mathematical Models of Load Optimization Problems for Elastic-Plastic Bodies under Repeated Loading", (in Russian), *Lit. Mekh. Sbornik*, Vol. 7, No. 2, 1970.

[10] ČYRAS, A. and ATKOČIUNAS, J., "Design of Elastic-Plastic Bodies under Variable Repeated Loading", (in Russian), *Lit. Mekh. Sbornik*, Vol. 8, No. 1, 1971.

[11] ČYRAS, A., "Optimization Theory of Limit Analysis of Deformable Solids", (in Russian with English summaries), *Mintis*, Vilnius, 1971.

[12] ČYRAS, A., BORKAUSKAS, A. and KARKAUSKAS, R., "Theory and Methods of Optimization of Elastic-Plastic Structures, (in Russian), *Stroiizdat*, Leningrad, 1974.

[13] ČYRAS, A., "Optimization Theory in the Design of Elastic-Plastic Structures", *CISM Lecture Notes on Structural Optimization*, Springer-Verlag, No. 237, 1975.

[14] ČYRAS, A. and VISLAVICIUS, K., "Optimization of Elastic-Plastic Structure under Movable Loading", *Mech. Res. Comm.*, Vol. 2, No. 4, 1975.

[15] ČYRAS, A., "Theory of Optimization of Elastic-Plastic Body Subjected to Moving Loads", (in Russian), *Lit. Mekh. Sbornik*,

[16] ČYRAS, A. and KALANTA, S., "Matrix of the Finite Element Yield Function", (in Russian), *Dep. Liet.*, MTII, No. 195-77, 1977.

[17] KOOPMAN, D. and LANCE, R., "On Linear Programming and Plastic Limit Analysis", *J. Mech. Phys. Solids*, Vol. 13, No. 12, 1965.

[18] DORN, W., GOMORY, R., and GREENBERG, H., "Automatic Design of Optimal Structures", *J. Mechanique*, Vol. 3, No. 1, 1964.

[19] MAIER, G. and ZAVELANI-ROSSI, A., "Finite Element Approach to Optimal Design of Plastic Discs, Part 1", Ist. Sci. Tecnica Costruzioni, Politecnido di Milano, *Publ. No. 482*, 1970.

[20] MROZ, Z. and GARSTECKI, A., "Optimal Design of Structures with Unspecified Loading Distribution", *J. of Optimization Theory and Applications*, Vol. 20, No. 3, 1976.

[21] PRAGER, W. and SHIELD, R.T., "General Theory of Optimal Plastic Design", *Trans. of A.S.M.E.*, Vol. 34, Ser. E., No. 4, 1967.

[22] PRAGER, W., "The Determination of Optimal Layout of a Structure", *CISM Courses*, Saint-Venant Session, October, 1974.

[23] PRAGER, W. and TAYLOR, J., "Problems of Optimal Structural Design", *J. Appl. Mech.*, Vol. 35, No. 4, 1968.

[24] REITMAN, M. and SHAPIRO, G., "Theory of Optimal Design in Structural Mechanics, Elasticity and Plasticity", (in Russian), *Itogi Nauki*, Viniti, 1966.

[25] REITMAN, M. and SHAPIRO, G., "Methods of Optimal Design of a Deformable Body", Moscow, *Nauka*, 1976.

[26] REITMAN, M. and SHAPIRO, G., "Optimal Design of a Deformable Body", (in Russian), *Itogi Nauki i Techniki*, Vol. 12, Moscow, Viniti, 1978.

[27] SAVE, M.A. and PRAGER, W., "Minimum-Weight Design of Beams Subjected to Fixed and Moving Loads", *J. Mech. Phys. Solids*, Vol. 11, 1963.

[28] SAVE, M.A., "Theory of Optimal Plastic Design of Structures", *CISM Courses and Lecture Notes*, No. 237, 1975.

[29] SAVE, M. and SHIELD, R., "Minimum-Weight Design of Sandwich Shell Subjected to Fixed and Moving Loads", *Proc. 11th Int. Congr. Appl. Mech.*, Munich, 1964, Springer-Verlag, 1966.

[30] SHIELD, R., "Optimum Design Methods for Structures", *Plasticity*, (editors, E.H. Lee and P.S. Symonds), London, Pergamon Press, 1960.

[31] *Optimum Structural Design, Theory and Applications*, (editors, R. Galagher and O.C. Zienkiewicz), John Wiley and Sons, 1973.

[32] *Optimization in Structural Design*, (editors, Z. Mroz and A. Sawczuk), IUTAM Symposium Warsaw, 1973, Springer-Verlag, 1975.

STRUCTURAL CONTROL, H.H.E. Leipholz (ed.)
North-Holland Publishing Company & SM Publications
© IUTAM, 1980

ON THE MORPHOLOGY OF CONTROLLED SYSTEMS

M.S. El Naschie and S. Al Athel

Faculty of Engineering
University of Riyadh
P.O. Box 800, Riyadh, Saudi Arabia

1. INTRODUCTION

An overriding consideration in the design of feedback control
systems is the question of stability and interest in this subject
is twofold. First, the knowledge of eigenvalues and eigenvectors
is by no means sufficient for the judgement of the actual stability
limit of many civil engineering structures such as thin walled
elastic shells [1]. Consequently, control methods based on the
knowledge of eigenvalues along could be questionable. Second,
the control itself plays a curious double role of stabilizing the
system in one way and destabilizing it in another way as may
happen in the case of a too large negative feedback stabilization.
From a control system one would usually require that it is stable,
at least in the large. It is clear that using linearized dynamical
equations, no information could be obtained about how large a perturbation, which is inevitable from the physical point of view,
can be tolerated before instability occurs. In addition to that,
only information about asymptotic stability can be obtained. How-
-ever, a system with an asymptotically unstable equilibrium may
still be stable with respect to regions that are quite small in

which case it could be ignored in practice.

The stability of stiffened elastic plates might be a good example for illustrating some of the dangers to which, for instance, linear optimization methods can lead. Relatively recent investigations have revealed that in enforcing the equality or near equality of the overall and the local eigenvalue critical load, topologically highly unstable singularities are generated by a nonlinear model coupling which unfolds to some of Thom's catastrophy surfaces with a most bizarre morphology [2, 3]. It seems, therefore, essential for a deeper understanding of the control theory of engineering structures to reconsider these problems in the light of the topological theory of structural stability in the presence of disturbance and nonlinearity.

In the present work we try to utilize some of the recent advances in global dynamics to achieve a better understanding of a particular mechanical and structural stability problem.

2. PHASE PORTRAIT, SYMPLECTIC MANIFOLDS AND GLOBAL DYNAMICS

The neoqualitative period starts with Poincaré's idea of proceeding directly to qualitative information using a qualitative method. A comparative study of analytical versus global methods was given in reference 12 in connection with Duffing's oscillator. This new method is characterized firstly by its global geometrical point of view which led Poincaré to use differentiable manifolds as a phase portrait. Thus modern mathematical models using Cartan's intrinsic calculus visualize a dynamical system as a vector field on a symplectic manifold. In this new model analytical methods are replaced by differential topology. Here, and as a main characteristic of this new model, a new question emerges, namely that of structural stability. This question was first posed by Andronov and Pontriagin [4].

3. LIAPUNOV STABILITY AND STRUCTURAL STABILITY

Definition

A vector field x on M is structurally stable if there is a neighbourhood O of $x \in \chi(M)$ in the Whitney C^r topology such that $\tilde{x} \in O$ implies that x and $\tilde{x}$ are topologically conjugate. The set of C^r structurally stable vector fields on M is denoted by

$$\Sigma_s^r(M).$$

From the preceeding definition we see an important difference between the more familiar notion of stability in the sense of Liapunov and structural stability. While the former is mainly concerned with the perturbation of a special trajectory, structural stability is a question concerning the globality of the trajectory of a vector field. We are thus using the word stability in two distinct senses and we will be asking, for instance, about the structural stability of an instability limit.

Structural stability is the most comprehensive of all notions of stability. Its great importance comes from the fact that all the qualitative information about a certain process is obtained from a phase portrait of an ideal unperturbed system and is applied in practice not to the ideal system, but to a perturbed, supposedly close system. The coefficients of a differential equation of a certain dynamical system is determined experimentally or by using simplified approximate models. Many systems of technical importance are also governed by parameters which change during the processes and include random design imperfection. To know the effect of small variations in the parameters of the system on the qualitative behaviour and structural stability is, therefore, of great practical relevance.

To sum up, one usually expects from a certain process that it remains stable against small variations of its variables and also small variations of the system itself. The first requirement is met by Liapunov's stability criteria while the second is fulfilled by the structural stability of Andronov-Pontriagin and Thom [4 - 7].

4. STRUCTURAL STABILITY AND BIFURCATION

There are by now a number of proven theorems connected with structural stability from which the following is of interest for the circle of problems discussed here.

Theorem

If M is a two dimensional compact and orientable manifold and $1 \leq r < \infty$ then:

(1) $\quad x \in \Sigma_s^r (M)$

if and only if x is a Morse-Smale system and

(2) $\quad \Sigma_s^r$

is an open and dense subset of $\chi(M)$ in the C^r topology.

From this theorem one would be inclined to conclude that the majority of systems which one may encounter have the property of being structurally stable. Nevertheless, for man-made systems this would be quite misleading since these systems are designed with a great deal of symmetry to operate at an optimum level and, therefore, very often at the very proximity of a point of bifurcation. A famous example for that in the field of civil engineering structures is a shell which carries its load mainly via compressive membrance forces. Another example is the Mitchell truss [8]. At the same time, man-made systems always deviate from their planned shape due to the inevitable imperfection which sums up many factors affecting the actual design and running process of a system. Therefore, there is no guarantee that small imperfections would not cause a qualitative change in the behaviour of the system at or near to a point of bifurcation which we may define generally as follows.

Definition

The bifurcation of a vector field is an instability within a parameterized family of vector fields.

In what follows, the importance of structural stability and bifurcation will be illustrated and discussed in connection with some problems of technical importance.

5. BIFURCATION AND CATASTROPHIES IN CONSERVATIVE SYSTEMS

In discussing a conservative system a few characteristics of the singular (equilibrium) points and of its phase portrait have to be considered. First a conservative system cannot have a focus (stable or unstable attractor) because there cannot be a source nor a sink in a conservative system. The only possibility left for instability is consequently the saddle singularity which is associated with a stationary point of the total potential energy that is not a minimum. The stable point, on the other hand, has to be a centre associated with a stationary point of the total potential energy which is a minimum [3]. It is clear that systems which display centres are structurally unstable since the addition of very small damping will qualitatively alter the phase portrait. At this stage it may be important to mention an illustrative example used by Andronov to show how, simply by varying the damping (resistance) of a feedback control system, we can make the system pass successively through five different regions corresponding to various types of motions and states of equilibrium. For large positive values to large negative values we get a stable node, stable focus, a centre, an unstable focus and an unstable node. The only type of singularity which cannot be attained by a simple variation of the friction damping in the system is the saddle point. We also note that the topology of phase trajectories in the neighbourhood of points of singularities in a nonlinear system remains the same as in a linear case.

In studying conservative scleronomic and holonomic mechanical systems it turns out that there are, from the *structural engineering* point of view, three distinct local bifurcation behaviours associated with three discrete branching points. A fourth "bifurcation" point, a point of instability associated with a limit point of an initially stable path, is not included in this classification. These points are the stable symmetric, the unstable symmetric which is topologically equivalent to the first and the asymmetric point of bifurcation [3]. These points have been extensively studied by Koiter and his followers [6, 8].

The variation of the phase portrait along the fundamental and the secondary path of these three points is shown in Figures 1, 2 and 3. Now we will discuss these points in connection with the structurally stable disappearance of an attractor of a field. In the course of doing so, it will be shown that it is not irrelevant to rephrase some of the results in the language of catastrophe theory [5, 6, 7, 12].

Morphology of Controlled Systems

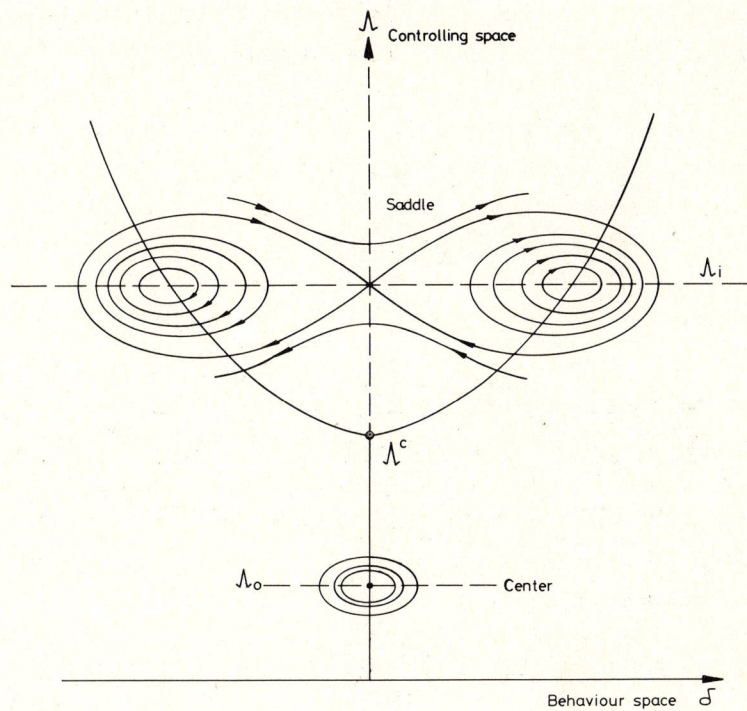

Figure 1 - Phase Space Transformation along the Fundamental and Secondary Path of the Symmetric Stable Point of Bifurcation [12]

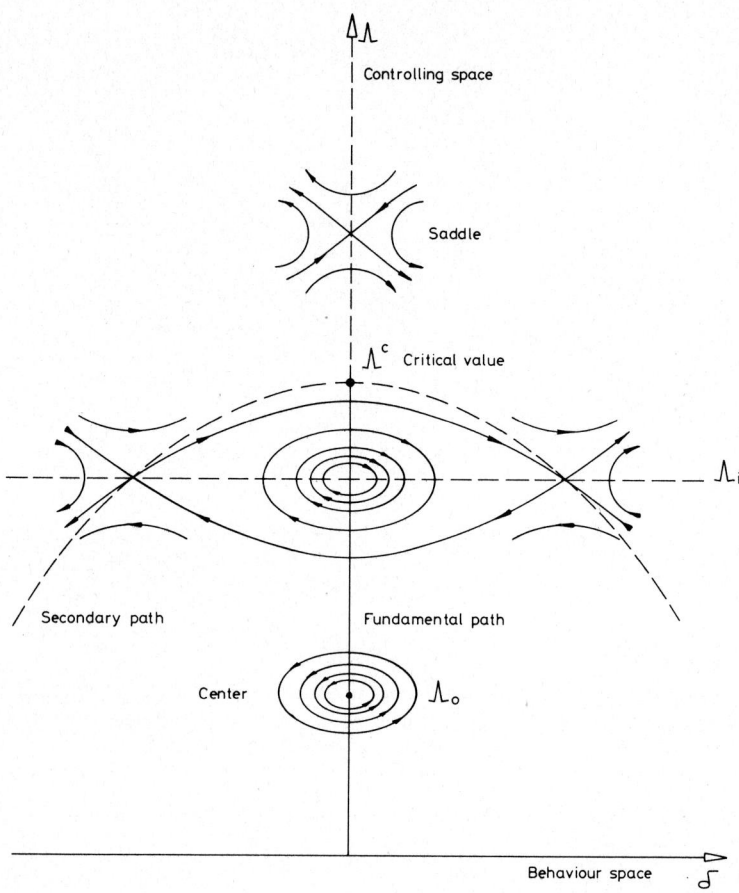

Figure 2 - Phase Space Transformation for the Unstable Symmetric Point of Bifurcation

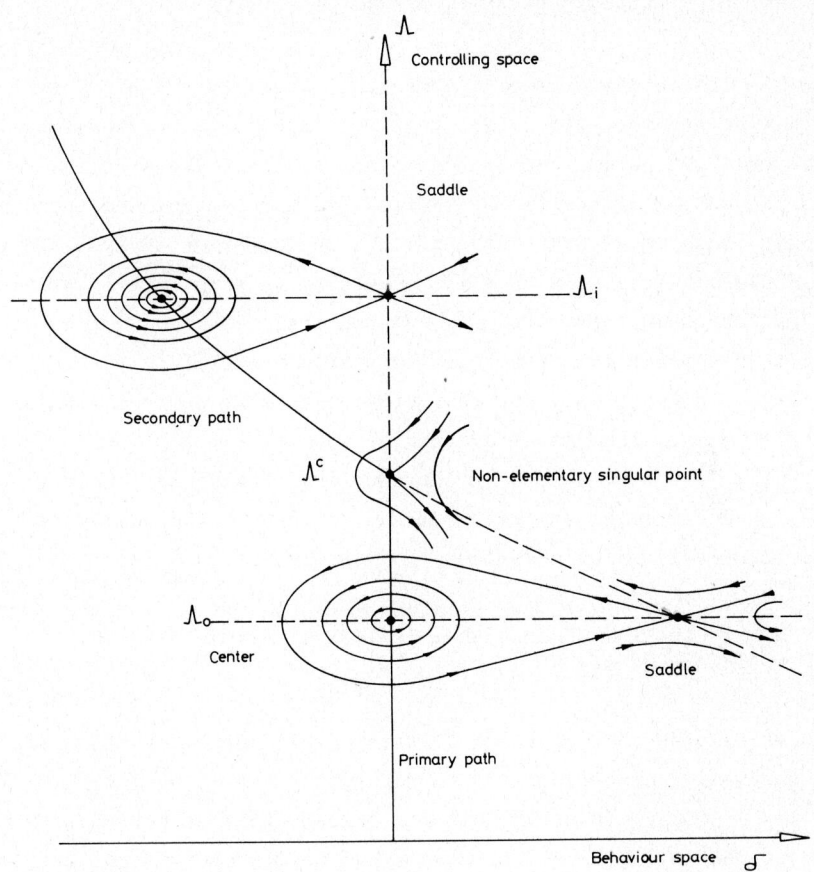

Figure 3 - Phase Space Transformation for the Asymmetric Point of Bifurcation [12]

6. THE DISAPPEARANCE OF AN ATTRACTOR OF A FIELD

The simplest attractors of a dynamical system (M, x) are first the point and then the closed generic trajectory. We consider here only the point. At such a point (P) the field X(P) is zero and the matrix of coefficients of the linear part of the components of X(P) in a local chart have all its characteristic values in the half plane $Re(z) < 0$. This is, of course, a structurally stable situation. Now there are only two possibilities for such an attractor to disappear in a structurally stable manner [5]:

(a) one and only one characteristic value of the Jacobian matrix becomes zero as a function of t,

(b) the real part of a characteristic value is zero for t = 0 and the imaginary part remains non-zero. The conjugate characteristic value undergoes the same evolution.

7. NONLINEAR DIVERGENCE - ANNIHILATION OF ATTRACTORS AND KOITER BIFURCATION

Now we examine the relevance of this result for bifurcation sets of the Hamiltonian type.

Clearly none of Koiter's three points of bifurcation correspond in any sense to these two fundamental types of structurally stable bifurcation. In fact, on perturbing the singularities of Koiter, one obtains a direct node-saddle singularity plus a nonbifurcating path. The node-saddle is, of course, the structurally stable annihilation. It corresponds to Case A where an attractor meets an unstable point and both are destroyed. The point at which this annihilation takes place can also be viewed as one of Thom's elementary catastrophies, namely the fold. It is also the type of instability prevailing in the real world of structural systems and is commonly found in arches and shells which snap under compressive loading. Incidentally, the gravitational collapse into a black hole can be viewed as such an annihilation, as suggeste by several authors in different versions.

It may be of interest to discuss some of the structural mechanical models used in illustrating the variation of a fold catastrophe with a second controlling parameter. Some such models are structurally unstable and are, stherefore, invalid as models for the Riemann-Hugoniot catastrophe of Thom, e.g., a von Mises truss with loading as a control parameter and variation of height as a second one [9]. The double cusp presented there is misleading, contrary to the genuine cusp given in [7]. One would notice this at once if the potential energy of the system is written down and compared with Thom's list of elementary catastrophies. (Note that the quadratic form has no influence on the form of the catastrophe.)

A valid model which correctly corresponds to Thom's Riemann-Hugoniot surface is actually any structure displaying a symmetric (stable or unstable) point of bifurcation in the presence of imperfection [3]. It is a frequent misunderstanding that Koiter's singularities are structurally unstable and, therefore, irrelevant to catastrophe theory. Koiter's singularities are structurally unstable. However, the unfolded three discrete branching points are structurally stable. The unfolding is achieved using linear imperfection.

An important result of the structural instability of Koiter's singularity is, of course, that a first order influence of perturbation will exist and consequently a first order so called imperfection sensitivity in the engineering sense can exist. This provides an alternative and global confirmation of Koiter's quantative analysis. Studying the phase portrait of Koiter's singularity may also reveal at a first glance that the type of unfolding in the case of the symmetric points, which are topologically equivalent but different from the engineering point of view, will be different from the asymmetric point. In fact, the asymmetric point unfolds as one may have expected from the phase portrait into a fold catastrophe while the other unfolds to a Riemann-Hugoniot catastrophe surface as already mentioned.

Finally, we might remark that the minimum dimension for a bifurcation of the type discussed is one. This might be termed statical creation. It is equally clear from the preceeding discussion that the stability in the sense of Liapunov of the conservative type cannot be influenced by vanishing damping in the case of a flexible structure which is a well known fact. However, in other cases this is naturally not true. We shall see in the following discussion that the so-called surprising effect of damping in a circulatory nonconservative stability set is in fact not surprising at all, as also noted by Thompson.

8. DYNAMICAL BIFURCATION AND NONCONSERVATIVE SYSTEMS

Contrary to conservative systems, in a nonconservative system stable and unstable attractors are no longer excluded a priori. Also closed orbits are no longer possible in a linear nonconservative system. We will see that some interesting dynamical instabilities are associated with these types of sets which are, of course, more real than the idealized Hamiltonian systems and which have no counterpart in linear conservative sets.

In the preceeding discussion we have seen that (statical) instability leads to new forms of bifurcation, the generalization of which may be terms after Taken's prototype statical creation. We now come to what may be termed dynamical creation. This is the case when the second possibility B of the structurally stable disappearance of an attractor occurs. As already mentioned, this is associated with the vanishing of the real part of a characteristic value for $t = 0$ while the imaginary part remains non-zero also for the conjugate characteristic value. Now this case is precisely the same subject of the classical study of Hopf and is termed after him, Hopf bifurcation (Abzweigung) or in modern terminology, Hopf catastrophe [10]. Qualitatively, the following happens. The

singular point becomes unstable but in the plane of the associated
characteristic vector, it is surrounded by small invariant attracting
cycles and the evolution is as if the initial point attractor
dilates into a two dimensional disc in this place while the centre
of the disc becomes a repellor point of the field. Finally, there
exists a stable regime but the associated attractor is no longer a
point but a closed trajectory. The new attractor is topologically
more complicated than the initial point attractor.

So far we have discussed the global qualitative picture
and it is important now to make clear the intuitive physical idea
behing the Hopf catastrophe. This is simply as follows: a system
is excited as a parameter λ is increased and beyond a certain critical value, excitation and dissipation balance to enable self
sustaining oscillation to occur. Hopf bifurcations are also local
in the sense that they can be detected by linearization about a
fixed point. It is also clear that vanishing damping will influence this bifurcation point significantly. However, damping cannot
lead to unfolding as in conservative sets.

It is now important to draw some diagrams for the Hopf
bifurcation which allows us some comparitive studies with the conservative Koiter types of bifurcation. In this way we meet two
types of Hopf bifurcation which we term stable and unstable in
analogy to the symmetric conservative branching points.

9. APPLICATION OF HOPF BIFURCATION TO POST FLUTTER BEHAVIOUR OF RODS UNDER FOLLOWER FORCES AND THE GALLOPING OF STRUCTURE IN WIND FLOW

(A) *The Stable Hopf Bifurcation (see Figure 4(a))*

Since this type of bifurcation is structurally stable the singularity is naked in the sense that no perturbation will change its
global character. Consequently, imperfection has a second order
effect similar to that of a limit point. The bifurcation point

is also detectable from a local linear analysis and no large
deformation far from the critical point is to be feared. From an
engineering point of view the critical value obtained using a
damped-linear analysis is completely sufficient as a design criterion. Now damping is important because instability is no longer
a saddle type divergence. This type of flutter instability
requires at least a two-degrees of freedom system and may be termed
soft flutter. The detector of such bifurcation to a stable limit
cycle can be achieved using the sign of the curvature of the load-
deflection-frequency curve. An application of a stable Hopf bifurcation is the post flutter behaviour of a two-degrees of freedom
link model under a tangential follower force, [12].

(B) *The Unstable Hopf Bifurcation (see Figure 4(b))*

This type of instability is also structurally stable. However, a
disturbance amplitude taking the system to a large deflection may
cause a subcritical instability whose limit is unknown since the
limit cycle itself is unstable. Thus from an engineering point of
view sufficient safety against disturbance near the critical
point is required. This type of bifurcation can also be detected
from the sign of the load-deflection-frequency curve which will
also give a first approximation (local) measure of tolerable
imperfection in amplitude.

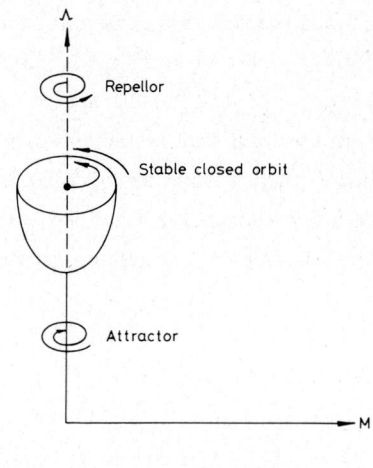

(a)

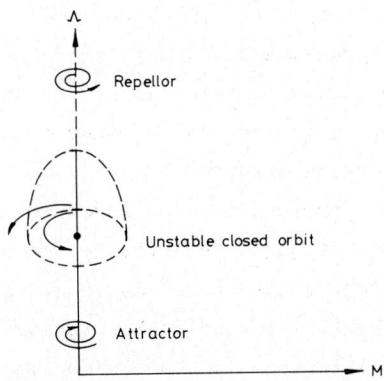

(b)

Figure 4 - Structurally Stable Bifurcation
(a) Super Critical Hopf Bifurcation (stable)
(b) Subcritical Hopf Bifurcation (unstable)

However, it seems that (A) and (B) are in general much more harmless for engineering structures than the corresponding divergence instability. We clearly see that the preceeding discussion also provides a global qualitative confirmation of the most important results obtained by Burgess and Levinson using quantitative methods [11]. Many applications of both the preceeding types of Hopf bifurcation can be found in civil engineering, for instance the galloping of structures under the action of wind flow, [13].

DISCUSSION AND CONCLUSIONS

The preceeding type of analysis seems to bridge the gaps between abstract mathematics and applied mechanics. It is shown here that some of the topological concepts of catastrophe theory can provide a surprisingly intuitive base for many analytically complicated mechanical problems. The notion of structural stability was found to be of considerable usefulness in placing Koiter's theory within a wider context and in giving a global picture for what may be happening in the case of many perplexing experimental observations of buckled structures.

We do not pretend to have revealed any radically new results in the present work. However, we hope that we have shown that qualitative conclusions based on quantitative analysis could be found directly via qualitative topological theorems. One of the major simplifications in this direction is the centre manifold theorem which enables one to restrict oneself to a finite dimensional problem.

Finally, for a class of nonconservative stability problems it is shown that the alternative methods of differential dynamics lead to the same qualitative conclusions as does the involved quantitative classical analysis.

ACKNOWLEDGEMENTS

The first author would like to express his indebtedness to countless stimulating discussions extending over many years with Professor J.M.T. Thompson who pioneered the application of catastrophe theory in structural engineering.

Both authors would like to thank Professor H.H.E. Leipholz whose work and encouragement made this paper possible.

REFERENCES

[1] EL NASCHIE, M.S., "High Speed Deformation of Shells", *Proc. IUTAM Symposium on High Velocity Deformation of Solids*, Springer, New York, 1979, pp. 363-376.

[2] THOMPSON, J.M.T., "Experiments in Catastrophe", *Nature*, Vol. 254, 1975, pp. 372-392.

[3] EL NASCHIE, M.S., "Durchschlagähnliches Stabilitätsverhalten von Rahmentragwerken", *der Stahlbau*, No. 11, 1977, pp. 338-340.

[4] ANDRONOV, A.A. and PONTRIAGIN, L., "Systems Grossiers", (Coarse Systems), *Dokl. Akad. Nauk., S.S.R.*, Vol. 14, 1937, pp. 247-251.

[5] THOM, R., *Structural Stability and Morphogenesis*, Addison-Wesley, New York, 1975.

[6] THOMPSON, J.M.T., "Imperfection Sensitivity Uninfluenced by Prestress", *Int. J. Mech. Sci.*, Vol. 20, No. 1, 1978, pp. 57-58.

[7] EL NASCHIE, M.S., ZAKY, A. and SOLIMAN, M., "A Mechanical Model for Multimodal Operations and Runaway Phenomena of Chemical Reactors", *J. Appl. Math. and Mech., ZAMM*, Vol. 59, 1979.

[8] THOMPSON, J.M.T. and HUNT, G., *A General Theory of Elastic Stability*, Wiley, London, 1973.

[9] TROGER, H., "Zur Einteilung von Sprungeffekten in mechanischen Systemen", *ZAMM*, Vol. 54, T177, 1974.

[10] HOPF, E., "Abweigung einer periodischen Losung eines Differential Systems", *Ber. Math.-Phys. Sachische Akademie der Wissenschaften Leipzig*, Vol. 94, 1942, pp. 1-22.

[11] BURGESS, I.W. and LEVINSON, M., "The Post-Flutter Oscillation of Discrete Symmetric Structural Systems with Circulatory Loading", *Int. J. Mech. Sci.*, Vol. 14, 1972, pp. 471-488.

[12] EL NASCHIE, M.S., "Stability and Catastrophe Theory in Applied Science", *Proc. of the Brazilian Congress of Mech. Eng.*, December 12-15, 1979, Campinas - SP, Brazil.

[13] NOVAK, M. and TANSKA, H., "Effect of Turbulence on Galloping Instability", *ASCE, J. of the Eng. Mech. Div.*, Vol. 100, 1974, pp. 252-258.

STRUCTURAL CONTROL, H.H.E. Leipholz (ed.)
North-Holland Publishing Company & SM Publications
© IUTAM, 1980

TWO EXTREME CASES OF ON-LINE
CONTROL OF STRUCTURES

M. Fanelli and G. Giuseppetti

Ente Nazionale per l'Energia Elettrica
Direzione Degli Studi e Richerche
Centro Ricerca Idraulica e Strutturale
Via Ornato 90/14, 20162 Milano, Italy

1.0 INTRODUCTION

Present practice in the field of safety check-up (passive control) of large structures, such as dams, is qualitatively viewed in the frame of automatic-control theory, evidencing the transfer function blocks, the comparison stages, the threshold gates and the feedback loops.

This part of the paper concerns static, man-aided control, in the sense that human judgement and action enter in the control loop.

In the second part of the paper, an attempt is made to envisage fully automatic, active control, during dynamic events such as earthquakes, for small-size critical pieces of equipment having failsafe requirements.

Such a proposal is conceptually appealing, but in order to assess its feasibility and reliability, a specialistic analysis should be undertaken. The authors, therefore, put this question to the attention of control-theory experts for more detailed and, if possible, quantitative discussion.

2.0 CASE I - STATIC, MAN-AIDED CONTROL OF DAMS

In this paper the possibility of continuous check-up of structural safety is viewed in the frame of the automatic control theory.

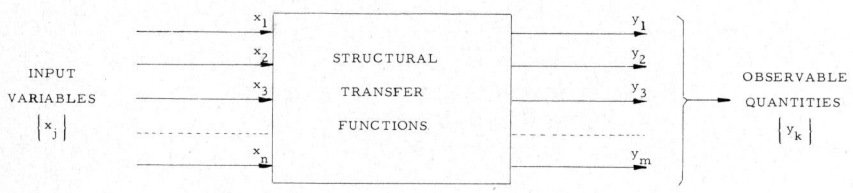

Figure 1 - The Structural Transfer Functions

It is supposed that it is possible to conceptualize a feedback loop such as in Figure 2.

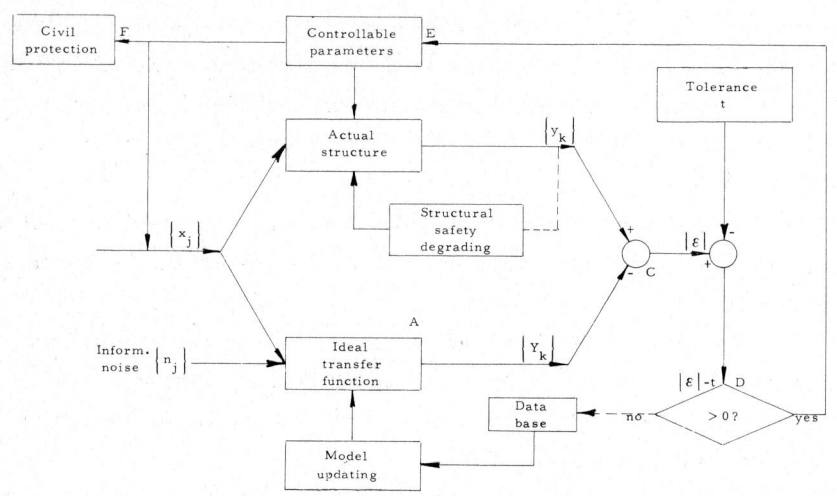

Figure 2 - The Basic Feedback Loop

The main building blocks of this diagram are:

- the input-variables vector, $\{x_j\}$;
- the actual structural response matrix, which transforms the vector $\{x_j\}$ into the "observable effects" vector $\{y_k\}$;
- the ideal transfer matrix, i.e., a mathematical "model" which translates the information about inputs into "expected" values $\{Y_k\}$ of the effects vector;
- a comparison stage, C, at which observed and expected effects are compared and a "norm" $|\varepsilon|$ of the differences is built up;
- a gate at which the difference norm, $|\varepsilon|$, is compared with a pre-set tolerance, t;
- a decision stage, D, at which the values of $|\varepsilon|$ in excess of t cause actions to be taken in one or more of three possible ways:

    (i) modification of controllable structural parameters, if any;

    (ii) modification, whenever possible, of some input variables;

    (iii) alert and activation of civil-protection procedures.

Other details also depicted in Figure 2 are:

- "information noise" affecting the estimates of the input variables that are fed into the mathematical model yielding expected values Y_k. In fact, not all the input variables are monitored, and even those that are will only be known with limited accuracy; this is characterized by a "noise vector" $\{n_j\}$; (for most structures, actual input variables will be infinite in number);
- storage of information about "normal" behaviour in a suitable "data base";
- updating of the mathematical model, drawing upon said data-base. The last two functions can be likened to "memory" and "experience learning" of an intelligent system.

It is obvious that the efficiency of this feedback loop in ensuring the attainment of "safety" in the most general sense (i.e., either structural safety proper or the safety of people and property that would be affected by structural failure) hinges on some critical "time constants" of the different blocks of the feedback loop itself:

- time constants of input-variables rates of variations, (in this connection, a distinction will have to be made between "static" and "dynamic" processes, the former being characterized by the fact that inertial forces are negligible. Typical dynamic process are seismic events);

- time constants of information processing stages A, C, D, (Figure 2);

- time constants of alert and corrective procedure activation; E, F, (Figure 2);

- time constants of structural safety deterioration up to failure. These latter time constants can be ideally characterized through an "inner loop" such as that visible in Figure 2 with the caption "structural safety degrading", which is made to depend on the magnitude of observable effects, and which operates by changing the actual structure response matrix.

It appears difficult to go much further along this sort of analysis without making definite hypotheses about each one of the building blocks. However, in the following it is proposed to attempt a schematization which avoids, as far as possible, any physical particularization of the system. Resort will be made, in particular for the "structural safety degrading" effect, to very simple "failure mechanism models":

(a) a "catastrophic" one, and

(b) a "soft-settling" one.

Both models can be derived from the general scheme of Figure 3, wherein a body M is tied to its support S by N identical bonds and is acted upon by a force F tending to tear it away from S. (F is for the moment assumed to be constant in time (static case)).

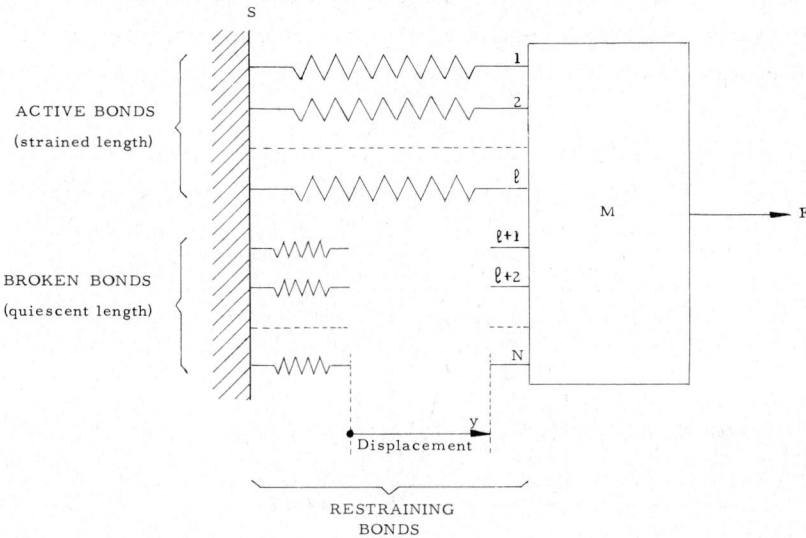

Figure 3 - The "Tear-Away" Process. Body M is gradually torn away from support S by acting force F; this meaning that a growing number of "in-parallel" bonds are failing.

In a given instant t there will be $1 \leq N$ bonds still active, and $N - 1$ broken bonds. The stress-strain characteristic of each bond will be of the elastic-brittle type depicted by Figure 4.

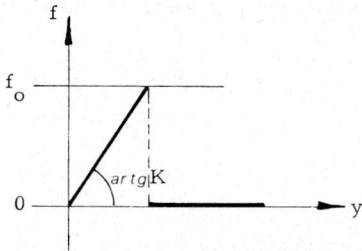

Figure 4 - Stress-Strain Characteristic of a Bond: $f = Ky$ until $f \leq f_0$, $f = 0$ for $y > f_0/K$

At this juncture, time is conveniently introduced through the "rate of mortality" of active bonds, depicted in Figure 5:

$$-\frac{d\ell}{dt} = \phi(\ell) .$$

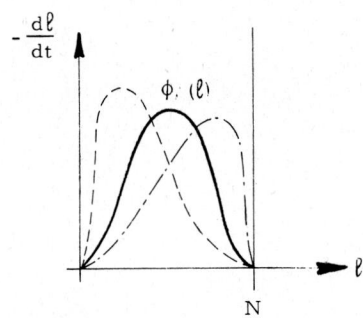

Figure 5 - *Failure Probability Distribution(s) for Bonds,* $\phi(\ell)$

According to the law assumed for $\phi(\ell)$, different types of failure will occur. By way of example, let us *first* assume that:

$$\phi(\ell) = \phi_i = \alpha\ell , \quad \text{with } \alpha = \text{constant.}$$

We shall obtain the following system of equations:

$$\ell \cdot Ky = F , \quad \ell = N \quad \text{for} \quad F \leq Nf_0 ,$$

$$\frac{d\ell}{dt} = -\alpha\ell \quad \text{for } F > Nf_0 .$$

In the latter case it will ensue:

$$\frac{d\ell}{\ell} = -\alpha dt , \quad \ln\frac{\ell}{N} = -\alpha(t-t_0) \quad \text{if } \ell = N \text{ at } t = t_0 ;$$

$$\ell = Ne^{-\alpha(t-t_0)} \quad , \quad y = \frac{F}{K\ell} = \frac{F}{KN} e^{\alpha(t-t_0)} \quad (\text{for } F > Nf_0) \ .$$

This represents a failure of the soft type (the structure exhibits an overall strain-softening; y tends to infinity only for an infinite time (see Figure 6). $\frac{\ell}{\alpha}$ is the time-constant of the process).

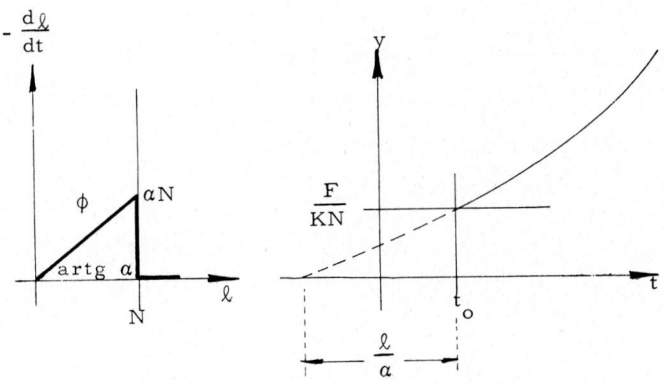

Figure 6 - "Soft" Failure

Let us now assume as a *second* example that

$$\phi(\ell) = \phi_2 = \frac{\beta}{\ell} , \quad \text{with } \beta = \text{constant}.$$

(In this case it is easy to see that the "mortality rate" $-\frac{d\ell}{dt}$ is proportional to displacement y.)

The system of equations will now be:

$$\ell \cdot Ky = F , \quad \ell = N \quad \text{for} \quad F \leq Nf_0 \ ,$$

$$\frac{d\ell}{dt} = - \frac{\beta}{\ell} \quad \text{for} \quad F > Nf_0 \ .$$

In the latter case it will obtain:

$$\ell \cdot d = -\beta dt \;,\; \frac{N^2-\ell^2}{2} = \beta(t-t_0) \quad \text{if } \ell = N \text{ at } t = t_0 ;$$

$$\ell = \sqrt{N^2-2\beta(t-t_0)} \;;\quad y = \frac{F}{K\ell} = \frac{F}{K\sqrt{N^2-2\beta(t-t_0)}} \quad (\text{for } F > Nf_0).$$

It is evident that this is a model of a castrophic failure; y goes to infinity for a finite time t_f

$$t_f = t_0 + \frac{N^2}{2\beta} \qquad (\text{see Figure 7}).$$

$\frac{N^2}{2\beta}$ is thus a characteristic time constant of the process.

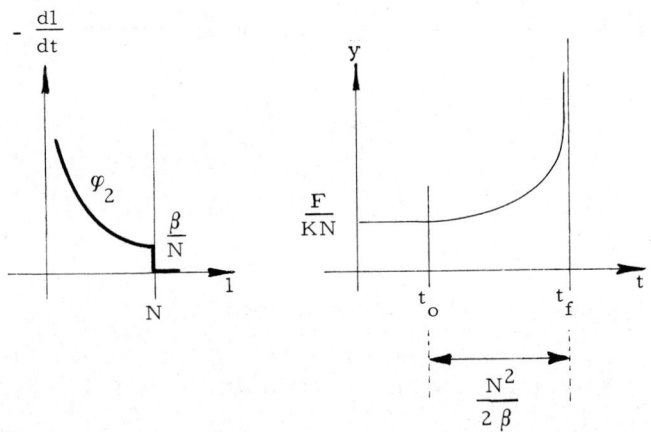

Figure 7 - "Catastrophic" Failure

The type of process, and the relevant time-constant, could in principle be identified through real-time analysis of observable variables, if the sampling rate is fast enough.

Thus, every time the "normality hypothesis" is falsified, such an identification process should be started. The scheme of

Figure 1 is accordingly modified, see Figure 8: it is obvious that the time-constants thus identified will influence the choice of different options for the type of alert and of correcting actions (a short time-constant will rule out long-range strategies and call for very quick emergency measures, etc.). The need for flexible strategies, according to contingencies that may arise, is brought into focus, if only in a qualitative way.

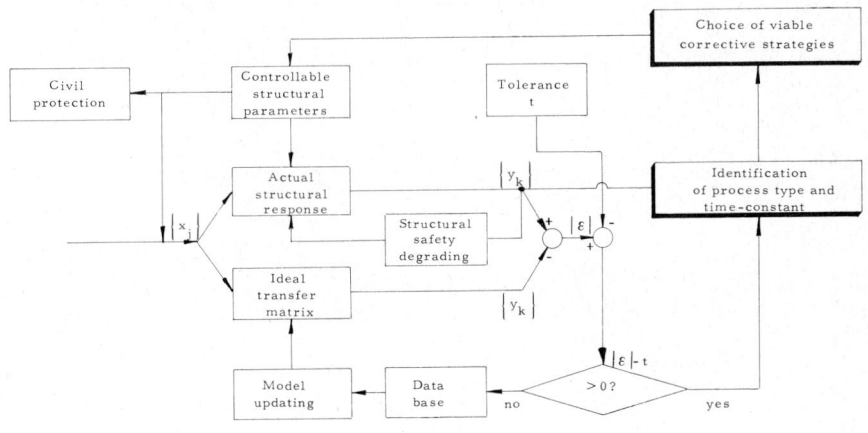

Figure 8 - Role of Identification in the Event of "Abnormal" Behaviour (see also Figure 1)

3.0 CASE II - DYNAMIC, REAL-TIME CONTROL OF FAILSAFE EQUIPMENT DURING EARTHQUAKES

Let us now consider the quite different case of a critical piece of equipment whose working has to be assured under dynamic conditions, in particular during violent earthquakes. (For instance a circuit breaker actuating the emergency shutdown of a nuclear power plant.)

If this piece of equipment (P henceforth) is not very massive, in principle the following line of approach could be outlined (see Figure 9, where for the moment the seismic excitation is supposed to happen in the horizontal direction).

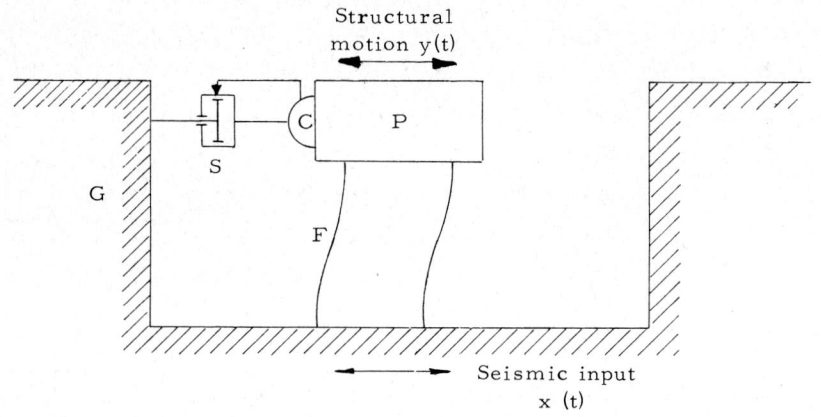

Figure 9

Let P be connected to the ground G [whose motion is x(t)] by an elastic support system F. Another connection between P and G is provided by a servo S controlled by a signal processor C whose input is the piece motion y(t). It is then possible in principle to design this processor so as to optimalize some prescribed "moment" of the statistical distribution of y or its time-derivatives.

Denoting by:

$F_1(s)$ the transfer function (force/displacement) of the support system F;

$F_2(s)$ the dynamic transfer function (force/displacement) of P;

$F_3(s)$ the servo (S) transfer function (actuating force/control signal);

$F_4(s)$ the processor (C) transfer function (control signal/displacement y(t));

the block diagram of the overall dynamic system will be like that depicted on Figure 10, and the following equation will apply for the Laplace transforms:

$$Y = F_2(s) [F_1(s)(X-Y) + F_3(s)F_4(s)Y] \text{, whence}$$

$$\frac{Y}{X} = F_s(s) = \frac{F_1(s)F_2(s)}{1+F_1(s)F_2(s)-F_2(s)F_3(s)F_4(s)} \text{ .}$$

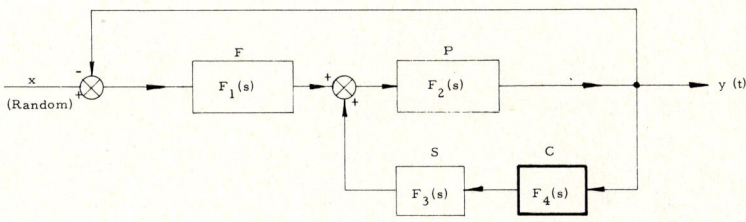

Figure 10

Let now the random input $x(t)$ be characterized by a power spectrum $\Phi(\omega)$:

$$A(\omega) = \frac{1}{2\pi} \int_{-\Theta/2}^{\Theta/2} x(t) e^{-i\omega t} dt \text{ ;}$$

$$\Phi(\omega) = \lim_{\Theta \to \infty} \frac{4\pi}{\Theta} |A(\omega)|^2 \text{ ;}$$

then we can compute the required "moment" of the statistical distribution of $y(t)$, or of its time-derivative of a given order.

Suppose, for the sake of argument, that it is required to minimize the root mean square $\overline{y''^2}$ of the acceleration of P, $y'' = \frac{d^2y}{dt^2}$: it is well-known [3] that the following relationship holds:

$$\overline{y''^2} = \int_0^\infty \omega^4 \Phi(\omega) |F_s(i\omega)|^2 d\omega =$$

$$= \int_0^\infty \omega^4 \Phi(\omega) \left| \frac{F_1(i\omega) F_2(i\omega)}{1+F_2(i\omega)[F_1(i\omega)-F_3(i\omega)F_4(i\omega)]} \right|^2 d\omega = \psi.$$

Now consider some design parameters Z_1, Z_2, ... of the signal processor transfer function, $F_4(s)$, as independent variables. Given the power spectrum $\Phi(\omega)$, $\overline{y''^2}$ will then be a function of these design parameters:

$$\overline{y''^2} = \psi(Z_1, Z_2, \ldots) ,$$

and ordinary minimum conditions can be imposed

$$\frac{\partial \psi}{\partial Z_1} = 0 ,$$

$$\frac{\partial \psi}{\partial Z_2} = 0 ,$$

$$\ldots, \text{ sub } \frac{\partial^2 \psi}{\partial Z_1^2} > 0 , \quad \frac{\partial^2 \psi}{\partial Z_2^2} \quad 0 , \ldots .$$

The design of $F_4(s)$, hence of C, can thus be finalized to obtain the required properties for $y(t)$.

What above can be repeated, if need be, to control other directions of motion (see Figure 11), by arranging a suitable intermediate body B connected to ground G by an elastic support system F_2 allowing some flexibility, e.g., in the vertical direction (or in an horizontal direction at right angles to the one previously considered).

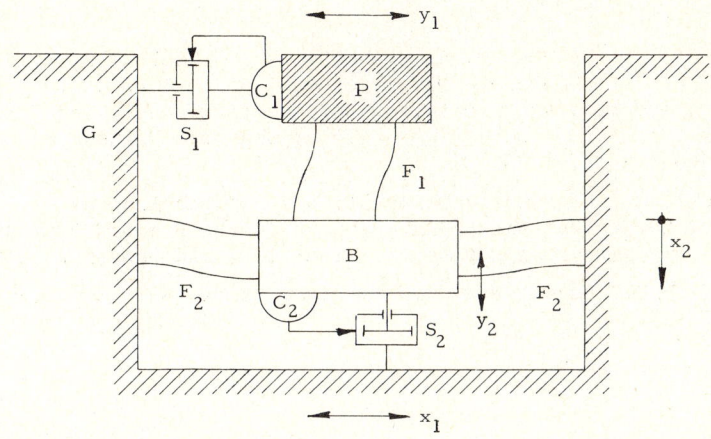

Figure 11

The "safety" problem would then be reduced to a question of proper design of the processors C_1, C_2, ... As such, it could be addressed by control-theory experts through well-known methodologies.

The authors feel that this possibility is a priori an interesting one, and that it should be investigated, at least at a feasibility-study level, so as to have a clearer idea of its potentialities and limitations.

Of course, several practical questions would arise if such a concept should prove to be technically viable. Among others, fail-safe features should be incorporated in the design of the servos, so as to prevent unchecked motion of P in the event, e.g., of power failure. (For this reason the servos have been depicted, in our schemes, to look like dashpots).

The problem has been only stated. The authors hope that some of the control theory specialists will be willing to take up from the point where they are obliged, for lack of specific knowledge, to lay down the pen.

REFERENCES

[1] BONALDI, P., DI MONACO, A., FANELLI, M., GIUSEPPETTI, G. and RICCIONI, R., "Concrete Dam Problems: An Outline of the Role, Potential and Limitations of Numerical Analysis", *Proc. of the Int. Symposium*, held at Swansea, United Kingdom, September 8-11, 1975.

[2] FANELLI, M., "Automatic Observation and Instantaneous Safety Control of Dams: An Approach to the Problem", *Water Power and Dam Construction*, November 1979 and December 1979.

[3] TSIEN, H.S., *Engineering Cybernetics*, McGraw-Hill Company, 1954.

CONTROL OF STRUCTURAL VIBRATION

Sabry F. Girgis

Aeronautical Department
Alfaatih University
Tripoli, Libya

1. INTRODUCTION

The full-slab tail is the result of some development and research work. It improves the flight control and performance of high speed aircraft. To utilize such type of control surface, flutter investigations were carried out. The mode shapes and the elasto-mechanical parameters were determined by ground resonance tests. These parameters were then introduced in correlated stability equations. Theoretical flutter analysis based on the strip theory was carried out. Flutter appeared at (1.18) Mach. This flutter was found to be the result of the rotational part in the main fundamental bending mode of such full-slab tail.

2. TECHNIQUES USED FOR FLUTTER INVESTIGATION

2.1 *Wind Tunnel Tests*

Dynamically similar models were built (Figure 1). The transonic model when excited at its fundamental bending mode, was destroyed at (0.98) Mach. The wind tunnel tests were carried out at Göttingen subsonic W.T. and at Bedford transonic W.T.

Figure 1

2.2 Flight Flutter Tests

Flight flutter tests were made on the original aircraft (Figure 2). During these tests, the full-slab tail was excited by rocket pulsers. The pulsers were properly located to excite the main fundamental bending mode of the full-slab tail. A pick-up located the node line of the main pitching mode, gave a clear response of the excited surface to the main fundamental bending mode. The flight flutter tests were performed at the same altitude and at different flight speeds. Each time the response of the pick-up was analyzed to determine the rate of decay of the recorded amplitude. The flight flutter tests showed that this rate of decay decreased with the aircraft speed, that meant approach of flutter. It could be forseen that flutter would occur at (1.01) Mach.

Control of Structural Vibration

Flight Flutter testing technique

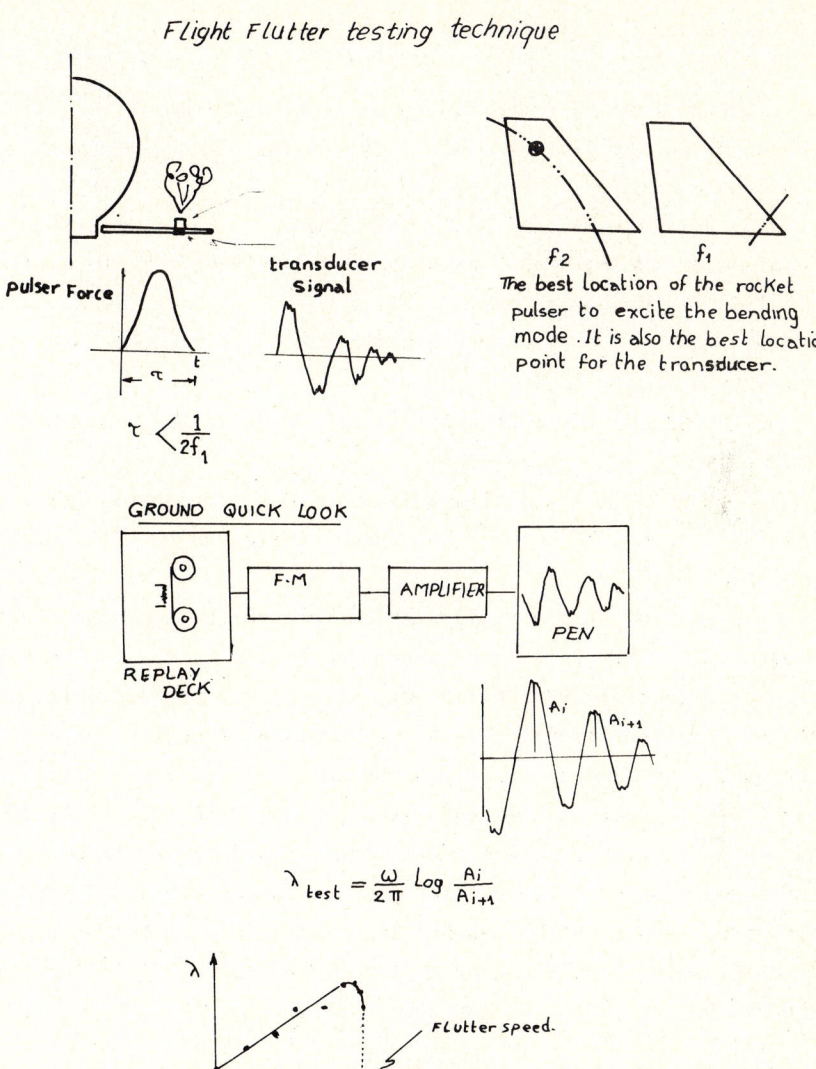

Figure 2 - Flight Flutter Testing Technique

2.3 *Techniques for Controlling Structural Vibrations*

The writer notices that the flutter analysts of industry have suggested antiflutter means. This means dealing with either of the following:

(a) Varying the stiffness distribution. This method should be applied at the early design stage of any aircraft. This method is applied in the design of MIG. 21.

(b) Adding a projected tip-chord mass. This method provides partial help in our case.

(c) Changing the root constraints. This method was found to be the worst in our case.

(d) Varying the position and direction of the pintle-axle. This method should be used in the early design stage. This method is applied to SU.7.

In view of the difficulty to foresee the suitable flutter improvement of this case, many ground resonance tests were carried out on dynamic similar model and on the original surface with different configurations. Tests were aimed at reducing this rotational part that exists in the fundamental bending mode. Theoretical flutter calculations were made for each case. From these investigations, it was found that fixing an Inertia-plate aft the root chord on the pintle-axle, would improve and suppress the flutter of such stabilizers. This method does not affect the aerodynamic cleanliness or airframe continuity. This Inertia-plate can be introduced at any design stage.

3. RESULTS

3.1 *Ground Resonance Test Results* (Figures 3, 4)

Figure 4 shows that the Inertia-plate reduces considerably the rotational part in the main fundamental bending mode.

Control of Structural Vibration 273

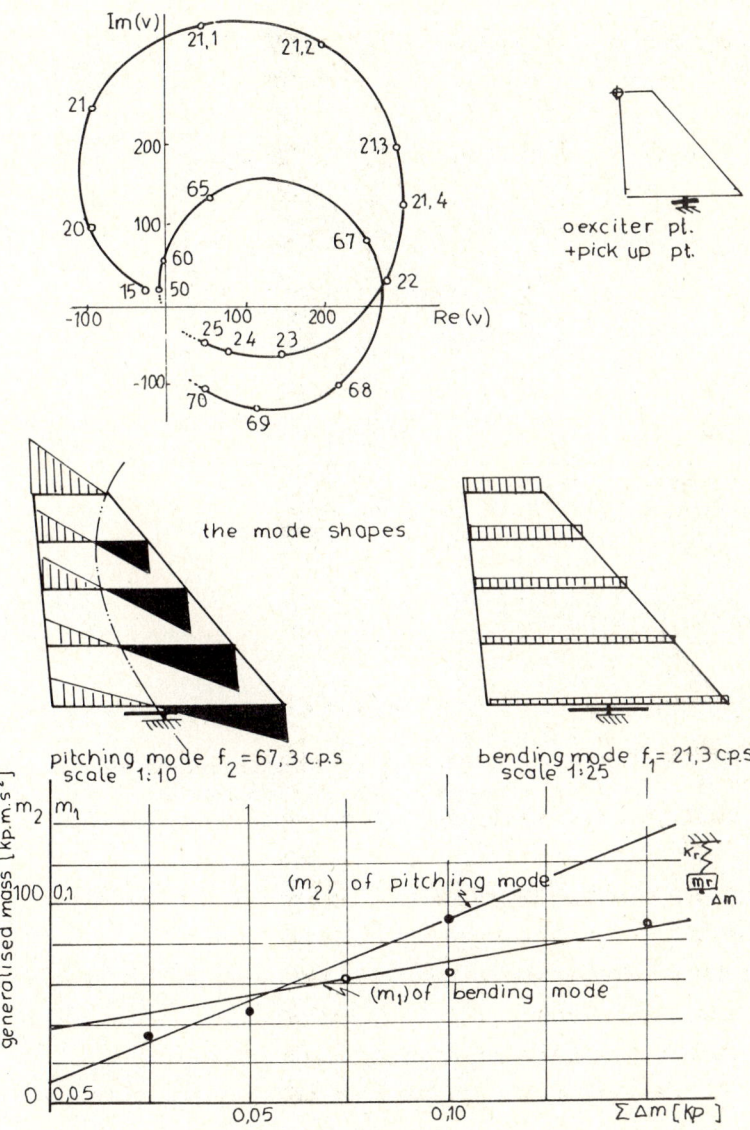

Figure 3 - The Experimental Determination of the Generalized Mass (m) and the Fundamental Modes

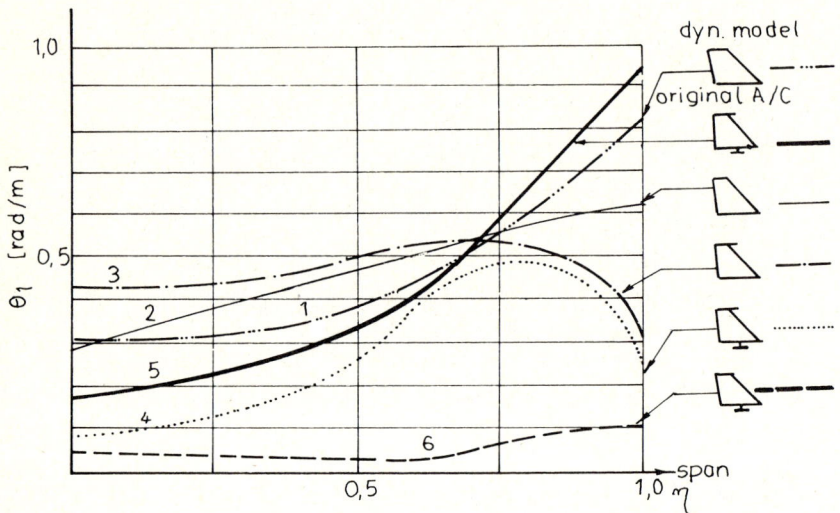

Figure 4 - Comparison of the Rotational Parts (θ_1) of the Bending Modes of the Full-Slab Tail with Different Configurations

(1) dynamical similar model
(2) original surface without attachment
(3) original surface equipped with projected tip-mass
(4) original surface equipped with projected tip-mass and inertia-plate
(5) original surface equipped with projected tip-mass and inertia-plate and one fixed point
(6) original surface equipped with inertia-plate

3.2 *Flutter Calculation Results* (Figures 5, 6, 7, 8, 9, 10)

(a) Figure 5 shows the flutter characteristics of the full-slab tail without attachment. No interference could be seen between the fundamental frequencies at the critical Mach number. The flutter is thus due to the fundamental bending mode only. The system can be considered as a single-degree of freedom flutter case. The coupling between rotation and translation in the fundamental bending mode is the cause of this flutter case.

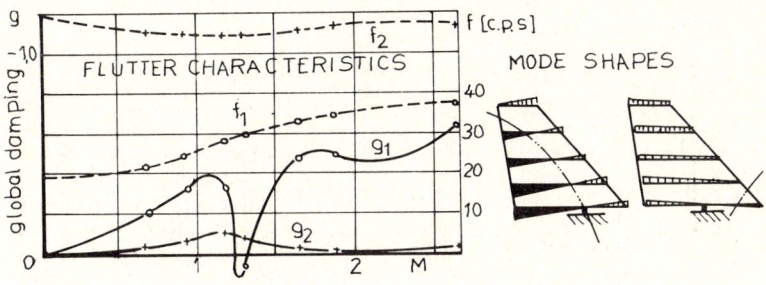

Figure 5 - Full-Slab Tail without Attachement

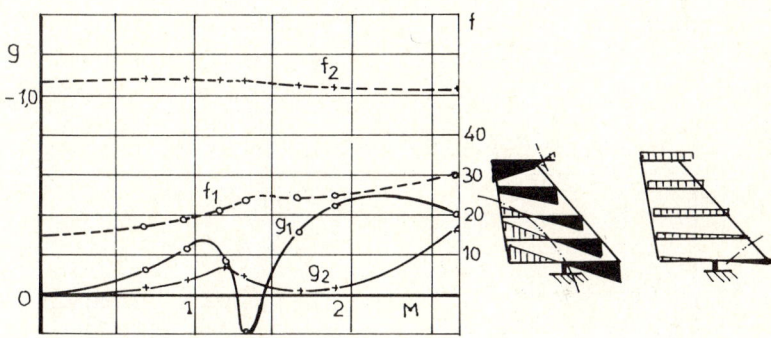

Figure 6 - Full-Slab Tail with Tip-Mass

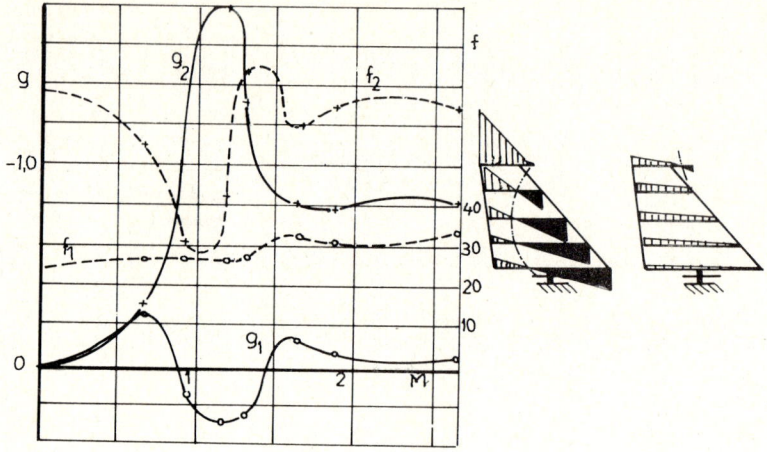

Figure 7 - Full-Slab Tail with Tip-Mass and Inertia-Plate

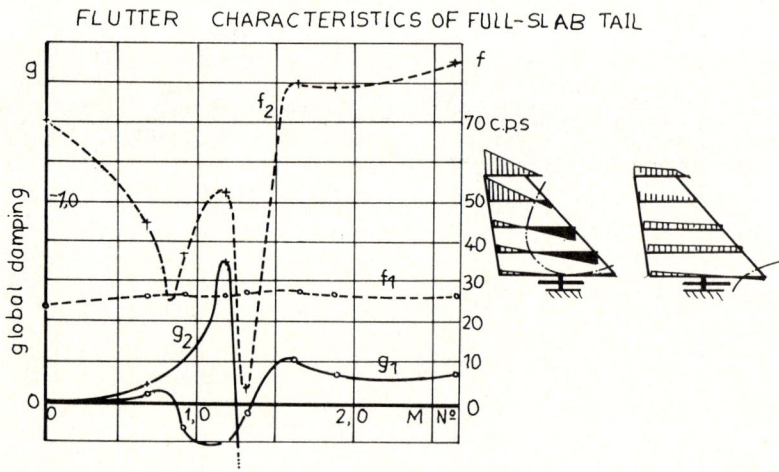

Figure 8 - Full-Slab Tail with Tip-Mass and Inertia-Plate and (1/4) Chord Fixed

(b) The flutter characteristics of the full-slab tail for the following configurations are shown:
- (i) with the projected mass at the tip chord;
- (ii) both the tip-mass and Inertia-plate at the root;
- (iii) the previous attachements together with one fixation point at quarter root chord.

(c) Figure 9 shows the original surface equipped with an appropriate Inertia-plate only. This represents 10 percent of the total weight of the tail surface. The results were found to be satisfactory, the flutter was suppressed within the flight range.

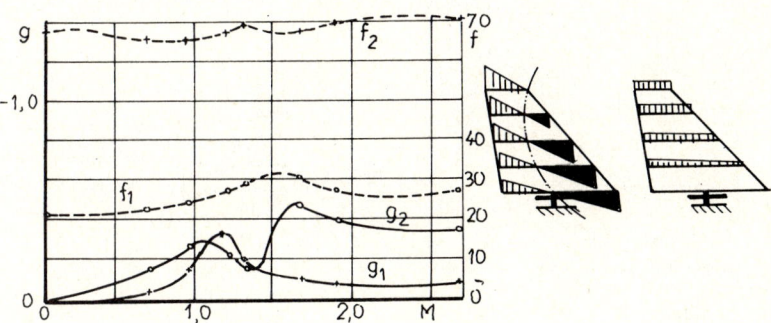

Figure 9 - *Full-Slab Tail with Inertia-Plate*

The attachment of this Inertia-plate does not require any structural modifications of the original main stream-lined surface. The structure is kept continuous and aerodynamically clean. This Inertia-plate can be mounted on the stabilizer pintle-axle inside the fuselage rear part. This method of suppressing the flutter of such a type of control surface can be made at any design stage.

(d) The bending and pitching normal modes of the full-slab tail with Inertia-plate were found to be of the following form: $h_1 = \eta^2$ which is the deflection in the first mode of the leading edge; $\theta_2 = e + 2e\eta$ is the rotation in the second mode about the leading edge; where (η) is the span ratio, and (e) is a constant that can be determined experimentally by measuring the transverse deflection (h) at two points at the reference chord ($\eta = 0.75$). The elaborate work for determining the deflections is considerably reduced. The theoretical flutter calculations based upon the deduced modes are in agreement with those obtained from the experimental ground resonance tests, (see Figure 10).

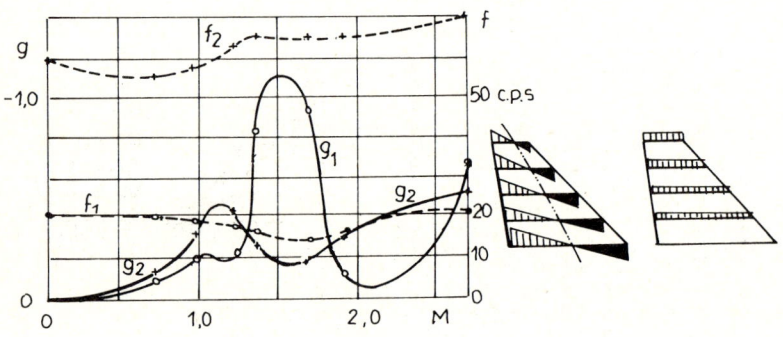

Figure 10 - *Theoretical Deduced Mode Shapes*

4. OTHER METHODS OF VIBRATION CONTROL

(a) A flap capable of inducing the appropriate unsteady aerodynamic force, is installed at the trailing edge of the full-slab tail. The flap oscillates by hydraulic oscillator. The oscillating pitching moment reduces the rotational part of the fundamental bending mode, near the flutter speed. The structure is considered as a part of the control loop. The closed loop system includes acceleration transducer on the structure and a circuit providing

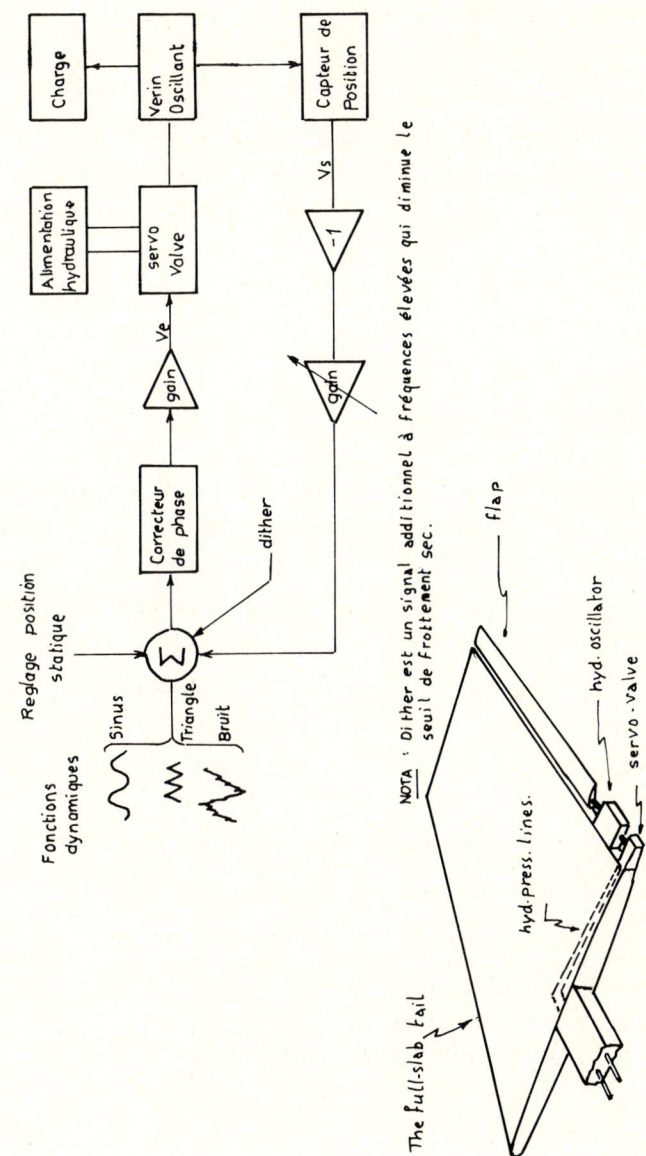

Figure 11

the control law relating the required flap oscillation (phase) to the signal of the transducer, near flutter speed, (see Figure 11). The design of a control loop for controlling the full-slab tail can be done systematically.

(b) Another method of controlling the structure to cure flutter consists of the following. A localized oscillating point control force is applied to the actual servo-jack that acts on the pintle-axle. This force exerts a suitable oscillating force on the structure. The structure will then be forced to vibrate in a stable mode with a predetermined frequency. This method eliminates the undesired mode.

ACKNOWLEDGEMENT

The author is grateful to Professor Dr. H. Försching, the Director of the Institute for Aeroelasticity in Göttingen, West Germany, for his sincere advice during this research work.
Thanks are also due to the Dean of the Faculty of Engineering at the University of Tripoli, Libya for giving me the opportunity for research work in the laboratory.

REFERENCES

[1] AFIFI, M.Y.M. and GIRGIS, S.F., "The Elasto-Mechanical Properties of Swept-Back, All-Moving Stabilizers", *Bul. Ain-Shams, Un.*, Vol. 5, 1970.

[2] GIRGIS, S.F., "Flutter Suppression of Aircraft Full-Slab Tail Control Surface", *Proc. of Euromech 107*, University of Edinburgh, September, 1978.

[3] FÖRSCHUNG, H., "Die Schwingungsanalysis Elastomechanischer Systeme mittels Vektorieller Ortskurven", *Z.V.D.I.*, 105, S. 1269-1278.

STRUCTURAL CONTROL, H.H.E. Leipholz (ed.)
North-Holland Publishing Company & SM Publications
© IUTAM, 1980

MINIMUM DEFORMABILITY DESIGN AND CONTROL OF CONSTRAINTS

Jan Grabacki

Institute of Structural Mechanics
Technical University of Cracow
ul. Warszawska 24, 31-145 Cracow, Poland

INTRODUCTORY REMARKS

The problem of optimum design for minimal deformability has been the subject of many papers. Among the various deformability measures used; [1, 2] the most interesting are the deflection of fixed point [3] and the absolute values of maximum deflection. These two measures have evident interpretation from the engineer's point of view, moreover, they can be immediately observed.

The other of the quoted above measures of deformability was discussed in detail in the paper by Komkov and Coleman [4]. However, some questions that arise during an investigation of the problem have never been discussed.

The answers to these questions are the subject of the first part of the presented paper. The second part refers to the problem of optimal control of constraints of the plates, by means of an optimal selection of the boundary conditions [5] or of optimal selection of the shape of the boundary. (The optimal control of the free part of the boundary was discussed by Mroz and Dems [6]).

Apart from an analytic approach more useful for numerical solutions, the results are presented here.

1. BAR STRUCTURES - AN ANALYTIC APPROACH

Denoting:

$M(t)$ - bending moment,

$m(t)$ - $EJ(t)$ - flexural stiffness,

$\phi(t) = \dfrac{-M(t)}{m(t)}$ - design variable,

Φ - set of admissible design variables,

the equation that governs the process takes the form:

$$\frac{d^2}{dt^2} x(t) = \phi(t) \quad ; \quad t \varepsilon <0,L> \quad ; \quad \phi \varepsilon \Phi \quad (1.1)$$

$$x(0) = x(L) = 0 \quad \text{or} \quad x(0) = x'(0) = 0 ,$$

$$x(t) = K(t,\tau) * \phi(\tau) \quad ; \quad x \varepsilon \mathcal{K} , \quad (1.2)$$

where:

$K(t,\tau)$ - Green's function,

$\mathcal{K}$ - the set of attainable trajectories.

In the following, (1.1) and (1.2) will be treated as mappings:

$$A: \mathcal{K} \to \Phi , \quad (1.3)$$

$$A^{-1}: \Phi \to \mathcal{K} , \text{ respectively.} \quad (1.4)$$

The problem that will be considered is:

- select the design: $\tilde{\phi} \varepsilon \Phi$,
- such that: $\sup\limits_{t} |x(\tilde{\phi},t)| = \min[\sup\limits_{x \varepsilon \mathcal{K}} |x|] = \min ||x||_c$.

Theorem 1

If:

1° - $\phi(t) \in \Phi \subset L^2(\Omega)$,

2° - $\int_\Omega |\phi(t)| dt \leq C$,

3° - $\forall_{t, \underline{\phi}, \overline{\phi}} \ \exists \ \underline{\overline{\phi}} \leq \phi \leq \overline{\overline{\phi}} : \quad \underline{\overline{\phi}}, \overline{\overline{\phi}}$ - real numbers ,

4° - $x(t, \phi) \in C^1(\Omega)$

then the unique optimal design exists.

Outline the proof:

Let $\mathcal{K}$ be equipped with the graph norm,

$$||x|| = ||x||_{C^1} + ||\phi||_{L^2} = ||x||_{W_2^1} \ ; \quad \mathcal{K} \subset W_2^1(\Omega) ,$$

then,

$$||Ax||_{L^2} \leq ||x||_{W_2^1} .$$

Hence the mappings:

$$A: \mathcal{K} \to \Phi \text{ and } A^{-1}: \Phi \to \mathcal{K}$$

are bounded. Since,

$$||x||_C = \sup \left| \int_\Omega K(t, \tau) d\tau \right| \leq K ||\phi||_{L^2(\Omega)} \ ; \quad K \in R ,$$

the mapping A^{-1} is bounded by means of the norm $|| \ ||_{C(\Omega)}$. From the inequality:

$$||x_n - x|| = \sup_\Omega |x_n - x| + \sup_\Omega |x_n' - x'| + ||\phi_n - \phi||_{L^2} \leq$$

$$\leq \left\{ K + [\mu(\Omega)]^{1/2} + 1 \right\} ||\phi_n - \phi|| = M ||\phi_n - \phi||_{L^2} ,$$

and

$$(\phi_n \to \phi) \to \phi\varepsilon\Phi \wedge x\varepsilon\text{Ж} \quad \text{follows} \quad \text{Ж} = \overline{\text{Ж}} \subset W_2^1(\Omega) ,$$

thus the set of attainable trajectories is closed.

By the theorem on embedding the W_2^1 space into the C space, [7] the compactness of Ж is shown.

Finally taking into account the continuity of the $||\ ||_C$ -norm by the Weierstrass theorem, the existence of the solution is proved. Let $\{x_n\}$ be such sequence that $||x_n|| > ||x_{n+1}||$. Since a convergent subsequence $\{x_{nk}\}$ can be chosen and $C(\Omega)$ is Haussdorff's, the solution is unique.

The next question concerns the relationship of optimal solutions in view of various types of the deformability measures. Partial answer to this question gives the following theorem: denote for the sake of concise notations:
- the problem which was defined above: *C problem*,
- the problem which arises by using the L_p - norm instead of the C - norm: *P problem*.

Theorem 2

Let $\{x_p\}$ be the sequence of solutions of P - problems, then:

1° $\lim\limits_{p\to\infty} x_p = x_c$ where x_c is the solution of the C - problem, a

2° $\lim\limits_{p\to\infty} ||x_p||_C = ||x_c||_C$.

Outline of the proof: assuming that for every p; $1 \leq p \leq \infty$; the uni solution x_p exists, the set:

$$\text{Ж}_p \stackrel{df}{=} \left\{ x_p \varepsilon \text{Ж} : \forall_p ||x_p||_p = \min_{x \varepsilon \text{Ж}} ||x||_p \right\} ,$$

equipped with order:

$$x_p < x_q \leftrightarrow ||x_p||_p < ||x_q||_q ,$$

can be defined. Let the sequence $\{||x_p||_p\} \subset \mathcal{K}_p$ be the sequence of P - problem solutions.

Taking into account well-known facts:

1° $\quad ||x||_p \leq ||x||q: \quad 1 \leq p \leq q \, \infty ; \quad \mu(\Omega) = 1 ,$

2° $\quad ||x||_p = \Psi(x,p): \quad \mathcal{K}_p \to R$ - continuous with respect to p as well as to x,

and the definition of the set, $\mathcal{K}_p$, it is easy to see that

$$\lim_{p \to \infty} ||x_p||_p = ||x_C||_C .$$

Under the additional assumption:

$$\underset{p>q}{\forall} \exists!_{p < r < q} ||x_p||_r = ||x_q||_r ,$$

the sequence $\{||x_p||_C\} \subset \mathcal{K}_p$ is monotonic and by property 2° - bounded, hence:

$$\lim_{p \to \infty} ||x_p||_C = ||x_C||_C ,$$

and the proof is completed.

If $\{R_i\}$ is the system of parameters and the design variable is defined as below:

$$\phi = \phi(R) \stackrel{df}{=} \frac{-M(R,t)}{m(t)} ,$$

then for every fixed system $\{R_i\}$ both of the theorems quoted above hold true. Particularly, if $\{R_i\}$ fulfills the system of compatibility equations, both theorems are valid for the hyperstatic system too.

Moreover introduce for convenience

(A) - Problem: Select a design $\tilde{m}$ that satisfies

$$\forall_t \; m_1 \leq m \leq m_2 \wedge \int_\Omega m \, dt = C ,$$

and minimizes the norm

$$||x(\tilde{m})||_C = \min_m ||x||_C .$$

(B) - Problem: Select a design $\tilde{m}$ that satisfies

$$\forall_t \; m_1 \leq m \leq m_2 \wedge ||x(m)||_C = x_0 ,$$

and minimizes the functional

$$\int_\Omega \tilde{m} \, dt = \min_m .$$

The relationship between the solutions of A and B - problems is subject to the following theorem.

Theorem 3

If $x_0 = ||\tilde{x}||_C$, where $\tilde{x}$ is the solution of an A - problem and $C = \tilde{C}$ is a solution of a B - problem, then both problems: A and B are mutual.

Outline of the proof. If:

$$m_a(t): \int_\Omega m_a(t) dt = C_a > C$$

and

$$\forall_t \; m_1 \leq m_a(t) \leq m_2 ,$$

then:

$$\forall_m \exists_{m_a} \forall_{t \in \Omega} \; m_a \geq m .$$

The solution of compatibility equations fulfills inequality:
$R \le ||A^{-1}|| \, ||b||$, where A - the matrix of coefficients, b - the vector of right-hand side terms.

Since $||A^{-1}||$ and $||b||$ are decreasing function of m it follows that:

$$\forall_{t \in \Omega}: \quad |\phi_a(R)| \le |\phi(R)| \, .$$

Taking into consideration that the solution of the equation governing the process is from a family of functions such that:

$$\forall_{\phi(R)} \quad x(\phi, t_k) = 0 \, ,$$

and t_k does not depend on m, the curvature of $x(\phi,t)$ is a decreasing function of m and finally:

$$m \le m_a \to \sup_t |x(\phi_a, t)| < \sup_t |x(\phi, t)| \, .$$

Let the solutions of problems A and B be different. Then,

1° Assumption $C > \tilde{C}$ and $||x||_C = ||\tilde{x}||_C$ leads to a contradiction because:

$$\exists_{x^*(t)} : \quad ||x^*||_C < ||x||_C \text{ and } ||x^*||_C < ||\tilde{x}||_C = x_0 \, .$$

2° Assumption $C < \tilde{C}$ and $||x||_C = ||\tilde{x}||_C$ implies that $\tilde{C}$ are not minimal values of the functional.
Thus, there remains as the last possibility only $C = \tilde{C}$ and the proof is completed.

Apart from the usefulness of the qualitative character of the proved theorems, they can be employed for the practical purposes. For example, theorem 1 allows to treat the necessary condition of optimality as sufficient.

According to theorem 2 the solution of the C - problem can be obtained as the limit of a sequence of solutions of P - problems (or the C - problem can be approximated by the P - problem with arbitrary accuracy.)

Finally, theorem 3 guarantees the solution of the C - problem to be economical by means of the global stiffness that cannot be less then calculated. However, an analytic solution can be obtained efficiently only in the simplest cases. Moreover, it is a solution which is useless for practice. This is the reason why one should seek an approximate solution. Assuming the design variable to be a piecewise constant function, the problem can be reduced to a finite dimensional one. Since the set of piecewise constant functions is dense in a space of quadratic integrable functions, the analytic solution (which exists) can be approximated with an arbitrary degree of exactness.

2. PLATES

The subject of this part of the presented paper is the discussion of the problem of the optimal selection of constraints by means of the optimal boundary conditions or the optimal shape of a boundary.

Apart from some analytic results, an approach will be presented which leads to a discretization of the problem and is useful for numerical calculations.

2.1 *The Boundary Conditions as Design Variable*

Notation:

Ω - the domain occupied by place,
$\Omega \subset \overline{\Omega}$ - the domain for which the Green's function is known,
G - Green's function
$f = \frac{q}{D}$ - the right-hand side of plate equation

$$\Delta_V = -\left[\frac{\partial^3}{\partial s^3} + (2-\nu)\frac{\partial^3}{\partial s \partial n^2}\right]; \quad \Delta_M = -\left[\frac{\partial^2}{\partial n^2} + \nu \frac{\partial^2}{\partial s^2}\right].$$

The solution of boundary value problems of the plates theory can be written as convolution:

$$w(x) = G(x,\xi)*f(\xi) + G(x,\xi)*v(\xi)\Big|_{\partial\Omega} + \frac{\partial G}{\partial n} * m(\xi)\Big|_{\partial\Omega}. \quad (2.1)$$

Here functions m and v are chosen in such a way that boundary conditions:

$$\left.\begin{array}{l} w\Big|_{\partial\Omega} = \Psi, \\[2ex] \dfrac{\partial w}{\partial n}\Big|_{\partial\Omega} = \phi \end{array}\right\} \quad (\partial\Omega - \text{the boundary of } \Omega) \qquad (2.2)$$

are satisfied.

In accordance with formulas

$$\left.\begin{array}{l} \Delta_V w\Big|_{\partial\Omega} = \Delta_V G*f + \Delta_V G*v + \Delta_V \dfrac{\partial G}{\partial n} * m = v_0 + v = V, \\[2ex] \Delta_M w\Big|_{\partial\Omega} = \Delta_M G*f + \Delta_M G*v + \Delta_M \dfrac{\partial G}{\partial n} * m = m_0 + m = M, \end{array}\right\} \quad (2.3)$$

the functions m and v represent the constraints that are applied.

The reactions of constraints can be written as internal-side limit

$$\left.\begin{array}{l} M^* = \lim\limits_{x \to \partial\Omega} M, \\[2ex] V^* = \lim\limits_{x \to \partial\Omega} V. \end{array}\right\} \qquad (2.4)$$

Since elastic properties of the constraints are described by the relations:

$$H_1 = \frac{\Psi}{-V^*} = \left.\frac{w(v,m)}{-V^*}\right|_{\partial\Omega},$$

$$H_2 = \frac{\phi}{-M^*} = \left.\frac{\frac{\partial w}{\partial n}(v,m)}{-M^*}\right|_{\partial\Omega},$$
(2.5)

functions m and v can be treated as the design variables. Now, the problem of optimal design can be defined as follows: select such a pair $\{\tilde{m},\tilde{v}\}$ that:

1° $\quad ||w(x,\tilde{m}_1\tilde{v})||_C = \min_{m,v} ||w(x)||_C$, or

2° $\quad ||w(x_0,\tilde{m}.\tilde{v})| = \min_{m,v} |w(x_0)|$; x_0 - fixed ,

subject to the restrictions:

(i) $\int_\Omega (f+v_0+v)d\Omega = 0$; D = 1 ,

$\int_\Omega [m+m_0+(v+v_0+f)\phi_0]d\Omega = 0$,

(ii) $\underline{\overline{H}}_1 \leq H_1 \leq \overline{\overline{H}}_1$; $\overline{\overline{H}}_1 > \overline{H}_1 > 0$,

$\underline{\overline{H}}_2 \leq H_2 \leq \overline{\overline{H}}_2$; $\overline{\overline{H}}_2 > \overline{H}_2 > 0$,

(for the whole boundary or its part), or

(iii) $V^*|_{\partial\Omega_V} = 0$; $M^*|_{\partial\Omega_M} = 0$, or

(iv) $\phi|_{\partial\Omega_\phi} = 0$; $\Psi|_{\partial\Omega_\Psi} = 0$.

The restrictions (ii), (iii) and (iv) can lead to an inconsistency so they should be appropriately combined.

For further discussion it is convenient to change the notation as follows:

$$w = w_0+w_1 \quad ; \quad w_0 = G*F \quad ; \quad w_1 = G*v + \frac{\partial G}{\partial n}*m;$$

(instead of (2.1), and

$$\frac{w}{w_0} \geq 0 \wedge \frac{v^*}{v_0} \geq 0 ,$$
$$\frac{\frac{\partial w}{\partial n}}{\frac{\partial w_0}{\partial n}} \geq 0 \wedge \frac{M^*}{m_0} \geq 0 , \qquad (2.7)$$

(instead of (2.5)). Let the set W be defined

$$W \stackrel{df}{=} \{w_1; w_1 = G*v + \frac{\partial G}{\partial n} * m \; ; \; \text{and restrictions (i) and}$$
(2.7) hold\} , \hfill (2.8)

then by the theorem on a support hyperplane [8] and convexity of the set w_0+W:

$$\inf ||w_0+w_1||_C = \max_{||w^*||\leq 1} |-h(w^*)| \; ; \; w^* \in NBV(\Omega) ,$$

and minimizer $\tilde{w}_1$ fulfills the condition:

$$<w_0+w_1|w^*> = ||w_0+\tilde{w}_1|| \; ||w^*||_{NBV} . \qquad (2.9)$$

Hence, the solution exists (but can be not "unique"). This means that more than our solution can exist.

The formula (2.9) represents the analytic optimality condition but is useless for practice.

Now, the finite dimensional version of the problem is proposed. It can be obtained in the following way. Dividing the boundary of the region Ω into subintervals Δs_j and assuming m and v to be piecewise constant functions, the functional takes the form:

$$w(x_0) = w_0 + \sum_j (a_{0j}v_j+b_{0j}m_j) , \qquad (2.10)$$

where x_0 is the point of local maximality or fixed, and the constraints respectively

$$\sum_j \Delta s_j v_j = 0 \quad ; \quad \underset{j}{\Delta}(m_j + v_j \cdot \rho_{0j})\Delta s_j = 0 , \qquad (2.11)$$

$$\begin{cases} 1 + \Sigma \overline{a}_{ij} v_j + \overline{b}_{ij} m_j \geq 0 \wedge 1 + \Sigma \overline{C}_{ij} v_j + \overline{d}_{ij} m_j \geq 0 , \\ \\ 1 + \Sigma \overline{e}_{ij} v_j + \overline{g}_{ij} m_j \geq 0 \wedge 1 + \Sigma \overline{h}_{ij} v_j + \overline{k}_{ij} m_j \geq 0 , \end{cases} \qquad (2.12)$$

where,

$$w_0 = \int_\Omega G(x \cdot \xi) f(\xi) d\Omega ,$$

$$a_{0j} = \int_{\Delta sj} G(x_0 1 \sigma) d\sigma \quad ; \quad b_{0j} = \int_{\Delta sj} \frac{\partial G(x_0 1 \sigma)}{\partial n} d\sigma ,$$

and the remaining coefficients respectively.

The same form is taken by the alternative constraints (iii) and (iv). The pair $\{\tilde{m}, \tilde{v}\}$ that solves the problem can be found by using the direct method of searching.

Finally, the elastic characteristics of optimal constraints are given by the formulas:

$$\left. \begin{array}{l} \tilde{H}_{1,i} = \dfrac{w_i(\tilde{m},\tilde{v})}{V^*(\tilde{m},\tilde{v})} , \\ \\ \tilde{H}_{2,i} = \dfrac{\frac{\partial w}{\partial n}(\tilde{m},\tilde{v})}{M^*(\tilde{m},\tilde{v})} . \end{array} \right\} \qquad (2.13)$$

The essential difficulty in such approach is, as usual, the dimension of the problem. In practice it is very often sufficient to search the boundary of the set of admissible control and the dimension of the problem can be considerably reduced.

If x_0 is the point of local maximality of a deflection, the most effective method is the iterative method.

In the case when more then one point of local maximality exists, additional restrictions:

$$|w(x_{0_i})| \geq |w(x_0)|,$$

should be taken into account. It is necessary for the existence of the optimal solution.

2.2 The Shape of Boundary as a Design Variable

Let the solution of the plate equation be given as previously in the form of convolution (2.1) and functions m,v which fulfill the boundary conditions (2.2) where ϕ,ψ are fixed.

Assume, in addition, that $\{\Omega_j\}$ is the set of such domains that:

(i) $\mu(\Omega_j) = $ where μ - measure,

(ii) Ω_j - star shaped; $r_j = r^*[1+g_j(\nu)]$,

(iii) $d_1 \leq \text{dist}(0,s) \leq d_2$; $s\epsilon\partial\Omega_j \wedge 0\epsilon\Omega_j$.

The problem of optimal design consists in the following: select such domain $\tilde{\Omega}$ with the boundary $\partial\tilde{\Omega}$ that:

$$||w(x,\partial\tilde{\Omega})||_C = \min_{\partial\Omega_j} ||w(x)||_C$$

or

$$|w(x_0,\partial\tilde{\Omega})| = \min_{\partial\Omega_j} |w(x_0)|, \quad \text{here } x_0 \text{ - fixed},$$

and restrictions (i), (ii) and (iii) hold.

Owing to assumption (ii) the convolution (2.1) can be integrated with respect to r and, as the result, the functional and additional equations (instead a boundary conditions (2.2)) depending on $g(\nu)$ only, will be obtained.

The analytic optimality condition can be derived in the usual way as the condition of stationarity of the Lagrangian functional. However, such an approach is useless in practice.

The approximate solution can be obtained by a discretization of the problem. Let the function $g(\nu)$ be expressed in the form of trigonometric series:

$$g_j(\nu) = \alpha_{jk} T_k(\nu) , \qquad (2.14)$$

and let $\Delta\nu_j$ be an angular subdomain corresponding to subinterval Δs_j and finally let m,v be piecewise constant. The problem under consideration takes the form:

$$w(x_0) = w_0(\alpha_{ik}) + \Sigma\, a_{0j}(\alpha_{ik}) v_j(\alpha_{ik}) + b_0(\alpha_{ik}) m_j(\alpha_{ik}) , \qquad (2.15)$$

$$\sum_i \int_{\Delta\nu j} [1+\alpha_{ik} T_k(\nu)] \rho_0 d\nu = \frac{1}{r^*}\Omega \;\; ; \;\; d_1 \leq r^*(1+\alpha_{ik} T_k) \leq d_2 . \qquad (2.16)$$

The optimality condition can be written as:

$$\nabla_{\alpha_{ik}} \phi(\lambda, \Lambda_j) = 0 , \qquad (2.17)$$

where ∇ - gradient and Λ - Lagrangian.

It is a system of algebraic equations and can be solved by using known methods.

Besides the usual difficulty of the discrete approach, an additional inconvenience of the proposed method is that for a numerical solution the problem should be hand-prepared. This inconvenience can be removed if the boundary conditions will be treated as additional constraints but it leads to a considerable limitation of the dimension of the problem which one is able to solve.

The double approximation and the mentioned limitation, reduce the exactness of the achievable solution.

In spite of all difficulties, the presented method allows to solve effectively a number of problems of optimal control of constraints which, according to the best knowledge of the author, have never been discussed. Moreover, it can be applied successfully to more general cases of optimal control of continuous media.

REFERENCES

[1] WASIUTYŃSKI, Z. and BRANDT, A., "The Present State of Knowledge in the Field of Optimum Design of Structures", *Appl. Mech. Rev.*, Vol. 16, 1963, pp. 341-350.

[2] PRAGER, W., "Optimality Criteria in Structural Design", *Proc. of the Nat. Acad. Sci.*, U.S.A., Vol. 61, 1968, pp. 194-196.

[3] SHIELD, R.T. and PRAGER, W., "Optimal Structural Design for a Given Deflection", *J. Appl. Math. Phys.*, Vol. 2, 1975, pp. 513-516.

[4] KOMKOV, V. and COLEMAN, N.P., "An Analytic Approach to Some Problems of Optimal Design of Beams and Plates", *Arch. of Mech.*, Vol. 27, 1975, pp. 565-575.

[5] MROZ, Z. and ROZVANY, G.I., "Optimal Design of Structures with Variable Support Conditions", *J. Opt. Theory and Appl.*, Vol. 15, 1975, pp. 85-101.

[6] MROZ, Z. and DEMS, K., "Multiparameter Structural Shape Optimization by the Finite Element Method", *Int. J. Num. Meth. in Eng.*, Vol. 13, 1978, pp. 247-263.

[7] СМИРНОВ, В.И., КУРС ВЫСШЕЙ МАТЕМАТИКИ Т.V. ГОС. ИЗД. ФИТ. МАТ. ДИТ. МОСКВА, 1959.

[8] ROLEWICZ, S., *Analiza Funkcjonalna i teoria Sterowania*, PWN, Warszawa, 1959.

STRUCTURAL CONTROL, H.H.E. Leipholz (ed.)
North-Holland Publishing Company & SM Publications
© IUTAM, 1980

ACTIVE FLUTTER CONTROL IN TRANSONIC CONDITIONS

A. Gravelle

Office National d'Études et de Recherches Aérospatiales
29, Avenue de la Division Leclerc
92320 Châtillon, France

1. INTRODUCTION

Two main reasons make active flutter suppression interesting in
the aeronautical domain:
- the design of lighter structures giving better performances of aircrafts, or more economical weight;
- the need to add external stores, generally on military
aircrafts, which change the vibration modes, sometimes very long
after the first flight.

In the second case, the cost of a structural modification is prohibitive and control is necessary. Therefore, theoretical studies and wind tunnel tests were performed at ONERA to design an active flutter suppression system on a fighter wing equipped with external stores, using aerodynamic forces induced by an existing aileron.

The difficulty of the problem is the poor accuracy in determining unsteady aerodynamic forces in these complex cases including external stores. Moreover the need to get a control law which applies in the whole flight domain (Mach number and altitude) leads to choose a rather simple system which is not very sensitive to aerodynamic force modifications and does not permit the use of optimal control.

2. PRINCIPLE OF THE METHOD

2.1 *Mathematical Formulation*

The classical flutter equation is:

$$\{-\omega^2 \mu + \gamma + \frac{1}{2} \rho V^2 A(\omega)\} q = Q . \qquad (1)$$

Q represents external forces and is zero in the case of flutter calculation since flutter is a stability problem.

Q may be considered as a control force, then it becomes:

$$Q = \frac{1}{2} \rho V^2 C(\omega) \beta , \qquad (2)$$

where $C(\omega)$ is the column matrix of unsteady generalized aerodynamic forces induced by the control surface rotation β on the structural modes of the wing.

In the general case, mass coupling between control surface and wing has to be considered, but it may be neglected in transonic conditions.

β is related to the displacement of different control points on the wing by means of different control transfer functions $T_i(\omega)$:

$$\beta = \sum_i T_i(\omega) z_i . \qquad (3)$$

In the modal representation, z_i is:

$$z_i = {}^t W_i q . \qquad (4)$$

Then, equations (2), (3) and (4) lead to:

$$Q = \sum_i \frac{1}{2} \rho V^2 C(\omega) T_i(\omega)^t W_i q , \qquad (5)$$

and the modified flutter equation becomes:

$$\{-\omega^2 \mu + \gamma + \frac{1}{2} \rho V^2 [A(\omega) + C(\omega) \sum_i T_i(\omega)^t W_i]\} q = 0 . \qquad (6)$$

Equation (6) has the classical form of the flutter equation, the unsteady aerodynamic forces being modified by the complex matrix $C(\omega) \sum_i T_i(\omega)^t W_i$ induced by the rotation of the control surface.

2.2 *Application*

In practical cases, mainly fighter wings with external stores configurations, the following assumptions may be made:
 - the flutter problem is a two degree of freedom problem, resulting from the coalescence of the frequencies of a bending mode and a store mode which appears as a torsion on the wing;
 - if the control surface is chosen so that its effect on the flutter modes is large enough, the C column is nearly purely real in the frequency range of interest.

Therefore, if it is assumed that the control can be realized with only one sensor, the control matrix is written as:

$$C(\omega) T(\omega)^t W = \begin{bmatrix} T(\omega) C_1(\omega) W_1 & T(\omega) C_1(\omega) W_2 \\ T(\omega) C_2(\omega) W_1 & T(\omega) C_2(\omega) W_2 \end{bmatrix} . \qquad (7)$$

It is possible to choose $T(\omega)$ so that the two diagonal terms $T(\omega)C_1(\omega)W_1$ and $T(\omega)C_2(\omega)W_2$ are nearly purely imaginary and positive for the flutter frequency, and we obtain a pure dissipative system.

Experiments showed that it is not the most efficient method.

It is also possible to make the two terms nearly purely real in a narrow band around the flutter frequence. The location of the sensor defines the values of W_1 and W_2. If the sensor is located between the node lines of the two modes, these quantities have opposite signs so that the control force is positive stiffness for one mode and negative stiffness for the other. Figure 1 shows that it is then possible to increase the convergence of frequencies, or on the contrary, to avoid the frequency coalescence and in both cases the flutter may disappear.

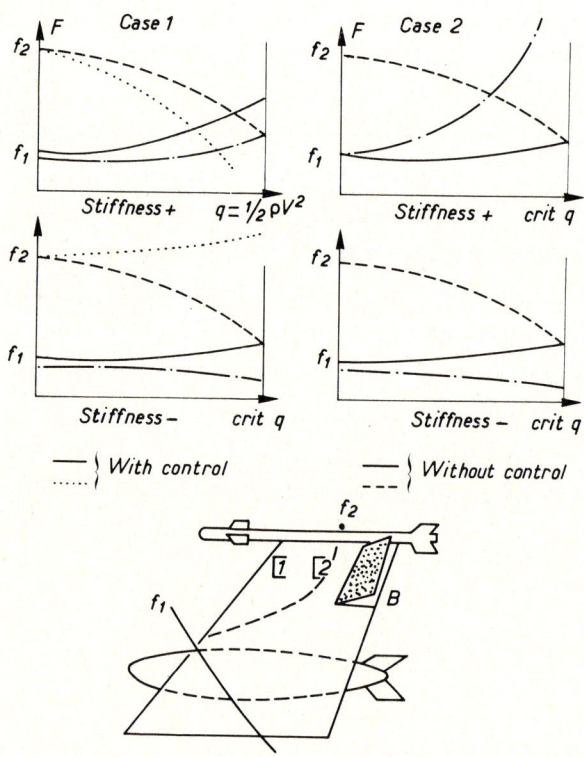

Figure 1 - Principle of Adding Stiffness

Of course the rigid body modes and the higher modes of the aircraft should not be destabilized by the control forces, therefore it is necessary to include a band-pass filter which is tuned to give no significant response at the corresponding frequencies.

The block diagram of the control law used during wind tunnel tests is given in Figure 2. The sensor is an accelerometer located in the wing, the global transfer function is:

$$\frac{\beta}{\ddot{z}} = T = T_1 T_2 (G_1 + G_2 T_2) G_3 T_3, \tag{8}$$

G_1, G_2 and G_3 are pure gains.

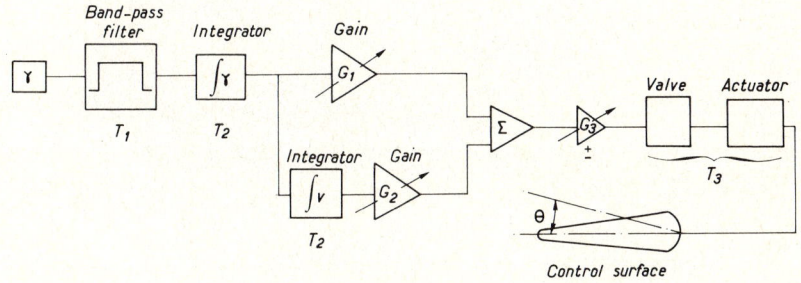

Figure 2 - Block Diagram of the Control Law

T_2 is a pseudo-integrator transfer function including a high pass filter which avoids an infinite response at zero frequency. The combination $(G_1+G_2T_2)$ is equivalent to a phase lag system and permits to compensate the phase due to the band-pass filter T_1 and the actuator and servo valve transfer function T_3.

3. EXPERIMENTAL CONDITIONS

The application of the flutter control was made on a similar dynamic half model of a modern fighter with two combinations of stores (Figures 3 and 10).

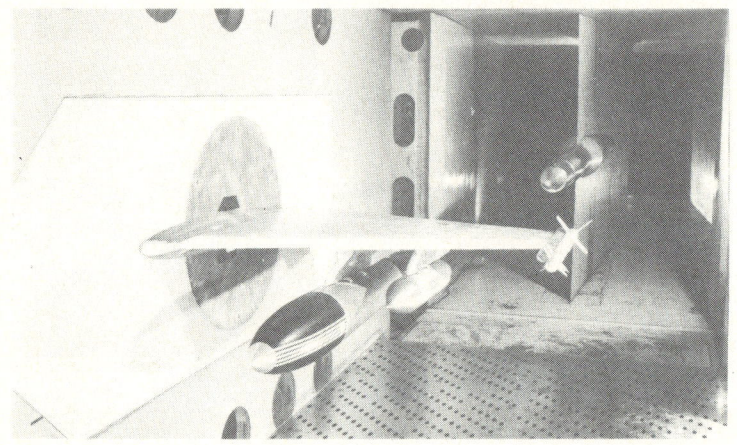

Figure 3 - *Aeroelastic Model Clamped at the Wind-Tunnel Wall*

In the first case, the wing was equipped with three stores, and has, at the root, a degree of freedom in rolling at low frequency (3.8 Hz), in order to represent an antisymmetric flutter case. The flutter modes were: the first bending at 20 Hz and the pitch of the tip store at 25.1 Hz. The first higher mode was rolling of the tip store at 34.3 Hz.

The second store configuration included only a big tank, and the model was clamped at the tunnel wall. The flutter modes were: the first bending at 24.7 Hz and the tank pitch at 30.5 Hz. The first higher mode was yawing of the tank at 35.5 Hz.

The control surface was a classical external aileron, the control system used only one accelerometer located at 30%

along the mid span of the aileron.

The hydraulic actuator was a small rotative actuator (Figure 4) designed and built at ONERA, whose transfer function is given in Figure 5.

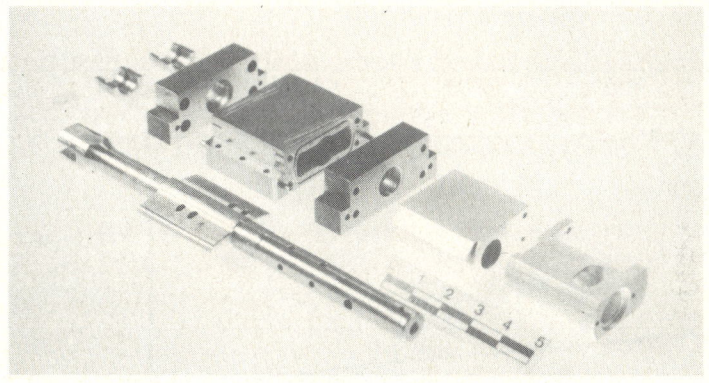

Figure 4 - Hydraulic Actuator (scale in cm.)

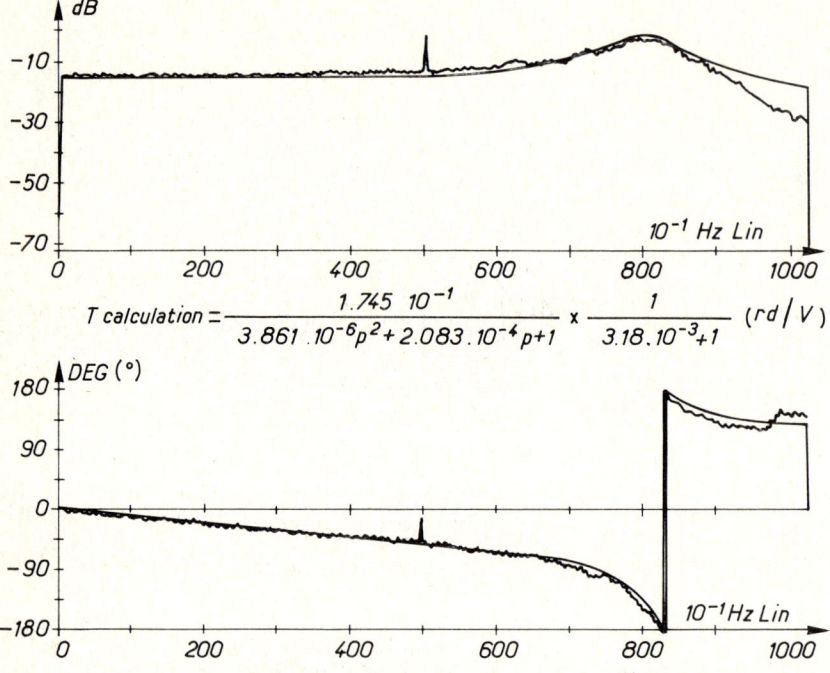

Figure 5 - *Actuator Transfer Function*

4. EXPERIMENTAL RESULTS

The unsteady pressure distribution induced by the rotation of the aileron was measured at a frequency near the flutter frequency (Figure 6) and compared to a theoretical pressure distribution calculated by means of a classical doublet lattice method. The measured pressure coefficients are smaller than the theoretical ones, and moreover the measured imaginary parts are nearly negligible.

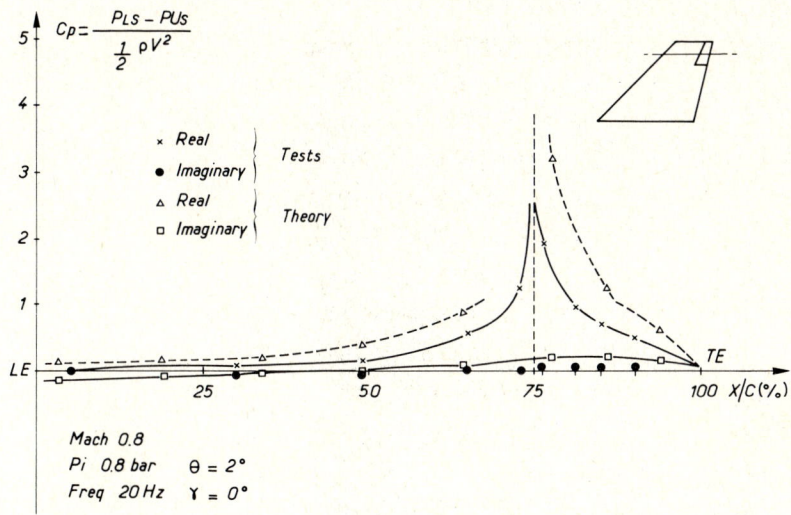

Figure 6 - Pressure Coefficient along the Middle Flap Chord

    Figure 7 shows theoretical and experimental frequency and damping evolutions of the model in the first configuration without any control. These results are in good agreement. The control (Figure 8) modifies the frequency evolution, the two modes (first bending and missile pitch) intersect very early, while coupling forces are small, and flutter disappears. The calculation accuracy is rather good.

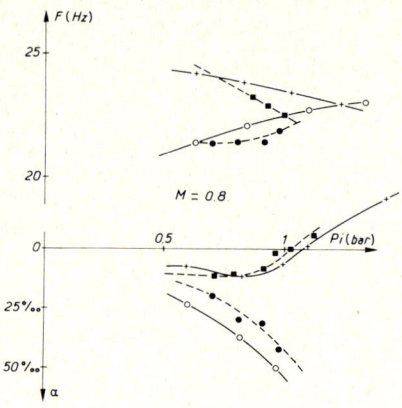

Figure 7 - Flutter Plot without Control

Theory without Control { + mode 3
                         o mode 2

Tests without Control { ■ mode 3
                        ● mode 2

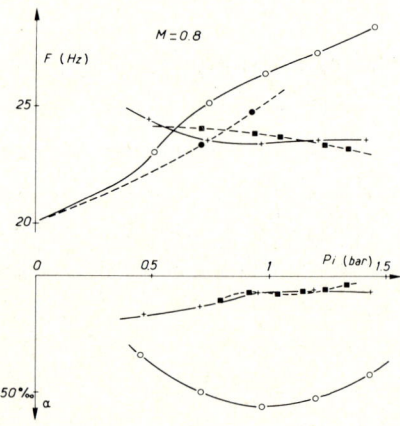

Figure 8 - Flutter Plot with Control

Theory with Control { + mode 3
                      o mode 2

Tests with Control { ■ mode 3
                     ● mode 2

Tests were performed up to M = 0.92 and the gain in critical stagnation pressure were very large, limitations being due to the wind tunnel ability.

A second loop was used at very low frequency in order to keep the wing in constant angle of roll, by means of the same aileron without any interference with the flutter control system.

The control efficiency is shown in Figure 9. The control loop is switched off above the nominal flutter point, the model starts to be unstable and for a given amplitude the control is switched on again. The flutter disappears after a few cycles.

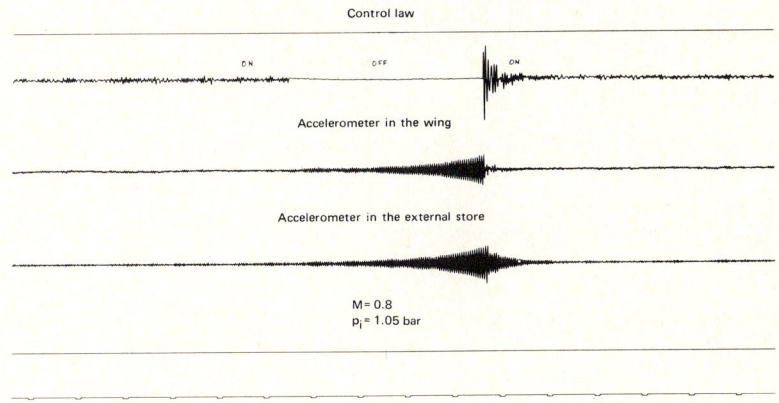

Figure 9 - *Efficiency of the Control*

For the second configuration (Figures 10 and 11), the control was also a stiffness control type, the sign of which was chosen in order to separate the flutter frequencies. The tank pitch mode frequency was not very much modified, but the bending mode frequency decreased and flutter disappeared in the whole wind tunnel domain.

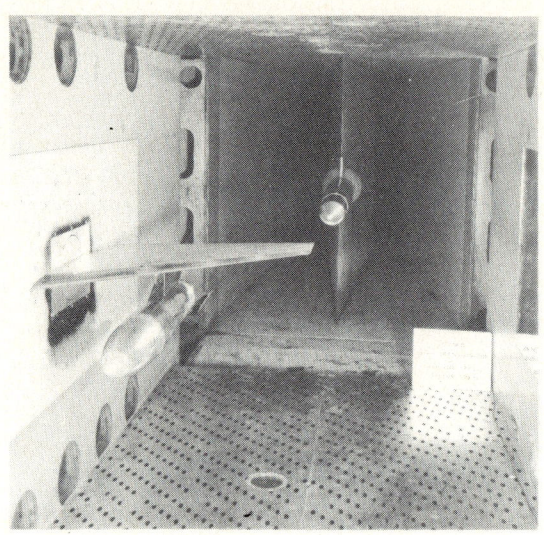

Figure 10 - Model with a Single Store

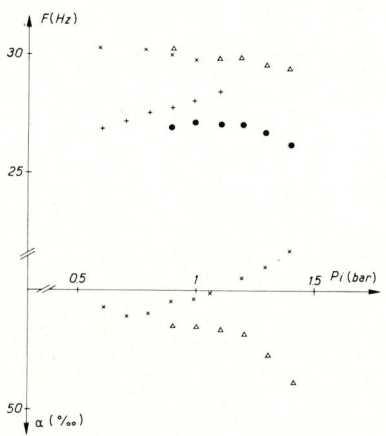

Figure 11 - Flutter Plot with and without Control (Tests)

Without Control { + Mode 1
                        x Mode 2

With Control { • Mode 1
                         △ Mode 2

The safety margins are shown in Figure 12, in a four modes calculation. The stability condition on the open loop Nyquist plot of the system with control loop is that the Nyquist plot turns around the -1 point counterclockwise. In this case, the gain margin for the control loop is ± 3db and the phase margin is greater than ± - 20° at a stagnation pressure of 1.4 bars. The theory shows that the controlled system becomes unstable at a higher stagnation pressure (1.6 bars). But this theoretical calculation is pessimistic because structural damping is not considered.

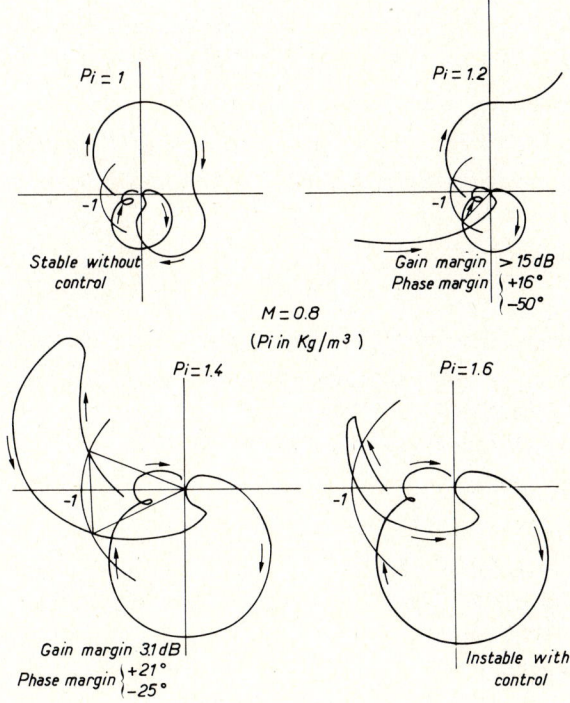

Figure 12 - Modified Nyquist Plots

5. CONCLUSIONS

These two wind tunnel examples prove the possibility to control flutter in the transonic range with a relatively simple control law using only one sensor per wing.

The control could be determined by means of theoretical calculation, improved by some experimental measurements, and successfully tested in spite of the poor accuracy of aerodynamic theories for such complex configurations in the transonic range.

Flight application of this method is possible, it will need the manufacturing of larger, fast hydraulic actuators. The main difficulty will be to avoid interactions between the different control systems, an aircraft in flight having a larger number of degrees of freedom than a wind tunnel model.

Very important improvements in aircraft weight and performances may result from active flutter control technology.

NOTATIONS

A	- unsteady generalized aerodynamic force matrix
C	- column of unsteady generalized forces induced by the control surface rotation
C_p	- unsteady pressure coefficient
M	- Mach number
V	- fluid velocity
T or T_i	- control transfer function
t_W or t_{Wi}	- transposed column of modal displacements at the control points
f	- frequency
p	- Laplace variables
p_i	- stagnation pressure
q	- column of generalized coordinates
z or z_i	- displacements at the control points
$\alpha (=2g)$	- reduced damping

β		- control surface rotation
γ		- structural stiffness matrix
μ		- generalized mass matrix
ρ		- fluid density
ω		- circular frequency

REFERENCES

[1] DESTUYNDER, R., "Aspects Structuraux du Contrôle Actif", *Proc. AGARD/FMP Symposium on Stability and Control*, Ottawa, Canada, September 25-29, 1978, AGARD CP-260.

[2] DESTUYNDER, R., "Problèmes d'Aérodynamique Instationnaire Posés par l'utilisation des Gouvernes dans le Contrôle Actif", *AGARD CPP-262*.

STRUCTURAL CONTROL, H.H.E. Leipholz (ed.)
North-Holland Publishing Company & SM Publications
© IUTAM, 1980

CRITICAL COMPARISON BETWEEN ACTIVE AND PASSIVE CONTROL
OF WIND INDUCED VIBRATIONS OF STRUCTURES
BY MEANS OF MECHANICAL DEVICES

G. Hirsch

Institut für Leichtbau
Technische Hochschule Aachen
Aachen, West Germany

1. INTRODUCTION

The importance of the problem of wind-induced oscillations of tall buildings and other slender structures is increasing with the growing tendency to build light structures. Because the inherent damping ability of a conventional structure is typically very low, the dynamic response according to vortex shedding, buffeting, gust load, galloping and flutter respectively is eminent in many cases.

The effects on the well-being of the inhabitants must be distinguished from the effects on the stability of the structure.

Some years ago various cases of wind-induced oscillations of tall buildings were discovered, which required additional measures. The 279 m-high Citicorp Center in New York City as well as the John Hancock Tower in Boston had to be equipped with damping-systems to reduce the movement of the buildings in strong wind to a tolerable size [1].

For some time other cases have been known where slender structures (steel chimney stacks) were damaged by wind-induced oscillations (vortex shedding resonance) and a reconstruction was only possible with additional damping measures [2], [3].

Therefore it seems advisable to discuss the possibilities of reducing wind-induced oscillations during the IUTAM Symposium "Structural Control", to demonstrate practical ideas, and to make critical comparisons.

This subject is also being treated in Europe. In September, 1978, the University of Edinburgh organized a EUROMECH-Kolloquium on the subject: "Control of Structure Vibrations".

2. A SURVEY ON WIND-INDUCED VIBRATIONS

It is necessary to give a summary of the mechanisms of wind-induced oscillations before talking about measures to reduce these motions. It is particularly to be seen how efficient these additional measures will be. On principle a distinction between forced and self-excited oscillations is sensible, although the boundary between these two forms of oscillations are effaced with increasing vibration amplitude.

2.1 *Forced Oscillations*

The following vibrations belong to this group: periodic excitation by vortex separation, by buffeting and the non-periodical stochastic excitation by gusts of winds.

2.1.1 Vortex-Shedding Excitation

Strong vibrations occur lateral to the wind-direction at the critical wind velocity from Strouhal relation ($V = D \cdot f/Sh$ where D = diameter of the cylinder or the lateral length of the rectangular cross-section, f = the natural frequency of the fundamental mode of vibration, Sh = The Strouhal number = 0.2 confirmed in many investigations).

It must be taken in consideration for a vortex excitation of bodies with a rectangular cross-section that the Strouhal number depends on the side ratio of the cross-section (S = 0.13 for a side ratio 1 : 1).

The dynamic load on the system depends mainly on the lift coefficient C_L and the resonance magnification factor π/δ with δ as the logarithmic decrement of the structure damping:

$$P_{cr} = C_L \cdot (\pi/\delta) \cdot (\rho/2) \cdot V^2 \cdot D \; ,$$

where ρ = the specific mass of air.

The oscillation-amplitude we can estimate [4] by means of the mass-damping-parameter [5] $M_\delta = 2m \, \delta/\rho \cdot D^2$ (where m = the mass per unit length):

$$(y_{max}/D) = 2 \cdot (C_L/M_\delta) \; .$$

In the subcritical Reynold number region we can assume $C_L \simeq 0.8$ and in the supercritical region $C_L \simeq 0.2$.

Seeing as the damping values can be extremely small (compare values in the tabulation) dangerous oscillations can be expected in the case of vortex-resonance.

Hence it appears that it is possible to reduce the dynamic response of the structure by reduction of the Lift-coefficient (aerodynamic measures) or increasing the structural damping.

2.1.2 Buffeting

In this case the wake from a steady wind passing over a bluff obstacle may provide turbulent flow in which eddy formation and fluctuations of wind speed and direction occur with a marked periodicity. A flexible structure in this wake may develop large amplitude oscillations when the eddy - and natural frequencies coincide [6]. The resonance magnification factor is: π/δ. There is also evidence that wind-excited oscillations of the leeward stack of a pair may be augmented by buffeting.

Structural damping has an important influence there and so damping measures can possibly reduce the wind-influence.

2.3.1 Non Periodic Excitation by Wind-Gusts

The unsteady nature of gusts may produce oscillatory displacements of tall buildings and other slender structures, especially when their frequencies coincide with a natural frequency, but large amplitudes are unlikely because of their random nature. Considering the estimation of the building's response the formula of the natural frequency is useful apart from the damping properties. This formula was found empirically using numerous test results [7]:

$$f = 0.4 \ (100/H)^{1.6} \qquad (c.p.s.)$$

where H = the height of the building (< 200 m). For extremely high buildings the formula can be simplified (f = 40/H).

2.2 *Self-Excited Oscillations*

The self-excited oscillations are motion-induced vibrations with continuous growing amplitudes. First of all, I would like to mention the galloping-type oscillations and moreover the instabilities in two or more degrees of freedom (flutter-instability).

2.2.1 Galloping-Oscillations

This type of oscillations can arise lateral to the wind direction of slender prismatic buildings with certain cross-sections and small damping-properties. The instability condition obtains when the slope of the lift versus incidence curve is more negative than the drag force is positive.

The onset-velocity of galloping-oscillations is:

$$V = 2 \cdot M_\delta \cdot f \ D/\sigma ,$$

where σ is equivalent to the gradient mentioned above and has the value minus 3.15 for a square cross-section for example. The structural damping plays an important role here.

Galloping vibrations also appear at cylinders standing tight together (e.g., rows of chimneys, steel-ropes of cable-stayed bridges, etc.). These oscillations were observed in model-tests as well as with original buildings [8], [9].

2.2.2 Flutter Vibrations

Two degrees of freedom flutter with bending and torsional motion is the classical case of aeroelastic instability. It is difficult to distinguish between the galloping and the flutter instability in the case of cross-sections varying from the aircraft wings.

After Klöppel-Thiele [10] and the van der Put [11] the critical wind velocity for slender streamline bridges - vibrations of this type can well be expected here - is:

$$V = c \ f_b b \left(2\pi + \left(\frac{f_T}{f_b} - 0.5 \right) \cdot 3.0 \left(\frac{m \cdot r}{\rho b^2} \right)^{0.5} \right) ,$$

with c as aerodynamic form-coefficient, 2 b as bridge-breadth in wind direction, f_b and f_T as natural frequencies of bending and torsional modes, m as mass/length and r as radius of inertia of the rotating mass. The form-coefficient can be relatively small (e.g., .2 for a double-T-section with a flange breadth of .1 to .2 times 2 b) and 1.0 (for stream-line sections).

By means of aerodynamic devices it is possible to extend the form-coefficient c. On the other side increasing the structural damping efficiently reduces the flutter instability, (see Figure 2).

3. REDUCTION OF WIND-INDUCED OSCILLATIONS BY DAMPING MEASURES

As shown above, the vibrations in the case of wind-induced oscillations with low structural damping can be reduced to allowable values by additional damping measures. It is to distinguish between active and passive methods (active and passive control). It can be by means of aerodynamic or mechanical action.

In the following treatment I would like to explain and give a critical look at these measures.

3.1 Active Measures

To limit the response of the structure within an acceptable range, active control devices can be used. The addition of active systems to tall buildings for the purpose of structural safety could enable future engineers to increase the present upper limit of the practical height.

3.1.1 Active Aerodynamic Measures

Active aerodynamic measures are being used successfully in aeronautical engineering to suppress flutter (e.g., wing flutter). Using additional control surfaces, as in Figure 1, forces that are anti-phase to the inducing forces are implied to prevent the vibration of the structure in the critical flow.

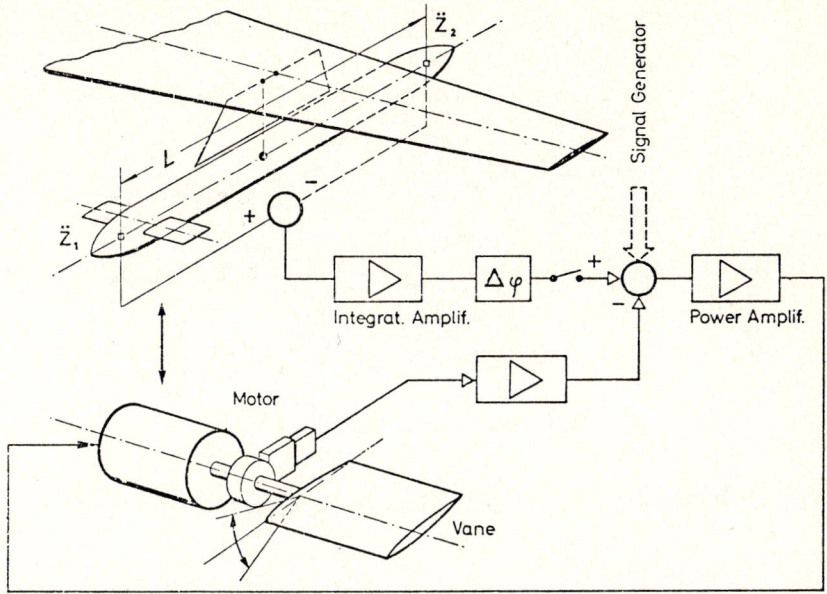

Figure 1 - Active Aerodynamic Measure to Suppress Wing-Store Flutter

The movement required for controlling this proceeding are measured with sensors and the additional control surfaces are regulated by integration-amplifiers, phase-shifters, power-amplifiers and electrical servo motors so that a damping effect is creases [12].

Figure 2 shows that the critical speed (e.g., of a cable-stayed bridge) can be increased this way. This measure can be converted to oscillating buildings without problems, if one can ensure that the aerodynamic forces induced by the additional surfaces are forcibly anti-phase to those inducing the vibration.

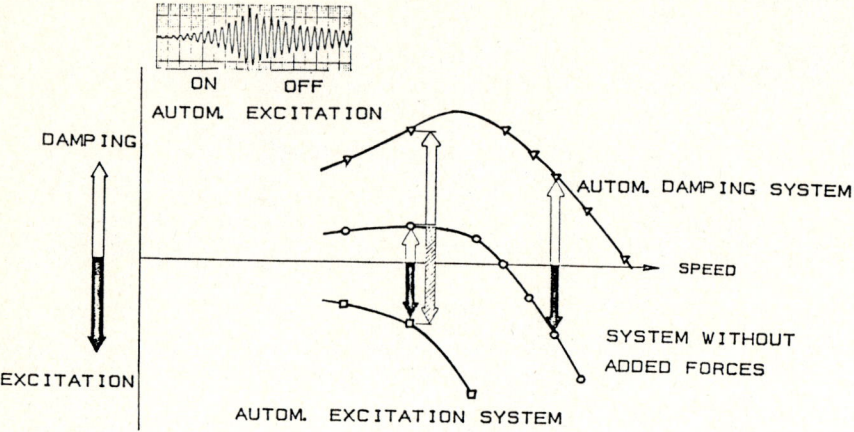

Figure 2 - Influence of the Aerodynamic Control

It should be very interesting to test active aerodynamic measures considering their use in civil engineering structures.

3.1.2 Active Mechanical Measures

The active control of building structures subjected to wind loads was discussed by Smanchai Sae-Ung, Yao, Yang, Giannopoulos et al, [13], [14].

Figure 3 shows the principle of an active control device to prevent wind-induced motions of tall buildings. The response of the structure is sensed by the accelerometer (for one direction). The signal from this is amplified and used to operate the hydraulic actuator, in turn, balances out the aerodynamic force by moving the mass m_2 (floating mounted) of the auxiliary system.

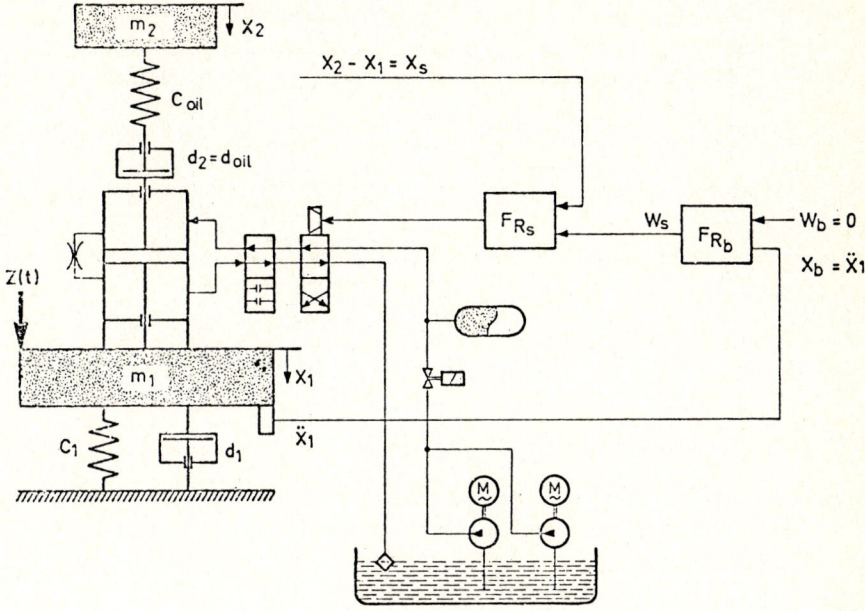

Figure 3 - Active Control Device to Prevent Wind-Induced Vibrations

The acceleration $\ddot{X}_1$, as a result of the random-like disturbation force $Z(t)$ is measured on the building and feed back as a controlled condition. $\ddot{X}_1$ is a unit for the well-being of the people in the skyscraper. The actual dimension X_b is compared with the normal dimension W_b and the deviation reaches the guidance regulator and returns to the auxiliary circuit.

Theoretically the building is reduced to a generalized system for a certain frequency (reduced mass m, reduced spring c, reduced damping d). These values are reduced to the top of the building. The spring-stiffness of the oil (c_{oil}) influences the frequency response of the complete system and is therefore especially important. The block-diagram of the system to be regulated is shown in Figure 4 and the block-diagram of the closed regulated circuit is shown in Figure 5.

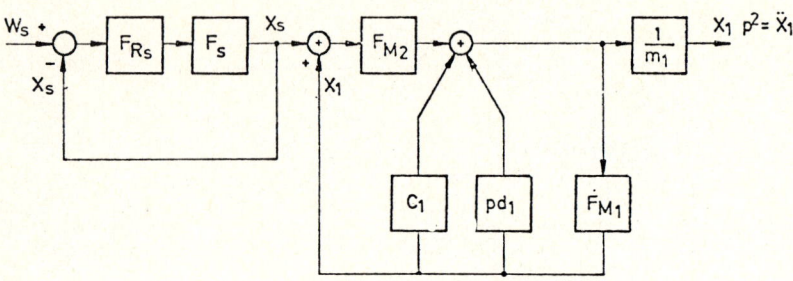

Figure 4 - Block-Diagram of the Active Mechanical Device

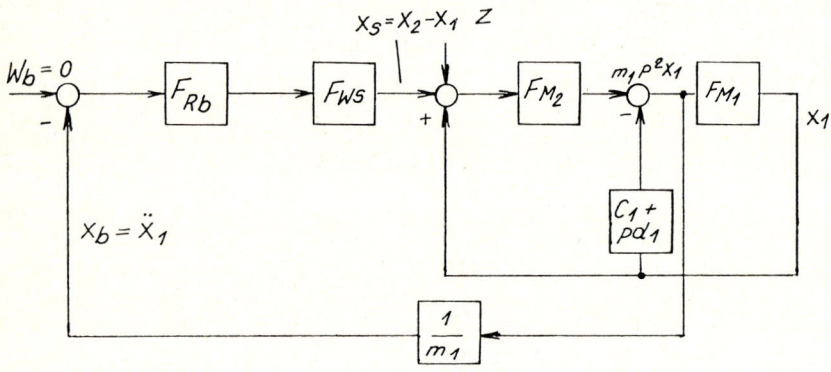

Figure 5 - Block-Diagram of the Closed Regulated Circuit

On the suggestion of the author, the CARL SCHENCK AG, Darmstadt, Germany [15], planned an active system of this kind and a system-example, basing on the Citicorp Building in New York City was tested. The result is shown in Figures 6 to 9. The effect of the additional system is obvious.

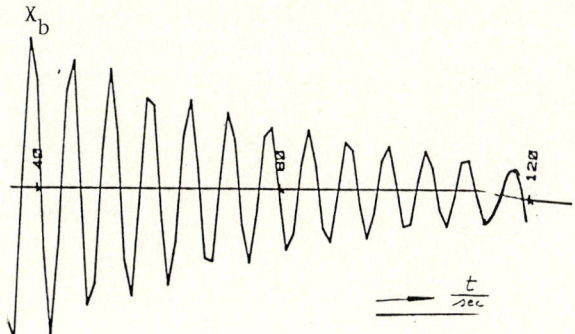

Figure 6 - Decay-Curve for Rectangular Impulse without Additional Damping

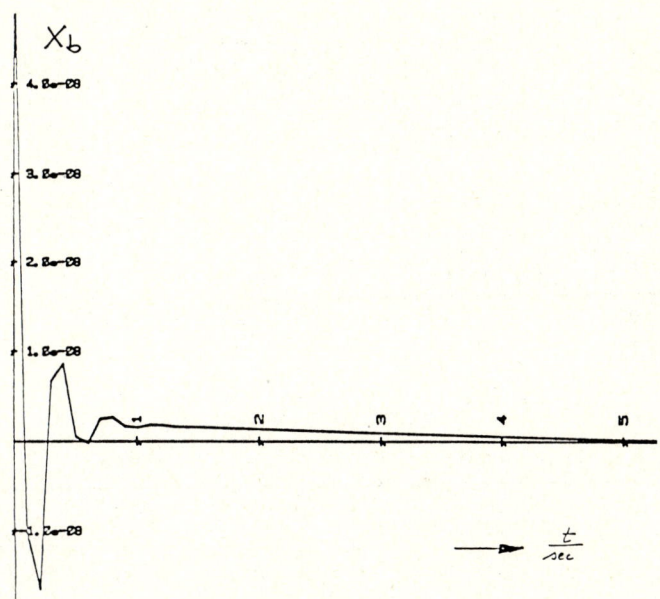

Figure 7 - Decay-Curve for Rectangular Impulse with Additional Damping

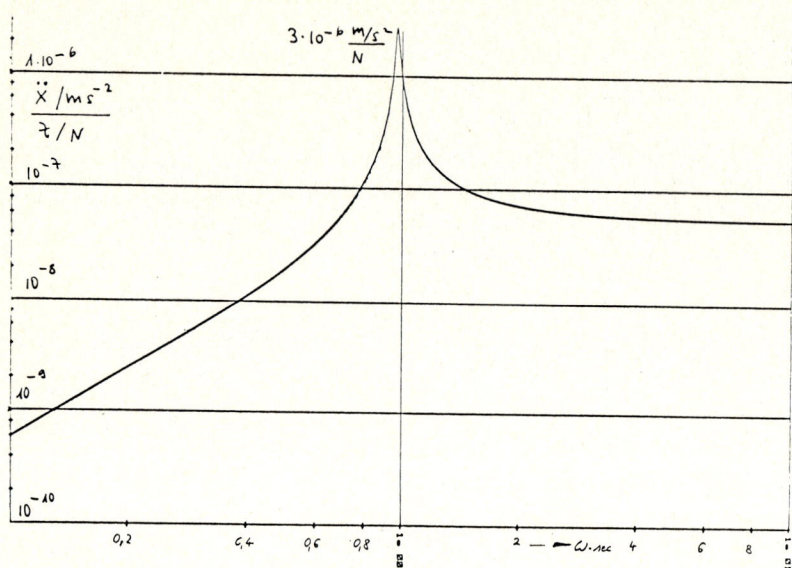

Figure 8 - Frequency-Response-Curve without Additional Damping

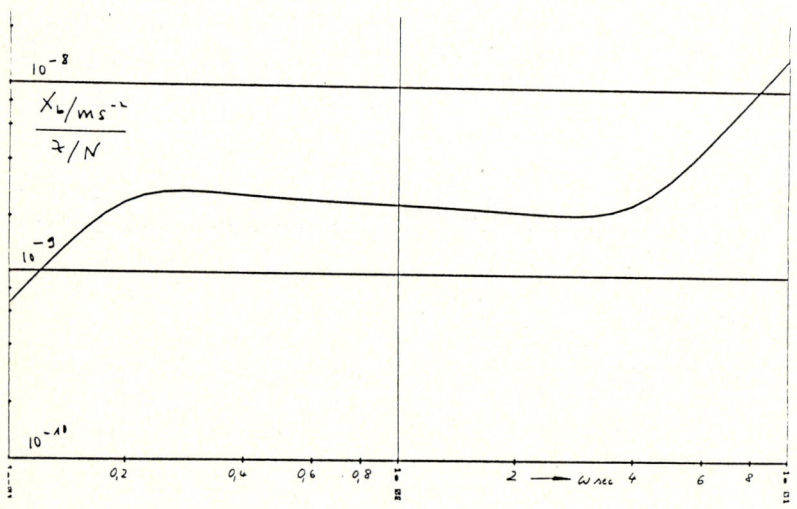

Figure 9 - Frequency-Response-Curve with Additional Damping

If the energy supply is disturbed additional valves keep up the passive effect of the damper-system. The advantage of this measure is that apart from the natural frequency the next frequencies of the building in wind-induced oscillations can be damped.

These measures could be used for buildings, where wind-induced oscillations must be reduced with regard to the well-being of the inhabitants and where constructive measures would be so much expenditure than the erection and maintenance of these additional dampers.

3.2 *Passive Measures*

A passive control of the wind-induced oscillations of structures is possible by certain measures basing on self-control. Here as well one must distinguish between aerodynamic and mechanical systems.

3.2.1 Passive Aerodynamic Measures

First I would like to mention the group of measures that reduce the extremely dangerous lateral oscillations of circular-cylindrical buildings (steel-chimneys, antennum-structures of communication towers, struts, cables, etc.). The goal is to prevent regular vortex separations respectively to shorten the correlation length. This measure consists of a row of disturbing bodies attached to the circumference of the cylinder. The efficiency of presently used systems is very varied.

H. Ruscheweyh gives a summary and a comparing view in [16]. It is quite obvious that the Scruton-Helical-Strakes are the most efficient [6]. The strakes must have the following features to obtain an optimum efficiency: triple start, width of 10% of the diameter and a pitch of five times the diameter.

The relative length must be 33% of the height. In such an optimum case the Lift-Coefficient only reaches values of .04. An increase in the drag coefficient is disadvantageous.

Vertical strips can suppress correlated delamination if their lateral extension is large enough. Ruscheweyh proved that the efficiency depends very much on the optimization (Figure 10). Increases must be mentioned (perforated metal-sheets). Shrouds are an optimum at a distance of .12 the diameter (from the cylinder wall) and with an opening-ratio of 30%. The openings are .07 D^2. The drag coefficient in relation to the outer diameter is .9. This measure is very expensive. Shrouds are of advantage for multiflow chimneys where the outer layer can be perforated.

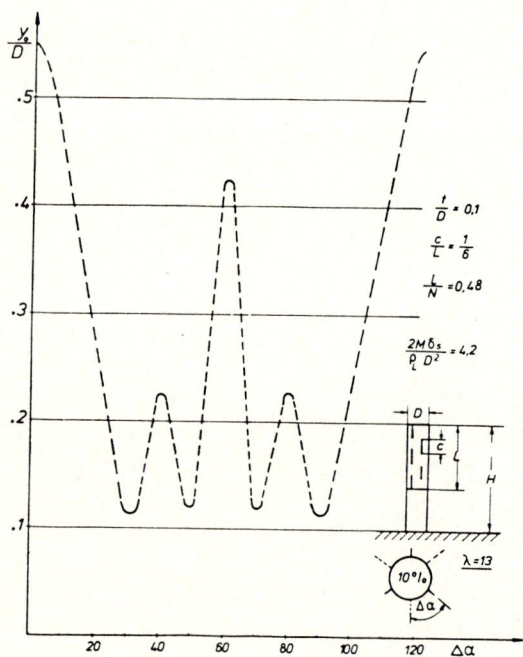

Figure 10 - *Vertical Strips for Aerodynamic Passive Damping*

The advantage of this case of aerodynamic measure mentioned above is that this system of additional construction is easy to realize. They can be constructed safely and kept without maintenance.

The increase in wind-drag is disadvantageous and this leads to a stronger load-bearing-structure respectively a stronger connecting structure. This immediately implies higher costs. Apart from that, measures of this kind are often of no use if they are fixed after the erection of the building as the structure was not built for the additional wind-forces.

The use of aerodynamic passive measures is not only common for circular-cylindrical bodies. It was discovered sometime ago that the oscillations of bridge-decks can be reduced efficiently by optimum coverings [17]. Figure 11 shows the influence of such systems (for example). Vortex formation is disturbed and the critical speed is increased. These facts have been taken into consideration for a long time with constructing bridge-profiles [10].

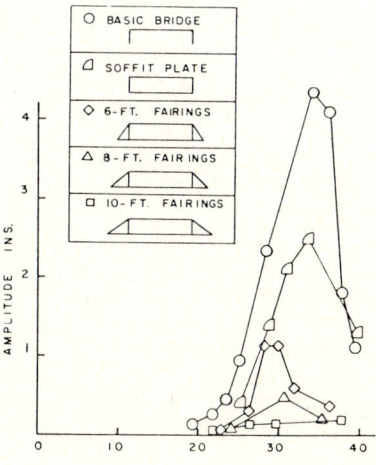

Velocity, M.P.H.
Figure 11 - Aerodynamic Measure on Bridge-Decks

3.2.2 Passive Mechanical Measures

First of all there are certain special cases in which there is a "fixed point" immediately next to the building that has to be damped [18]. Furthermore, cables are a classical measure to give additional damping. The nonlinearity of the cables can be used in addition [19]. Another case of fix-point damping can be obtained for bridges (e.g., under construction) by mounting a body so that it will reach below the water-surface [20].

Passive measures that reduce building-vibrations with one or more additional systems may be of special interest.

This method of tuned mass dampers (dynamic vibration absorbers) is based on Den Hartog's Theory in which an additional system constructed to an optimum reduces the resonance oscillations of an undamped (or extremely small damped) one-mass-oscillator.

Figure 12 shows this kind of system with the main-system (a tower-like building) generalized to a one-mass-oscillator. At optimum construction of the additional damper the effective damping depends on the mass-ratio $\mu = m_1/m_2$,

$$D_{eff} = .5(1 + 2/\mu)^{-0.5} .$$

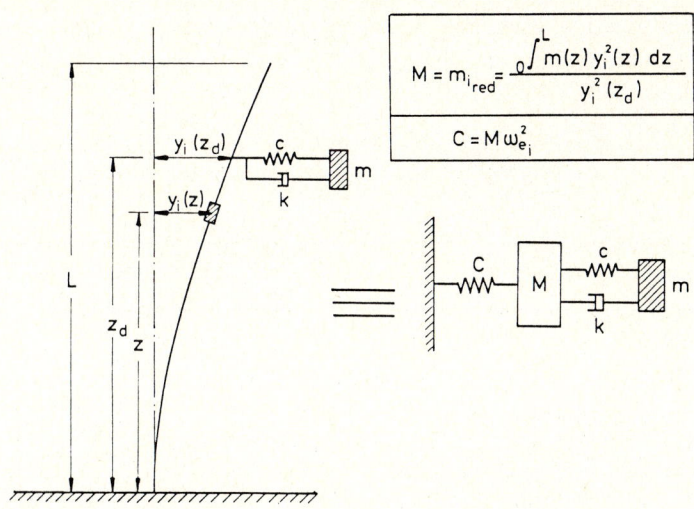

Figure 12 - *Generalized Main-System with Dynamic Vibration Absorber*

Harmonic induction is assumed here (vortex resonance). The optimization requires the following frequency-ratio

$$f_2/f_1 = 1/1 + \mu .$$

Apart from that the damping must be designed as follows,

$$D_2 = (3 \mu/8 (1 + \mu)^3)^{0.5} .$$

A mass ratio of only 0.5 results in an effective damping of 7.8% of the critical damping. The log. decrement is 0.5 and the resonance-magnification factor is only 6.4.

Seeing as the main-system's reduced mass for the natural frequency is only 20% of the complete mass, the mass of the additional system is only 1% of the total mass of the building to be damped. This is often called the "one-percent-damper".

The resulting damping is many times larger than the value of the damping used for buildings with small damping (see Table 1).

Table 1 - *Damping and Frequency Characteristics of Typical Structures (measured by means of autocorrelation-function)*

Structure	Material	Sizes	Nat. Frequ. c.p.s.	Log. Decr.
Cables of a Cable-Stayed Bridge	Steel	100 m 50 mm ⌀	2.5 (5th)	0.005
Chimney-Stack Unlined	Steel, Welded	140 m 6 m ⌀	0.5	0.01
Chimney-Stack Unlined	Steel, Welded	40 m 0.6-1.3 m ⌀	0.73	0.01
Antennum-Carrier	Glasfiber-reinforced	50 m 0.2-1.2 m ⌀	0.53	0.01
Tall Building	Concrete	150 m	0.16	0.04
Chimney-Stack Lined	Steel, Welded	140 m 6 m ⌀	0.28	0.1
Cable-Stayed Bridge	Steel, Welded	350 m 16 m (1. vertical)	0.52	0.075

That the relative-movement between the main-system and the damping-mass depends also on the mass-ratio that must be taken into consideration. For a mass-ratio of 0.05 the relative movement is 36-times the size of the static amplitude under the influence of the inducing forces (without dynamic magnification).

The additional system reacts more or less sensible to derivations from optimum construction. The derivations from the optimum damping are less critical (20% result in 10% deterioration. If the optimum frequency tuning is missed by 5% the effective damping decreases by 27%.

For stochastic vibration excitation the additional system is less efficient. The effective damping is 40% less for white noise excitation than for harmonic induction.

Figure 13 shows a classical dynamic vibration absorber (system Reutlinger, Germany) for circular-cylindrical buildings. A combination of spring and damping elements is shown in Figure 14. This kind of system was developed by the author in cooperation with the KA-BE-Company, Germany. The prototype is being tested at the present time. It is of great advantage that this damping system required no maintenance.

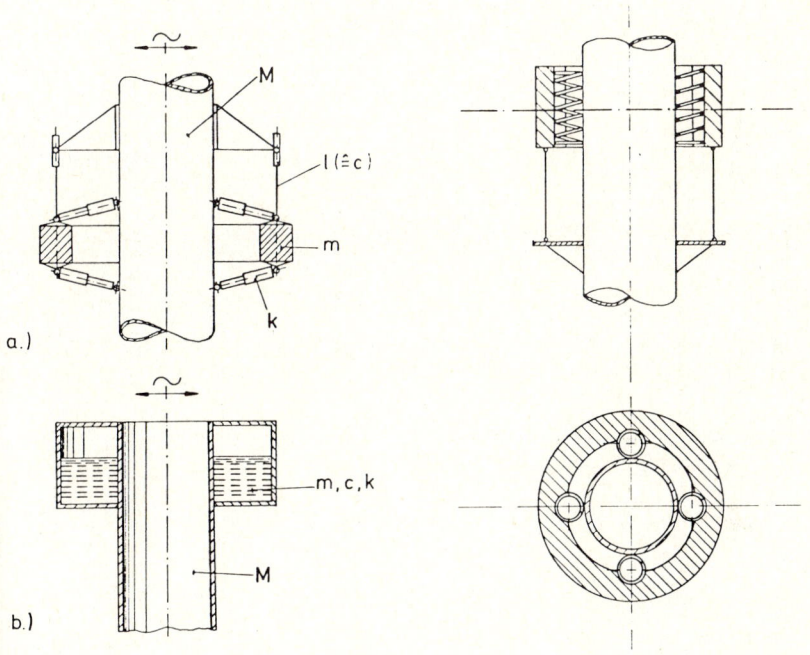

Figure 13 - Dynamic Vibration-Absorber

Figure 14 - Dynamic Vibration Absorber (with combined elements)

Dynamic vibration absorbers have now been, used successfully for some time to reduce wind-induced vibrations of structures. I would like to show some examples now. Figures 15 and 16 show a cable-stayed bridge with two pylons that oscillated violently at the critical wind-velocity.

$f_x = 1.93$ cps $\qquad f_y = 1.1$ cps
$\delta = 0.02$ $\qquad a_x = \pm 40$ cm
$a_y = \pm 20$ cm

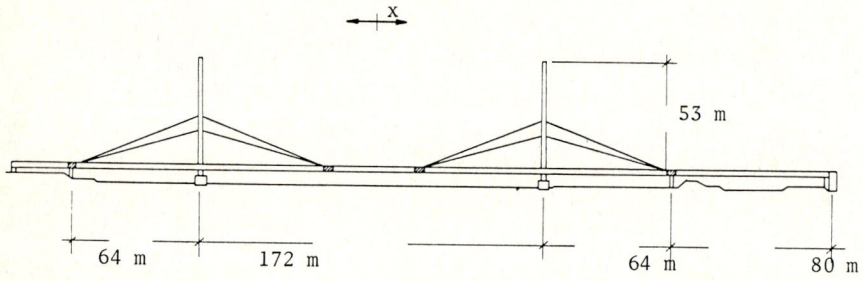

Figure 15 - *Cable Stayed Bridge*

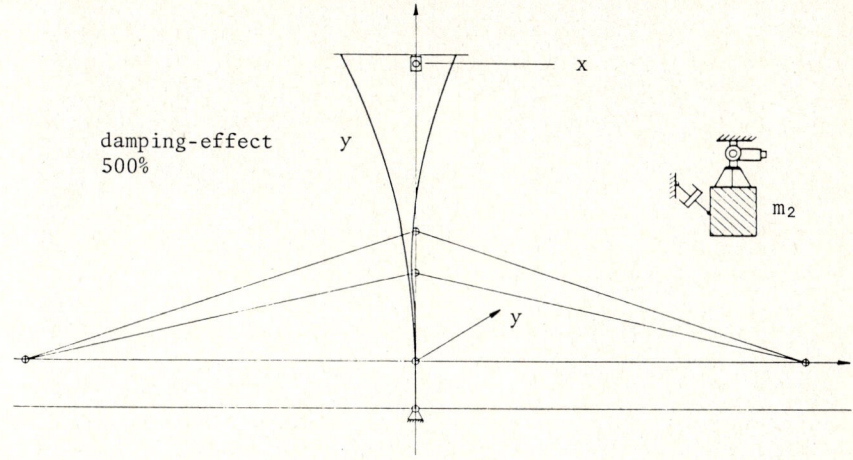

Figure 16 - Fundamental-Vibration Mode in x and y of the Pylons
$m_1 = 7\%$ of $93 \cdot 10^3$ kg
$m_2 = 490$ kg
$m_2/m_1 = 0.076$

With using the dynamic vibration absorbers shown in Figure 17 the oscillations transferred to the bridge-deck could be efficiently suppressed. The damping mass is fixed by means of two torsional springs connected in series and shows the natural frequency in both x- and y-direction.

(a)

Figure 17 - Details of the TMD for a Cable - Stayed Bridge
(a) Torsional Springs
(b) Complete System

(continued)

(continued)

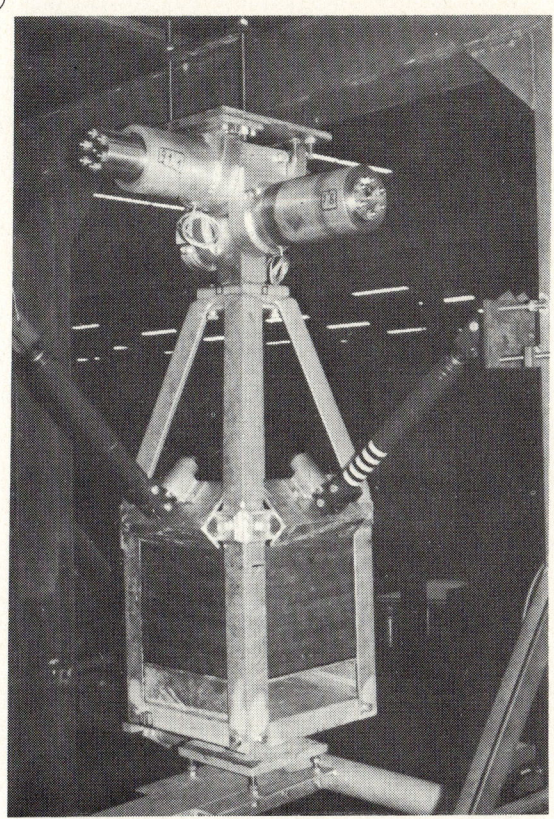

(b)

Figure 17 - Details of the TMD for a Cable - Stayed Bridge
  (a) Torsional Springs
  (b) Complete System

Figure 18 shows a telecommunication-tower with an antennum-carrier (glasfiber reinforced plastic). Considering the suppression of the third eigenvalue a pendulum-type-absorber was attached to the top of the carrier. The damping was obtained by means of shock-absorbers. It is noticeable that the oscillating mass of the tower influences the calculation of the absorber considerably. The generalized mass of the third vibration mode was 60% of the antennum-carrier-mass. In addition it must be taken in consideration that to be precise the damper system cannot be designed with the Den Hartog's Theory in the case of higher frequencies - this is what occurred here. There is nonproportional damping, because the additional damping system is fixed pointlike to the continuum (main-system) [21]. In this particular case the damping matrice is not diagonal and the equations of motion are coupled by the damping forces. The calculation for the optimum design requires much more expenditure. For the fundamental mode Den Hartog's Theory can be used to obtain satisfied results. In the example mentioned above the third mode is the quasi-fundamental mode with regard to the antennum-carrier. For other buildings (e.g., high circular-cylindrical buildings with multiple cable-staying) this can be completely different [22].

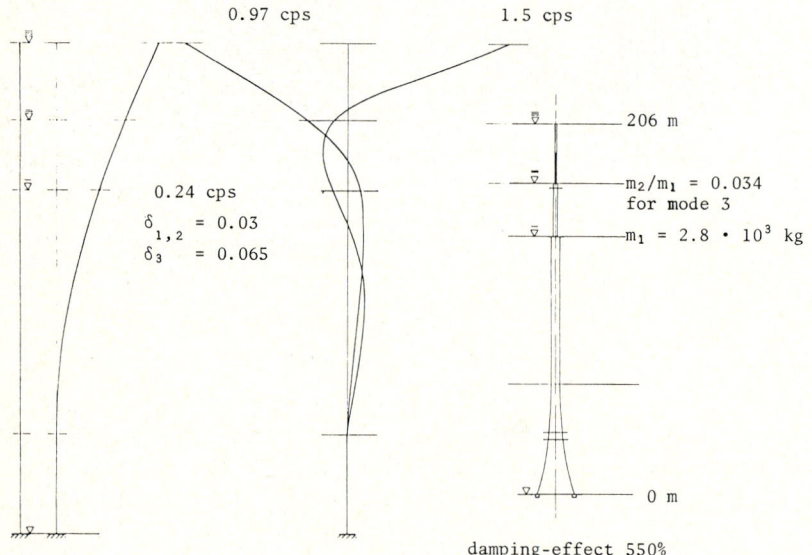

Figure 18 - Telecommunication-Tower with Antennum-Carrier (glasfiber reinforced plastic)

The advantage of using passive mechanical measures it that the calculations require less expenditure. There is no increase in drag caused by the wind. Also this system as an additional measure does not cause any problems. The fact that the dampers require maintenance is disadvantageous (therefore the use of combined elements should be attempted).

There must be a guarantee that the spring-elements do not change considerably during their life-time (this can happen with rubber elements for example).

Passive damping systems are increasingly being used for smaller buildings to obtain stability in an economical way. For larger buildings they are disadvantageous compared with active measures, because one system can only influence one frequency.

4. FINAL REMARKS

It was shown from the present point of view which possibilities there are to control wind-induced oscillations of buildings using damping-devices.

The advantages and disadvantages of the various active and passive possibilities were mentioned, noting that active measures have not found practical use in building constructions and that it seems promising to enter further examinations considering the growing importance of "Control of Structures".

REFERENCES

[1] "Tuned Mass Dampers Steady Sway of Skyscrapers in Wind", *Engineering New Record*, August, 1977, pp. 28-29.

[2] HIRSCH, G., RUSCHEWEYH, H. and ZUTT, H., "Failure of a 140 m High Steel Chimney Caused by Wind-Excited Oscillations Transverse to the Wind Direction", *Proc. Int. Symp. Steel Plated Structures*, London, Dowling-Harding-Frieze, 1977.

[3]   PETERSEN, C., "Tilgung der Querschwingungen Zylindrischer Bauwerke durch Mechanische Dämpfer unter besonderer Berücksichtigung eines Schadensfalles an einem Kraftwerks-Kamin", *HdT-Veröffentlichungen*, Vol. 347, 1976, S.93-101.

[4]   NOVAK, M., "The Wind-Induced Lateral Vibration of Guyed Masts with Circular Cross-Section", *Proc. Symp. on Tower-Shaped Steel and Reinforced Concrete Structures, IASS*, Bratislava, June, 1966, pp. 169-188.

[5]   VICKERY, B.J. and WATKINS, R.D., "Flow Induced Vibrations of Cylindrical Structures", reprinted from *Proc. First Australia Conf. on Hydraulic and Fluid Mechanics, 1962*, Pergamon-Press, London, 1963.

[6]   SCRUTON, C. and FLINT, A.R., "Wind-Excited Oscillations of Structures", *The Inst. of Civil Engineers*, Paper No. 6758, London, 1964, pp. 673-702.

[7]   HIRSCH, G. and RUSCHEWEYH, H., "Newer Investigations of Nonsteady Wind Loadings and the Dynamic Response of Tall Buildings and Other Structures", *Proc. Third Int. Conf. on Wind Effects on Buildings and Structures*, Tokyo, 1971, pp. 811-823.

[8]   RUSCHEWEYH, H., "Winderregte Schwingungen Zweier Engstehender Kamine", *Proc. Third Coll. on Industrial Aerodynamics, II*, Aachen, 1978, FH-Aachen, Fluid-Laboratorium.

[9]   GAUBE, E., "Schwingungen im Chemie-Betrieb", *Chem.-Ing. Techn.*, Vol. 5, 1979, No. 1, S. 14-22.

[10] KLÖPPEL, K. and THIELE, F., "Modellversuche im Windkanal zur Bemessung von Brücken gegen die Gefahr Winderregter Schwingunge *Der Stahlbau*, 1967, S. 353-365.

[11] van der PUT, J., "Rigidity of Structures against Aerodynamic Forces", Assoc. Intern. des Ports et Charpents, *Mémoires*, 36-1, Zurich, 1976.

[12] HAIDL, G. and STEININGER, M., "Excitation and Analysis Technique for Flight Flutter Tests", *AGARD Report No. 672*.

[13] SAE-UNG, S. and YAO, J.T.P., *"Active Control of Building Structures"*, Civil Engineering, Purdue University, West Lafayette, Indiana, 1975.

[14] YANG, J.N. and GIANNOPOULOS, F., "Dynamic Analysis and Active Control of Two Cable-Stayed Bridges", *Tech. Report No. 1 and 2* School of Engineering and Applied Science, The George Washington University, 1978.

[15] JUNG, G., "Geregelter Schwingungsdampfer für ein Hochhaus", unveröff. Bericht der Carl Schenck AG, Darmstadt, 1978.

[16] RUSCHEWEYH, H., "Massnahmen zur Verhinderung gefährlicher Schwingungen von Kaminen", *Koll. Aer. Probleme 1978*, Curt-Risch-Institut, TU Hannover.

[17] WARDLAW, R.J., "Some Approaches to Improving the Aerodynamic Stability of Bridges and Road Decks", *Proc. Third Conf. on Wind Effects*, Tokyo, pp. 931-940.

[18] HIRSCH, G. and RUSCHEWEYH, H., "Full-Scale Measurements on Steel Chimney Stacks", *J. of Ind. Aerodyn.*, Vol. 1, 1975/76, pp. 341-347.

[19] KOLLAR, L., "Dämpfung der Schwingamplituden Seilverspannter Systeme infolge des nichtlinearen Verhaltens der Seile", *Acta Technica Academiae Scientiarum Hungaricae*, Tomus 75, 1973, S. 203-217.

[20] HIRSCH, G. and RUSCHEWEYH, H., "Vibration Measurement on a Cable-Stayed Bridge under Construction", *J. of Ind. Aerodyn.*, Vol. 1, 1975/76, pp. 297-300.

[21] HIRSCH, G. and WAHLE, M., "Design Criteria for Dynamic Vibration Absorbers with regard to the Nonproportional Damping", *Proc. Third U.S. National Conf. on Wind Engineering*, Gainesville, Florida, 1978, IV-17.

[22] PACHT, H., "Schwingungsuntersuchungen an stählernen Turmbauwerken wie Maste und Schornsteine aus der Sicht der Praxis", *VDI-Berichte Nr. 221*, 1974, S. 127-133.

STRUCTURAL CONTROL, H.H.E. Leipholz (ed.)
North-Holland Publishing Company & SM Publications
© IUTAM, 1980

FINITE ELEMENT ANALYSIS OF CONTACT PROBLEMS BASED ON THE UNILATERAL CONSTRAINTS FORMULATION

Nguyen Dang Hung and Gery de Saxcé

Department of Structural Mechanics
and Stability of Constructions
University of Liège, Belgium

G.M.L. Gladwell

Department of Civil Engineering
University of Waterloo
Waterloo, Ontario, Canada

1. INTRODUCTION

Even when the deformation is small, the contact of elastic bodies poses a nonlinear problem because of the complication introduced by the presence of the unknown surface contact. In the literature one may find two distinct numerical approaches to this problem: the iterative method and the direct method. The first method consists of calculating the increment of loading and verifying the contact condition at each step. The following authors may be cited among the users of this procedure: Chan and Tuba [1], Gaertner [2], Zienkiewicz and Francavilla [3] and Fredriksson [4].

The second method consists of reducing the elastic contact problem to a mathematical programming problem. In this way, Feng and Huang [5] have examined the contact problem of an inflated plane membrane. The variational form of the contact problem is studied by Fremond [6] who has presented some numerical examples of contact between elastic bodies and rigid foundations. Panagiotopoulos [7] has generalized this approach to an inelastic foundation and presented some dual forms of the variational inequalities for the contact problem.

The present work belongs to the second type of formulation where an appropriate linearization of the contact condition is adopted. It is assumed that no friction exists between the solid bodies so that the contact condition may be expressed uniquely in term of displacements. It is supposed also that the deformation is small and the material obeys the linear elastic constitutive equations. The numerical examples concern only plane strain or plane stress problems. The generalization of the formulation to three dimensional bodies would present no major difficulties apart from a considerable increase in the number of variables. This paper may be divided into two sections. In the first section, the contact between an elastic body and a rigid foundation is considered. It is shown that there is good agreement between the numerical results obtained using the present finite element algorithm and the super-element technique.

In the second section, the formulation is generalized to the two elastic bodies. We represent the space separating the contacting surfaces as a fictitious perfectly locking material.

This concept is accompanied by a new and general non-interpenetration condition. In the particular case of rigid-elastic contact under small deformations, the general non-interpenetration condition is reduced to the usual constraint condition.

2. CONTACT OF AN ELASTIC BODY AND A RIGID FOUNDATION

2.1 *Kinematically Admissible Displacements for Contact Problems*

Let Ω be the rigid foundation, $\partial\Omega$ its boundary. Let V be the elastic body; its boundary is composed of three portions: Γ_σ where surface tractions $\bar{t}_i$ are prescribed, Γ_u where displacements $\bar{u}_i$ are prescribed and Γ_c where contact may happen (see Figure 1).

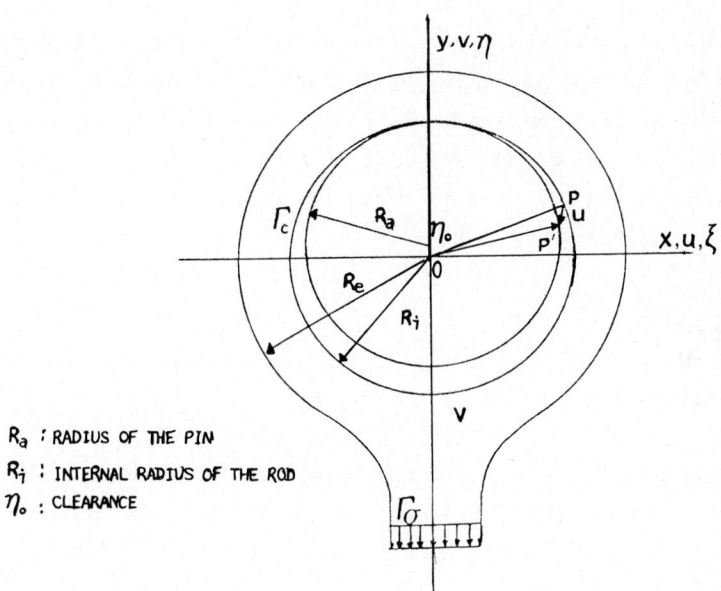

R_a : RADIUS OF THE PIN
R_i : INTERNAL RADIUS OF THE ROD
η_o : CLEARANCE

Figure 1 - *Contact Surface Γ_c and External Surface Γ_σ Contact Between a Pin and a Piston Rod*

The initial configuration of the body V is determined by a position vector $\vec{OP} = \vec{x}$. The components x_i are the Lagrangian coordinates of the material point. Let $\vec{u}(\vec{x})$ be the displacement vector of the point P under loading. The new position P' of P after deformation is described by the new coordinates

$$\vec{OP'} = \vec{\xi} = \vec{x} + \vec{u} , \qquad (2.1)$$

where the components ξ_i of $\vec{\xi}$, are Eulerian coordinates of the material point P'. We assume that the surface of the foundation

$\delta\Omega$ is regular, so that it may be defined by the following equation

$$\Pi(\vec{\xi}) = 0 . \tag{2.2}$$

In practice, equation (2.2) defines only the possible contacting surface but not necessary the whole surface $\partial\Omega$. By convention, we suppose that the region where $\Pi(\vec{\xi})$ is negative belongs to the foundation Ω. The displacement field $\vec{u}$ is kinematically admissible for the contact problem if it satisfies the following non-interpenetration condition

$$\Pi(\vec{x} + \vec{u}) \geq 0 . \tag{2.3}$$

2.2 *Displacement Variational Principle*

The classical total potential energy of an elastic body V subjected to imposed body forces f_i and prescribed traction $\bar{t}_i$ on Γ_σ is:

$$\Phi(u_i) = \int U(\varepsilon_{ij}) dV - \int f_i u_i dV - \int \bar{t}_i u_i d\Gamma_\sigma , \tag{2.4}$$

where U is the strain energy density.
ε_{ij} is a kinematically admissible strain field such that

$$\varepsilon_{ij} = \frac{1}{2} (D_i u_j + D_j u_i) . \tag{2.5}$$

The contact problem [6] is reduced to a minimization of the functional (2.4) subject to the constraint (2.3) with respect to an arbitrary kinematically admissible displacement field u_i.

The displacements must therefore satisfy

$$u_i \in H_K ,$$

$$H_K = \{u_i | u_i = \bar{u}_i \text{ on } \Gamma_u \text{ and } \Pi(x_i + u_i) \geq 0 \text{ on } \Gamma_c\} . \tag{2.6}$$

Introducing a slack variable ω we may replace the inequality (2.3) by the following equality:

$$\Pi(x_i + U_i) - \omega^2 = 0 . \tag{2.7}$$

Following Friedrich, we introduce a Lagrange parameter λ and a dislocation potential

$$\phi(u_i, \lambda, \omega) = - \int \lambda [\Pi(x_i + u_i) - \omega^2] d\Gamma_c . \tag{2.8}$$

If a modified functional Φ^* is adopted such that:

$$\Phi^*(u_i, \lambda, \omega) = \Phi(u_i) + \phi(u_i, \lambda, \omega) , \tag{2.9}$$

the variational problem is

$$\delta\Phi^*(u_i, \lambda, \omega) = 0 , \tag{2.10}$$

with $u_i \in H : \{u_i | u_i = \bar{u}_i \text{ on } \Gamma_u\}$.

It is not difficult to see that the Euler-Lagrange equations of the problem (2.10) are:

internal equilibrium:	$D_j \sigma_{ij} + \bar{f}_i = 0 ,$	(2.11)
surface equilibrium on Γ_σ:	$n_j \sigma_{ij} = \bar{t}_i ,$	(2.12)
contact equilibrium on Γ_c:	$n_j \sigma_{ij} = \lambda D_i \Pi ,$	(2.13)
displacement constraint on Γ_c:	$\Pi(x_i + u_i) - \omega^2 = 0 ,$	(2.14)
complementarity condition on Γ_c:	$\lambda \omega = 0 .$	(2.15)

It appears from (2.15) that if $\omega \neq 0$, then $\lambda = 0$. Equation (2.14) shows that $\Pi > 0$, so that there is no contact. Equation (2.13) shows that there is no contact pressure. If $\omega = 0$, then $\lambda > 0$ and there is contact. So the complementarity condition (2.15) restores the Kuhn-Tucker optimality conditions. Figure 2 illustrates the situation discussed above. An interpretation of the Lagrange parameter λ may be performed in the following way. There is no friction, so the surface tractions must be normal to the contacting surface Γ_c:

$$\vec{t} = \lambda \nabla \Pi .$$

Consequently

$$\lambda = \frac{||\vec{t}||}{||\nabla \Pi||} . \qquad (2.16)$$

If we had chosen the normalized constraint function

$$\Pi_0 = \Pi / ||\nabla \Pi|| , \qquad (2.17)$$

instead of Π, then λ would be identified exactly as the contact pressure.

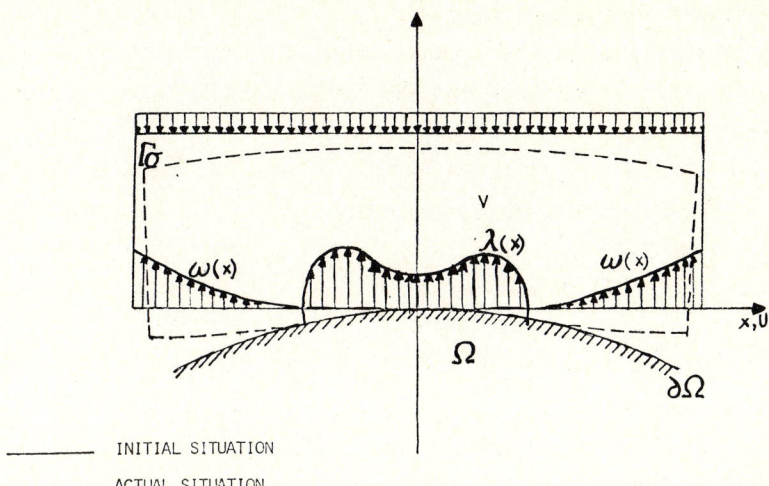

INITIAL SITUATION
ACTUAL SITUATION

Figure 2 - Geometric Situations Before and After Contact of a Solid Body and a Rigid Foundation. Distribution of λ(x) and ω(x)

2.3 *Finite Element Formulation*

Classical discretization using the finite element and compatible displacements leads to the following total potential energy of the whole structure:

$$\Phi = \frac{1}{2} q^T K q - g^T q ,\qquad(2.18)$$

where: q is the nodal displacement vector,
         g is the generalized force vector,
         K is the stiffness matrix.

The finite element discretization of the contact potential

$$\Phi_c = \int \lambda \, \Pi(x_i + u_i) d\Gamma_c ,\qquad(2.19)$$

may be realized in the following way.

Assuming the displacements are small, we may retain only the first order terms of the Taylor expansion of the function $\Pi(x_i+u_i)$:

$$\Pi(x_i+u_i) \simeq \Pi(x_i) + D_k\Pi(x_i)u_k + O(u_i u_k) ,$$

so that inequality (2.3) becomes:

$$D_k\Pi(x_i)u_k \geq - \Pi(x_i) . \tag{2.20}$$

Now, let us consider an arbitrary finite element V_e with an interface Γ_e belonging to Γ_c. Let s be the current coordinate defined on Γ_e, and q_{re} a system of nodal displacements chosen on Γ_e and $M_e(s)$ a shape function for $u(s)$:

$$u(s) = M_e(s)q_{re} . \tag{2.21}$$

Let $N_e(s)$ be another system of shape functions corresponding to the nodal system λ_e of Lagrange's parameter λ:

$$\lambda(s) = \lambda_e^T N_e(s) . \tag{2.22}$$

According to (2.20), (2.21) and (2.22) the contact potential (2.19) of element V_e may be reduced to the matrix form:

$$\Phi_{ce} = \lambda_e^T (A_e q_{re} + h_e) , \tag{2.23}$$

with

$$h_e = \int N_e \Pi \, d\Gamma_{ce} , \tag{2.24}$$

$$A_e = \int N_e (\nabla\Pi)^T M_e d\Gamma_{ce} . \tag{2.25}$$

The assembling of the contact elements may be performed by using Boolean matrices L_e, G_e

$$q_{re} = L_e q_r , \qquad (2.26)$$

$$\lambda_e = G_e \lambda ,$$

$$h_e = G_e^T h , \qquad (2.27)$$

and writing the total contact potential as the sum of the contact potential of each element:

$$\Phi_c = \sum_e \Phi_{ce} = \lambda^T (Aq_r + h) , \qquad (2.28)$$

where,

$$A = G_e^T A_e L_e , \qquad (2.29)$$

is the constraint matrix of the contact problem. Taking the variation of (2.28) with respect to λ in problem (2.10) we find the following linear constraint in terms of the nodal displacements defined on Γ_c:

$$Aq_r \geq - h . \qquad (2.30)$$

It should be emphasized that the constraint (2.3) on the convex function Π has been linearized to the form (2.20), i.e., to the finite element form (2.30). This is fundamental to the procedure. The convexity of Π ensures the existence of a solution.

2.4 *Superelement Technique*

Let

$$g_0 = A^T \lambda , \qquad (2.31)$$

be the generalized contact force vector. The constraint affects only the nodal displacement defined on Γ_c. In order to reduce the size of the problem before optimization, the superelement technique is used. Application of the principle of virtual work for the entire structure:

$$\delta \left[\frac{1}{2} q^T K q - g^T g - g_0^T q_r - \lambda^T h \right] = 0 , \qquad (2.32)$$

gives the nodal equilibrium equation:

$$\begin{aligned} K_{cc} q_c + K_{cr} q_r &= g_c , \\ K_{rc} q_c + K_{rr} q_r &= g_r + g_0 , \end{aligned} \qquad (2.33)$$

where q_c are the condensed displacements which must be eliminated before the optimization step. Thus

$$q_c = K_{cc}^{-1} [g_c - K_{cr} q_r] . \qquad (2.34)$$

Substituting this in (2.33) one obtains:

$$\overline{K}_{rr} q_r = \overline{g}_r + g_0 , \qquad (2.35)$$

with

$$\overline{K}_{rr} = K_{rr} - K_{rc} K_{cc}^{-1} K_{cr} , \qquad (2.36)$$

$$\overline{g}_r = g_r - K_{rc} K_{cc}^{-1} g_c . \qquad (2.37)$$

Now the size of the problem (2.32) is considerably reduced:

$$\delta \left[\frac{1}{2} q_r^T \overline{K}_{rr} q_r - \overline{g}_r^T q_r - \lambda^T (A q_r + h) \right] = 0 , \qquad (2.38)$$

which is equivalent to the following quadratic programming problem:

Minimize $\frac{1}{2} q_r^T \bar{K}_{rr} q_r - g_r^T q_r$,
q_r

Subjected to: $A\, q_r \geq - h$. (2.39)

Once the optimal solution q_r and the associated Lagrange parameters λ are known, the internal displacements and stresses of the whole structure are obtained from (2.34) by back-substitution, and the contact force vector is found from (2.31).

2.5 *Example*

It seems useful to illustrate the above finite element formulation of the contact potential Φ_c by taking a simple example shown by Figure 3. If V_e is a linear finite element having an interface 1-2 belonging to Γ_c, one has the following interpolation matrix M_e and displacement q_{re} of formula (2.21).

$$M = \begin{bmatrix} 1-\eta & 0 & \eta & 0 \\ 0 & 1-\eta & 0 & \eta \end{bmatrix} \qquad \eta = s/\ell$$

$$q_{re}^T = [q_{11}, q_{12}, q_{21}, q_{22}] ,$$

where s is the current coordinate on the interface 1-2 and l is its length. If linear interpolation is taken for $\lambda(s)$ in (2.22) one has:

$$\lambda_e^T = [\lambda_1, \lambda_2] ,$$

$$N_e(s) = [1-\eta,\ \eta] .$$

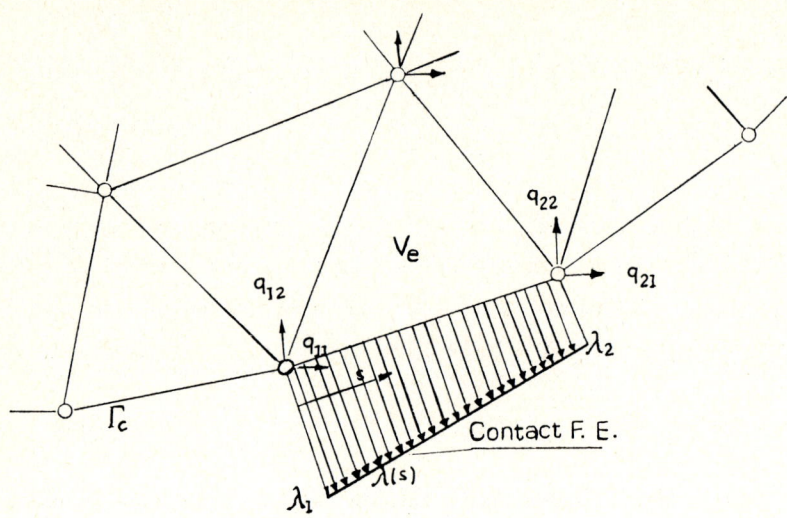

Figure 3 - Elastic Finite Element and Interface Contact Finite Element

In these conditions the constraint matrix A_e defined by (2.25) takes the following form

$$A_e = \ell \begin{bmatrix} A_{11} & A_{12} & A_{13} & A_{14} \\ A_{21} & A_{22} & A_{23} & A_{24} \end{bmatrix},$$

with

$A_{11} = \int_0^1 (1-\eta)^2 D_1 \Pi(\eta) d\eta$, $\qquad A_{12} = \int_0^1 (1-\eta)^2 D_2 \Pi(\eta) d\eta$,

$A_{13} = \int_0^1 (1-\eta)\eta D_1 \Pi(\eta) d\eta$, $\qquad A_{14} = \int_0^1 (1-\eta)\eta D_2 \Pi(\eta) d\eta$,

$A_{21} = \int_0^1 (1-\eta)\eta D_1 \Pi(\eta) d\eta$, $\qquad A_{22} = \int_0^1 (1-\eta)\eta D_2 \Pi(\eta) d\eta$,

$A_{23} = \int_0^1 \eta^2 D_1 \Pi(\eta) d\eta$, $\qquad A_{24} = \int_0^1 \eta^2 D_2 \Pi(\eta) d\eta$.

In the case of Figure 4 where the rigid foundation is the circle

$$\Pi(x_1,x_2) = x_1^2 + x_2^2 + 2x_2 R ,$$

one has

$$D_1\Pi = 2 \ell(x_{11}/\ell + \eta) ; \quad D_2\Pi = 2R .$$

Hence:

$$A_e = \frac{\ell^2}{3} \begin{bmatrix} 2(x_{11}/\ell + \frac{1}{4}) & 2R/\ell & x_{11}/\ell + \frac{1}{2} & R/\ell \\ (x_{11}/\ell + \frac{1}{2}) & R/\ell & 2(x_{11}/\ell + \frac{3}{2}) & 2R/\ell \end{bmatrix} ,$$

and the matrix h_e defined by (2.24) is

$$h_e^T = \ell^3 \left[\frac{1}{2}\left(\frac{x_{11}}{\ell}\right)^2 + \frac{1}{3}\frac{x_{11}}{\ell} + \frac{1}{2} \quad \frac{1}{2}\left(\frac{x_{11}}{\ell}\right)^2 + \frac{x_{11}}{3\ell} + \frac{1}{4} \right] .$$

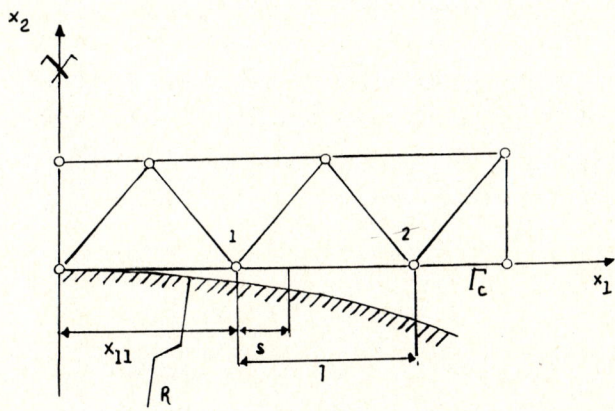

Figure 4 - *Example of Discretization in the Neighbourhood of the Contact Surface*

3. CONTACT OF TWO ELASTIC BODIES

3.1 *Formulation*

The major difficulty for the extension to two elastic bodies consists in the fact that not only the contact zone is unknown here, but the contact surface may deform during loading. We can take the boundary of one body as reference like the rigid foundation in the previous section but by doing so we lose the symmetry of the problem. We do not treat the two bodies equally. For these reasons, we propose a new and general formulation described as follows. Let V_A, V_B (Figure 5) be two elastic bodies in contact with Γ_{AC}, Γ_{BC} as the possible contacting boundaries and $\Gamma_{A\sigma}$, $\Gamma_{B\sigma}$ the portions of the boundaries where loading $\bar{t}_i$ is prescribed. Let $V = V_A \cup V_B$ be the total volume of the two elastic bodies and $\Gamma_\sigma = \Gamma_{A\sigma} \cup \Gamma_{B\sigma}$ the portion of its boundary where $\bar{t}_i$ are prescribed. Let V_C be the region adjacent to the contact surface. This region is bounded by $\Gamma_c = \Gamma_{AC} \cup \Gamma_{BC}$ and $\Gamma = \Gamma_1 \cup \Gamma_2$. It may be multiply connected if the V_A and V_B are initially in contact. The basic assumption of the formulation is that the displacement field existing in V may be extended into the fictitious material in the region V_C. Let x_i be the Lagrangian coordinates of the material point. Under loading the configuration changes:

$$x_i \to \xi_i = x_i + u_i , \qquad (3.1)$$

and V_A, V_B, V_C become respectively the deformed Z_A, Z_B, Z_C. The associated Jacobian transformation matrix of (3.1) is

$$J_{ij} = D_i \xi_j = \delta_{ij} + D_i u_j , \qquad (3.2)$$

with the so-called determinant, or Jacobian

$$J = \det [J_{ij}] . \qquad (3.3)$$

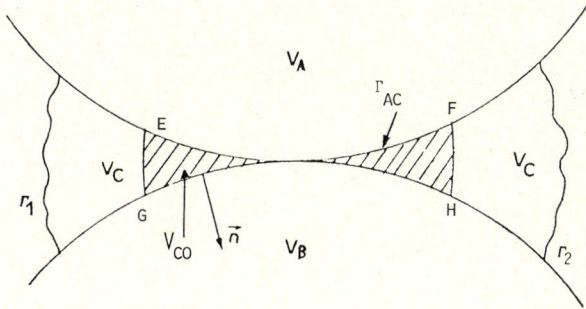

Figure 5 - *Contact of Two Elastic Bodies. Initial Geometric Situation*

The volume of the fictitious material is reduced to:

$$Z_C = \int_{V_C} d\,Z_C = \int_{V_C} J\,d\,V_C \ . \tag{3.4}$$

The narrow volume V_{Co} near the elastic bodies and limited by E, F, G, H is a region of high strain and tends to zero as the loading increases. It is evident that for a small element of volume in V_{Co} the contact between V_A and V_B is realized when

$$J = 0 \ . \tag{3.5}$$

In the remainder of the region V_C the condition of non-interpenetration holds, namely

$$J \geq 0 \ . \tag{3.6}$$

This is the form which the constraint (2.3) takes in this case. A displacement is called kinematically admissible for the contact problem it is satisfies (3.6).

The same formulation (2.9), (2.10) may be used here with the dislocation potential

$$\phi(u_i,\lambda,\omega) = - \int \lambda(J-\omega^2)d V_C , \qquad (3.6)$$

and the new contact potential

$$\Phi_C = \int \lambda J d V_C . \qquad (3.7)$$

Noting that

$$I_{ij} = \frac{DJ}{DJ_{ij}} , \qquad (3.8)$$

is the minor of element J_{ij} of the Jacobi matrix $[J_{ij}]$, one has

$$\delta\Phi_C = \int \lambda \frac{DJ}{DJ_{ij}} \delta J_{ij} d V_C ,$$

and because of (3.2)

$$\delta\Phi_C = \int \lambda I_{ij} D_j \delta u_i dV_C = \int D_j(\lambda I_{ij} \delta u_i)dV_C - \int D_j(\lambda I_{ij})\delta u_i dV_C , \qquad (3.9)$$

and according to the Gauss formulae

$$\delta\Phi_C = \int \lambda n_j I_{ij} \delta u_i d\overline{\Gamma} - \int D_j(\lambda I_{ij})\delta u_i dV_C ,$$

with

$$\overline{\Gamma} = \Gamma_c \cup \Gamma ,$$

the total boundary of V_C.

Taking the variation of the modified functional

$$\Phi^*(u_i,\lambda,\omega) = \Phi(u_i) + \phi(u_i,\lambda,\omega) , \qquad (2.9)$$

where (u_i) is defined by (2.4), we may find the following Euler-Lagrange equations:

$$\text{in } V \quad -- \quad D_j \sigma_{ij} + \bar{f}_i = 0 \ , \tag{3.10}$$

$$\text{on } \Gamma_\sigma \quad -- \quad n_j \sigma_{ij} = \bar{t}_i \ , \tag{3.11}$$

$$\text{on } \Gamma \quad -- \quad n_j \sigma_{ij} = 0 \ , \tag{3.12}$$

$$\text{on } \Gamma_C \quad -- \quad t_i = \lambda n_j I_{ij} \tag{3.13}$$

$$\text{in } V_C \quad -- \quad D_j (\lambda I_{ij}) = 0 \ . \tag{3.14}$$

3.2 Physical Interpretation

The three first equations are classical equilibrium equations. Focusing our attention on the last two we recognize in (3.13) the presence of the normal ν_i (Figure 6) of the strained surface Γ_C

$$\nu_i = n_j I_{ij} \ , \tag{3.15}$$

so that it may be written:

$$t_i = \lambda \nu_i \ . \tag{3.16}$$

It appears from this that the tractions are still normal to the strained surface. This is the consequence of the frictionless contact condition. Again, as in Section 3, λ may be identified with the contact pressure on the contact surface.

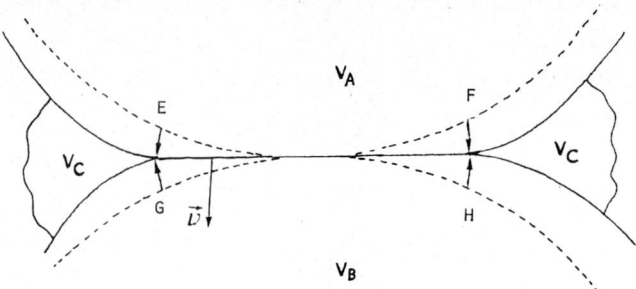

Figure 6 - Contact of Two Elastic Bodies Actual Geometric Situation

Besides, in the virtual work equation (3.9) the conjugate quantity of the strain $\alpha_{ij} = D_j u_i$ is the Piola stress tensor:

$$\sigma_{ij} = \lambda I_{ij} , \qquad (3.17)$$

therefore, (3.13) and (3.14) may be written respectively:

$$\text{external equilibrium on } \Gamma_C \quad -- \quad t_i = n_j \sigma_{ij} , \qquad (3.18)$$

$$\text{internal equilibrium in } V_C \quad -- \quad D_j \sigma_{ij} = 0 . \qquad (3.19)$$

It is very interesting to point out that relation (3.16), (3.17) is just the normality law of an ideal locking material defined by the potential J which constitutes a dual material of the rigid perfectly plastic material (8),

$$\sigma_{ij} = \lambda \frac{\partial J}{\partial \alpha_{ij}} \quad \begin{array}{l} \lambda = 0 \quad \text{if } J > 0 , \\ \lambda \neq 0 \quad \text{if } J = 0 . \end{array} \qquad (3.20)$$

3.3 *Relation Between the Constraint Function J and* Π

In the case of $V_B \equiv \Omega$, $\Gamma_{BC} \equiv \delta\Omega$, the body V_B is identified as the rigid foundation examined in Section 2. It is interesting to relate the constraint function J defined by (3.6) to the constraint function Π defined by (2.3).

Due to the fact that in the rigid foundation there is no strain, $\Phi_B = 0$, and the modified functional (2.9) is reduced to form:

$$\Phi^*(u_i,\lambda,\omega) = \Phi_A(u_i) + \phi(u_i,\lambda,\omega), \qquad (3.21)$$

with $\Phi_A(u_i) = \int U(\varepsilon_{ij})dV_A - \int f_i u_i dV_A - \int \bar{t}_i u_i d\Gamma_{\sigma A}$.

The dislocation potential $\phi(u_i,\lambda,\omega)$ is defined by (3.6), so that we focus our attention on the contact potential

$$\Phi_C = \int \lambda J dV_C. \qquad (3.22)$$

Now to relate Π of (2.19) to J of (3.22) it is necessary to transform the volume integral (3.22) to a surface integral along $\Gamma_{AC} \equiv \Gamma_C$. Let $A \in \Gamma_C$, $B \in \delta\Omega$ be two points which must be in contact during the loading process. We introduce (Figure 7) a special curved orthogonal coordinate system

$$r = r(x_1,x_2),$$
$$s = s(x_1,x_2),$$

such that for $r = r_B = 0$, the line $r = r(B)$ coincides with the boundary of the foundation $\delta\Omega$ and the line $r = r(A)$ coincides with the boundary of the elastic solid Γ_C. In these conditions the frontiers Γ_1, Γ_2 are chosen to coincide respectively with the line $s = s_1$ and $s = s_2$. Therefore the domain V_C of the perfect locking material is

$$V_C \equiv (0,r_A) \times (s_1,s_2), \qquad (3.23)$$

and the integral (3.22) has the more precise form:

$$\Phi_C = \int \lambda J dr ds \ . \tag{3.24}$$

As the contact pressure λ depends uniquely on s we may write

$$\Phi_C = \int_{r_C} \lambda ds \int_0^{r_A} J dr \ . \tag{3.25}$$

Putting

$$\Pi = \int_0^{r_A} J dr \ , \tag{3.26}$$

we may write (3.25) in the form

$$\Phi_C = \int \lambda \Pi d\Gamma_C \ , \tag{3.27}$$

and this is exactly the constraint potential (2.19),

$$J = 1 + \frac{\partial u_\rho}{\partial \rho} + \frac{\partial u_\theta}{\partial \theta} = 1 + \varepsilon_{\rho\rho} + \varepsilon_{\theta\theta} \ .$$

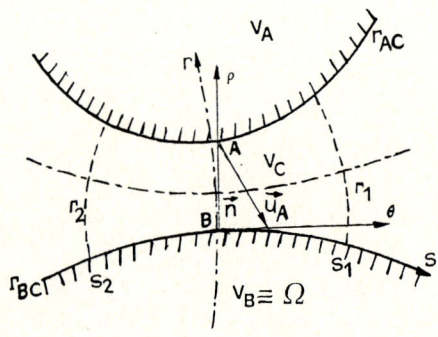

Figure 7 - Linearization of the Non-Interpenetration Condition $J \geq 0$. Body V_B is Identified with the Rigid Foundation

We denote that in V_C the material is identified as a perfect locking material with r, s as principle stress directions. Near B the curved principal directions coincide with cartesian axes ρ, θ. Near B and in the domain V_C we must have

$$\sigma_{\rho\rho} = \lambda = \frac{\partial J}{\partial \varepsilon_{\rho\rho}} \,,$$

$$\sigma_{\theta\theta} = 0 = \frac{\partial J}{\partial \varepsilon_{\theta\theta}} \,.$$

The term $\varepsilon_{\theta\theta}$ in (3.28) must disappear so that the integral (3.26) becomes:

$$\Pi(\rho, u) = \rho(A) + u_\rho(\rho_A) = \rho(A) + \vec{n} \cdot \vec{u}(A) \,, \qquad (3.29)$$

where $\vec{n}$ is the normal vector to $\partial\Omega$. This is the linear form of Π which is used in the non-interpenetration constraint. Returning to the system x_1, x_2 which is the Lagrangian system linked to the rigid foundation one has:

$$\Pi(\vec{u} = 0) = \rho(A) = \rho(x_1, x_2) \,, \qquad (3.30)$$

and due to the fact that the displacements are small:

$$\vec{n} = \nabla\Pi(\vec{u} = 0) = D_k \Pi(\vec{u} = 0) \,. \qquad (3.31)$$

According to (3.26), (3.29), (3.30) and (3.31) the non-interpenetration condition (3.6) leads to:

$$D_k \Pi(\vec{u} = 0) u_k > -\Pi(\vec{u} = 0) \,, \qquad (3.32)$$

which is just the linearized inequality (2.20) already obtained in Section 2.

The unity of the two formulations has been demonstrated.

4. NUMERICAL RESULTS

Computation of the rigid-elastic contact formulation described in Section 2 is realized by connecting the finite element algorithm ADELEF to the mathematical programming code called ACDPAC developed by Best and Bowler [9]. The types of finite element used in the numerical examples are the isoparametric quadrangle proposed by Ergatoudis [10], Argyris and Fried [11] and the hybrid triangle proposed by Dang Hung [12]. The latter element has the following displacement field on the interfaces:

$$u = \alpha_1 + \alpha_2 x + \alpha_3 y + \alpha_4 x^2 + \alpha_5 xy + \alpha_6 y^2 ,$$

$$v = \alpha_7 + \alpha_8 x + \alpha_9 y + \alpha_{10} x^2 + \alpha_{11} xy + \alpha_{12} y^2 ,$$

and the stress field may have arbitrary degree derived from the Airy stress function:

$$F = \sum_{i=0}^{n} \sum_{j=0}^{i} \beta_{ij} x^{i-j} y^i .$$

This element belongs to a very efficient class of hybrid elements named the mongrel elements discussed in [13] and [14]. Externally, this element possesses the same number of degrees of freedom as the classical conforming quadratic triangle.

4.1 *Hertz's Problem*

The classical contact problem of an infinite elastic cylinder on a rigid semi-infinite medium (Figure 8) is tested. The analytical solution of the problem has been given earlier by Hertz [15]. Geometric and mechanic data of the example are:

$R = 1000$ mm , $E = 21.000$ kg mm^{-2} , $\nu = 0.3$, $P = 600$ kg mm^{-1} .

Taking advantage of the symmetry, only half of the infinite cylinder
is discretized into triangular elements. The mesh is shown successively by Figures 8, 9, 10 and 11. For this pattern 80 mongrel
triangles are necessary with 394 degrees of freedom. The number
of interface constraint elements is 16 with 36 degrees of freedom.
Figure 12 illustrates the distribution of the pressure along the
contact surface AB. The dotted curve is Hertz' analytical result.
The points indicate the present results. Agreement between analytical and finite element results is satisfactory. The CPU of this
problem is 26 sec. on IBM 370-158 at the University of Liège.
Figure 13 presents the nodal forces g_r obtained directly in the output. We see that the coincidence with Hertz's ellipse is more
regular because the forces, being obtained by integration of the
interpolation shape function, are thereby smoothed. One may distinguish the upper curve formed from the forces taken from the
mid-points of the contact elements and the lower curve formed from
the connected vertices of the same elements.

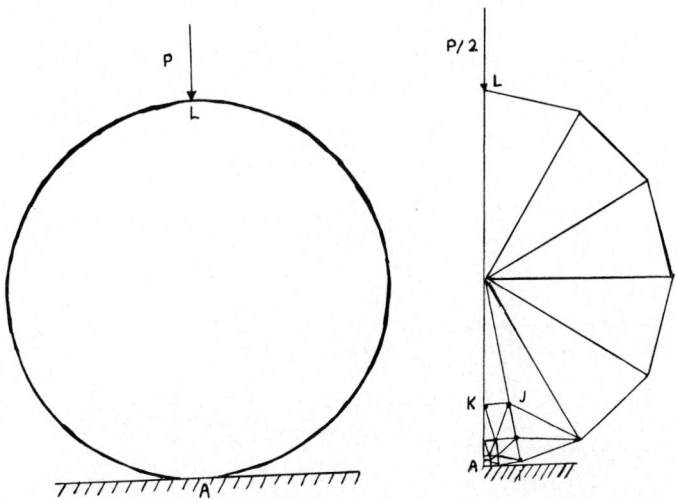

Figure 8 - *Hertz's Problem. Discretization of a Semi-Area of the Cylinder*

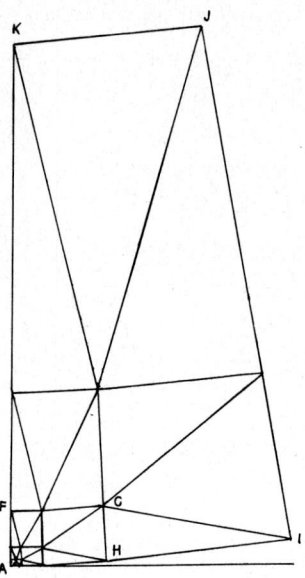

Figure 9 - Finite Element Mesh of Region AKJI

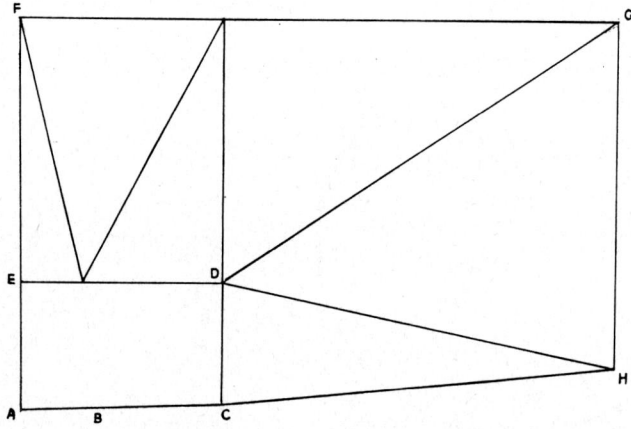

Figure 10 - Finite Element Mesh of Region AFGH

Finite Element Analysis of Contact Problems 365

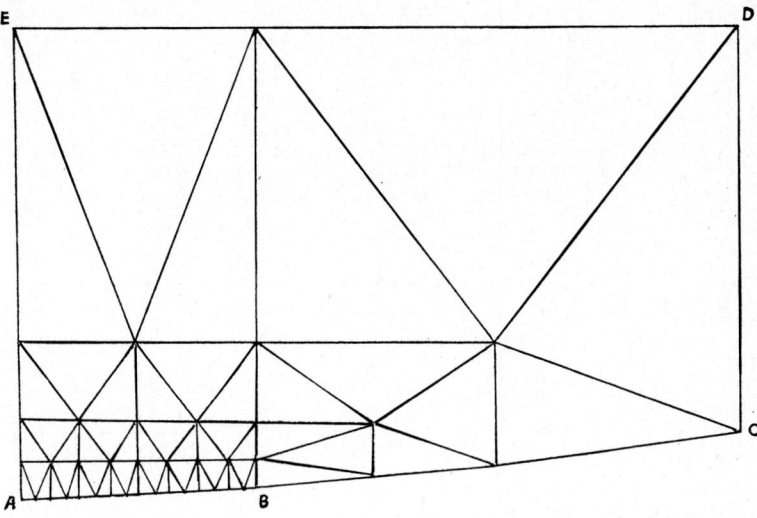

Figure 11 - Finite Element Mesh of Region AEDC

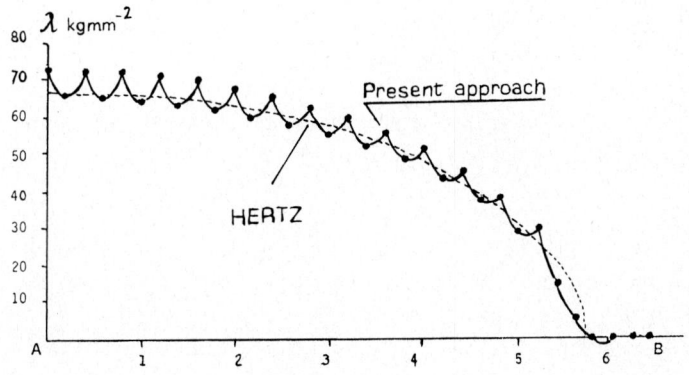

Figure 12 - Comparison of the Distribution of the Pressure Along the Contact Surface

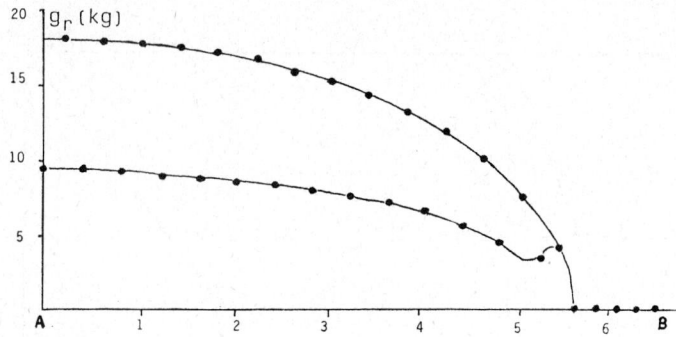

Figure 13 - Distribution of the Nodal Forces Along the Contact Surface

4.2 *Piston Rod*

Figure 14 shows a piston rod acting on a pin under an external force F distributed along Γ_σ. A clearance exists before loading and the two solids are initially in contact on the top:

$$c = \eta_0(1-\cos\theta) ,$$

where

$$\eta_0 = R_i - R_e ,$$

is the initial gap between the radius of the hole on the rod and of the pin. The following geometric data are chosen for the present test:

$E = 7288$ Kg mm^{-2} , $\nu = 0.32$, $R_i = 41$ mm , $R_e = 55.5$ mm ,

$t = $ (thickness) $= 19.8$ mm , $F = 612$ kg (Figure 14) .

Again the symmetry allows us to discretize only half of the structure, using 48 isoparametric elements. The connecting rod is considered elastic, but the shaft as rigid. The constraint function is the equation of the boundary of the pin:

$$\Pi(\xi,\eta) = \xi^2 + (\eta-\eta_0)^2 - R_a^2 \geq 0 .$$

The pattern (Figure 14) needs 394 degrees of freedom with 15 contact elements corresponding to 62 degrees of freedom.

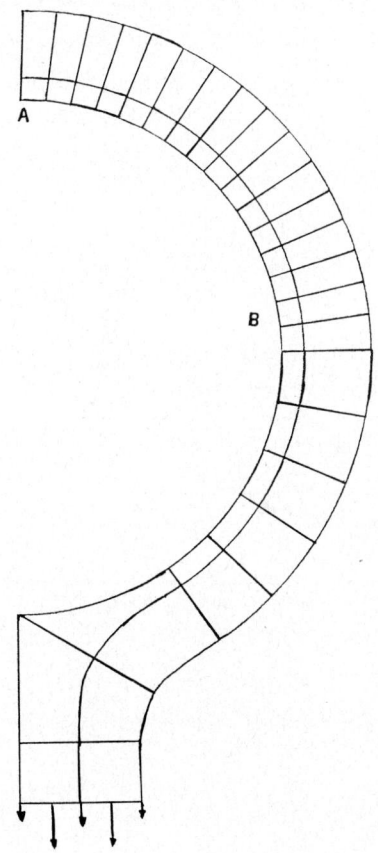

Figure 14 - Finite Element Mesh. The Semi-Area of the Piston Rod

Figures 15, 16 and 17 illustrate some results for the case $\eta_0/R_i = 0.003$. Distributed pressures $\lambda(\theta)$ along the contact surface are shown by Figure 15. Figure 16 presents the distribution of the nodal forces g_r of two kinds of nodes of contact elements. Figure 17 gives the comparison of the circumferential stress distribution derived from the present results and the numerical and experimental results obtained by Gaertner [2]. Agreement is good despite the nonrealistic assumption that the shaft is perfectly rigid. Figure 18 shows the angular contact and the distributed pressure in the 3 cases $\eta_0/R_i = 0.008$, $\eta_0/R_i = 0.003$, $\eta_0/R_i = 0.001$. It is seen that for large clearance the pressure distribution tends rapidly to the elliptic form predicted by Hertz.

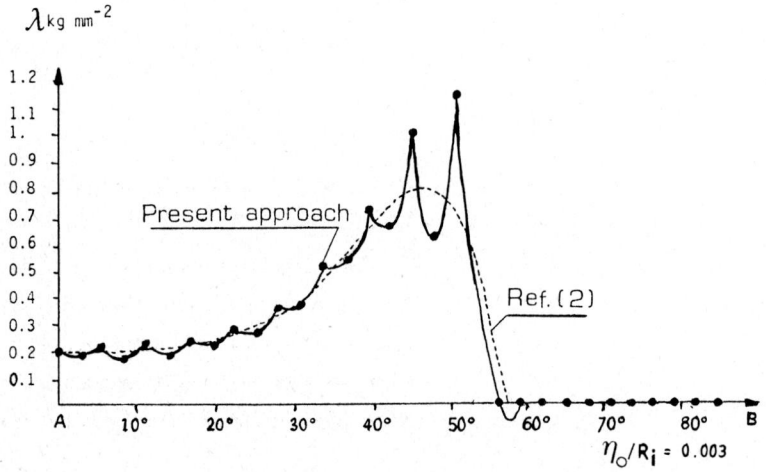

Figure 15 - *Comparison of the Distribution of the Pressure along the Contact Surface*

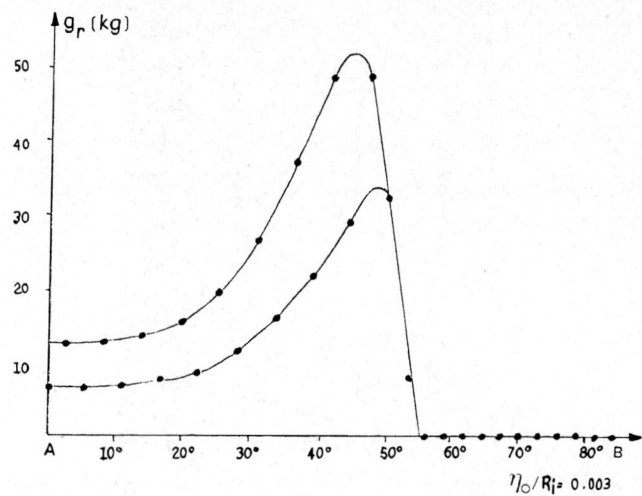

Figure 16 - Distribution of the Nodal Forces along the Contact Surface

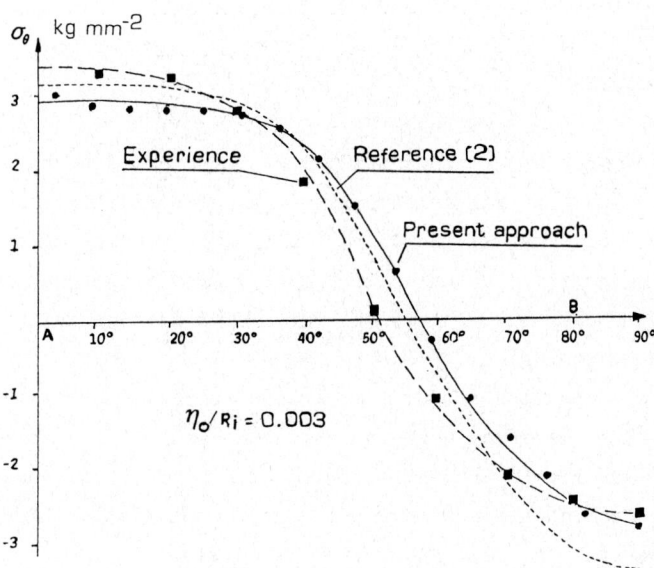

Figure 17 - Comparison of the Distribution of the External Circumferential Stress

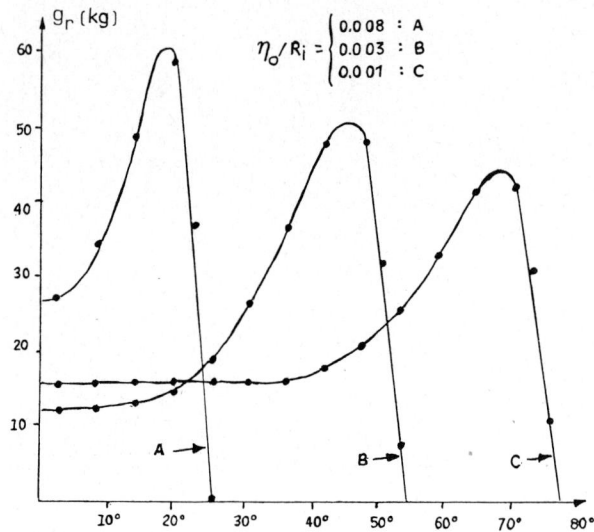

Figure 18 - Nodal Forces Distribution for Different Clearance for $F = 612$ KG

5. CONCLUSIONS

The paper presents a linearization procedure for the non-interpenetration condition for frictionless contact of elastic bodies. This allows a modelling of the contact surface by finite elements. This method renders the optimization problem convex and convergence is assured. Numerical experiences for the rigid-elastic contact case are very encouraging.

Extension to the elastic contact problem is achieved by considering the space near the contact surfaces as a perfectly locking material. The introduction of a symmetrical non-interpenetration condition leads to an elegant formulation of the frictionless contact problem of two elastic bodies.

This formulation may be reduced to the rigid-elastic contact problem for small displacements conditions.

NOTATION

x_i		
ξ_i	–	Lagrangian coordinates
V, V_A, V_B	–	geometric domains of elastic bodies
$\Gamma_\sigma, \Gamma_{\sigma A}, \Gamma_{\sigma B}$	–	portion of boundary of V, V_A, V_B where $\bar{t}_i$ are prescribed
$\vec{n}, n_i$	–	normal vector defined on the boundaries
Ω	–	rigid domain
$\partial\Omega$	–	boundary of Ω
$\varepsilon_{ij}, \varepsilon$	–	strain tensor, strain field
$U(\varepsilon_{ij})$	–	density of the strain energy
σ_{ij}, σ	–	stress tensor, stress field
$\bar{f}_i$	–	body forces
$\bar{t}_i$	–	tractions prescribed on the boundaries
q	–	nodal displacement vector
g	–	nodal generalized force fector
K	–	global stiffness matrix
a	–	displacement parameters
Φ	–	total potential strain energy
Φ^*	–	modified total potential energy
Φ_c	–	contact potential
ϕ_c	–	dislocation potential
Π	–	contact constraint function
λ	–	Lagrange's parameter
ω	–	slack variable
A	–	constraint matrix
D_i, ∇	–	Lagrangian gradient operator
∂_i	–	Eulerian gradient operator
$[J]$	–	Jacobi matrix
J	–	Jacobian: determinant of $[J]$
I_{ij}	–	minor of J_{ij} in $[J]$
J_1, J_2	–	first and second invariant of the deformation tensor

REFERENCES

[1] CHAN, S.H. and TUBA, I.S., "A Finite Element Method for Contact Problems of Solid Bodies", *Int. J. Mech. Sci.*, Vol. 13, 1971, pp. 615-639.

[2] GAERTNER, R., "Investigation of Plane Elastic Contact Allowing for Friction", *Computers and Structures*, Vol. 7, 1975, pp. 59-63.

[3] FRANCAVILLA, A. and ZIENKIEWICZ, O.C., "A Note on Numerical Computation of Elastic Contact Problems", *Int. J. for Num. Methods in Eng.*, Vol. 9, 1975, pp. 913-924.

[4] FREDRIKSSON, B., "Finite Element Solution of Surface Nonlinearities in Structural Mechanics with Special Emphasis to Contact and Fracture Mechanics Problems", *Computers and Structures*, Vol. 6, 1976, pp. 281-290.

[5] FENG, W.W. and HUANG, P., "On the General Contact Problem of an Inflated Nonlinear Plan Membrane", *Int. J. Solids and Structures*, Vol. 11, 1975, pp. 437-448.

[6] FREMOND, M., "Etude des Structures Visco-Élastiques Stratifiées soumises à des charges harmoniques et de solides élastiques reposant sur des structures", *Thèse*, Université de Paris, 1971. "Formulation duale des énergies potentielles et complémentaires Application à la méthode des éléments finis", *C.R. Ac. Sc. Paris*, 273, Serie A, pp. 775-777.

[7] PANAGIOTOPOULOS, P.D., "A Nonlinear Programming Approach to the Unilateral Contact and Friction-Boundary Value Problem in the Theory of Elasticity", *Ing. Archives*, Vol. 44, 1975, pp. 421-432.

[8] DANG HUNG, N., "Modèles mathématiques et calcul numerique du comportement inélastique des matériaux", Notes de cours, Laboratoire de Mécanique des Materiaux et de Statique des Constructions, Université de Liège, 1978.

[9] BEST, M.J. and BOWLER, A.T., "ACDPAC, A FORTRAN IV Subroutine to Solve Differentiable Mathematical Programmes User's Guide", Level 2.0, University of Waterloo, Waterloo, Ontario, Canada, 1976.

[10] ERGATOUDIS, J.G., "Isoparametric Elements in Two and Tridimensional Analysis", *Ph. D. Thesis*, University of Wales, Swansea, 1969.

[11] ARGYRIS, J.H. and FRIED, I., "The LUMINA Element for the Matrix Displacement Method", *Aero. J.*, Aero. Soc., Vol. 72, 1968, pp. 514.

[12] DANG HUNG, N.; "Direct Limit Analysis via Rigid-Plastic Finite Elements", *Comp. Meth. in Appl. Mech. and Eng.*, Vol. 8, 1976, pp. 81-116.

[13] DANG HUNG, N., "Sur une classe particulière des éléments finis hybrides: les éléments métis', Communication présentée au "Congrès Int. sur les Méthodes Numériques dans les Sciences de l'Ingénieur", organisé par le GAMNI, Paris, France, November 27 - December 1, 1978, le compte rendu est à paraître dans les Editions Dunod.

[14] DANG HUNG, N., "On the Monotony and the Convergence of a Special Class of Hybrid Finite Elements: the MONGREL Elements", *Proc. of the IUTAM Sym. on Variational Methods in Mechanics*, held at Northwestern University, Evanton, Illinois, U. S. A., September 11-13, 1978.

[15] HERTZ, H., "Über die Beruhrung fester elastischer Körper", *J. für Math.*, Vol. 92, 1882, pp. 156-171.

STRUCTURAL CONTROL, H.H.E. Leipholz (ed.)
North-Holland Publishing Company & SM Publications
© IUTAM, 1980

COMPLEMENTARITY CONDITIONS IN ENGINEERING OPTIMIZATION

I. Kaneko

Departments of Industrial Engineering and Computer Sciences
University of Wisconsin-Madison
Madison, Wisconsin 53706, U.S.A.

1. INTRODUCTION

Two nonnegative n-vectors w and z are called *complementary* if for each $j \in \{1, \ldots, n\}$ either w_j and/or z_j is zero, i.e., $w_j > 0$ implies $z_j = 0$ (and vice versa). Since both vectors are nonnegative, w and z are complementary if and only if their inner product equals zero; i.e.,

$$w^T z = 0 . \qquad (1)$$

(A vector is always assumed to be a column; transposition of vectors and matrices is indicated by T.) The condition (1) is known as a *complementarity condition* (between w and z). In the recent years certain mathematical programming models involving complementarity conditions have been found remarkably versatile and useful in formulating and solving various problems in engineering and physics. The nature of this paper is largely expository; its purposes are (i) to present a few examples of engineering problems formulated as these mathematical programming models, (ii) to explain some numerical techniques to solve these problems and (iii) to indicate some computational results using these numerical techniques.

In this paper, we shall consider the following two mathematical programming models. For a given n-vector q and n by n matrix M, the *linear complementarity problem* (LCP) is that of finding n-vectors w and z satisfying the conditions:

$$w = q + Mz, \quad w \geq 0, \quad z \geq 0, \qquad (2.1)$$

$$w^T z = 0. \qquad (2.2)$$

The *linear programme with a complementarity constraint* (LP with a CP constraint) is given by:

$$\text{minimize:} \quad (c^1)^T x + (c^2)^T w + (c^3)^T z \qquad (3.1)$$

$$\text{subject to:} \quad A^1 x + A^2 w + A^3 z = b, \qquad (3.2)$$

$$x \geq 0, \quad w \geq 0, \quad z \geq 0, \qquad (3.3)$$

$$w^T z = 0. \qquad (3.4)$$

Here, c^1 is a k-vector, c^2 and c^3 are n-vectors, A^1 is an m by k matrix, A^2 and A^3 are m by n matrices and b is an m-vector. Clearly any linear inequality constraints and/or variables with any sign restrictions can be incorporated in the above model.

Note that (2) is merely a system of linear equations (in nonnegative variables) except for the complementarity condition (2.2). Similarly, the problem (3) would be an ordinary linear programme if not for (3.4). In general, a complementarity condition provides a compact way of expressing often complex nonlinearities which prevail in many engineering problems.

The merit of considering these mathematical programming models to deal with engineering problems is two-fold. First, one may be able to apply theoretical results obtained for the mathematical models to establish some analytical properties (such as existence and uniqueness of solutions) of the engineering problems. Also, numerical solutions of these engineering problems may be obtained by using some solution procedures devised for the mathematical programming models. In this paper, we shall be concerned with only the computational aspect.

Extensive research has been done on the LCP in the past decade or so and powerful theoretical and computational results are available. The LCP provides a unifying framework to study a variety of subjects including quadratic programming, game theory, finance, regression and actuarial graduation. For instance, the Kuhn-Tucker optimality conditions for the quadratic programme: minimize: $c^T x + \frac{1}{2} x^T D x$; subject to: $Ax \geq b$, $x \geq 0$ can be viewed as a LCP (2) with

$$q = \begin{bmatrix} c \\ -b \end{bmatrix} \quad \text{and} \quad M = \begin{bmatrix} D & -A^T \\ A & 0 \end{bmatrix}.$$

2. OPTIMAL EXCAVATION OF SEABED

As the first example, we shall consider in this section the problem of determining the shape of an elastic pipeline in equilibrium laid on a seabed (see Figure 1). Under a certain practical set of assumptions, this problem is formulated (see [1, 2]) as a LCP (2).* We refer to [1, 2] for details but here we only point out that for the purpose of analysis, pipeline is considered as an assembly of rigid elements connected with elastic hinges ("nodes"). Assume that there are n nodes.

---

* The paper by N.D. Hung presented at the IUTAM Symposium deals with a similar "contact problem" in a more general setting.

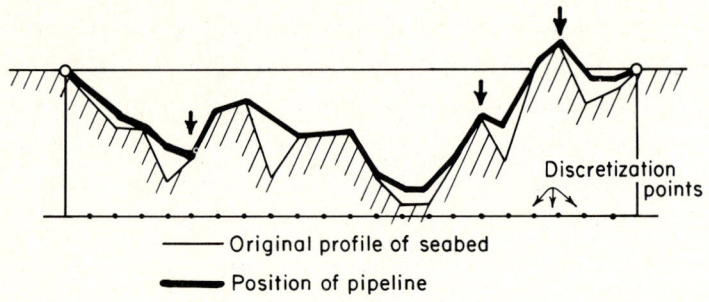

Figure 1 - Pipeline Problem

In the LCP formulation, w denotes the n-vector such that its jth component is the reaction force generated by the seabed at the jth node and z denotes the n-vector whose j^{th} component is the (vertical) distance between the seabed and the pipeline in equilibrium at node j. The matrix M denotes the "assembled" stiffness matrix and q is given by q = F - Mr, where F is the n-vector of loads and r is the n-vector denoting the depth of the seabed. In this formulation, the complementarity condition represents the following nonlinearity; the reaction force may be positive only at points of contact.

Having formulated the problem as a LCP one can obtain the solution by using some efficient numerical techniques in Operations Research (see Section 5).

In many cases, the shape of the seabed is such that the pipeline in equilibrium presents some excessive curvature (for example, in Figure 1, much curvature is present at those places indicated by arrows). If one considers that a curvature beyond a certain allowable limit causes serious difficulties, then one may need to excavate some of the "peaks" and/or embank some of the "valleys" to make the seabed profile smoother (see Figure 2).

Since the excavation/embankment is usually quite expensive, one naturally wants to find an excavation/embankment scheme with the minimum possible cost, subject to the (additional) condition that the curvature of the pipeline be within certain limits.

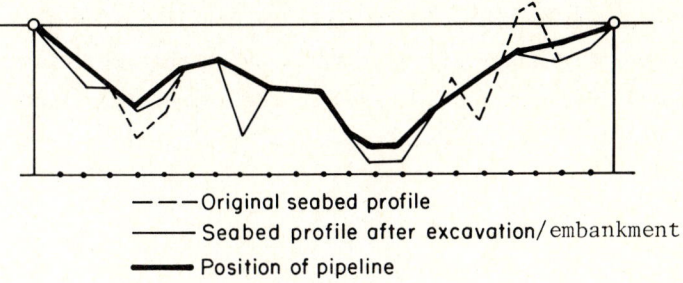

– – – Original seabed profile
——— Seabed profile after excavation/embankment
▬▬▬ Position of pipeline

Figure 2 - Pipeline Problem

Letting x and y be the n-vectors such that x_j and y_j denote the depths of excavation and embankment, respectively, at node j, the optimal excavation/embankment problem is formulated as:

$$\text{minimize:} \quad c^T x + a^T y \tag{4.1}$$

$$\text{subject to:} \quad w = q + M(z-x+y) , \tag{4.2}$$

$$l \leq C(x-y-z) \leq u , \tag{4.3}$$

$$w \geq 0 , \quad z \geq 0 , \quad x \geq 0 , \quad y \geq 0 , \tag{4.4}$$

$$w^T z = 0 . \tag{4.5}$$

Here, c and a are n-vectors such that c_j and a_j denote, respectively, the excavation and embankment costs per unit depth. The matrix M and n-vector q have the same meanings as before: the condition (4.3) ensures that the value of the vector of relative rotations, $C(x-y-z)$, is within the allowable lower/upper limits, where C is the n by n tri-diagonal matrix. The complementarity condition (4.5) is as explained above.

We would like to remark on a technical point about the mathematical programming problem (4). After converting inequality constraints into equalities, the problem (4) contains 3n linear constraints and 6n variables. Applying a certain manipulation, we can reduce the number of constraints by n; this reduction leads to some significant saving in terms of the computation time to solve the problem. More specifically, we replace (4.3) with

$$v = C(x-y-z) - l, \qquad (5)$$

and require

$$0 \leq v \leq u - l. \qquad (6)$$

The resulting model (in equality form) will have 2n linear constraints and 6n variables. (The computational efficiency of algorithms based on the Simplex Method depends critically on the number of constraints; the number of variables and whether or not variables have lower and/or upper bounds tend to have less influence on the computation time required to solve the problem.)

3. OPTIMAL LIMIT DESIGN OF STEEL FRAMES

Next example, taken from the paper [3], is the problem of finding a minimal cost limit design of a plastic steel framed structure. Some of the assumptions made for the formulation include the following (see [3] for details). The structure is decomposed into prismatic rigid elements connected with plastic hinges. The unit cost of a beam (or column) segment depends only on the yield moment for the segment. Let n be the number of "critical sections" (corresponding to the plastic hinges), of which only t ($\leq$ n) correspond to distinct beam/column segments. Let m denote the t-vector of yield moments at these t critical sections; the yield moments at all n critical sections are given by the n-vector Qm, where Q is the geometric matrix.

The problem is formulated as:

minimize: $G(m)$ (7.1)

subject to: $m = Tx$, $x \geq 0$, (7.2)

$-Qm \leq m^E + Zr \leq Qm$. (7.3)

Here, $G(m)$ is the total cost, x is the s-vector ($s \leq t$) of design variables which determines m through (7.2) (T is a t by s "technology" matrix), m^E is the n-vector of linear responses to the load, Z is the n by k "influence" matrix and r is the k-vector of redundant forces (so $m^E + Zr$ is the n-vector of moments at all critical sections).

Now, the total cost $G(m)$ is given by

$$G(m) = \sum_{j=1}^{t} l_j \cdot g_j(m_j),$$ (8)

where l_j and $g_j(m_j)$ are, respectively, the length and the cost per unit length of the beam/column segment with respect to the jth critical section. In [3] it is mentioned that $g_j(m_j)$ is usually given by:

$$g_j(m_j) = \beta_j \cdot (m_j)^{\alpha_j},$$ (9)

for some positive scalar β_j and some scaler α_j with $0 < \alpha_j < 1$. See Figure 3 for an example of such a cost function.

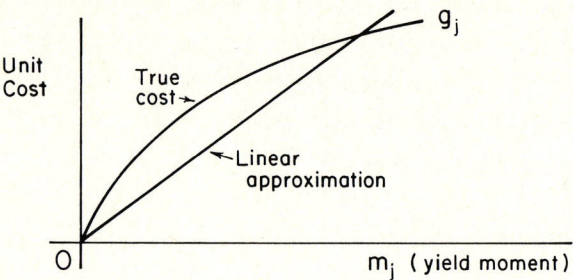

Figure 3 - Unit Cost Curve

One approach would be to replace $g_j(m_j)$ with a linear function (of m_j); if one does so, the problem (7) becomes a linear programme. To improve the accuracy of the assessment of the cost, here we consider a tri-linear approximation of the cost curve such as that shown in Figure 4. For each $j \in \{1, \ldots, t\}$ let c_j^i be the slope of the ith linear piece, $i = 1, 2, 3$, and let f_j^i be the length of the interval as shown in the figure, $i = 1, 2$.

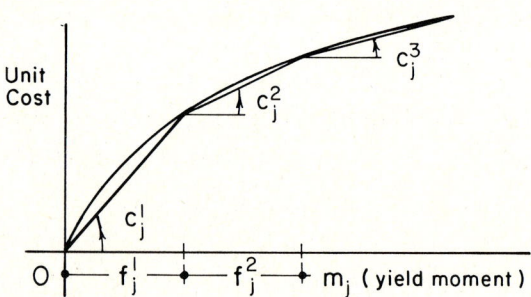

Figure 4 - Trilinear Approximation

As explained in an appendix of Kaneko [4], the tri-linear function $g_j(m_j)$ can be expressed as

$$g_j(m_j) = c_j^1 m_j^1 + c_j^2 m_j^2 + c_j^3 m_j^3 , \qquad (10)$$

where $m_j = m_j^1 + m_j^2 + m_j^3$ with m_j^i, $i = 1, 2, 3$, satisfying

$$0 \leq m_j^1 \leq f_j^1, \quad 0 \leq m_j^2 \leq f_j^2, \quad m_j^3 \geq 0, \quad (11.1)$$

$$(f_j^1 - m_j^1) \cdot m_j^2 = 0, \quad (f_j^2 - m_j^2) \cdot m_j^3 = 0. \quad (11.2)$$

Let m^i be the t-vector whose j^{th} component is m_j^i, $i = 1, 2, 3$, and let f^i be the t-vector such that its jth component is f_j^i, $i = 1, 2$. Then the optimal limit design problem with the trilinear cost function is the following LP with CP constraints

$$\text{minimize:} \quad \sum_{j=1}^{t} \ell_j \left(\sum_{i=1}^{3} c_j^i m_j^i \right) \quad (12.1)$$

subject to:

$$\sum_{i=1}^{3} m^i = Tx, \quad x \geq 0, \quad (12.2)$$

$$-Q \sum_{i=1}^{3} m^i \leq m^E + Zr \leq Q \sum_{i=1}^{3} m^i, \quad (12.3)$$

$$0 \leq m^i \leq f^i, \quad i = 1, 2, \quad m^3 \geq 0, \quad (12.4)$$

$$(f^i - m^i)^T m^{i+1} = 0, \quad i = 1, 2. \quad (12.5)$$

In closing this section we would like to point out the following. The above expression of $g_j(m_j)$ (given by (10) and (11)) is somewhat different from that used in the original paper [3]. Our formulation has a technical advantage in that it involves fewer number of constraints and thus leads to a more efficient solution as noted above; the model given in [3] has $2n+2t$ linear constraints, while ours involves $2n+t$ of them.

4. MAIER'S ELASTIC-PLASTIC ANALYSIS

The third and final example we discuss in this paper is the LCP mehtod initiated by G. Maier (see e.g., [5, 6]) of analyzing elastic-plastic structures. (This "Maier's LCP method" has led to much research by mathematical programmers; see e.g., Cottle [7, 8] and Kaneko [4, 9, 10]).

Omitting the details (see [4, 5, 6]) we only mention that the problem of determining the plastic strains of an elastic-plastic structure may be formulated as that of finding n-vectors w and z satisfying:

$$w = q + F(z), \quad w \geq 0, \quad z \geq 0, \quad (13.1)$$

$$w^T z = 0. \quad (13.2)$$

Here n is the number of distinct yielding modes under consideration, q is a constant n-vector, F is a possibly nonlinear function from Euclidean n-space to itself. If F is indeed nonlinear, then the problem of the form (13) is called a *nonlinear complementarity problem*. The complementarity condition (13.2) in this particular problem enbodies the following situation.

In formulating the problem as (13) we employ a piecewise linear approximation of the "elastic domain" (in the generalized stress space), each of whose n boundaries corresponds to a distinct yielding mode. According to the theory of plasticity, no plastic activity takes place as long as the "stress point" lies in the interior of the elastic domain. In (13) w_j denotes the distance between the stress point and the jth boundary of the elastic domain and z_j denotes the intensity of plastic activity with respect to the jth yielding mode; thus we must have that z_j cannot be positive unless w_j is zero, or the complementarity between w and z.

Under certain additional assumptions and after some manipulations, the nonlinear complementarity problem (13) is transformed into a LCP (using Maier's scheme as in [6]) or what is known as an n by dn LCP (using Kaneko's scheme in [4]). The n by dn LCP is a variation of the LCP. According to the computational experiment conducted by the author, the n by dn LCP formulation leads to a more efficient computation of the solution.

5. NUMERICAL SOLUTION METHODS

We have illustrated, in the preceeding three sections, that mathematical programming models involving complementarity conditions (LCP, LP with a CP condition and n by an LCP) provide a useful framework for some engineering problems. In this final section we shall mention some of the solution methods by which one can obtain numerical solutions to these problems. We shall also refer to the availability of computer codes and give some indications of the computational efficiency of these methods.

An appropriate choice of a solution method for a LCP depends on the nature of the matrix M defining the problem. If one knows, a priori, that M has one or more of certain properties, then one should use some special-purpose algorithms taking advantage of the properties. Otherwise, a general-purpose algorithm must be used. In most cases, special-purpose algorithms, if applicable, produce much saving in the computation time.

The properties of M particularly relevant to engineering applications are positive definiteness (p.d.) and positive semi-definiteness (p.s.d.). If the LPC has p.d. (or p.s.d.) M, then, one of the most effective solution methods is Lemke's algorithm (Lemke [11]). Fortran code, LCPL, for this algorithm is available at the Department of Operations Research, Stanford University.

To give some indication of the computational performance of LCPL solving a LCP with p.s.d. matrix we quote from Cottle [12] the computation times required to solve LCPs of various sizes, as shown in Table 1 below.

Table 1 - Lemke's Algorithm for a LCP with a p.s.d. Matrix

Number of Constraints	CPU Time
15	0.23 sec.
36	0.93 sec.
105	2.08 sec.
520	16.21 sec.

All the results cited in the above table were obtained by using IBM 370/168. Note that the problem size of the LCP is specified by the number of constraints. (In the LCP the number of variables is always twice as many as that of the constraints.) The computer code LCPL is generally considered to be reliable and fast.

If the matrix M in (2) does not have the required properties, then Lemke's algorithm need not yield a solution and some general purpose methods must be used. One of the general-purpose methods is the implicit enumeration algorithm developed and coded by the author and Hallman (see [13]). Fortran code, GLCP, for the algorithm is available through this author, at the University of Wisconsin-Madison.

The author has conducted some substantial tests using GLCP and found its performance satisfactory. It should be pointed out that a general LCP (i.e., LCP where M does not possess required properties such as positive semi-definiteness) is inherently a "difficult" problem and much computational efforts may be needed if the problem has a large size.

Some of the results from the computational tests are recorded in Table 2 (see [13] for more details). All the computations were done on UNIVAC 1110. It should be noted that UNIVAC 1110 is much (about 30-50 times) slower than IBM 370.

Table 2 - Kaneko/Hallman Algorithm for a General LCP

Problem Type	Number of Constraints	CPU Time
A	20	0.18 sec.
B	20	1.09 sec.
C	20	2.36 sec.
A	30	1.07 sec.
B	30	0.95 sec.
C	30	44.38 sec.
B	50	108.88 sec.
C	50	16.85 sec.
A	90	45.85 sec.

Remarks:

(a) The above results indicate that the difficulty of the problem depends critically on the type of problem as well as the size.

(b) The computation time in each row of Table 2 is the average time of solving several (2 to 10) problems of the same type and same size.

The LP with a CP constraint (3) is also considered as a "difficult" problem and solving it tends to take much longer time than solving a linear programme of a comparable size. This is understandable considering the fact that the complementarity condition makes it possible to deal with various complex non-linearities. Among several numerical methods potentially applicable to solving this problem, it seems that the branch-and-bound algorithm developed and coded by the author and Hallman is one of the most effective procedures. Fortran code, LPCPC, for this algorithm is available through the author. (Some computational experience is reported in [3] using a forerunner of LPCPC; LPCPC is about ten times faster than the early version used in [3].)

We have obtained a set of numerical data from Professor Maier for the optimal excavation problem discussed in Section 2, and solved the problem by LPCPC. The problem has 42 constraints, 84 variables and the dimension of w and z in (31) was 21. It took 54.6 seconds on UNIVAC 1110. Giannessi, Jurina and Maier reported in [2] that by using an iterative method they devised, the same problem was solved in 41 seconds on IBM 370/168. (There were some differences in the mathematical problems dealt with in [2] and in our experiment. In particular, we employed the "simpler" version of the model using (5) and (6) in place of (4.3) and so our model has 21 fewer constraints than that dealt with in [2]. This size reduction is believed to be one of the main reasons for the improved computation time.) Considering that UNIVAC 1110 is about 30-50 times slower than IBM 370/168, it appears that LPCPC solved this particular problem about 30 times faster than the method used in [2]. The following table shows some of the other computational results using LPCPC.

Table 3 - LPCPC for LP with a CP Constraint

Number of Linear Constraints	Dimension of w and z	CPU Time (UNIVAC 1110)
11	10	1.11 sec.
20	10	8.38 sec.
25	15	14.87 sec.
27	20	31.28 sec.
35	15	3.00 sec.
40	20	30.14 sec.

Finally, we would like to refer to the author's paper [4] for results of some systematic computational tests using Lemke's algorithm and the algorithm the author has developed to solve the elastic-plastic problem mentioned in Section 4. Fortran code, NUGRV, for this algorithm specially developed for the analysis of reinforced concrete frames (based on the n by dn LCP formulation) is available through the author.

ACKNOWLEDGEMENTS

This research was supported, in part, by the National Science Foundation under Grant No. ENG77-11136.

REFERENCES

[1] MAIER, G. and ANDREUZZI, F., "Elastic and Elastoplastic Analysis of Submarine Pipelines, An Unilateral Contact Problem", Tech. Report I.S.T.C., Politecnico di Milano, September, 1977; also to appear in *J. of Computers and Structures*.

[2] GIANNESSI, F., JURINA, L. and MAIER, G., "Optimal Excavation Profile for a Pipeline Freely Resting on the Sea Floor", paper presented at the 4th Congress of AIMETA, October, 1978.

[3] LO BIANCO, M., MAZZARELLA, C., PANZECA, T. and POLIZZOTTO, C., "Limit Design of Frame Structures with Piecewise Linear Cost Functions", paper presented at the 4th Congress of AIMETA, October, 1978.

[4] KANEKO, I., "Piecewise Linear Elastic-Plastic Analysis", *Int. J. for Numerical Methods of Engineering*, Vol. 14, 1979, pp. 757-767.

[5] MAIER, G., "A Quadratic Programming Approach for Certain Classes of Nonlinear Structural Problems", *Meccanica*, Vol. 3, 1968, pp. 121-130.

[6] MAIER, G., "A Matrix Structural Theory of Piecewise Linear Elastoplasticity with Interactive Yield Planes", *Meccanica*, Vol. 5, 1970, pp. 54-66.

[7] COTTLE, R.W., "Monotone Solutions of the Parametric Linear Complementarity Problem", *Mathematical Programming*, Vol. 3, 1972, pp. 210-224.

[8] COTTLE, R.W., "Complementarity and Variational Problems", *Symposia Mathematica*, Vol. 19, 1976, pp. 177-208.

[9] KANEKO, I., "A Linear Complementarity Problem with n by 2n "P" - Matrix", *Mathematical Programming Study*, Vol. 7, 1978, pp. 120-141.

[10] KANEKO, I., "Complete Solutions of a Class of Elastic-Plastic Structures", *Computer Methods in Applied Mechanics and Engineering*, to appear.

[11] LEMKE, C.E., "Bimatrix Game Equilibrium Points and Mathematical Programming", *Management Science*, Vol. 11, 1965, pp. 681-689.

[12] COTTLE, R.W., "Fundamentals of Quadratic Programming and Linear Complementarity", *Tech. Report SOL 77-21*, Department of Operations Research, Stanford University, August, 1977.

[13] KANEKO, I. and HALLMAN, W., "An Enumeration Algorithm for a General Linear Complementarity Problem", *Tech. Report WP 78-11*, Department of Industrial Engineering, University of Madison-Wisconsin, June, 1978.

CONTROL DEVICES FOR EARTHQUAKE-RESISTANT
STRUCTURAL DESIGN

James M. Kelly
Department of Civil Engineering
University of California
Berkeley, California, U.S.A.

1. INTRODUCTION

The research work to be reported here concerns experimental studies on the use of control devices in structural systems designed to resist earthquake loading. The purpose of the devices is *au fond* to absorb the energy induced in a structure by earthquake ground motion. It is generally accepted that a structure designed to resist earthquake attack must have some capacity to dissipate energy; this capacity is normally provided by detailing beam-column connections so that they can accept a certain amount of plastic deformation. The inherent ductility of a structural system so designed assures its ability to survive, even if damaged, the largest foreseeable earthquake. However, the provision of ductility in a structure the primary purpose of which is to carry vertical loads means that if this energy-absorbing capacity is used, some damage to the structure will result. Thus, the question arises as to whether it is possible to incorporate into a structure a set of replaceable devices with the specific purpose of absorbing energy, and the consequent damage, that would, under conventional design methods, be absorbed at beam-column connections.

The experimental work that we have carried out on the shaking table at the Earthquake Simulator Laboratory of the Earthquake Engineering Research Center, University of California, Berkeley, has involved two applications of specially designed control devices, the first to a stepping frame project and the second to a natural rubber bearing base isolation system. This experimental work is described below.

2. CONTROL DEVICES IN A STEPPING FRAME

While stepping structures are unusual, there are good reasons to allow a very tall structure which could be subject to earthquake loading to step off its foundation. If the structure is to be designed conventionally, then very high foundation tensions would have to be designed for, possibly leading to a very expensive foundation. An example of a stepping structure is a very high railway viaduct in New Zealand. The reinforced concrete piers were allowed to step off the foundation to reduce the very high tensions that could be induced in the concrete [1]. However, when stepping, the inherently low damping in the system meant that the motion could continue with large excursions at the rail level for many minutes after the ground motion had ended. A set of control devices were therefore designed to introduce damping [2 - 4]. The bridge is now under construction. Another structure in which this principle has been used, a chimney at the Christchurch Airport, has been built in New Zealand.

The control devices used in the experimental programme described here are based on the cyclic plastic torsion of hot-rolled, low carbon, mild steel [5]. Torsion appears to be a mode of deformation that is highly resistant to low-cycle fatigue in this material; localized plastic instability, such as develops in other materials subject to alternating compression and tension or alternating bending, is not severe. Further, the devices have

Earthquake-Resistant Structural Design

been designed so that welding is either well away from highly stressed regions or is not present. This is achieved in one type of device by using welds to hold the device together, but not to transmit stress, and in another type of eliminating welds altogether, the pieces being held together by pressed fits. The energy-absorbing element in the device is a rectangular torsion bar to which torque is applied through levers (Figure 1).

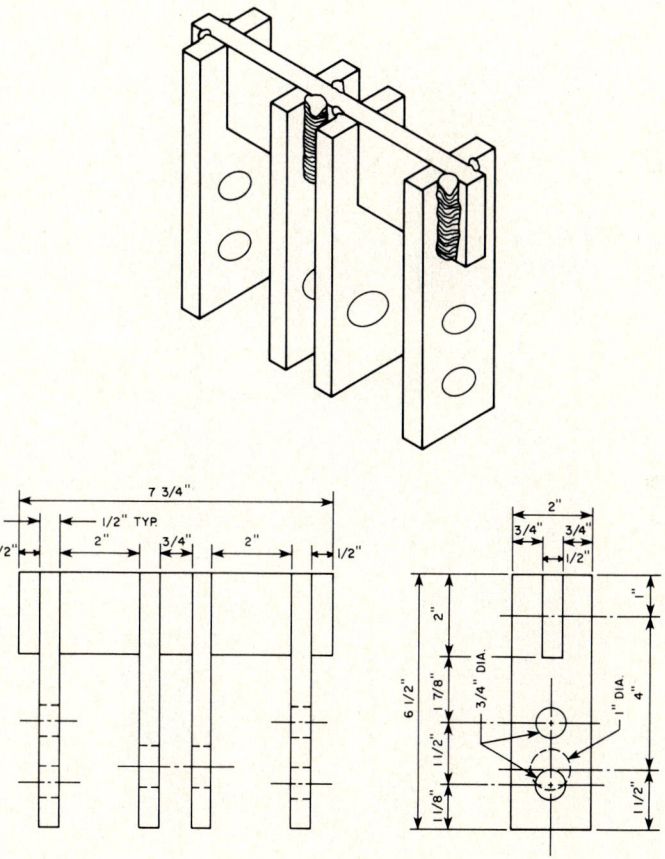

Figure 1 - Dimensions of Control Devices

For tests of the concept in the laboratory, a three-story, single-bay test frame was modified to allow its feet to step off the table and a set of devices was made and attached so that one device was attached to each of the four column feet. The device does not operate until the foot lifts off, but produces downward forces when the column lifts off and an upward force when the foot moves back down. The test frame and the arrangement of the devices are shown in Figures 2 and 3.

Figure 2 - Frame Modified to Allow Uplift

Figure 3 - Column Foot/Device Integration

Results of tests run on the shaking table [6] where the El Centro and Pacoima Dam ground motion records were used as input are discussed below for three base conditions: fixed, free to uplift, and free to uplift with the control devices installed. For all earthquake intensities used in the tests, the uplife of the frame footings was significantly lower when the frame was unanchored. The rocking motion of the frame is shown in Figures 4 and 5; the uplift of the north side of the frame is shown on the top grid and that of the south side on the bottom grid. The effect of the control devices on frame response is clear in these figures.

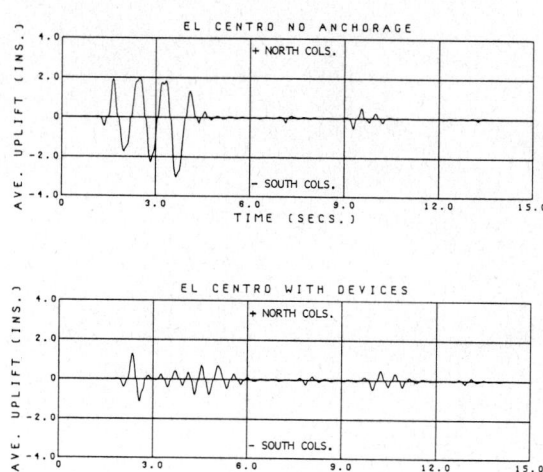

Figure 4 - El Centro Vertical Uplift Displacement Comparisons

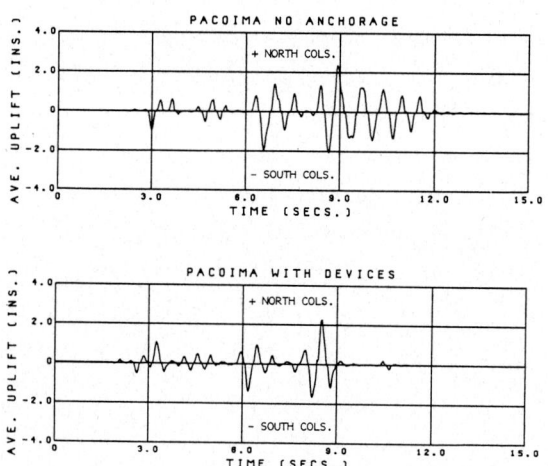

Figure 5 - Pacoima Vertical Uplift Displacement Comparisons

For the El Centro motion with a peak input acceleration of 0.786g, third floor displacement was substantially lower with the devices installed than when the frame was free to uplift (Figure 6). The relative story displacements of the frame with the devices installed were similar to those when the base frame was fixed except that the peak displacements were slightly larger with the devices. The influence of the devices on the overall displacement history is, however, apparent in that considerably more damping of the motion occurred.

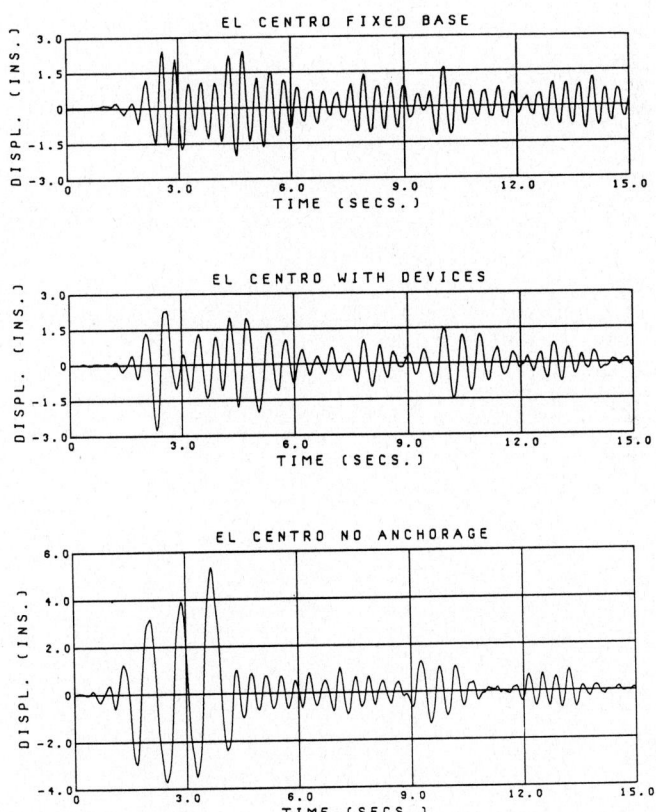

Figure 6 - El Centro Relative Third Floor Horizontal Displacement Comparisons

For the Pacoima Dam test, in which the peak input acceleration was 0.955g, the largest third floor displacement occurred when the frame with the devices installed was tested, the next largest in the unanchored frame, and the least in the fixed-base frame (Figure 7). While this result is perhaps not in favour of the control devices, the displacement of the top story was reduced with the control devices installed, especially for the latter portion of the time-history record which followed the most intense portion of the input. First floor column tension in the frame with the control devices was greater for both earthquake series than that in the unanchored frame, but substantially less than that in the fixed-base frame, as shown, for example, in Figure 8. Test results for the El Centro and Pacoima Dam test series are summarized in Tables 1 and 2 where the peak responses for average column uplift, relative third floor displacement, first floor column axial force, base shear, and base overturning moment are indicated for the three test conditions.

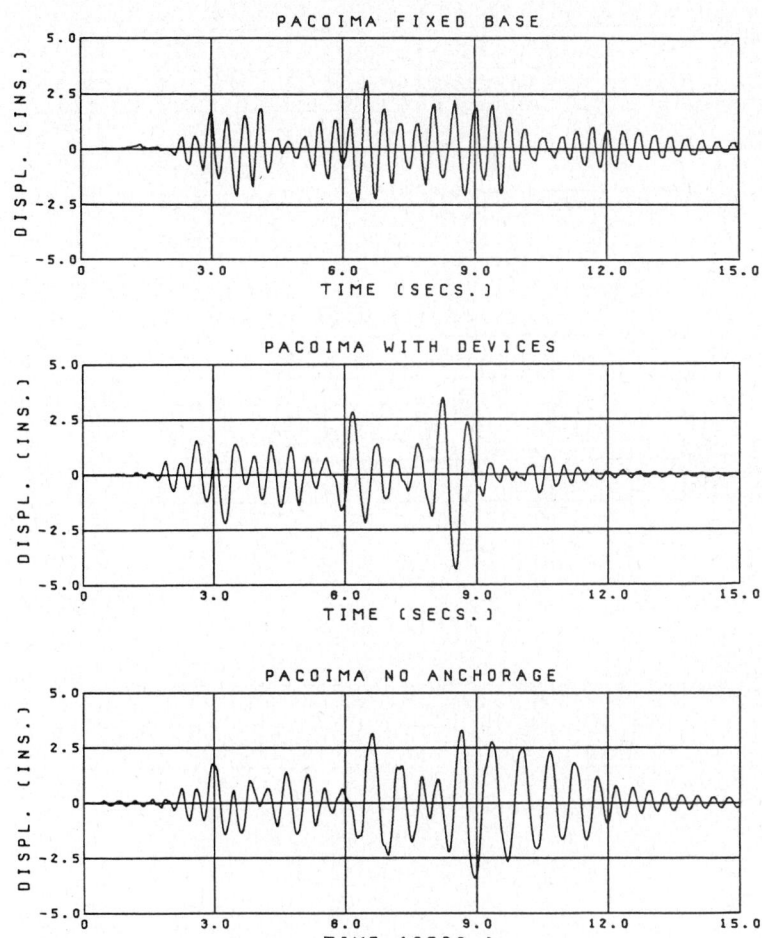

Figure 7 - Pacoima Relative Third Floor Horizontal Displacement Comparisons

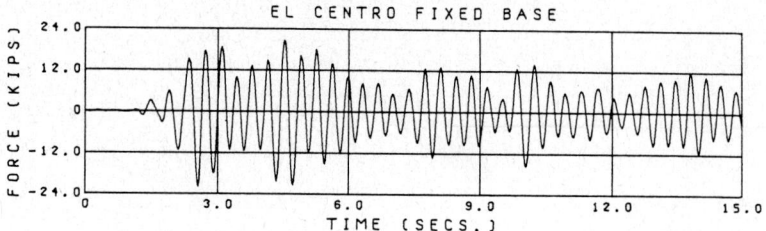

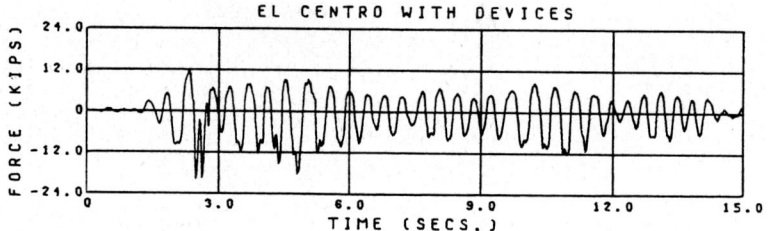

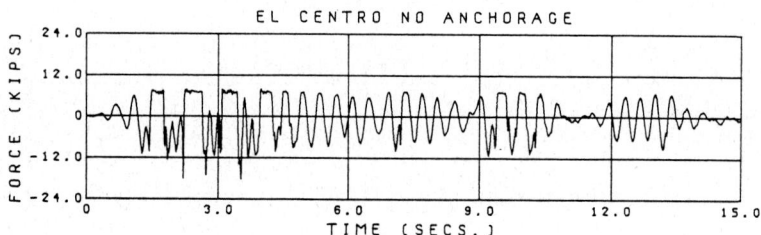

Figure 8 - El Centro First Floor North Column Axial Force Comparisons

Table 1 - Summary of El Centro Peak Responses

	FIXED BASE	WITH DEVICES	WITHOUT ANCHORAGES
AVE. UPLIFT			
NORTH COLS.	0"	1.26"	1.99"
SOUTH COLS.	0"	1.10"	3.03"
REL. 3RD. FLR. DISPL.	+2.43" -2.02"	+2.29" -2.74"	+5.34" -3.72"
1ST. FLR. COL. AXIAL FORCE			
NORTH COL.	+20.76 K -22.14 K	+11.64 K -19.93 K	+7.85 K -18.44 K
SOUTH COL.	+22.10 K -18.93 K	+14.42 K -13.33 K	+8.99 K -20.50 K
BASE SHEAR	+35.21 K -35.53 K	+19.57 K -24.11 K	+26.90 K -25.25 K
BASE OVERTURNING MOMENT	+471.3 K-FT -434.3 K-FT	+278.7 K-FT -297.7 K-FT	+323.2 K-FT -296.1 K-FT

Table 2 - Summary of Pacoima Peak Responses

	FIXED BASE	WITH DEVICES	WITHOUT ANCHORAGES
AVE. UPLIFT			
NORTH COLS.	0"	2.24"	2.40"
SOUTH COLS.	0"	1.65"	2.05"
REL. 3RD. FLR. DISPL.	+3.12" -2.36"	+3.47" -4.31"	+3.27" -3.44"
1ST. FLR. COL. AXIAL FORCE			
NORTH COL.	+22.96 K -25.15 K	+14.20 K -22.59 K	+7.94 K -21.03 K
SOUTH COL.	+25.72 K -20.89 K	+15.73 K -31.36 K	+10.20 K -21.18 K
BASE SHEAR	+37.58 K -32.72 K	+27.80 K -31.35 K	+24.32 K -30.72 K
BASE OVERTURNING MOMENT	+563.8 K-FT -481.1 K-FT	+358.3 K-FT -431.8 K-FT	+343.3 K-FT -398.4 K-FT

Hysteresis loops for each device taken before and after testing showed that no detectable deterioration of the devices had occurred during testing. The concept of control devices associated with a form of base isolation has been demonstrated by this test series to be feasible, and should, as such, be considered as a design concept for tall structures subject to earthquake loading.

3. BASE ISOLATION EXPERIMENTS

The control devices used in the stepping frame experiments were also tested in conjunction with a natural rubber bearing base isolation system [7]. This is a natural use of such devices since large relative displacements are needed to dissipate the large amounts of kinetic energy induced in a structure during an earthquake attack. Large displacements, say between floors, would lead to unacceptable damage to nonstructural components, but small displacements would require large forces so that the points of attachment of the devices to the structural system would have to be specially strengthened.

The devices play a number of roles in the response of the isolation system to earthquake loading. Since they are elastic for small displacements and their elastic stiffness is high relative to that of the rubber bearing system, they act as mechanical fuses and cause the structure to behave as if rigidly based for small excitation. Thus, under small excitation the structure typically amplifies ground acceleration. As the excitation increases in intensity, the devices yield, producing large hysteresis loops as the structure oscillates. The tangent stiffness of the devices when yielded is around 5% of their elastic stiffness. Thus the fundamental frequency of the structure drops, and the system acts as an isolator with very high damping. The accelerations induced in the structure are, of course, somewhat greater than they would be if the rubber bearing system alone were used, but much higher earthquake intensities can be sustained [8].

The base isolation system to which the devices were added has been extensively tested [7, 8]. It uses multilayer natural rubber bearings which are a development of bearings currently used as vibration isolators in buildings constructed in areas of high traffic disturbance such as above underground railway systems. These vibration isolation systems have been in use since the mid-sixties and are a logical extension of bearings used in highway bridges. The requirements for bearings for an earthquake isolation system differ, however, from those for the above applications and a completely new type of bearing had to be developed. The lateral stiffness of such a bearing must be very low, it must be able to accept large lateral deflections but remain stable, and it must perform well under long-term high vertical loading. The bearings used in these tests (Figure 9) were developed and constructed at the Malaysian Rubber Producers' Research Association, Hertford, England.

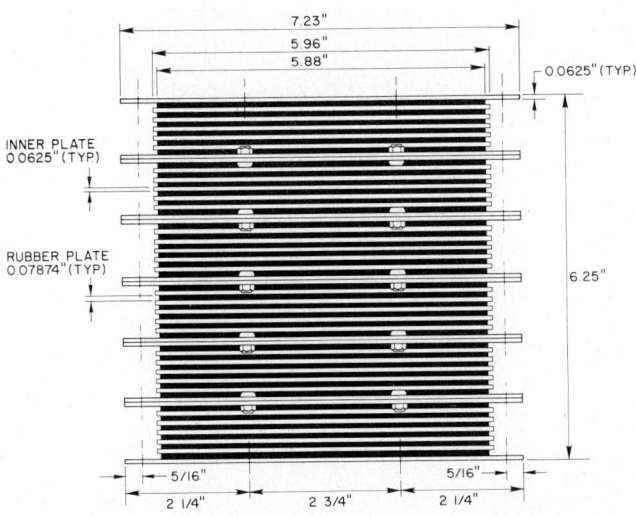

Figure 9 - Typical Natural Rubber Bearing

The isolation system was tested on the shaking table at the Earthquake Engineering Research Center. The model steel frame in which the system was incorporated is shown in Figure 10. Except for the base floor and associated isolation system, this model is identical to that used by Clough and Tang [9]. The model weighs 39.5 kips and is twenty feet high and twelve-by-six feet in plan dimension. Each of the three floors and the base of the model were loaded with 8-kip concrete blocks to simulate dead weight and to provide a period of vibration in the range appropriate to steel and concrete structures. A single control device was mounted on a load cell and attached to the frame as shown in Figure 11.

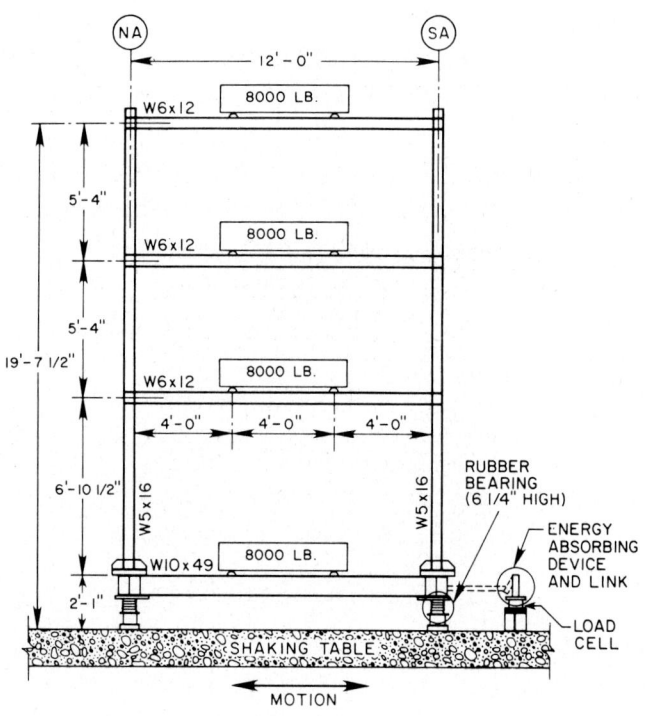

Figure 10 - Dimensions of Structure

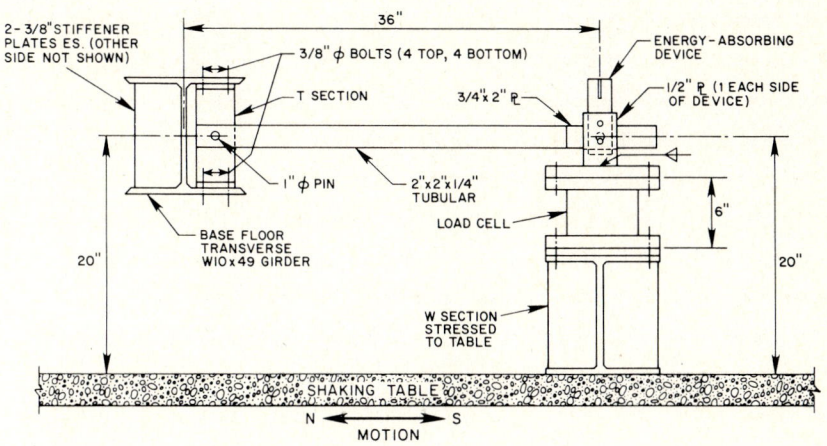

Figure 11 - Detail of Control Device Connection

For minor earthquake loading, the structural behaviour of the base-isolated model did not differ significantly from that of the structure on a conventional rigid foundation. For moderate to severe ground motion, the first story yielded and the effective first mode period of the structure increased. The torsion devices then absorbed large amounts of energy equivalent to from 30% to 35% of critical viscous damping. Both maximum base deflection and peak acceleration of the model in response to the scaled earthquake motions were reduced considerably. The higher mode frequencies and response of the structure with the control devices and the bearings incorporated did not change significantly after the devices had yielded. When mounted on a conventional rigid foundation and subjected to the input used to test the base-isolated model, the response of the structure was at least 50% greater.

The displacement of the system when fixed and when base-isolated and subjected to a scaled El Centro ground motion record are compared in Figure 12; the acceleration response is compared in the following figure. The fixed frame responded primarily at its fundamental frequency, with displacement and acceleration increasingly amplified the higher the story. In contrast, the base-isolated structure with the control devices produced a fairly uniform response in each floor and responded with a low frequency during periods of high-intensity excitation and with a higher frequency during less intense portions of the input record. The ground motion of the first seven seconds of the El Centro record is intense, followed by a period of relatively low intensity between seven and twelve seconds, after which a second high-intensity pulse occurs. The frequency shift at these points in the record is particularly clear in the displacement record (Figure 13). This result illustrates one of the important features of the action of the control devices, namely the drop in frequency associated with large displacements, a significant factor in reducing acceleration in the structure. The damping action provided by the control devices is also clear in Figure 14 which shows hysteresis loops for the type A device for a portion of the force-displacement time history during the most intense phase of the table input. This damping action, which for the record shown is equivalent to 35% of critical viscous damping, has an additional and substantial effect in reducing acceleration.

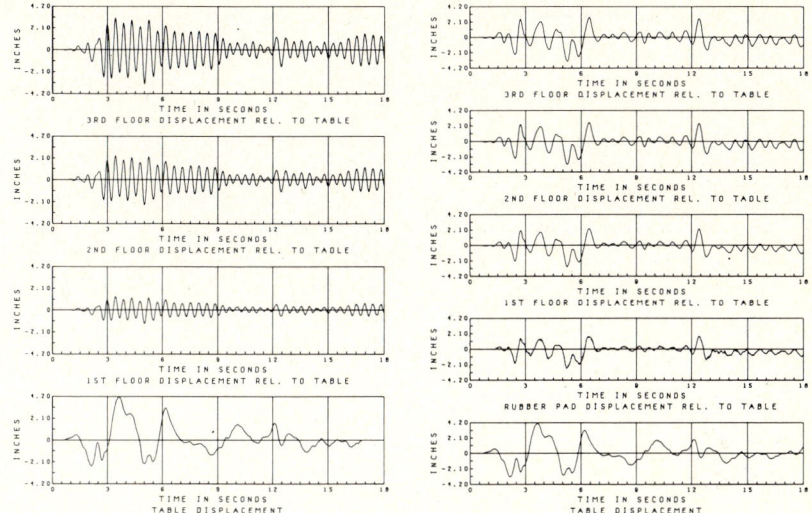

Figure 12 - Displacement Response, Fixed Foundation (Left) and Type A Control Device (Right), El Centro Motion

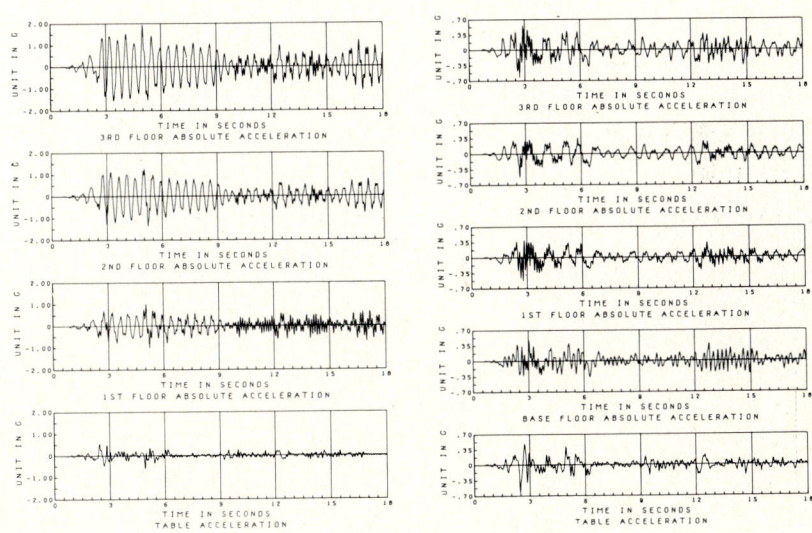

Figure 13 - Acceleration Response, Fixed Foundation (Left) and Type A Control Device (Right), El Centro Motion

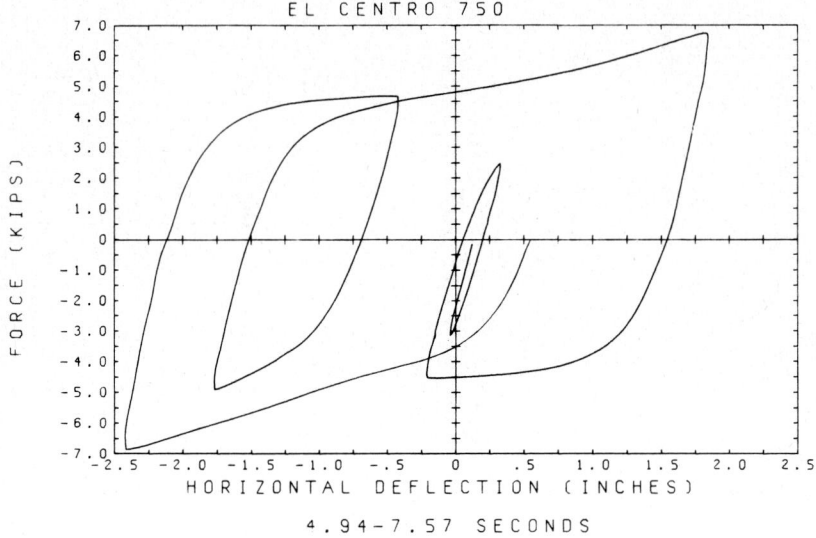

Figure 14 - Hysteresis Loops for Type A Control Device, El Centro Motion

The features of the response of the base-isolated structure just described for the El Centro ground motion used in the testing programme held true for a scaled Parkfield motion. The most intense portion of the Parkfield record is between three and six seconds; the reduction in frequency during this period is clear from Figures 15 and 16.

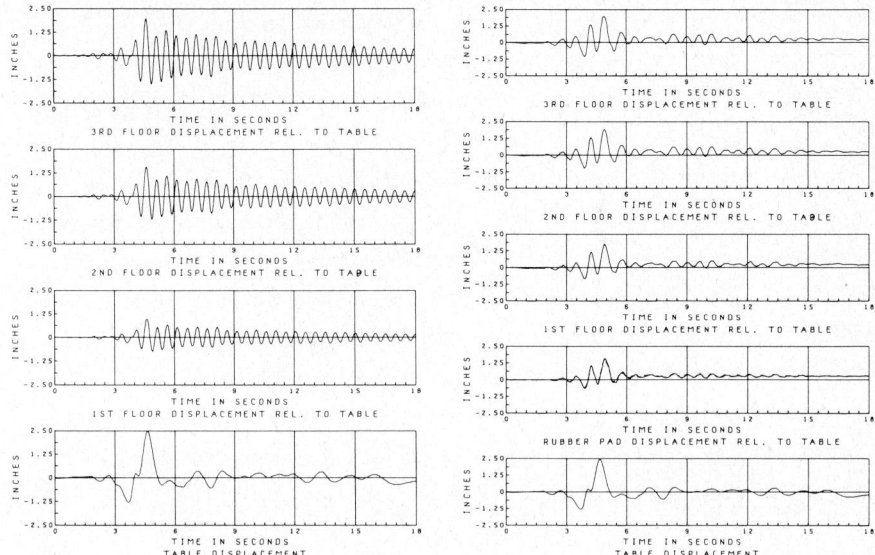

Figure 15 - *Displacement Response, Fixed Foundation (Left) and Type A Control Device (Right), Parkfield Motion*

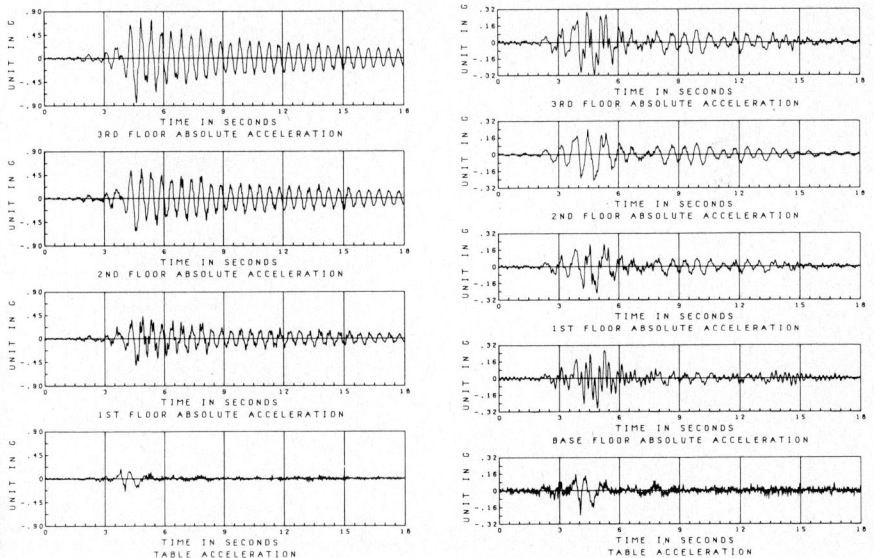

Figure 16 - *Acceleration Response, Fixed Foundation (Left) and Type A Control Device (Right), Parkfield Motion*

The response of the frame foundation in which different types of control device were incorporated is illustrated in Figures 17 and 18. The post-yield stiffness of the type A device is lower than that of the type B device and, when subjected to a scaled Pacoima Dam excitation, this device exhibited a period of drift in the foundation. This drift did not appear when the type B device, with a significantly higher post-yield stiffness, was tested. The hysteresis loops for the type B device (Figure 19) differ markedly from those for the type A device (Figure 14). The displacement and acceleration maxima for the type B device are somewhat higher than those for the type A device, but the former may be preferable where permanent drift at the end of ground motion is considered to be a problem.

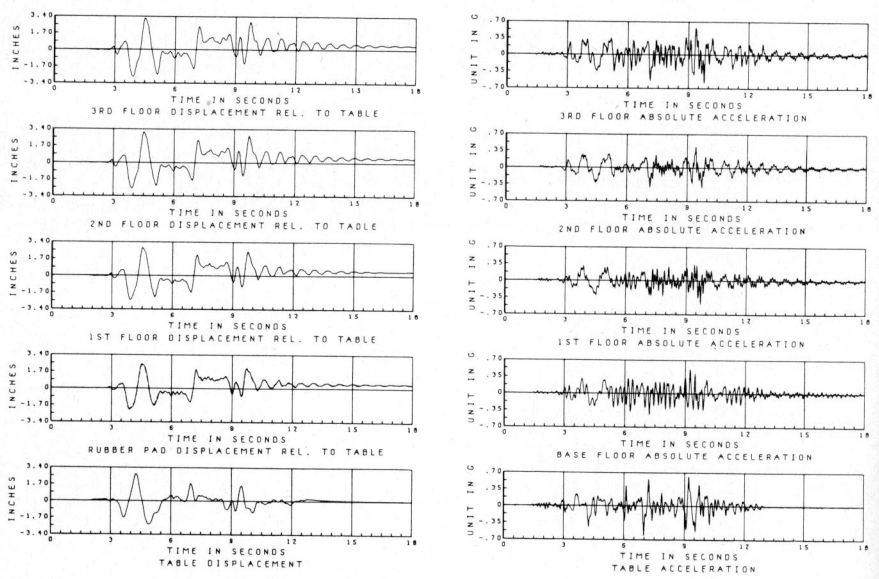

Figure 17 - *Displacement and Acceleration Responses, Type A Control Device, Pacoima Dam Motion*

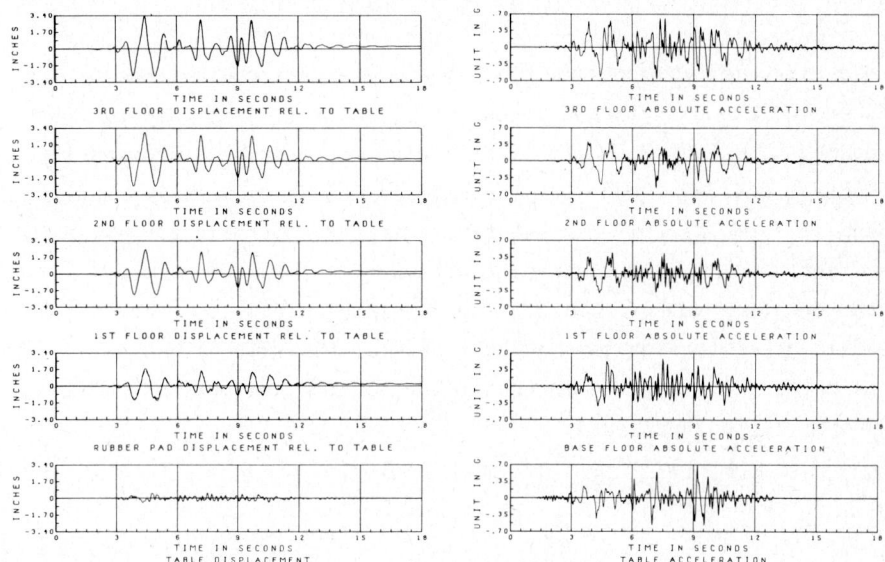

Figure 18 - *Displacement and Acceleration Responses, Type B Control Device, Pacoima Dam Motion*

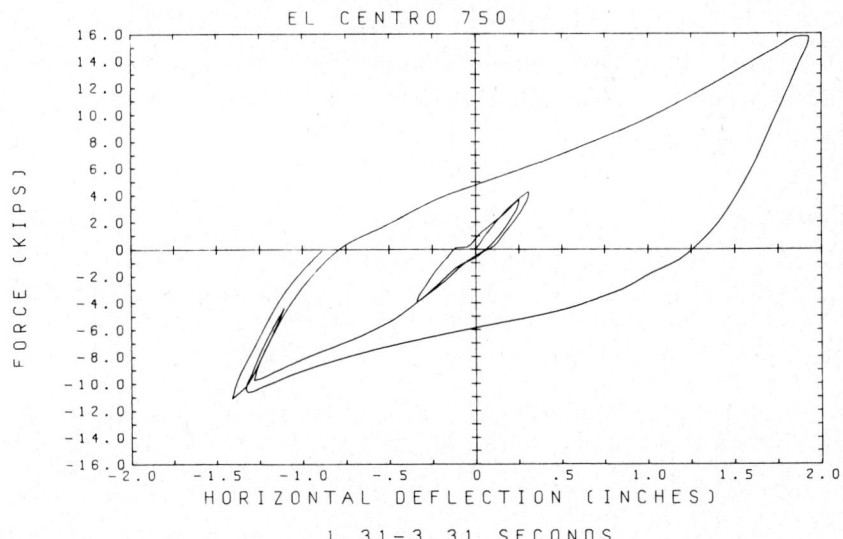

Figure 19 - *Hysteresis Loops for Type B Control Device, El Centro Motion*

4. CONCLUDING REMARKS

The predominant theme of this symposium on structural control has been the use of active devices in the control of structural response. The emphasis here is different. The devices are passive control devices. However, the application is to seismic loading and in such cases it is essential that the control devices remain effective over many years of inactivity, but act when needed in the event of an earthquake. Thus, mechanical devices such as the solid steel torsion bars and the natural rubber bearings have advantages over active devices which require sensors and on-line computers to function.

5. ACKNOWLEDGEMENTS

The Malaysian Rubber Producers' Research Association, Hertford, England, developed and constructed the natural rubber bearings described in this report and helped support the testing of the base isolation system; financial support for the development and testing of the control devices was provided by the National Science Foundation under Grant No. ENV-76-04262. This support is gratefully acknowledged.

REFERENCES

[1] BECK, J.L. and SKINNER, R.I., "The Seismic Response of a Reinforced Concrete Bridge Pier Designed to Step", *Int. J. of Earthquake Engineering and Structural Dynamics*, Vol. 2, 1974, pp. 343-358.

[2] KELLY, J.M., SKINNER, R.I. and HEINE, A.J., "Mechanisms of Energy Absorption in Special Devices for Use in Earthquake Resistant Structures", *Bulletin of the New Zealand National Society for Earthquake Engineering*, Vol. 5, 1972, pp. 63-68.

[3] SKINNER, R.I., KELLY, J.M. and HEINE, A.J., "Hysteretic Dampers for Earthquake Resistant Structures", *Int. J. of Earthquake Engineering and Structural Dynamics*, Vol. 3, 1975, pp. 287-296

[4] SKINNER, R.I., KELLY, J.M. and HEINE, A.J., "Energy Absorption Devices for Earthquake-Resistant Structures", *Proc. of the Fifth World Conference on Earthquake Engineering*, Rome, Italy, 1973.

[5] KELLY, J.M. and TSZTOO, D.F., "The Development of Energy-Absorbing Devices for Aseismic Base Isolation Systems", *Report No. UCB/EERC-78/01*, Earthquake Engineering Research Center, University of California, Berkeley, 1978.

[6] KELLY, J.M. and TSZTOO, D.F., "Earthquake Simulation Testing of a Stepping Frame with Energy-Absorbing Devices", *Bulletin of the New Zealand National Society for Earthquake Engineering*, Vol. 10, 1977, pp. 196-207.

[7] DERHAM, C.J., EIDINGER, J.M., KELLY, J.M. and THOMAS, A.G., "Natural Rubber Foundation Bearings for Earthquake Protection - Experimental Results", *Natural Rubber Technology*, Vol. 8, Part 3, 1977, pp. 41-61.

[8] KELLY, J.M. and EIDINGER, J.M., "A Practical Soft Story Earthquake Isolation System", *Report No. UCB/EERC-77/27*, Earthquake Engineering Research Center, University of California, Berkeley, 1977.

[9] CLOUGH, R.W. and TANG, D.T., "Earthquake Simulator Study of Steel Frame Structure, Vol. I: Experimental Results", *Report No. UCB/EERC-75/6*, Earthquake Engineering Research Center, University of California, Berkeley, 1975.

STRUCTURAL CONTROL, H.H.E. Leipholz (ed.)
North-Holland Publishing Company & SM Publications
© IUTAM, 1980

THE TIME-OPTIMAL CONTROL OF WIND INDUCED
STRUCTURAL VIBRATIONS USING ACTIVE APPENDAGES

R.E. Klein and H. Salhi

Department of Mechanical and Industrial Engineering
University of Illinois at Urbana-Champaign
Urbana, Illinois 61801, U.S.A.

1.0 INTRODUCTION

The need for active control of civil engineering structures is becoming increasingly apparent. Various methods for increasing the energy damping of structures are now being employed. Examples include the energy absorbing dampers installed on the World Trade Center Towers in New York City [1], the tuned mass Hula-Hoop type lead ring dampers used to damp the second and the fourth natural modes of the Canadian National Tower in Toronto [2], and the tuned mass dampers installed on the Citicorp Center in Manhattan and the Hancock Center in Boston [3].

As the state of the art progresses, it is becoming increasingly evident that effective tuned mass damping devices on structures often require substantial increases in added mass, and, in the case of the tuned mass damper for buildings [3], external power is often required. In fact, the analysis by Petersen [3] indicates that a fully active mass damper, for a building similar in size to the Citicorp Center, would require an energy input capability of 3,000 Hp (2240 KW) of hydraulic power. This high level of predicted energy input for a fully active system for the

Citicorp Center was considered prohibitive and thus the design used primarily a passive approach, with hydraulic power assist to overcome friction in the device.

The notion of using active aerodynamic appendages or control surfaces has been suggested by Klein [4, 5, 6]. The purpose of this paper is to further discuss and explore the design possibilities and consequences of active aerodynamic structural control methods. Many of the concepts concerning active aerodynamic appendage control have been previously stated in the literature. See references [4-6]. In addition, the thesis by Salhi [7] presents some preliminary modelling and time-optimal control results. The balance of this paper consists of a brief physical description of an appendage system, a collection of remarks, and then a section dealing with conclusions.

2.0 PHYSICAL DESCRIPTION

An active aerodynamic appendage system can take many forms, however, it is easy to visualize it as a small manipulatable control surface attached to the structure. The horizontal force on a structure, say, in the alongwind direction, is known to be approximated by

$$F(t) = C_d \, A \, V(t)^2 \, , \qquad (1)$$

where C_d is the coefficient of drag, A is the projected frontal area, $V(t)$ is the fluid or wind free stream velocity relative to the structure and t is time. The control surface or appendage can, conceivably, be designed to alter, or vary, either the coefficient of drag term C_d or the projected frontal area term A, or both of them. The desired objective then is to use this manipulated control surface, varied according to a negative feedback velocity signal from the structure, to alter the net force on the structure.

Thus, with the appendage, the horizontal force is given by

$$F(t,\dot{x}) = C_d(\dot{x}) \, A(\dot{x}) \, V(t)^2. \qquad (2)$$

It is well known that a variety of aerodynamic phenomenon may result in wind induced structural excitation of a structure. Some of these mechanisms are periodic vortex shedding, buffeting by wakes from other structures, gust loading, galloping, and flutter, for example. One effect of these mechanisms is to permit wind created energy to enter the structure and to be exhibited as vibrational energy. The appendage control system considered herein works on the principle of dissipating that vibrational energy once it is present in the structure. Irrespective of the particular aerodynamic mechanism for input of energy, a negative velocity feedback signal controlling a moveable appendage will dissipate energy. Furthermore, because the energy input to a structure is often due to broadband excitation and also becuase it is distributed spatially over the structure, the energy in the structure, as seen as an oscillation of the natural frequencies, is slow to accumulate. However, the manipulated appendage can, hypothetically, be positioned at a location which avoids all zero crossings of the modal vibrations. Consequently, a small manipulated additional force due to increased drag and controlled by the feedback signal can be quite effective in dissipation of energy because of the preciseness of its modulation and due to its select location on the structure.

The preliminary work by Klein et al., [3] demonstrates that a total differential in projected frontal area, assuming that C_d remains constant, of a structure of one percent is capable of reducing ambient vibrational energy by ninety percent.

3.0 REMARKS

Remark 1. An active appendage can, in certain circumstances, have the ability to minimize the magnitude of the aerodynamic disturbance. Specifically, in the case of aero-elastic excitations such as galloping, flutter, and aero-elastic lock in, the stabilization of the structure has the secondary effect of preventing the onset of the excitation mechanism. Similar results using active control surfaces on aircraft wings prevent, inhibit, or shift to higher velocities, the onset of high speed flutter.

Remark 2. The aerodynamic approach to structural control for wind induced excitation permits the designer to exploit the energy in the mean value of the wind to control the structure which is being excited by the very same wind. Any requirement for externally supplied energy can be obviated.

Remark 3. In the case of a cantilevered type structure, such as a building, the top and/or the upper sides are the best selection site for an appendage controller. This occurs because all modes of vibration are typically large at the free boundary condition and this provides an excellent location for dissipating energy. This occurs because the average or mean velocity of the wind at that location is at a maximum, the controller mechanism is capable of simultaneously dissipating all modal vibrations of interest, and the upper corners have an enhancing or magnified effect on the flow field around the structure.

Remark 4. It is conjectured that torsional motion of a cantilevered structure may be controlled using, for example, two controllers located at the outer upper edge of the structure so as to create appropriate moments.

Remark 5. Unlike tuned mass dampers, tendons, and other similar devices, the appendage device increases its effectiveness in a wind storm because of the increase in the average wind velocity impacting on the appendage. This is in considerable contrast to other methods, such as the tuned mass damper, which in certain buildings, are designed to shut down if predetermined travel limits of the mass are exceeded.

Remark 6. The selection of an appendage location in structures having non-cantilevered structure or geometry is a complex matter. Types of structures which are nontrivial include bridges, arches, antenna dishes, and cable stayed towers. In these cases, the selection of a controller location, if only one is used for example, requires a location selection which satisfies a suitable design objective. A suitable objective might be to maximize the minimum energy dissipation per mode.

A suitable formulation, or index of performance IP, could be

$$IP = \max_{z \in [0,L]} \{\min[F(z)A_1|\psi_1(z)|w_1, \ldots, F(z)A_n|\psi_n(z)|w_n]\},$$

where:

- z - structural axial coordinate indicating length;
- L - maximum axial length of structure (assumed to be one dimensional for simplicity in this formulation);
- $F(\cdot)$ - average available incremental aerodynamic force at each location on the structure;
- $\psi_j(z)$ - normalized modal mode for the jth mode;
- A_j - an *a priori* bound on the maximum amplitude of the jth mode in the absence of control;
- w_j - frequency of the jth mode in radians/second.

The above IP is formulated to maximize the minimum energy dissipation per mode. It is predicted on force available, the amplitude of dynamic sway, and the rate of cycling. Criterion analogous to this have been discussed by Cannon and Klein [8]. Other criteria to minimize fatigue load and/or maximum stress are also capable of formulation. It is worthy to note that the control systems community has expressed an interest or concern in finding the best or most optimal location for controllers and sensors in a spatially distributed structure. While such a quest is of mathematical interest esoterically, the opinion of the senior author is that a vast number of locations are suitable in engineering terms and emphasis might be better placed upon avoiding bad locations. Bad locations are those on or adjacent to zero crossings of mode shapes and locations with economic or architectural shortcomings. Goodson and Klein [9] discuss general observability requirements in distributed systems, and Cannon and Klein [8] provide an analysis of one method for determining optimal sensor location in a particular distributed parameter system. The opinion of the senior author, being a contributor in that area, is that a rigorous search for the very best sensor of controller location is not necessarily warranted in view of the availability of a good number of "suitable" locations which assure observability and controllability conditions. As an additional note, the degree of degradation in performance as one departs from an optimal or best location is not serious so long as reasonable prudence is used. In short, many suboptimal locations can serve quite adequately.

Remark 7. The use of an aerodynamic appendage necessarily requires a slight increase in the average or static horizontal load or drift of the structure. However, this allows the designer to achieve other goals such as:

(i) suppression of "lock in";
(ii) lower peak deflections (static plus dynamic combined);
(iii) reduction of fatigue cycle damage;
(iv) reduction of ambient levels of building acceleration and jerk so as to improve occupant comfort, and thus reduction in creaking movement, and shaking of the building's interior, such as lamp fixtures, hanging curtains, water in open bowls, and the like.

Remark 8. The nature of externally fitted light weight appendages on a structure holds open the feasibility of economical retrofit on troubled structures.

Remark 9. The aerodynamic appendage control method for structures increases the effective damping ratio of a structure. However, the percentage amount of increase varies approximately as the inverse of amplitude. Thus, the controller is capable of driving the motion to a stop in finite time similar to the action of Coulomb friction in a simple second order mechanical system.

Remark 10. The implementation of the aerodynamic control can be visualized as an appendage, however, it is conceivable that it can take on other forms. Other possibilities include:

(i) bleeds and/or suction devices;

(ii) a variation in effective surface roughness (such as a system of movable shrouds, strakes, and the like);

(iii) a design involving variable porosity changes in the interior of the structure.

It should be noted that the control can exploit either changes in projected frontal area (such as with an appendage) or changes in the coefficient of drag C_d of the structure via geometry of flow changes. Work in other subsonic, high Reynolds number applications indicate that the edge geometry of leading edges on bluff bodies is crucial to drag determination. In addition, the use of bleeds, suctions, and dimples can be, in general, effective in determining the shape of the turbulent boundary layer and trailing wake. It is well to recall that the dimension or size of the trailing turbulence is useful as an approximate measure of drag force on a structure. Successes in other areas such as frontal drag reduction on semi-trailer trucks or lorries of the order of 26% have been achieved by means of deflectors although the projected frontal area is unchanged.

Remark 11. Simple water table flow studies by the author (see Figures 1, 2 and 3) illustrate the possible changes in turbulent boundary layers for a rectangular object in a flow stream. Figure 1 is with no appendage. Figure 2 is with an upright appendage and Figure 3 indicates the reduced size of the trailing wake resulting from a rotated appendage. Conjecturally, it is felt that achievement of changes in C_d using modest geometry changes are worthy of further study.

Figure 1 - Flow Visualization of Structure without Appendage

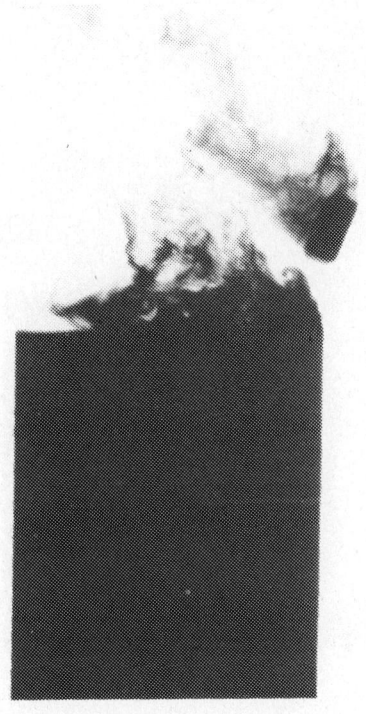

Figure 2 - Flow Visualization with Appendage in the "On" Position

Figure 3 - Flow Visualization with Appendage in the "Off" Position

Remark 12. Because of the hesitance on the part of designers to consider vibration control devices that require a sizable level of external power, such as electricity to power hydraulic pumps, consideration has been given to design of low energy requirement aerodynamic appendages. A number of configurations are available, however it is sufficient to think of appendages which rotate using wind energy upon a specific input or release command.

The only energy from external sources to permit devices to function is that required by the latching devices. By proper mechanical design that can be held small. It should be noted that:

(i) energy input requirements are limited to the energy required for the command signals to latch and delatch a wind activated appendage;

(ii) the cycle time is restricted in size, but size can be kept small enough to meet any timing requirement by means of multiple devices which can be arranged in stacks.

Remark 13. The control of wind induced oscillations in the crosswind direction represents an interesting challenge. One alternative is to use an angled appendage to create a controlled force component in the crosswind directions. The ancient Phonecians did this *millenia* ago, and its potential application to structures is a distinct possibility. A second alternative is to exploit any existing coupling between the various fundamental vibrational modes of a structure. It is axiomatic that if energy can be dissipated from any mode of a coupled dynamic process, then the energy level in all modes is diminished due to the nature of the coupling. A fundamental requirement of control by means of either alternative above is that the structure be well conditioned in that the various natural frequencies of concern be sufficiently distinct.

Remark 14. The structural control of any building, tower, or bridge is a significantly easier task if the structure has been designed in advance to be compatible with control concepts. Several major requirements may often be met in advance by use of proper design, and thus greatly simplify the control problem. First, the various natural frequencies should be separated or distinct which can be achieved by proper design. Second, the coupling between various modes should be increased and encouraged which is also possible by proper design. For example, Garland [10] illustrates that if the mass axis and the elastic axis of a structure do not coincide, then significant coupling occurs between the torsional mode and one or more of the lateral modes. Third, the design should be such that the proper control location, e.g., the best position for a controller, has sufficient structural strength to handle the forces which would be created by a controller.

Remark 15. The oscillation of a natural frequency or mode shape yields a specific period of oscillation which decreases inversely with the corresponding frequency. Clearly, the aerodynamic appendage control method will be limited to control of those vibrations having periods of sufficient duration to be compatible with the ability to cycle and appendage on and off. While this can be construed as a limitation of sorts, it is worthy to note that as structures become larger, the periods of oscillation will become longer. For example, the World Trade Center Towers in New York have a fundamental lateral period of approximately eleven seconds. As height and size increase, and as construction methods yield "softer" buildings, then the periods will increase. In fact, it could be speculated that the advent of technology for structural vibrational control will lessen the rigidity requirements of buildings. This, in turn, will make control easier due to the longer periods involved.

Remark 16. It should be recalled that over design or under design of any product, including large structures, is costly. Heretofore, building design, in order to handle dynamic loads, such as those which are wind induced, has been based largely upon static design and safety factors. The establishment of acceptable and economical design codes for design of structures with respect to dynamic loading is a nontrivial task. It is worthy to note, conjecturally, that structural control devices could circumvent, in part, the dynamic loading problem and hence reduce the design criterion to that of static considerations.

Remark 17. Buildings and other large structures require an immense commitment of capital. Space vehicles and warships are often funded from a public treasury, with an air of urgency and accepted uncertainty, and without conventional "insurance" coverage in the event of loss. Large buildings, however, are constructed from risk capital and the investors then routinely protect themselves with insurance. Furthermore, the nature of the building as an investment is largely dictated by tax laws in many nations and a major incentive for the structure is the creation of a secure and low risk tax shelter. Thus, the investors in a building are seldom looking for spectacular profits (or losses). This strengthens the requirement of insurability of the structure as a prerequisite for construction. On the other hand, insurance underwriters and companies often require either a statistical base or history and also adherence to conventional construction codes. When a new technology, such as structural control by whatever means is introduced, it is reasonable to assume that a building with those innovations will be uninsurable due to the absence of a historical base of a publically accepted confidence level. The difficulty in meeting insurance underwriting standards or in obtaining municipal code variances, for example, can often deter buildings with novel design approaches, especially when dynamic behaviour or its control is concerned.

4.0 CONCLUSIONS

This paper has discussed the active control of large structures using feedback methods and manipulated aerodynamic appendages. The investigations to date have been theoretical or computer simulation based. Some preliminary two dimensional flow visualization studies have been done at the University of Illinois in the Department of Mechanical and Industrial Engineering using a conventional water table. At the time of this writing, wind tunnel model prototypes illustrating the concepts are being conducted at both the University of Illinois at Urbana-Champaign, by the senior author and at the State University of New York at Buffalo under direction of Dr. T.T. Soong in conjunction with CALSPAN Corporation's wind tunnel facilities.

The remarks presented in the paper are often conjectural or theoretical in nature, but they are based upon the senior author's past decade of study and theorizations of active control of tall buildings using active appendages, and upon a rigorous training and study in the observability and controllability of spatially distributed parameter systems.

REFERENCES

[1] Industrial Specialties Division, 3M Company, 3M Center, St. Paul, Minnesota, 55101, U.S.A.

[2] "Lead Hula-Hoops Stabilize Antenna", *Engineering News Record*, Vol. 197, No. 4, July 22, 1976, p. 10.

[3] PETERSEN, N.R., "Design of Large Scale Tuned Mass Dampers", *Proc. of the IUTAM Symposium on Structural Control*, held at the University of Waterloo, Waterloo, Ontario, Canada, June 4-7, 1979.

[4] KLEIN, R.E., CUSANO, C. and STUKEL, J.J., "Investigation of a Method to Stabilize Wind Induced Oscillations in Large Structures", presented at the 1972 ASME Winter Annual Meeting, New York, November, 1972, Paper #72-WA/AUT-11.

[5] KLEIN, R.E., "Methods for Vibrational Energy Dissipation", *Proc. of the Third U.S. National Conference on Wind Engineering Research*, University of Florida, Gainesville, Florida, U.S.A., 1978.

[6] KLEIN, R.E., "The Potential for Application of Closed-Loop Control Concepts in Structures", presented at the ASCE Convention and Exposition, Boston, April, 1979, Paper #2, Session EM-3.

[7] SALHI, H., "The Time Optimal Control of Wind Excited Structures Using Controllable Aerodynamic Appendages", *M.S. Thesis*, Department of Mechanical and Industrial Engineering, University of Illinois at Urbana-Champaign, January, 1979.

[8] CANNON, J.R. and KLEIN, R.E., "Optimal Selection of Measurement Locations for Approximate Determination of Temperature Distributions", *ASME, J. of Dynamic Systems, Measurement and Control*, September, 1971.

[9] GOODSON, R.E. and KLEIN, R.E., "A Definition and Some Results for Distributed System Observability", *IEEE Trans. on Automatic Control*, Vol. AC-15, No. 2, April, 1970, pp. 165-174.

[10] GARLAND, C.F., "The Normal Modes of Vibrations of Beams Having Noncollinear Elastic and Mass Axes", *ASME, J. of Appl. Mech.*, September, 1940, pp. A-97-A-105.

STRUCTURAL CONTROL, H.H.E. Leipholz (ed.)
North-Holland Publishing Company & SM Publications
© IUTAM, 1980

OPTIMAL PROJECT OF A CYLINDRICAL SHELL FOR
MODERATELY LARGE DEFLECTIONS

J. Lellep

Tartu State University
Tartu, Estonia, U.S.S.R.

A. Sawczuk

Institute of Fundamental Technological Research
Polish Academy of Sciences
Warsaw, Poland

1. INTRODUCTION

Plastic structures, optimal under the condition of minimum weight, are usually designed for the collapse load, thus, under the requirement of incipient plastic flow. Such design appears to be sensitive to geometrical changes, the structures undergo in the post-yield range [1]. The necessity thus arises to account for configuration variation in the optimal plastic design.

This involves first the question of optimality criteria as the incipient collapse load is no longer suitable to the purpose since the load carrying capacity varies when the structure deforms. Nonlinearity of the strain-displacement relations and the displacement dependence of the equilibrium equations make the load carrying capacity depend on the actual configuration, and stable as well as unstable behaviours can occur. An optimization for a given deflected shape, associated with the load carrying capacity of a uniform structure, seems reasonable.

On the other hand, a suitable method of optimization to account for nonlinearities involved is to be applied. A possible approach is that of the optimal control theory, thickness of the structural elements being the control variable.

In the present note optimal design of a plastic cylindrical shell for moderately large deflections is considered as a problem of the optimal control theory. The deflection is required to be that of the constant thickness shell in the post-yield range, and the optimal thickness variation is sought for under the requirement of minimum material consumption. Applicability of the maximum principle to plastic structures optimization at configuration changes is shown.

2. PROBLEM FORMULATION

We consider a closed, sandwich cylindrical shell of length 2L and radius A, hinged at the end sections and allowed to displace in the axial direction. The shell wall consists of a constant layer H carrying shears and of two layers of thickness $h(x)$ carrying membrane forces and moments. The coordinate system with its x-axis coinciding with the shell's undeformed generator has the origin at the median cross-section of the tube. The structure consisting of a cylindrical shell and of two end plates is subjected to internal pressure P considered as a dead load at small configuration changes.

We are looking for a shell with variable thickness $h(x)$, whose deflection $w(x)$ coincides with the deflections of a shell of constant thickness h_* but whose volume takes a minimum value. As a justification for such a requirement can serve the conclusion that for a rigid-plastic material, the condition [2, 3]

$$\int_V (Q_1-Q_2)(\dot{q}_1-\dot{q}_2)dv = 0 \qquad (1)$$

holds,

where 1 and 2 denote two different solutions at the incipient plastic flow. Q_i stands for generalized stresses and $\dot{q}_i$ denote generalized strain rates. For $\dot{q}_1 = \dot{q}_2$ the situation might thus exist such that $Q_1 \neq Q_2$. The stress distribution in an optimal structure will be different in comparison with that corresponding to the constant thickness case and therefore it seems reasonable to assume (1) at moderately large deflections of a shell.

For the considered shell and loading, the internal forces contributing to the energy dissipation are: the axial force $N_x = AP/2$, the circumferential force N_ϕ and the bending moment M_x. The axial and transverse displacements are U and W respectively. We attempt to find the optimal project within the theory of moderately large deflections assuming the shell takes a specified shape.

The equilibrium equation for a shell element is

$$\frac{d^2 M_x}{dx^2} - \frac{AP}{2} \frac{d^2 W}{dx^2} + \frac{N_\phi}{A} - P = 0 , \qquad (2)$$

whereas the strain-displacement relations have the form

$$\varepsilon_x = \frac{dU}{dx} + \frac{1}{2}\left(\frac{dW}{dx}\right)^2 , \qquad \varepsilon_\phi = \frac{W}{A} ,$$

$$K_x = \frac{d^2 W}{dx^2} , \qquad K_\phi = 0 . \qquad (3)$$

For convenience we introduce the following dimensionless variables

$$\xi = \frac{X}{L} , \quad v = \frac{h}{h_*} , \quad w = \frac{W}{H} , \quad u = \frac{UL}{H^2} , \quad \alpha = \frac{2L^2}{AH} ,$$

$$n_x = \frac{N_x}{2\sigma_0 h_*} , \quad n_\phi = \frac{N_\phi}{2\sigma_0 h_*} , \quad m_x = \frac{M_x}{\sigma_0 H h_*} , \quad p = \frac{PA}{2\sigma_0 h_*} , \qquad (4)$$

where σ_0 denotes the material yield point and h_* relates to the reference shell of constant wall thickness.

In view of (4) the equilibrium and geometrical relations take the form

$$m_x'' - pw'' + \alpha(n_\phi - p) = 0 ,\qquad(5)$$

and

$$\varepsilon_x = \frac{H^2}{L^2}\left(u' + \frac{1}{2}w'^2\right), \qquad \varepsilon_\phi = \frac{H}{A}w ,$$

$$\kappa_x = \frac{H^2}{L^2}w'', \qquad\qquad \kappa_\phi = 0 ,\qquad(6)$$

where prime denotes differentiation with respect to ξ.

The shell material is assumed to obey the Tresca yield condition [4, 5]. For a sandwich shell of variable thickness, a yield surface is used as defined by the planes

$$n_\phi = \pm v , \quad n_\phi - n_x = \pm v , \quad 2n_\phi - n_x - m_x = \pm 2v ,$$
$$-n_x - m_x = \pm v , \quad -n_x + m_x = \pm v , \quad 2n_\phi - n_x + m_x = \pm 2v .\qquad(7)$$

As we are interested more in developing a design procedure for plastic structures at large deflection range employing the optimal control theory rather than in a specific solution, we assume that the plastic potential law applies to the strain components (6). A sort of deformation theory of plasticity is therefore employed, according to which the strain components are orthogonal to the yield surface (7).

3. REFERENCE SOLUTION

For the intended optimization, the deflected shape is needed for a shell of constant thickness in the large deflection range. The load-deflection relation concerning sandwich shells when the equilibrium equation (5) holds was studied by Duszek [6]. For completeness the appropriate relation regarding shells obeying the employed deformation law will be determined and will serve as a reference.

For a shell of constant wall thickness, $v = 1$, two different stress regimes among those given by (7) occur when the structure goes plastic and continues to deform in the post-yield range. In the central zone, half of which is $0 \leq \xi \leq \xi_1$, the stress corresponds on the yield surface to the edge given by $n_\phi = 1$, $n_x - m_x = 1$ whenever $n_x \geq 1/2$. Then the deflections are subjected to the requirements $\varepsilon_\phi \geq 0$, $\varepsilon_x = -\kappa_x$, $\kappa_x \leq 0$. This stress state is given by the point E' in Figure 1 representing the part of the yield surface belonging to the plane $n_\phi = 1$. In the edge zone $\xi_1 \leq \xi \leq 1$ the stress profile belongs to the plane $n_\phi = 1$, the line EF, Figure 1, and $\varepsilon_\phi \geq 0$, $\varepsilon_x = \kappa_x = 0$.

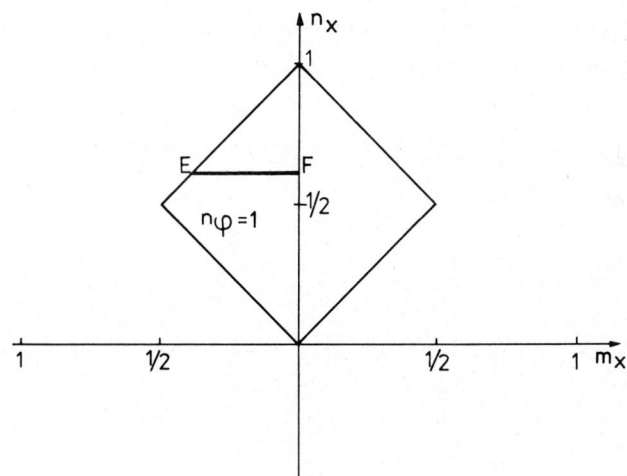

Figure - 1

Making use of the stress profile in the equilibrium equation (5) and accounting for the geometrical relations (6) the load-displacement relation can be derived. In view of the boundary conditions $w(o) = w_o$, $w'(o) = 0$, $m_x(1) = 0$, $w(1) = 0$ and the continuity imposed at $\xi = \xi_1$ on u, w, w', m_x and m'_x, the following relation is eventually obtained

$$w = \begin{cases} w_o - \dfrac{\alpha(p-1)}{2p} \xi^2 , & 0 \leq \xi \leq \xi_1 , \\ \\ \dfrac{\alpha(p-1)}{p} \xi_1(1-\xi) , & \xi_1 \leq \xi \leq 1 , \end{cases} \qquad (8)$$

whereas

$$w_o = \frac{1}{2p} [p(1+\alpha) - 2 - \alpha] , \quad \xi_1 = 1 - \sqrt{\frac{2-p}{\alpha(p-1)}} . \qquad (9)$$

The relations (8) and (9) apply in the case of moderately large deflections, namely those of the order of shell wall thickness as the equations (5) and (6) are limited to such situations. They apply therefore to short shells, $\alpha \leq 8$ as indicated in [6].

The reference solution (9) will be used in deriving an optimal thickness distribution for a shell sensitive to geometrical changes in the post-yield range.

4. MATHEMATICAL SETTING

We look for the optimum design of a short shell. The problem consists in finding the minimum of the functional

$$I = \int_0^1 v(\xi) d\xi \qquad (10)$$

under the requirement that the shell deflections are as given by (8) and (9). It means that when minimizing the functional (10) there have to be satisfied: the equilibrium requirement (5), the geometrical relations (6) as well as the potential law associat with the yield surface (7). It appears that the requirement $v \leq 1$ throughout cannot be satisfied for the considered problem and therefore the optimal project will be looked at without such a constraint.

The problem will now be set in terms of the optimal control theory. The equilibrium equation (5) and the requirements (8) lead eventually to the following relations

$$m_x = \begin{cases} \alpha(1-v), & 0 \leq \xi \leq \xi_1, \\ \alpha(p-v), & \xi_1 \leq \xi \leq 1. \end{cases} \quad (11)$$

Let us consider $x_1 = m_x$ and $x_2 = m_x'$ to be the state variables whereas v is the control. Then, equations (11) can be written in the form

$$\begin{aligned} x_1' &= x_2, & x_2' &= \alpha(1-v), & \xi &\in [0,\xi_1], \\ x_1' &= x_2, & x_2' &= \alpha(p-v), & \xi &\in [\xi_1,1]. \end{aligned} \quad (12)$$

To the system (12) are associated the boundary conditions

$$x_1(1) = 0, \quad x_2(0) = 0. \quad (13)$$

In the central zone, the deflection (8) is associated with the stress profile E on the yield surface, Figure 1. Two yield equations are then satisfied, namely $n_\phi = v$ and $-m_x + n_x = v$. Since $n_x' = 0$ in the considered theory and $n_x = p/2$ for the considered case of loading, the following additional constraint is obtained

$$v = \frac{p}{2} - x_1, \quad 0 \leq \xi \leq \xi_1. \quad (14)$$

In the second part of the trajectory, thus $\xi_1 \leq \xi \leq 1$, the stress profile belongs solely to the plane $n_\phi = v$. Assuming that for the considered short shells the axial bending moment does not change its sign, so that the stress profile EF is as indicated in Figure 1, the following constraint on the control variable is obtained

$$v \geq \frac{p}{2} - x, \qquad \xi_1 \leq \xi \leq 1. \tag{15}$$

This allows to disregard in the subsequent analysis the requirement $v \geq p/2$ imposed by the equilibrium condition. It will appear later that the equality sign is associated with the solution.

The considered problem consists thus in minimization of functional (10) among trajectories of the system (12) and (13) under the requirements (14) and (15).

5. NECESSARY OPTIMALITY CONDITION

The problem defined by the relations (10) and (12) to (15) is characterized by discontinuities. Therefore it is a problem "with discontinuous right sides". As the discontinuity point $\xi = \xi_1$ is assumed to be known, the dependent variables and their adjoint quantities should be continuous [7, 8].

Let us denote the adjoint coordinates by $\psi_1^\pm$, $\psi_2^\pm$, $\psi^\pm$, where the minus sign corresponds to the zone $(0, \xi_1)$ and analogously the plus sign means that the respective variable belongs to the zone $(\xi_1, 1)$ of the trajectory. As the state variables satisfy the boundary requirements (13), the transversality conditions result in

$$\psi_1^-(0) = 0, \qquad \psi_2^+(1) = 0. \tag{16}$$

For the trajectory in the zone $(0, \xi_1)$, the Hamiltonian function becomes

$$H^- = -v + \psi_1^- x_2 + \psi_2^- \alpha(1-v) + \psi^-\left(v - \frac{p}{2} + x_1\right), \tag{17}$$

Optimal Project of a Cylindrical Shell

and the adjoint coordinates satisfy the following relations

$$\bar{\psi}_1{}' = -\bar{\psi}, \qquad \bar{\psi}_2{}' = -\bar{\psi}_1, \qquad (18)$$

$\bar{\psi}$ is a non-zero valued function.

The maximum principle yields the equation

$$\frac{\partial \bar{H}}{\partial v} = -1 - \alpha \bar{\psi}_2 + \bar{\psi} = 0. \qquad (19)$$

In usual cases of control problems, condition (19) yields the optimal control. In the considered case, the control variable is governed, however, by expression (14). Hence, equation (19) allows to determine $\bar{\psi}$ as the control (13) in this zone is singular, [9].

Solving (18) and (19), when taking into account the boundary conditions (16), the adjoint coordinates are found to have the form

$$\bar{\psi}_1 = -\beta A_1 \operatorname{sh}\beta\xi, \qquad \bar{\psi}_2 = -\frac{1}{\alpha} + A_1 \operatorname{ch}\beta\xi, \qquad \bar{\psi} = \alpha A_1 \operatorname{ch}\beta\xi, \qquad (20)$$

where $\beta = \sqrt{\alpha}$ and A_1 denotes the integration constant.

In the right part of the trajectory, hence in the range $(\sqrt{\xi_1}, 1)$ the control variable v is subjected to the inequality constraint (15). Thus we shall introduce an additional control parameter t such that the constraint in question satisfies

$$v - \frac{p}{2} + x_1 + t^2 = 0. \qquad (21)$$

The Hamiltonian function for the considered part of the trajectories is therefore

$$\overset{+}{H} = -v + \overset{+}{\psi}_1 x_2 + \overset{+}{\psi}_2 \alpha(p-v) + \overset{+}{\psi}\left(v - \frac{p}{2} + x_1 + t^2\right), \qquad (22)$$

and the adjoint variables ψ_1^+, ψ_2^+, ψ^+ satisfy the conditions

$$\psi_1^{+\prime} = -\psi^+, \qquad \psi_2^{+\prime} = -\psi_1^+ . \tag{23}$$

According to the maximum principle, in this case,

$$-1 - \alpha\psi_2^+ + \psi^+ = 0, \qquad \psi^+ t = 0 . \tag{24}$$

From the second of equations (24) it follows that two cases have to be considered, namely: (a) $\psi^+ = 0$, $t \neq 0$ and (b) $\psi^+ \neq 0$, $t = 0$. The first case is, however, inadmissible since then, according to (24), it is $\psi_2^+ = -\frac{1}{\alpha} = \text{const.}$, which contradicts the boundary requirements (16). Hence, the case $t = 0$, $\psi^+ \neq 0$ applies. It follows from (21) that the necessary optimality condition for the second part of the trajectory is

$$v = \frac{p}{2} - x_1, \qquad \xi_1 \leq \xi \leq 1, \tag{25}$$

and therefore the equality sign applies in (15).

The maximum principle applies if the adjoint variables are non-zero valued functions. Solving the system (23) to (24) under the boundary conditions (16) and taking into account the continuity requirements imposed on ψ_1 and ψ_2 at $\xi = \xi_1$ the following relations are obtained

$$\psi_1^- = \psi_1^+ = -\frac{1}{\beta} \frac{\sh\beta\xi}{\ch\beta} ,$$

$$\psi_2^- = \psi_2^+ = -\frac{1}{\alpha} + \frac{\ch\beta\xi}{\alpha\ch\beta} , \tag{26}$$

$$\psi^- = \psi^+ = \frac{\ch\beta\xi}{\ch\beta} ,$$

where relations (20) are taken into account.

It appears that in spite of the fact that the right hand sides of equation (12) have discontinuities at $\xi = \xi_1$ the adjoint variables are given by (26) for both parts of the trajectory.

6. OPTIMAL PROJECT

It was shown that the optimal project requires the relations (25) to be satisfied. This allows to integrate the set (12). The integration yields

$$x_1 = \frac{p}{2} - 1 + A_1 ch\beta\xi + A_2 sh\beta\xi ,$$
$$x_2 = \beta A_1 sh\beta\xi + \beta A_2 ch\beta\xi ,$$
(27)

in the zone $0 \leq \xi \leq \xi_1$, whereas for $\xi_1 \leq \xi \leq 1$,

$$x_1 = -\frac{p}{2} + B_1 ch\beta\xi + B_2 sh\beta\xi ,$$
$$x_2 = \beta B_1 sh\beta\xi + \beta B_2 ch\beta\xi .$$
(28)

A_1, A_2, B_1, B_2 appearing in (27) and (28) are integration constants. They are determined by the boundary conditions (13) and by the continuity requirements imposed on m_x and m_x' at $\xi = \xi_1$. The calculations yield the optimal trajectory x_1, x_2, and eventually the bending moment $m_x = x_1$,

$$m_x = \begin{cases} \frac{p}{2} - 1 + \left\{\frac{p}{2} - (p-1)ch[\beta(1-\xi_1)]\right\} \frac{ch\beta\xi}{ch\xi} , \\ \\ -\frac{p}{2} + \frac{1}{ch\beta} \left\{\frac{p}{2} ch\beta\xi + (p-1)sh\beta\xi_1 sh[\beta(1-\xi)]\right\} \end{cases}$$
(29)

According to (25), the optimal control, the shell wall thickness is:

$$v = \begin{cases} 1 - \left\{\dfrac{p}{2} - (p-1)\text{ch}[\beta(1-\xi_1)]\right\} \dfrac{\text{ch}\beta\xi}{\text{ch}\beta}, & 0 \leq \xi \leq \xi_1, \\ \\ p - \dfrac{1}{\text{ch}\beta}\left\{\dfrac{p}{2}\text{ch}\beta\xi + (p-1)\text{sh}\beta\xi_1\text{sh}[\beta(1-\xi)]\right\}, & \xi_1 \leq \xi \leq 1. \end{cases} \quad (30)$$

7. DISCUSSION

The optimal thickness of a moderately deforming shell is shown in Figure 2 at $\xi_1 = 0{,}3$. It can be remarked that in the central part the thickness differs, but insignificantly, from that corresponding to the reference shell of constant thickness. Specific values at different extents of the central zone corresponding to the stress regime E on a ridge of the yield surface are given in Table 1. The thickness in that zone exceeds unity.

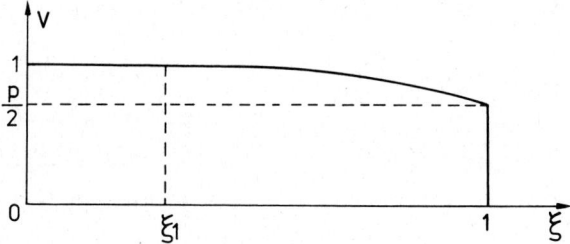

Figure 2

Table 1 - Optimal Shapes

α =	4	4	8	8
ξ_1 =	0,3	0,7	0,4	0,7
ξ	v	v	v	v
0	1,0153	1,00107	1,0116	1,00151
0,1	1,0156	1,00109	1,0120	1,00158
0,2	1,0166	1,00115	1,0134	1,00176
0,3	1,0182	1,00127	1,0160	1,00209
0,4	1,0137	1,00143	1,0198	1,00259
0,5	0,9963	1,00165	1,0148	1,00330
0,6	0,9651	1,00193	0,9902	1,00427
0,7	0,9190	1,00230	0,9442	1,00559
0,8	0,8561	0,98800	0,8728	0,98395
0,9	0,7738	0,94371	0,7704	0,91419
1	0,6689	0,86765	0,6289	0,79070

Economy of the optimal project established can be assessed by the ratio $\theta = I(v)/I(1)$, where the integrals I are calculated according to (10) both for the uniform thickness and the optimal one (30). The ratio is

$$\theta = \xi_1 + p(1-\xi_1) + \frac{1}{\beta ch\beta}\left[-\frac{p}{2}sh\beta + (p-1)sh\beta\xi_1\right]. \qquad (31)$$

Its values are given in the last column of Table 2 corresponding to the case $\alpha = 4$.

Table 2 - Volume Ratios at $\alpha = 4$

ξ_1	ξ_2	p	θ_1	θ
0	0,1186	1,2000	0,9172	0,9107
0,1	0,1911	1,2358	0,9267	0,9207
0,2	0,2673	1,2809	0,9369	0,9313
0,3	0,3473	1,3378	0,9478	0,9426
0,4	0,4312	1,4098	0,9591	0,9544
0,5	0,5188	1,5000	0,9705	0,9665
0,6	0,6100	1,6097	0,9812	0,9782
0,7	0,7043	1,7353	0,9903	0,9884
0,8	0,8013	1,8621	0,9966	0,9958
0,9	0,9002	1,9615	0,9995	0,9993
1	1,0000	2,0000	1,0000	1,0000

It is worthwhile to mention that the relations (12) to (15) are satisfied also by the following project

$$v = \begin{cases} 1, & 0 \le \xi \le \xi_1, \\ p, & \xi_1 \le \xi \le \xi_2, \\ \frac{p}{2} - x_1, & \xi_2 \le \xi \le 1, \end{cases} \quad (32)$$

where

$$\xi_2 = 1 - \frac{1}{\beta} \ln \left[\frac{p}{2(p-1)} + \sqrt{\frac{p^2}{4(p-1)^2} - 1} \right].$$

The ratio θ_1 specifying the material consumption of such a stepwise continuous shell is

$$\theta_1 = \xi_1 + p(1-\xi_1) - \frac{p-1}{\beta} \text{sh}[\beta(1-\xi_2)] . \quad (33)$$

Its values as well as the parameters specifying the position of thickness discontinuities are given in Table 2 for $\alpha = 4$. The economy ratio θ_1 differs slightly from that corresponding to the optimal project θ.

The solution procedure regarding optimization of rigid-plastic shells within the post-yield range has been developed. The optimal project of a shell assuming a required shape beyond the incipient collapse load was found employing the optimal control theory to singular control problems. For one part of the optimal trajectory the control variable assumes values exceeding those of the reference shell of constant thickness. This effect, and specific features of control theory problems appearing in optimization of plastic structures in the large deflection range, are discussed in [10].

ACKNOWLEDGEMENT

The study was started when the first writer was visiting the Institute for Fundamental Problems of Technology in Warsaw under a co-operative research programme operating between the respective governments.

REFERENCES

[1] MRÓZ, Z. and GAWĘCKI, A., "Post-Yield Behaviour of Optimal Plastic Structures", *Proc. IUTAM Symposium on Optimization in Structural Design*, Warsaw, 1973, Springer, Berlin, 1975, pp. 518-540.

[2] PRAGER, W., *Introduction to Plasticity*, Addison-Wesley, Reading, Mass., 1959.

[3] SAWCZUK, A. and JAEGER, Th., *Grenztragfähigkeits-Theorie der Platten*, Springer, Berlin, 1963.

[4] HODGE, P.G., Jr., *Limit Analysis of Rotationally Symmetric Plates and Shells*, Prentice-Hall, Englewood Cliffs, New Jersey, 1963.

[5] OLSZAK, W. and SAWCZUK, A., *Inelastic Behaviour in Shells*, Noordhoff International Publishers, Groningen, The Netherlands, 1967.

[6] DUSZEK, M., "Plastic Analysis of Cylindrical Sandwich Shells Accounting for Geometrical Changes", (in Polish), *Rozpr. Inz.*, Vol. 15, 1967.

[7] BRYSON, A.E., Jr., and HO, Y-C., *Applied Optimal Control*, Blaisdell, Waltham, 1969.

[8] LEPIK, U. and LELLEP, J., "Foundations of the Optimal Control Theory", (in Russian), Tartu University Press, Tartu, Estonia, U.S.S.R., 1978.

[9] BELL, D.J. and JACOBSON, D.H., *Singular Optimal Control Problems*, Academic Press, New York, 1975.

[10] LELLEP, J., "Application of the Optimal Control Theory to Optimum Design of Plastic Structures in the Post-Yield Range", (in preparation).

STRUCTURAL CONTROL, H.H.E. Leipholz (ed.)
North-Holland Publishing Company & SM Publications
© IUTAM, 1980

APPLICATION OF THE CONTROL THEORY FOR OPTIMAL DESIGN
OF NONELASTIC BEAMS UNDER DYNAMIC LOADING

Ülo Lepik

Tartu State University
Estonian S.S.R., U.S.S.R.

1. OPTIMAL DESIGN OF VISCOUS BEAMS UNDER IMPULSIVE LOADING

The first problem of optimal design, regarded in this paper, is the following. We shall take a simply supported beam of variable thickness. The material of the beam is nonlinearly viscous with a constitutive equation

$$\sigma = c\, \dot{\varepsilon}^n, \quad c > 0, \quad 0 \leq n \leq 1 \quad . \tag{1}$$

The beam is subjected to impulsive loading with a given kinetic energy K_o. The mean residual deflection of the beam is also given. The thickness distribution $h = h(x)$, for which the beam volume is minimal, is to be determined.

Now let us put this problem into a mathematical form. The equations of motion are

$$\overline{M}' = \overline{Q}, \quad \overline{Q}' = \rho h(x)\, B\, \ddot{w} \quad , \tag{2}$$

where ρ, B, $\overline{M}$ and $\overline{Q}$ stand correspondingly for the density, the width of the beam, the bending moment and the shear force; primes and dots denote differentation with respect to the coordinate x and time. For materials with the constitutive equation (1) permanent mode form motions exist and we can assume that

$$\dot{w}(x, t) = \Phi(t)\, v(x)\,. \qquad (3)$$

Let us introduce the quantities

$$M(x) = \overline{M}(x,t)\, \Phi^{-n}(t),\ Q(x) = \overline{Q}(x,t)\, \Phi^{-n}(t)\,. \qquad (4)$$

By substituting (3) into the equations (2) and by separating the variables, we obtain

$$\dot{\Phi} + \lambda^2 \Phi^n = 0\,,\ M' = Q\,,\ Q' = -\rho \lambda^2 B\, h\, v\,, \qquad (5)$$

where λ^2 is an eigenvalue of the mode form solution.

The first equation of (5) can be integrated easily. By taking $\Phi(0) = 1$, we get

$$\Phi^{1-n} = 1 - (1-n)\,\lambda^2 t\,. \qquad (6)$$

Let us assume that $B = $ const and the Bernoulli hypothesis $\dot{\varepsilon} = -z\dot{w}''$ holds good. Multiplying equation (1) by z and integrating over the area of the cross-section of the beam, we find

$$M = \frac{1}{A^n} h^{n+2}(-v'')^n\,,\quad A = 2\left[\frac{2(n+2)}{cB}\right]^{1/n}\,. \qquad (7)$$

If l is a half of the beam's length, then the initial kinetic energy is

$$K_0 = \int_0^l \rho B\, h\, v^2 dx\,. \qquad (8)$$

We shall define the mean residual deflection of the beam as follows:

$$\overline{w} = \left[\int_0^1 \rho B h \dot{w}^2 (x, t_f) \, dx \right]^{1/2} . \tag{9}$$

Making use of (3), (6) and (8), the formula (9) can be put into the form

$$\overline{w} = \frac{\sqrt{2K_o}}{(2-n)\lambda^2} . \tag{10}$$

Since K_o and n are given constants, it follows from (10), that a specified eigenvalue λ^2 corresponds to a given value of the mean residual deflection.

Now we shall introduce the following state variables

$$y_1 = v , \quad y_2 = v' , \quad y_3 = M , \quad y_4 = Q . \tag{11}$$

The last two equations of (5) and the isoperimetric constraint (8) can now be presented in the form

$$y_1' = y_2 , \quad y_2' = - A y_3^{\frac{1}{n}} h^{-\frac{n+2}{2}} , \quad y_3' = y_4 ,$$
$$y_4' = - \rho \lambda^2 B h y_1 , \quad y_5' = \rho B h y_1^2 . \tag{12}$$

The boundary conditions are

$$y_2(0) = y_4(0) = y_1(1) = y_3(1) = 0 , \quad y_5(0) = 0 , \quad y_5(1) = K_o . \tag{13}$$

The functional

$$V = \int_0^1 B h \, dx , \tag{14}$$

subjected to the constraints (12) and the boundary conditions (13), has to be minimized. To solve this mathematical problem, we shall use the methods of the optimal control theory. First we shall make up the Hamiltonian

$$H = \psi_0 B h + \psi_1 y_2 - \psi_2 A y_3^{\frac{1}{n}} h^{-\frac{n+2}{2}} + \psi_3 y_4 - \psi_4 \rho \lambda^2 B h y_1 + \psi_5 \rho B h y_1^2 . \qquad (15)$$

The adjoint system is

$$\psi_1' = -\frac{\partial H}{\partial y_1} = \psi_4 \rho \lambda^2 B h - 2\psi_5 \rho B h y_1, \quad \psi_2' = -\frac{\partial H}{\partial y_2} =$$
$$= -\psi_1, \quad \psi_3' = -\frac{\partial H}{\partial y_3} = \frac{A}{n} \psi_2 y_3^{-\frac{1-n}{n}} h^{-\frac{n+2}{n}}, \qquad (16)$$
$$\psi_4' = -\frac{\partial H}{\partial y_4} = -\psi_3, \quad \psi_5' = -\frac{\partial H}{\partial y_5} = 0 ,$$

with the boundary conditions $\psi_1(0) = \psi_3(0) = \psi_2(1) = \psi_4(1) = 0$.

Since there are no constraints for the control $h = h(x)$, the necessary condition for an extremum is $\partial H/\partial h = 0$. This equation allows us to eliminate the variable h from our equations.

In the following part is is convenient to go over to the new variables

$$x = \xi 1 , \quad y_1 = \frac{x_1}{\sqrt{\psi_5}} , \quad y_2 = \frac{x_2}{1\sqrt{\psi_5}} , \quad y_3 = 0.5 \rho \lambda^2 1^4 B N x_3 ,$$
$$y_4 = \rho \lambda^2 1^3 B N x_4 , \quad \psi_1 = \psi_5 1^3 N x_5 , \quad \psi_2 = \psi_5 1^4 N x_6 ,$$
$$\psi_3 = \frac{\sqrt{\psi_5}}{\rho \lambda^2 B 1} x_7 , \quad \psi_4 = \frac{\sqrt{\psi_5}}{\rho \lambda^2 B} x_8 , \quad N = \left(\frac{\rho \lambda^2 B}{1 \psi_5}\right)^{\frac{1}{n+1}} . \qquad (17)$$

The equations (12) and (16) can be put into the form

$$\dot{x}_1 = x_2, \quad \dot{x}_2 = -\frac{\beta}{\alpha} T s , \quad \dot{x}_3 = 2x_4, \quad \dot{x}_4 = -x_1 T ,$$

$$\dot{x}_5 = (x_8 - 2x_1) T , \quad \dot{x}_6 = -x_5 , \quad \dot{x}_7 = \frac{2\beta}{\alpha n} \frac{s T}{x_3} x_6 , \quad \dot{x}_8 = -x_7 ,$$

where

$$\alpha = \sqrt{n+2}\left(\frac{2}{n}\right)^{\frac{n}{2(n+1)}}, \quad \beta = \frac{2}{\sqrt{n+2}}\left(\frac{n}{2}\right)^{\frac{n+2}{2(n+1)}},$$

$$s = \frac{1}{x_6}(1 + x_1 x_8 = x_1^2), \quad T = \alpha\, x_3^{\frac{1}{2(n+1)}} s^{-\frac{n}{2(n+1)}}.$$

Dots in equations (18) denote now differentiation with respect to ξ.

The system (18) is accompanied by the following boundary conditions $x_2(0) = x_4(0) = x_5(0) = x_7(0) = x_1(1) = x_3(1) = x_6(1) = x_8(1) = 0$. So we have to solve a nonlinear boundary value problem. In paper [1] it has been done by using the methods of nonlinear programming, some results for different values of n are shown in Figure 1. It follows from the condition $\partial H/\partial h = 0$ that $h = 1^2 N \sqrt{\psi_5}\, T$; since ψ_5 = const, the function $T = T(\xi)$ determines the thickness distribution of the beam.

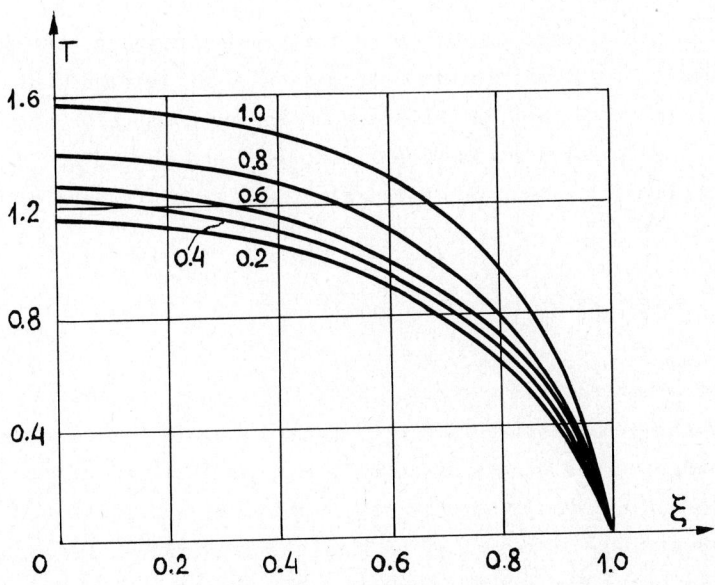

Figure 1

The method, which was described above, is also applicable for problems with two or three space coordinates or for non-linear elastic structures. For these aspects consult papers [1-2].

2. OPTIMAL DESIGN OF NONELASTIC BEAMS WITH ADDITIONAL SUPPORTS

In dealing with practical problems it is often important to increase the stiffness of structures. One way of doing this is to provide some points of the structure with additional supports. The location of these supports must be selected so as to maximize the global structural stiffness or minimize its compliance. Such a problem for beams was set up and solved by Mróz and Rozvany [3]. Taking the global complementary energy for the measure of the structure compliance, they derived the following optimality condition

$$W(s-) - W(s+) - R v' = 0 , \qquad (19)$$

where s is the location of the support, W - complementary energy of the beam, R - reaction of the support, v' - deflection's slope at x = s. The condition (19) holds good for nonelastic or viscous beams. The case of rigid-plastic material was solved by Prager and Rozvany [4]. For such materials the optimality condition has the form

$$Q(s-) v'(s-) = Q(s+) v'(s+) , \qquad (20)$$

where Q denotes the shear force. Making use of the optimal control theory the problem of optimal locations of beams for an arbitrary compliance criterion was discussed in paper [5].

In papers [3-5] only static problems were dealt with. It is interesting to make clear whether the criterions (19) and (20) are valid also for dynamic loading. This problem has been studied by Lepik [6]. It was shown that in the case of implusive loading, when we use the method of mode form solutions, the conditions

(19)-(20) hold good. In the following subsection some results, given in paper [6], will be reviewed. For the sake of conciseness we shall consider only rigid-plastic beams with rigid supports (as to other cases the reader is advised to make use of the original paper [6]).

Let us consider a rigid-plastic beam of constant thickness h, which has an additional support in the cross-section $x = s$. Similarly to the preceding part, the quantities K_o and λ^2 are given. We must find out such a location of the support where the volume of the beam will be minimal.

In the case of a rigid-plastic material we have $n = 0$ in the equation (1). The equations (2)-(6) and (8)-(11) remain valid. The condition (7) must be replaced by the inequality $|M| \leq M_o$, where $M_o = 0.25\, \sigma_o B\, h^2$ and σ_o denotes the yield stress. The equality $|M| = M_o$ holds good only for some discrete values $x = x_i$ where $i = 1, 2, \ldots, m$. In these cross-sections plastic hinges appear. For the other cross-sections, where $|M| \leq M_o$, the beam remains rigid and we have $\dot{w}'' = 0$ or correspondingly $v'' = 0$.

The state variables shall be taken again in the form (11). The formulae (12) will be valid if we replace $y_2' = 0$. The boundary conditions can be presented in the form

$$a_j y_j(0) + b_j y_j(1) = 0, \quad j = 1, 2, 3, 4,$$

where the constants a_j, b_j have values 0 or 1. Besides in the cross-section $x = s$ the following discontinuity condition must be fulfilled

$$y_4(s+) - y_4(s-) = R(s) \quad . \tag{21}$$

For solving the problem of optimal design we construct the augmented functional

$$L = \int_0^1 \sum_{j=1}^5 (\psi_j y_j' - H)dx + \sum_{j=1}^4 \nu_j [a_j y_j(0) + b_j y_j(1)] +$$

$$+ \nu_5 y_5(0) + \nu_6 [y_5(1) - 2K_o] + \nu_7 [y_4(s+) - y_4(s-) - R(s)] +$$

$$+ \sum_{j=1}^m \mu_i [y_3^2(x_i) - M_o^2] \quad ,$$

where $\psi_1, \ldots, \psi_5$, $\nu_1, \ldots, \nu_7$, μ_i are multipliers of Lagrange and the Hamiltonian H is defined by the equation (15), if we omit the third term. Further on we must calculate the total variation of L. From the equality $\Delta L = 0$, where the independence of the variations of state and control variables is taken into account, we see that the adjoint system (16) will remain valid if we take $\psi_3' = 0$. In addition to that we get the following groups of equations

1)
$$\psi_1(x_i-) = \psi_1(x_i+) \;,\; \psi_2(x_i-) = \psi_2(x_i+) = 0 \tag{23}$$

$$\psi_3(x_i-) - \psi_3(x_i+) = -2\mu_i y_3(x_i) \;,\; \psi_4(x_i-) = \psi_4(x_i+)$$

2)
$$\psi_i(s-) = \psi_i(s+) \;,\; i = 2, 3 \text{ for } |y_3(s)| < M_o \tag{24}$$

$$\psi_2(s-) = \psi_2(s+) = 0 \;,\; \psi_3(s-) - \psi_3(s+) = 2\mu_i y_3(s)$$

$$\text{for } |y_3(s)| = M_o$$

$$\psi_4(s-) = \psi_4(s+) = 0 \;.$$

3)
$$H(s-) = H(s+) \tag{25}$$

4)
$$\int_0^1 \left[\frac{\partial H}{\partial h} + 2 \left(\sum_{i=1}^m \mu_i \right) M_o \frac{\partial M_o}{\partial h} \right] dx = 0 \;. \tag{26}$$

If we take

$$\psi_1 = 0, \quad \psi_2 = 0, \quad \psi_3 = -y_2, \quad \psi_4 = y_1, \quad \psi_5 = 0.5\lambda^2 \tag{27}$$

the adjoint system (16) and the equations (23)-(24) will be satisfied; consequently the problem which was considered above is selfadjoint. By making use of the equations (15) and (27) we get

$$H = \psi_0 B h - y_2 y_4 - 0.5 \rho \lambda^2 B h y_1^2 .$$

In this formula only the state variable y_4 is discontinuous at $x = s$. Fulfilling the continuity condition (25), we get the optimality condition (20).

In an analogical way it can be shown, that in case of nonelastic or viscous materials the optimality condition (19) holds good as well.

The solution, which was given in the previous section, is based on the method of mode form motions. In order to estimate the exactness of such solutions it is important to obtain a class of more exact solutions. For this purpose the partial differential equations (2) must be integrated, keeping in mind the proper boundary and initial conditions. The optimization problem can be solved by the aid of the optimal control theory with distributed parameters. In connection with it some special effects (e.g., moving plastic hinges in the case of a rigid-plastic material) should be accounted. Some results of this field of investigation have recently been obtained by the author. The limited extent of this paper does not allow to present them here; they will be published in the "Transactions of Tartu State University" (Tartu Riikliku Ülikooli Toimetised). Therefore here we shall confine only to some brief remarks.

It turns out that in case of an exact solution the problem is generally not selfadjoint and it is impossible to get a concise optimality condition. In order to get a selfadjoint problem we must use particular forms of initial and terminal conditions. So in the case of a rigid-plastic beam it can be shown, that the solution is selfadjoint if the functional, to be minimized, has the form $F[x(x_*,t), w'(x_*,T), h]$ and for the final instant $t = T$ a constraint $g[w(x,T), w'(x,T), h] \leq S$ is valid (here F and g are arbitrary functions; S and x are given constants, $x \in [0,1]$). Now the optimality condition (20) takes an integral form

$$\int_0^T [Q(s+,t) w'(s+,t) - Q(s-,t) w'(s-,t)]dt = 0 \ . \qquad (28)$$

In order to get some estimates for the error due to the method of mode form solutions we shall consider an example. A rigid-plastic beam on two variable supports is subjected to a constant initial velocity field (Figure 2a). The beam volume is given. Such locations of supports, in which residual deflections in the centre and in the free ends of the beam are equal, will be considered optimal.

This example was solved both exactly and approximately, by using the mode form solution. In the second case the initial distribution cannot be uniform, but it must have the form as shown in Figure 2b. The location of the support s and the maximal

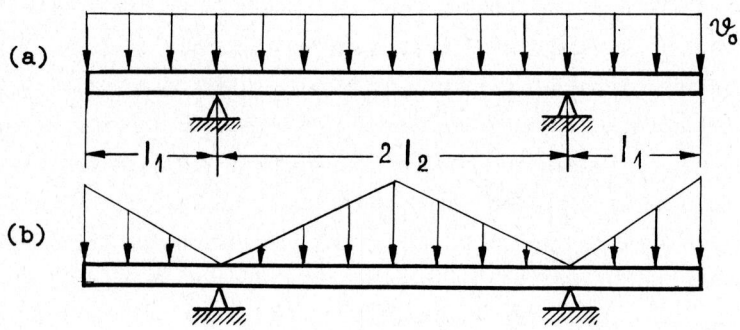

Figure 2

residual deflection w_{max} were calculated in both cases. By introducing the auxiliary variables

$$s = \frac{2}{l_1 + l_2}, \quad \mu = \frac{m\, l_1^2}{M_o},$$

where m is the mass of the beam unit, we get

(i) $s = 0.597$, $w_{max} = 0.367\mu$ for the exact solution,

(ii) $s = 0.586$, $w_{max} = 0.375\mu$ for the mode form solution.

Hence it follows that the method of modal form solutions guarantees good exactness: the error in calculating the location of the support and the residual deflection in the centre of the beam is only about 2 per cent.

REFERENCES

[1] MRÓZ, Z. and LEPIK, Ü., "Optimal design of viscous structures under impulsive loading," *Mekch. Polimerov N° 6*, 1977, pp. 1021-1028, (in Russian).

[2] MRÓZ, Z., "Mode approach to rational synthesis of structures under impulsive and dynamic pressure loading," *Trans. 4th Int. Conf. Struct. Mech. React. Tech.*, California, 1977, Vol. L, Amsterdam e.a., 1977, pp. L2.4/1-L2.4/8.

[3] MRÓZ, Z. and ROZVANY, G.I.N., "Optimal design of structures with variable support conditions," *J. Optimization Theory Appl.*, 15, 1975, pp. 85-101.

[4] PRAGER, W. and ROZVANY, G.I.N., "Plastic design of beams: optimal locations of supports and steps in yield moment," *Int. J. Mech. Sci.*, 17, 1975, pp. 627-631.

[5] LEPIK, Ü., "Optimal design of beams with minimum compliance," *Int. J. Non-Linear Mechanics*, 13, 1978, pp. 33-42.

[6] LEPIK, Ü., "Optimal design of nonelastic beams with additional supports in the case of dynamic loading," *Tartu Riikliku Ülikooli Toimetised*, 430, 1977, pp. 132-143, (in Russian).

STRUCTURAL CONTROL, H.H.E. Leipholz (ed.)
North-Holland Publishing Company & SM Publications
© IUTAM, 1980

ACTIVE DAMPING OF LARGE STRUCTURES IN WINDS

Richard A. Lund

MTS Systems Corporation
Minneapolis, Minnesota, U.S.A.

1. INTRODUCTION

Mass dampers have been installed in large buildings to reduce building motion during high winds. These systems are presently designed to operate as passive tuned mass dampers [1, 2]. An investigation of the benefits of active control in such systems is presented.

It is easily shown that it is not feasible to connect a suitable large damper mass to a tall building entirely with servohydraulics. The forces and velocities necessary can require thousands of hydraulic horsepower [1]. The large systems presently installed use a pneumatic spring to provide a majority of the force between the mass block and building. A hydraulic actuator provides the damping force as well as a "trim" force to insure smooth operation as a linear passive tuned mass damper.

It is of interest to investigate the possibility of an actuator force as a function of system motion variables that achieves a degree of active control in conjunction with the spring. The amount of force than can be achieved is severly penalized by the cost of the actuator, servovalve and hydraulic power supply required. Mass block motion (i.e., system stroke requirement) is similarly penalized and in addition, is limited by the cost of the mass block bearing surface and the available space. A desirable goal is to use active control to reduce the mass of the mass block while keeping the above variables within acceptable limits.

A simple approach to achieving active control is presented which, although not optimized, illustrates the nature of the possible improvement. Optimization of the design is beyond the scope of this effort. Results of the application of optimization techniques to similar systems can be found in the literature [3, 4]. These results are not directly applicable since the actuating device applies the total force between the structure and damper mass. Optimum design of a system which includes a passive spring is currently being investigated by Soong [5].

2. MODEL FOR ANALYSIS

A schematic of the system is shown in Figure 1. M_B represents the first mode modal mass of the building as seen by the damper. C_D represents the effective linear damping in the damper support bearings. The actuator force F represents the compressive force (minus on M_B and plus on M_D).

A pair of equations can be used to describe this model as:

$$M_B \ddot{y}_B + C_B \dot{y}_B + K_B y_B = C_D \dot{z} + K_D z + F_W - F , \qquad (1)$$

$$M_D \ddot{y}_D + C_D \dot{z} + K_D z = F , \qquad (2)$$

where $z = y_D - y_B$ is the relative motion between the mass block and building.

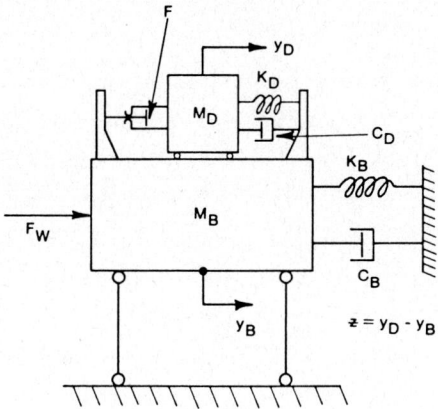

Figure 1 - Schematic of the System Model

3. CONTROL TECHNIQUE

It is desirable from a stability point of view to operate the servohydraulic actuator in stroke control. One way to do this is to compute a stroke reference from an ideal model of the desired mass block motion (1). A linear passive tuned mass damper model is obtained from equation (2) by setting the actuator force to zero obtaining:

$$M_D \ddot{y}_D + C_D \dot{z} + K_D z = 0 . \qquad (3)$$

This is put in a more convenient form by subtracting $M_D \ddot{y}_B$ from both sides and using $z = y_D - y_B$,

$$M_D \ddot{z} + C_D \dot{z} + K_D z = -M_D \ddot{y}_B . \qquad (4)$$

Now both sides are divided by M_D and the result put in transfer function form in terms of reference damping ratio ζ_R and reference natural frequency $\omega_{N,R}$ as:

$$\frac{z}{\ddot{y}_B} = \frac{-1}{p^2 + 2\zeta_R \omega_{N,R} p + \omega_{N,R}^2} \;, \qquad (5)$$

where $p = d/dt$, the derivative operator.

Equation (5) provides a simple way of computing the reference damper stroke from the building acceleration for a desired damping ratio and natural frequency. Combining this is equations (1) and (2) provides a complete description for analysis of required motions and forces for a mass damper controlled to operate as a passive tuned mass damper.

A simple way of achieving active control is to incorporate a gain factor (K_A) into equation (5) to allow the damper stroke reference level to be increased above that of a passive system. The stroke reference becomes:

$$\frac{z}{\ddot{y}_B} = \frac{-K_A}{p^2 + 2\zeta_R \omega_{N,R} p + \omega_{N,R}^2} \;, \qquad (6)$$

where $K_A = 1.0$ provides a passive system reference.

The nature of the active control achieved by this technique is not optimized. However, the simplicity of the approach might outweigh any additional gains in performance that are possible. Additional gains appear to be quite limited, principally due to the fact that performance is determined by the nature of the system at or very near the structure's natural frequency. The phase angle of the mass block motion relative to the structure is limited if good use is to be made of the spring to avoid large forces in the actuator. With these restrictions, the majority of the performance improvement available with active control might be achievable with the approach in equation (6). The forthcoming results from Soong [5] will clarify this point.

4. METHOD OF ANALYSIS

Wind force inputs to a building have been shown to be representable by a broad band random process [6]. An example of real wind spectra can be found in Amyot et al [7]. Since system response is mainly influenced by the spectral content of the wind force at the first mode natural frequency of the structure, it is reasonable to approximate the wind force spectrum as white noise. Extensive analysis of a passive tuned mass damper as applied to a large building with random wind inputs can be found in Wiesner [2] and McNamara [8].

Similar analysis is performed here with the exception that the spectral density of the wind input is limited in frequency content. This allows mean-square and RMS estimates to be calculated by numerically integrating the spectral density functions. Since the spectral density of the wind input is assumed constant, response spectral densities are that constant times squares and products of the frequency response functions. The frequency response functions are calculated using a technique similar to that described by McNamara [8].

An energy flow or power balance analysis of the system can also be performed from cross-spectral density considerations [9]. For example, the average wind power input to the building can be estimated from:

$$\text{wind power input} = \int_{-\infty}^{\infty} S_{f,v}(f) \, df , \qquad (7)$$

where $S_{f,v}(f)$ is the cross-spectral density between wind force input and building velocity $v_B = \dot{y}_B$. Wind power input must be dissipated by the building damping (C_B), the damper friction (C_D) or by the hydraulic actuator. These quantities can be estimated from similar cross-spectral calculations.

If a Gaussian probability distribution is assumed, the peak values of stroke, velocity and force can be estimated. In addition, the minimum hydraulic power required at the servovalve can be estimated as the average of the absolute value of the velocity times the peak force. Peak values are taken as three times the RMS for this purpose. Minimum hydraulic power is approximately equal to the actual required hydraulic power at the servovalve since actuator pressure drop does not inhibit flow through the servovalve in any of the systems analyzed.

5. RESULTS OF ANALYSIS

An analysis was performed on a system with approximately the same base-line parameters as that installed in the Citicorp Center in New York City [1]. A full scale analysis was performed to indicate the magnitudes of the quantities involved. Four specific cases were highlighted as shown in Table 1. The wind input used has a constant spectral density to a frequency of 1 Hz and a level which produces an RMS building motion of 4 inches with 4% structural damping. This level represents at least a 50 year return period storm [10].

Table 1

	M_B	$P_{N,B}$	ζ_B	M_D/M_B	$P_{N,D}$	ζ_D	K_A	ζ_R	$P_{N,R}$
Case I Building with No Damper	20,000 Ton (18.2E6 kg)	6.75 secs	.01	0.0	-	-	-	-	-
Case II Passive Damper	20,000 Ton (18.2E6 kg)	6.75 secs	.01	0.02	6.89 secs	.036	1.0	.11	6.89 secs
Case III Active Damper	20,000 Ton (18.2E6 kg)	6.75 secs	.01	0.02	6.89 secs	.036	1.5	.20	6.89 secs
Case IV Active Damper	20,000 Ton (18.2E6 kg)	6.75 secs	.01	0.015	6.89 secs	.031	2.0	.20	6.89 secs

System stroke and actuator force are of primary importance and so graphs of these parameters versus effective damping were made. Effective damping is defined as the damping required to achieve the same RMS level in a structure without a mass damper. The two curved plots in Figure 2 are the stroke requirements versus effective damping for a passive system. The reference damping is varied to trade off stroke versus effectiveness.

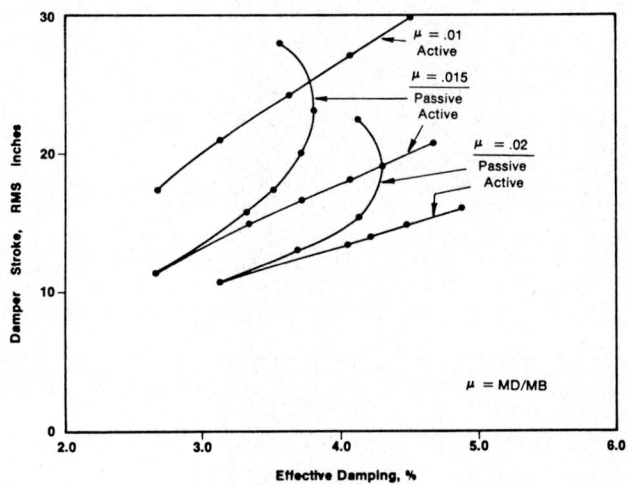

Figure 2 - Damper Stroke Versus Structure Effective Damping

The three relatively straight lines are the stoke requirements with active control. The reference damping was fixed at .2 while K_A was varied to achieve the stroke versus effectiveness tradeoff. It is clear that active control can be used to reduce the stoke requirement and to achieve damping levels with a given mass ratio that are beyond that achievable with a passive system.

This gain does not come free as can be seen in Figure 3, which shows the actuator force required versus effective damping. The two curves that are concave to the left are for the passive systems, while the remaining curves are for the active cases. It should be noted that the force primarily resists velocity for the passive systems while the force can approach ± 90° of phase with respect to velocity for the active systems. Thus the active systems require a larger servovalve or greater system pressure.

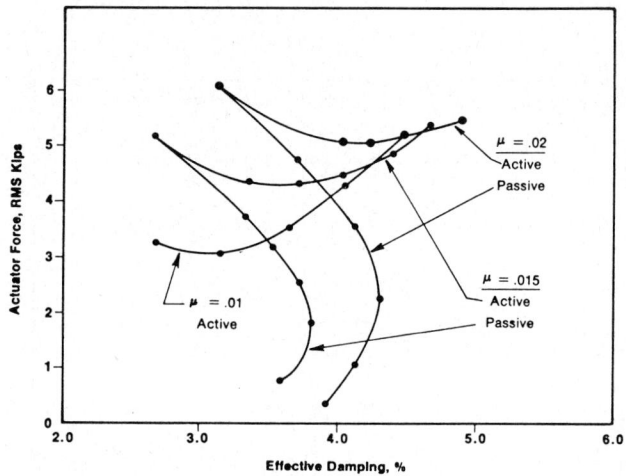

Figure 3 - Actuator Force Versus Structure Effective Damping

A study of Figures 2 and 3 points out three interesting cases for comparison. There are three combinations which achieve 4% equivalent damping in a very feasible manner. These are listed in Table 1 as Cases II, III and IV. The pertinent performance parameters are presented in Table 2.

Table 2

	Case I Building with No Damper	Case II Passive Damper μ = .02	Case III Active Damper μ = .02	Case IV Active Damper μ = .015
Wind Force Level	22,160 Kip^2/Hz (4.38E5 kN^2/Hz)	22,160 Kip^2/Hz (4.38E5 kN^2/Hz)	22,160 Kip^2/Hz (4.38E5 kN^2/Hz)	22,160 Kip^2/Hz (4.38E5 kN^2/Hz)
Building Displacement	8.0 in (20.32 cm)	4.0 in (10.16 cm)	4.0 in (10.16 cm)	4.0 in (10.16 cm)
Equivalent Damping	.01	.04	.04	.04
Damper Stroke RMS	-	15.0 in (38.1 cm)	13.6 in (34.5 cm)	18.1 in (46.0 cm)
Damper Relative Velocity RMS	-	13.9 ips (35.3 cm/sec)	12.7 ips (32.3 cm/sec)	17.0 ips (43.2 cm/sec)
Actuator Force RMS	-	3.9 Kip (17.3 kN)	5.3 Kip (23.6 kN)	5.1 Kip (22.7 kN)
Wind Power Input	16.4 Hp (12.2 kW)	16.2 Hp (12.1 hW)	16.3 Hp (12.2 kW)	16.4 Hp (12.2 kW)
Building Damping Dissipation	16.4 Hp (12.2 kW)	4.0 Hp (3.0 kW)	3.9 Hp (2.9 kW)	4.0 Hp (3.0 kW)
Damper Friction Dissipation	-	4.0 Hp (3.0 kW)	3.4 Hp (2.5 kW)	3.9 Hp (2.9 kW)
Actuator Absorption	-	8.2 Hp (6.1 kW)	9.0 Hp (6.7 kW)	8.5 Hp (6.3 kW)
Hydraulic Power Required at Servovalve	-	19.7 Hp (14.7 kW)	24.4 Hp (18.2 kW)	31.5 Hp (23.5 kW)
Minimum Oil Cooling Capacity	-	1354 BTU/min	1562 BTU/min	1863 BTU/min

All magnitudes are shown in Table 2 as RMS with peak values of three times the RMS normally used for a Gaussian distribution. All power values are calculated from the cross-spectral density between the associated force and velocity. The wind power input must be absorbed by the three elements listed below it in the table as dictated by an energy balance on the structure.

The hydraulic power required at the servovalve assumes a normal four-way flow-control servovalve which is sized to flow the $\sqrt{2}$ times the peak flow at supply pressure drop. This allows the actuator differential pressure to be within the region in phase angle of 180° ± 90° of velocity. The oil cooling capacity is obtained as the sum of hydraulic power, the actuator absorption, and the damper friction dissipation. This assumes that an oil bearing support system introduces the significant friction in the damper system. The relationship between power required at the hydraulic power supply and the power required at the servovalve depends on the hydraulic configuration and so is not considered.

In addition to the above analysis, each configuration was tested for stability. Frequency response and transient response analyses were performed to insure that strokes, velocities, and forces responded in a reasonable fashion. The results were consistent with the random process analysis and also provided insight on servovalve sizing.

6. CONCLUSIONS

An active control technique provides many new possibilities for achieving the required structural damping. The required stroke can be decreased with a given mass ratio (M_D/M_B) or the mass ratio can be reduced if additional mass block travel is allowed. Although additional force and hydraulic power are required to achieve significant performance improvement, this is shown to be well within reason. Thus active control should be considered a serious contender in future mass damper installations for motion control of large structures.

7. ACKNOWLEDGEMENT

The author wishes to thank Niel R. Petersen for many helpful suggestions in support of this work.

NOTATIONS

C_B — modal damping coefficient for the first mode of the building as seen at the damper

C_D — damping coefficient for the mass damper (estimated as the equivalent linear damping in the mass support system using a coefficient of friction of .003 and sinusoidal motion of 24 in. (61 cm).

F — actuator compressive force between the mass block and building

F_W — wind force input to the building

K_A — reference filter gain, 1.0 for passive control and > 1.0 for active control

K_B — modal spring rate for first mode of the building as seen at the damper

K_D — spring rate of the passive spring connecting the mass block to the building

M_B — modal mass of the first mode of the building as seen at the damper

M_D — mass of the damper mass block

$P_{N,B}$ — building first mode natural period

$P_{N,D}$ — damper natural period

$P_{N,R}$ — reference filter natural period

y_B — displacement of the first mode of the building at the damper

y_D — displacement of the damper mass block

z — relative displacement $y_D - y_B$

μ — mass ratio, damper mass to building mass (M_D/M_B)

ζ_B — building damping ratio, fraction of critical

ζ_D — damper damping ratio due to inherent friction, fraction of critical

ζ_R — damping ratio of the reference filter, fraction of critical

REFERENCES

[1] PETERSEN, N.R., "Design of Large-Scale Tuned Mass Dampers", *ASCE National Convention*, Boston, April, 1979.

[2] WIESNER, K.B., "Tuned Mass Dampers to Reduce Building Wind Motion", *ASCE National Convention*, Boston, April, 1979.

[3] KARNOPP, D. and MORISON, J., "Comparison of Optimized Active and Passive Vibration Absorbers", *14th Annual Joint Automatic Control Conference*, Ohio State University, Columbus, Ohio, June, 1973, pp. 932-938.

[4] YANG, J.N., "Application of Optimal Control Theory to Civil Engineering Structures", *Trans. ASCE, J. of the Engineering Mechanics Division*, Vol. 101, No. EM6, December, 1975, pp. 819-838.

[5] SOONG, T.T., Private Communication, State University of New York at Buffalo, February, 1979.

[6] DAVENPORT, A.G., "The Application of Statistical Concepts to the Wind Loading of Structures", *Proc. of The Institution of Civil Engineers*, Vol. 19, August, 1961, pp. 449-472.

[7] AMYOT, J.R., WARDLAW, R.L., COOPER, K.R. and VAN BLOKLAND, G.P., "Computer Studies of a Vibration Damper for Wind-Induced Motion of a Tall Building", *National Research Council*, Research Report, June, 1977.

[8] McNAMARA, R.J., "Tuned Mass Dampers for Buildings", *Trans. ASCE, J. of the Structural Division*, Vol. 103, No. ST9, September, 1977, pp. 1785-1798.

[9] PAPOULIS, A., *Probability, Random Variables and Stochastic Processes*, McGraw-Hill, 1965, 339 p.

[10] ISYUMOV, N., HOLMES, J.D., SURRY, D. and DAVENPORT, A.G., "A Study of Wind Effects for the First National City Corporation Project - New York, U.S.A.", *Boundary Layer Wind Tunnel Laboratory Special Study Repord*, BLWT-SS1-75, University of Western Ontario, London, Ontario, Canada, April, 1975.

STRUCTURAL CONTROL, H.H.E. Leipholz (ed.)
North-Holland Publishing Company & SM Publications
© IUTAM, 1980

ON-LINE PULSE CONTROL OF TALL BUILDINGS

S.F. Masri, G.A. Bekey and F.E. Udwadia

School of Engineering
University of Southern California
Los Angeles, California 90007, U.S.A.

1. INTRODUCTION

In the last decade or two an increasing amount of attention has been given by scientists and engineers all over the world to the mitigation of damage caused by strong earthquake ground shaking. A significant portion of this effort has been devoted to improvements in the analysis, design and construction of tall building structures built in seismically active areas. Though these advances will, no doubt, lead to structures which will be safer in the earthquake environment, the analyses for such seismic designs basically have to cope with major uncertainties in the time history of ground shaking at a site, and in the dynamic modelling of the building structural system.

An alternative approach to the exclusive reliance on such analyses is to investigate the possibility of actively controlling the structure during an episode of strong ground shaking. It is with this alternative method that this paper concerns itself.

Among the major difficulties encountered in the application of modern control techniques to building structural systems are the following:

(1) active control requires the ability to generate and apply large controlled forces to the structure,

(2) modern control theory often leads to feedback control laws, thus requiring on-line measurement (or estimation) of all the system state variables,

(3) on-line control requires that both measurement and control be performed in real time.

The work of Yang [1], Soon and Leipholz [2, 3] are but a few examples of feasibility studies concerned with the optimal control of civil engineering structures. From a practical standpoint, while the application of large control forces on a structure does not pose insurmountable problems, the generation of such forces over sustained periods of time (as often dictated by continuous optimal feedback control theory) may cause the concept of active control to be driven outside the realm of today's possibilities. To bypass this possible drawback, this paper attempts to utilize pulses of relatively short duration to control the structural system.

Moreover, it has been demonstrated that a sequence of force pulses applied to a structure can be selected in such a way that the power spectral density of the displacement at a particular point within the structure matches the spectral density produced by earthquake ground motions as closely as desired [4]. This result naturally suggests the possibility of using the force-pulses to counteract or reduce displacements produced by earthquakes. This paper presents an approach to this problem. A heuristic algorighm is introduced and shown (by computer simulation) to be effective in reducing the motion of a structure in response to a stochastic disturbance that resembles earthquake loads.

1.1 *Some Drawbacks of Standard Optimal Control Theoretic Methods*

In order to illustrate potential approaches and their difficulties, assume, for the moment, that the structure to be considered is a linear system with n degrees of freedom. We further assume that the modal approach is not applicable, so that the system is described by the 2n first-order differential equations

$$\dot{\underline{x}}(t) = A(t)\underline{x}(t) + B(t)\underline{p}(t) + \underline{V}(t) , \qquad (1)$$

where $\underline{x}(t)$ is a 2n-dimensional vector of state variables representing all the positions and velocities, $\underline{p}(t)$ is the m-dimensional vector of control forces produced by the actual controllers, $\underline{V}(t)$ is a 2n-dimensional vector of disturbances representing the earthquake (or other excitations), A is a (2n x 2n) coefficient matrix and B is a (2n x m) coefficient matrix. The optimal control problem consists of finding the control vector $\underline{p}(t)$ such that an appropriate cost functional is minimized. When the disturbance is Gaussian, A and B constant matrices, and the cost function quadratic, we obtain the classical Linear Quadratic Gaussian (LQG) formulation. For example, if the cost function is given by

$$J = E \int_{t_0}^{t} (\underline{x}^T Q \underline{x} + \underline{u}^T R \underline{u}) dt , \qquad (2)$$

then its minimization is equivalent to minimization of the covariance of the state variables and of the control energy. Q and R represent weighting matrices that make it possible to select or emphasize certain elements of the state or control vector. If the system is given by equation (1) and the performance index by equation (2), the optimal control $\underline{u}^*(t)$ is given by

$$\underline{u}^* = - F(t)\underline{x}(t) , \qquad (3)$$

where F(t) depends on the matrices B and R and on the solution of a matrix Riccati equation [5]. We note that the solution is a feedback solution requiring (1) the state variables to be measureable, and (2) if control is to be on line, extremely fast computation to solve the necessary Ricatti equation. In the absence of measurements, the state variables may be estimated using, say, Kalman Filters, but this would further increase on-line computational task. It is the on-line solution of the Ricatti equation, however, which may prove to be difficult when the number of degrees of freedom of the system is large.

Furthermore, the building structures during strong ground shaking show strongly nonlinear and time variant (degrading) behaviour. Even in the absence of nonlinearities, we obtain a linear differential euqation as in equation (1) with time varying coefficient matrices. The use of such a representation would prevent the system from being decomposed, in general, to normal modes, thereby making the concepts of modal control difficult to apply.

It is thus evident that the available standard methods of control system design are not well suited to the structural control problem because they either fail to take into account the statistical nature of the earthquake disturbances, or they do not lend themselves easily to on-line implementation, or both. Also, the problem of time invariant systems complicates the issue considerably. Perhaps the most serious problem of the standard design techniques is that they do not lend themselves readily to restricting the class of control signals to relatively narrow, high-energy pulses of the type suitable for active control of buildings.

1.2 *Scope of the Paper*

Based upon the actual recorded structural responses during strong ground shaking, this paper presents an algorithm designed to overcome the limitations of the existing controller design techniques. The algorithm requires a continuous monitoring of the state variables and the estimation of the energy content of the earthquake ground motion record. Following determination that some specified threshold has been exceeded, an open-loop pulse control is applied. The determination of the optimum pulse magnitude is based on a performance criterion which depends in a nonlinear way on both the deterministic and stochastic components of the response. Since both the threshold and the performance index contain nonlinearities, the analytical solution of the control problem is not possible and the usefulness of the algorithm is evaluated using simulation techniques.

2. PULSE CONTROL METHOD

We consider first the nature of typical building structural responses during strong ground shaking.

The Millikan Library at the California Institute of Technology is a nine-story RC building which has undergone extensive ambient and forced vibration testing [6]. Figure 1(a) shows the "transfer function" of the building in the E-W direction obtained by dividing the Fourier transform of the input record (obtained at the roof level of the structure) by that of the input record (obtained at the base of the structure). These transfer functions are obtained for two different events - the Lytle Creek Event and the San Fernando Event. We notice that the San Fernando earthquake resulted in a transfer function with reduced peak amplitude around the natural frequencies of the system, and a strikingly increased "equivalent viscous damping." The broadening of these resonance peaks can be explained as being caused by the gradual increase in the structural periods during the duration of strong ground shaking, the system behaving in a time variant nonlinear fashion. Though the larger event causes shifts in the natural frequencies towards the lower values, the estimates of the fundamental frequencies obtained from vibration tests (vertical arrows) appear to be reasonably good.

Figure 1(b) shows the actual time history of responses to the San Fernando event. Perhaps the most striking feature of this response is the oscillatory behaviour of the structure in a markedly single mode of vibration. Also, one observes that the build-up of oscillation occurs gradually, the oscillations gradually increasing in size at each period from about four to twelve secs. The structural response gradually changes with each oscillation - a period of time involving the characteristic time of the system.

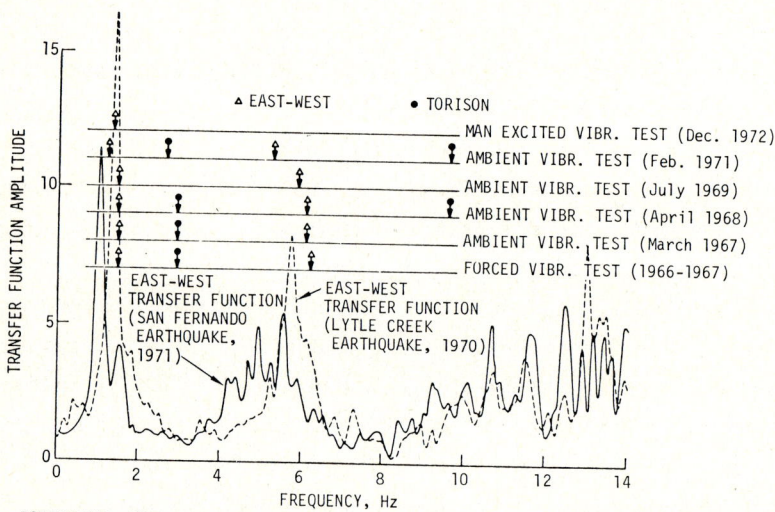

(a)

Figure 1 - *Response of a Reinforced Concrete Building to Earthquake Excitation* [6]

(continued)

(continued)

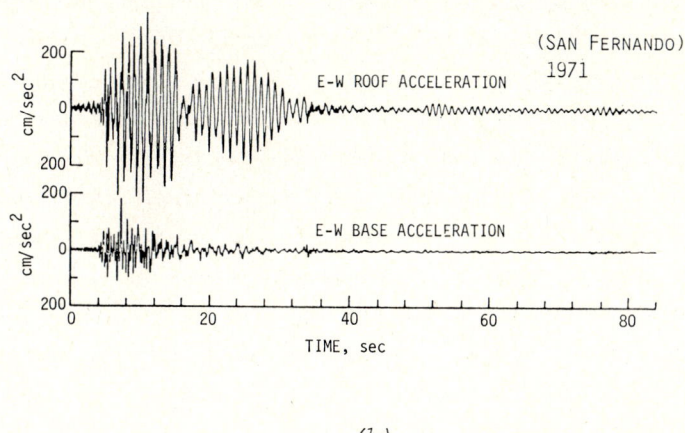

(b)

Figure 1 - *Response of a Reinforced Concrete Building to Earthquake Excitation* [6]

Turning to the nature of ground shaking, a considerable amount of effort has been devoted by many investigators (see, e.g., [7 - 9]) to modelling earthquake ground motions. Among the simplest evolutionary models of typical earthquakes is the one in which the process is assumed to consist of modulated white noise with zero mean:

$$n(t) = g(t)w(t) , \qquad (4)$$

where,

$w(t)$ is Gaussian white noise,

$g(t)$ is a deterministic envelope function,

$n(t)$ is a nonstationary stochastic process representing the earthquake.

Typically, the spectral content of the excitation is much wider than the lower natural frequencies of the systems. Therefore, to first order approximation, during the period of time comparable to the system time constant (fundamental period, T_1, of the structure) the earthquake can be modelled as a stationary white process. Thus over each period of time of $O(T_1)$, the base point input can be thought of as a sample function excised from a stationary process having a certain power spectrum. One can imagine the whole input record as made up of a patchwork of such sample functions, each created by excision of a portion of time length $O(T_1)$ from suitable samples of stationary ergodic processes with each of these processes having different power spectra. We shall assume that these power spectra over two adjacent intervals of time change slowly and that during each interval of $O(T_1)$, the ground motion can be defined as having a constant equivalent power spectrum (Figures 2 and 3).

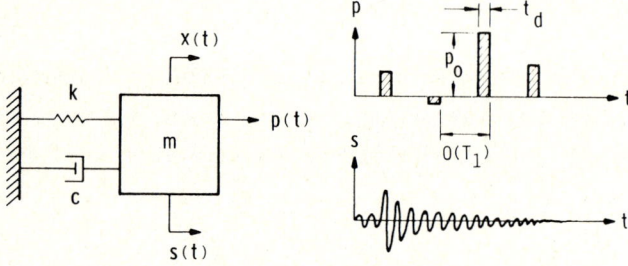

Figure 2 - Equivalent SDOF System to be Controlled

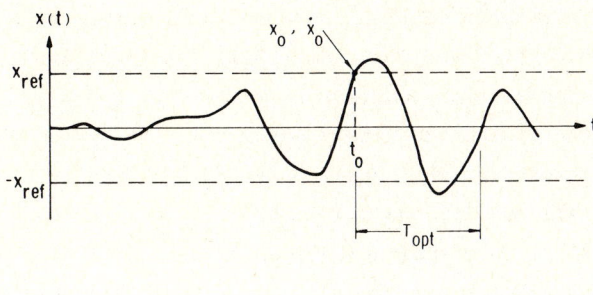

(a)

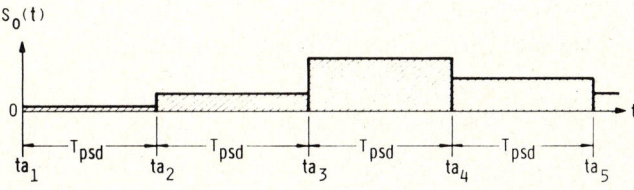

(b)

Figure 3 - Response and Excitation Characteristics
(a) System Displacement and Threshold Levels,
(b) Approximate Stepwise Variation of Typical Earthquake Excitation PSD Level

2.1 Control Algorithm

The heart of the control algorithm lies in the physical realization that the gradual rhythmic build-up of the structural response (the vibrational energy) can be destroyed by applying a pulse of suitable magnitude in the proper direction.

Furthermore, in order to minimize the amount of control energy utilized, the control should be applied only when the structural response exceeds a certain threshold related to the resistance of the structure.

As a first approximation, we model a structure as a linear single-degree-of-freedom system (SDOF), as in Figure 2, and our strategy for control requires that we:

(1) pulse the system adequately every time the system response crosses a threshold (which can be determined from full scale vibration tests) (Figure 1), and

(2) keep the minimum spacing between pulses to be of the order T_1 - the fundamental period of the structure (Figure 2). The time duration of pulsing, t_d, as well as the acutal minimum interpulse duration would be controlled primarily by the response time of the pulsing equipment at hand. We shall assume for simplicity that rectangular pulse forms will be generated.

2.2 *Determination of Pulse Height*

The pulse height is related to the threshold level used, the specific response parameter(s) (displacement, velocity, acceleration) which are not allowed to be exceeded, and the cost function to be minimized.

As the engineer is interested in limiting the interstory base shears and the story bending moments in a structure during ground shaking, the typical quantities that should be prevented from exceeding certain values would be related to the relative displacements and the weighted accelerations. The relative interstory displacements control to a major extent the interstory shear forces while the accelerations weighted with respect to the story heights control the bending moments. For the simpler case of a single-degree-of-freedom oscillator undergoing almost periodic motion, the accelerations are approximately constant multiples of the displacements. In the sequel, the threshold variable used is the relative displacement.

As the interpulse interval is of the order of T_1 (say, equal to the optimum interval T_{opt}) we shall endeavour to find the pulse height for which the energy of the oscillator is minimized over the time period T_{opt}, under the constraint that the energy of the pulse not exceed a certain predetermined value. This value would of course depend on the pulsing mechanism used. Thus, at time t_0 (Figure 3(a)), the oscillator's velocity and displacement are known, and an optimal pulse height is required to be ascertained. However, as our pulse is purely deterministic, the system model linear, and our control open-loop, this pulse would have no effect on controlling that part of the motion of the system which is in response to the stochastic excitation. The response of the system prior to time t_0 is contained in "initial conditions" at time t_0, but the pulse cannot be tailored to take into account the interpulse stochastic excitation. In fact, the response $x(t)$ of the system to all but the stochastic excitation can be explicitly written for $t \varepsilon [t_0, t_0+T_{opt}]$ as

$$x(t) = x_0 u(t-t_0) + \dot{x}_0 v(t-t_0) + \int_t^{T_{opt}} h(t-\tau) P(\tau) d\tau , \qquad (5)$$

where

$$u(t) = \exp(-\zeta\omega t)\left(\cos\omega_d t + \frac{\zeta\omega}{\omega_d}\sin\omega_d t\right),$$

$$v(t) = \frac{1}{\omega_d}\exp(-\zeta\omega t)\sin\omega_d t ,$$

$h(t)$ is the impulse response of the system; in this case $h(t) = v(t)$, $\omega_d = \omega\sqrt{1-\zeta^2}$,

$\omega = \sqrt{k/m}$, $\zeta = c/(2\sqrt{km})$, and $P(t)$ is the control force.

Furthermore, if $P(t)$ is restricted to the class of pulses of very short duration t_d compared to the system time constant, equation (5) reduces to

$$x(t) = x_0 u(t-t_0) + \left(\dot{x}_0 + \frac{I}{m}\right) v(t-t_0) , \qquad (6)$$

where

I is the impulse, created by the pulse, of magnitude $P_0 t_d$.

Though the response of the system to the interpulse portion of the earthquake excitation is unknown at the time of pulsing, the variance of the response to the stochastic excitation can be obtained. Assuming that the power spectral density of the piecewise stationary stochastic input in $t\varepsilon[t_0, t_0+T_{opt}]$ is S_0 (see Figure 3(b)) and assuming that the mean of the excitation in that period of time is zero, the variance is given by [10]

$$\begin{aligned}\sigma_x^2(t) &= E[\{x(t) - E[x(t)]\}^2] \\ &\approx \frac{\pi S_0}{4\zeta\omega^3}\left\{1 - \frac{e^{-2\zeta\omega t}}{\omega_d^2}\left[\omega_d^2 + \frac{(2\omega\zeta)^2}{2}\sin\omega_d t + \omega\omega_d \zeta \sin 2\omega_d t\right]\right\},\end{aligned} \qquad (7)$$

which in the case of negligible damping, reduces to

$$\sigma_x^2(t) \approx \frac{\pi S_0}{4\omega^3}(2\omega t - \sin 2\omega t) . \qquad (8)$$

The value of $\sigma_x^2(t)$ is a measure of the fluctuations of the response about the mean value, which in the presence of the "initial" conditions x_0 and $\dot{x}_0$, is given by $x(t)$. Clearly the value of S_0 is not known at time t_0; however, a close enough approximation to it can be obtained by monitoring the power levels in the time intervals $(0(T_1))$ prior to time t_0.

A cost function which incorporates the effects of these possible fluctuations about the mean and which accounts for a measure of the system energy as well as the constraints on impulse levels can then be stated as

$$J(I) = \frac{1}{T_{opt}} \int_{t_0}^{t_0+T_{opt}} \{[C_1|x(t)| + C_2\sigma_x(t)]^2$$

$$+ [C_3|\dot{x}(t)| + C_4\sigma_{\dot{x}}(t)]^2\}dt + C_5 I^2 , \qquad (9)$$

where C_i, $i = 1,2,\ldots,5$ are weighting factors which can be selected to emphasize the importance of the deterministic portion of the response (C_1 and C_3), the stochastic part of the response (C_2 and C_4) or the control energy used (C_5).

3. SIMULATION OF CONTROL STRATEGY

Figure 4(a) shows an artificial accelerogram generated by using a time modulated white noise signal. The building structure has been modelled as an SDOF system with period $T_1 = 1$ sec. and ζ, the damping ratio, to be 0.05. The uncontrolled response of the system is shown in Figure 4(b) over a segment of time equal to 40 periods of the system time.

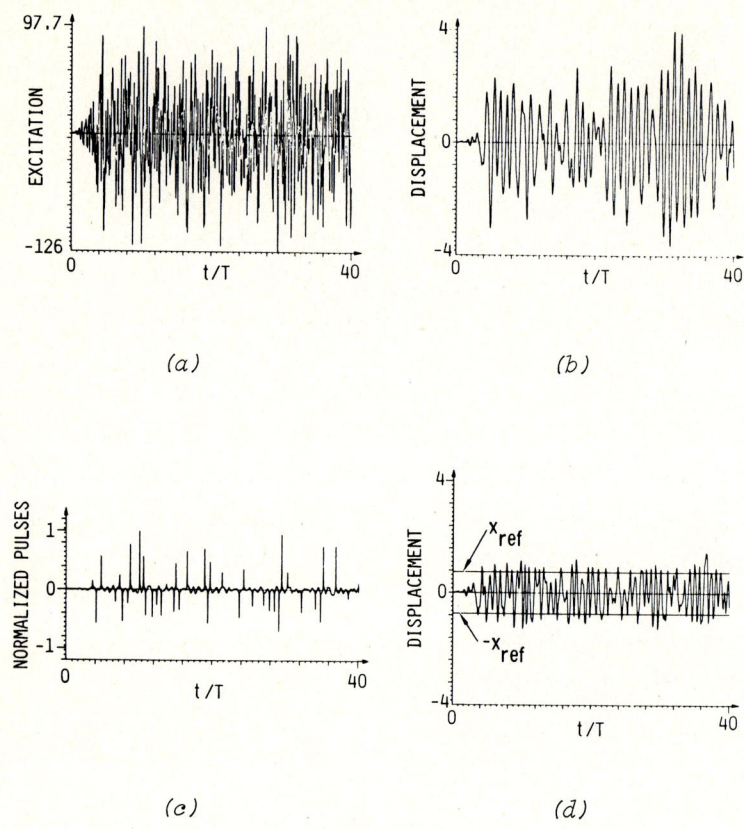

Figure 4 - Normalized Controlled Response
(a) Excitation,
(b) No Control,
(c) Normalized Pulses,
(d) Controlled Response

The system was controlled using the strategy discussed in Section 2. An estimate of the power spectrum over a period of time $T_{psd} = T_1$ was carried using the approximate relation

$$S_0 = \frac{\Delta t_s}{2\pi} \sigma_s^2 , \tag{10}$$

where Δt_s is the sampling interval and σ_s^2 is the variance of the input about the mean. The trigger thresholds were maintained at the displacement response equal to ≈ 0.8 cm. The maximum allowable impulse was restricted to 25 units and the value of t_d was taken to be 1/20 sec. The constants C_1 and C_2 were chosen to be 1 and 4 respectively, while $C_3 = C_4 = C_5 = 0$. The control force time history is shown in Figure 4(c). The controlled response as shown in Figure 4(d) indicates the ability of the method under consideration to control the structural motions. Figure 5 shows the components of the cost function with and without control over one period T_1 of time. As observed from Figure 4(d), the component $\bar{x} = E[x(t)]$ of the response has been significantly reduced by the application of the control. Additional sample cases with and without control are shown in Figures 6 and 7.

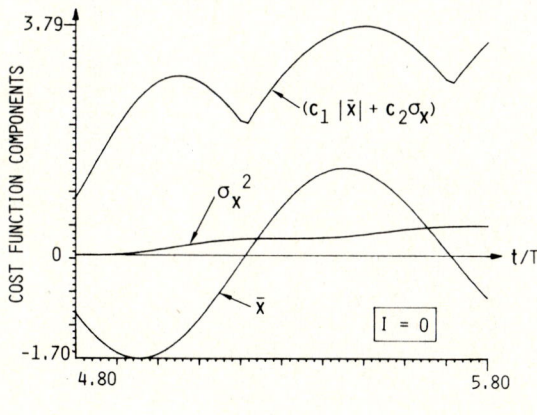

(a)

Figure 5 - Components of Cost Function

(continued)

(continued)

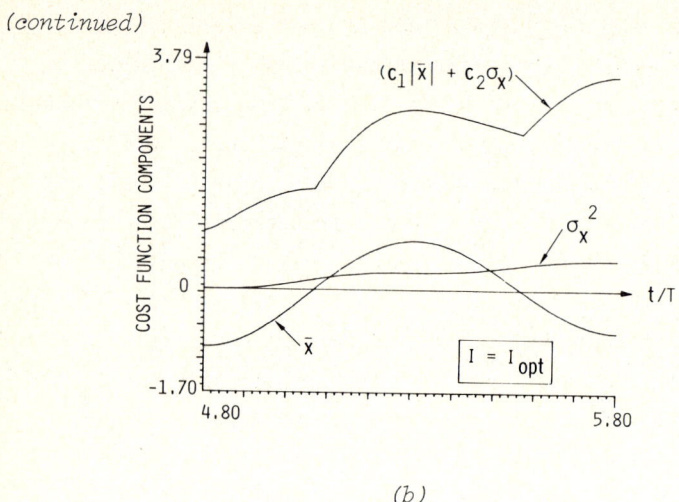

(b)

Figure 5 - Components of Cost Function

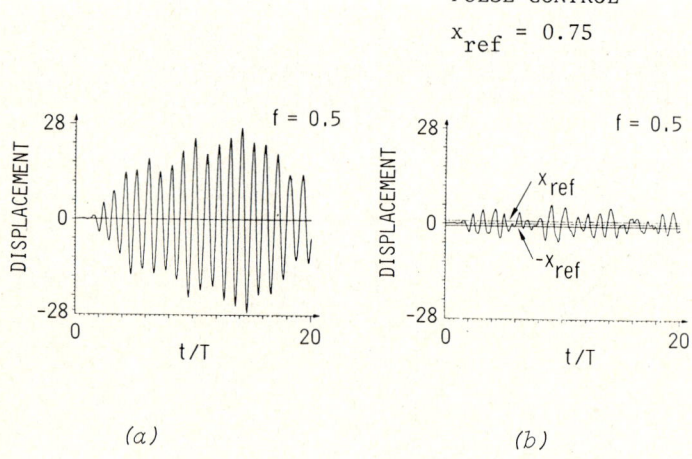

Figure 6 - Effects of Oscillator Frequency f (Hz)

(continued)

(continued)

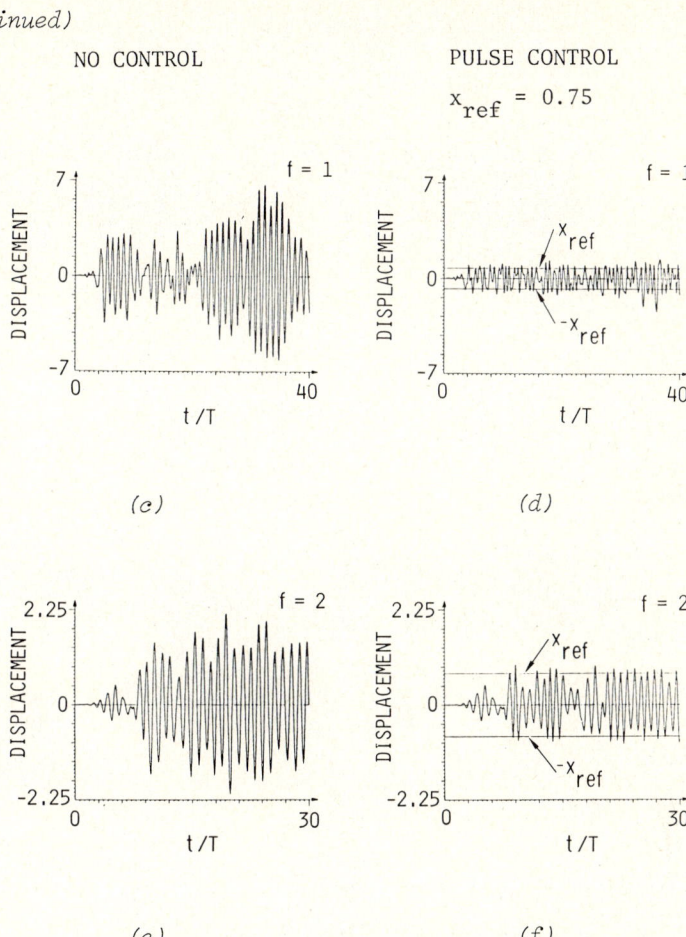

Figure 6 - *Effects of Oscillator Frequency f (Hz)*

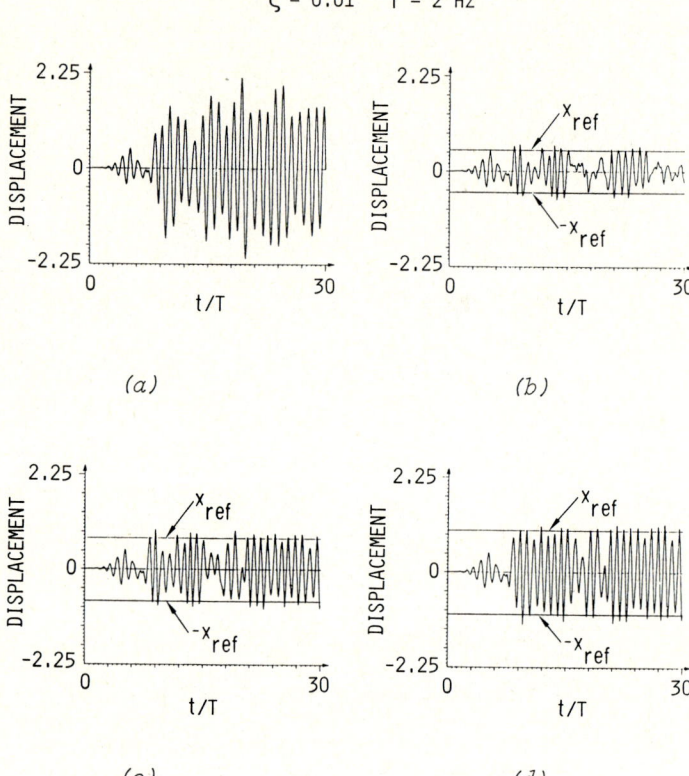

Figure 7 - Effects of Threshold Level x_{ref}
(a) No Control,
(b) $x_{ref} = 0.5$,
(c) $x_{ref} = 0.75$,
(d) $x_{ref} = 1.0$

4. REDUCED ORDER MODELLING EFFECT

Whereas the actual structure is a continuous system, the control pulses P(t) have been obtained on the assumption that the system could be adequately modelled as an SDOF system. However, the control pulses having been so ascertained, a physical realization of the control technique would indeed involve the application of these forces to the actual continuous system. The question then remains whether these control pulses could have an adverse (possibly destabilizing) effect on the higher modes of vibration of the structure.

Assuming the structure to have normal classical modes (experimental studies on tall buildings validate this assumption) it can be shown that the energy E_n of the nth mode of a pulse-controlled structure is given by

$$E_n(t) \lesssim \left(\frac{\lambda_1}{\lambda_n}\right) \frac{I_{max}^2}{4\pi\zeta} \left[1 - e^{-2\lambda_n \zeta_n t}\right],$$

where I_{max} is the impulse upper bound, and λ_i is the ith eigenvalue of the system.

Thus, for a very long earthquake record ($t \to \infty$), not only are the energies of oscillation corresponding to the higher modes bounded, but they fall off inversely as the modal frequencies.

5. NONLINEAR AND DEGRADING SYSTEMS

The adaptive open-loop concept developed promises to be especially useful for systems which undergo degradation (or time variations) during the actual period through which the excitation occurs. Preliminary simulation results suggest that improvements in performance as dramatic as those shown in Section 4 for the linear time-invariant SDOF system are possible for both time varying and nonlinear systems. Nonlinear systems can be handled by representing them by linearized systems over intervals of time of the order of T_1. Also, extensions of the method to the control of several modes of vibration of linear systems (with classical modes of vibration) can be carried out.

6. DISCUSSION AND CONCLUSIONS

This paper shows the feasibility of using pulsed open-loop adaptive control for reducing the oscillations of tall building structures (or similar distributed-parameter systems) subjected to strong ground shaking (or arbitrary nonstationary disturbances). The method is adaptive in order to take into account the varying nature of the system; it uses pulse control to get around our inability to produce large control forces over sustained periods of time, and it is open-loop to reduce computing time.

Perhaps, the primary asset of this control method lies in its basic simplicity - a factor which foreshadows the reliable functioning of a system. In years to come, we may find ourselves not asking how to control structures that have already been designed, but rather how best to design structures so that they can be adequately controlled.

REFERENCES

[1] YANG, J.N., "Application of Optimal Control Theory to Civil Engineering Structures", *ASCE, J. of the Eng. Mech. Div.*, Vol. 101, No. EM6, December, 1975, pp. 819-838.

[2] ABDEL-ROHMAN, M. and LEIPHOLZ, H.H.E., "Structural Control by Pole Assignment Method", *ASCE, J. of the Eng. Mech. Div.*, Vol. 104, No. EM5, October, 1978, pp. 1159-1175.

[3] MARTIN, C. and SOON, T.T., "Modal Control of Multistory Structures", *ASCE, J. of the Eng. Mech. Div.*, Vol. 102, No. EM4, August, 1976, pp. 613-623.

[4] MASRI, S.F., BEKEY, G.A. and SAFFORD, F.B., "Optimum Response Simulation of Multidegree Systems by Pulse Excitation", *J. of Dynamic Systems, Measurement and Control, Trans. ASME*, Vol. 97, No. 1, 1975, pp. 46-52.

[5] KWAKERNAAK, H. and SIVAN, R., *Linear Optimal Control Systems*, Wiley-Interscience, 1972.

[6] UDWADIA, F.E. and TRIFUNAC, M.D., "Time and Amplitude Dependent Response of Structures", *Int. J. of Earthquake Engineering and Structural Dynamics*, Vol. 2, 1974, pp. 359-378.

[7] KANAI, K., "An Empirical Formula for the Spectrum of Strong Earthquake Motions", *Bull. Earthquake Research Institute*, University of Tokyo, Tokyo, Japan, Vol. 39, 1961, pp. 85-95.

[8] JENNINGS, P.C., HOUSNER, G.W. and TSAI, N.C., "Simulated Earthquake Motions", *EERL*, California Institute of Technology, Pasadena, California, April, 1968.

[9] KAMEDA, H., "Evolutionary Spectra of Seismogram by Multifilter", *ASCE, J. of the Eng. Mech. Div.*, Vol. 101, No. EM6, December, 1975.

[10] CAUGHEY, T.K. and STUMPF, H.J., "Transient Response of a Dynamic System under Random Excitation", *Journal of Applied Mechanics*, 1961, pp. 563-566.

STRUCTURAL CONTROL, H.H.E. Leipholz (ed.)
North-Holland Publishing Company & SM Publications
© IUTAM, 1980

EFFECTS OF REDUCED ORDER MATHEMATICAL MODELS ON DYNAMIC
RESPONSE OF FLEXIBLE AIRCRAFT WITH CLOSED-LOOP CONTROL

D. McLean

Department of Transport Technology
University of Technology
Loughborough, Leicestershire LE11 3TU, England

1. INTRODUCTION

To achieve worthwhile reductions in future in the fuel consumption of, say, transport aircraft will involve, inter alia, a reduction in the weight of the aircraft. It is commonly expected that such reductions will be made possible by the application of a group of techniques which have come to be known in world aviation circles as Active Control Technology. Two particular members of that group are generally considered to offer the most promise of achieving the aim of reducing the aircraft's weight without impairing its performance. These techniques consist of:

(1) relaxing the static stability of the aircraft reducing thereby the size of the empannage needed;

(2) using a Structural Load Alleviation Control system so that a lighter, but inevitably more flexible, wing can be used although the aircraft must still be able to respond to the same manoeuvre commands from either the pilot or his path control systems.

It is with the latter type that this paper is concerned.

2. MATHEMATICAL MODELS OF SUBJECT AIRCRAFT

The aircraft used as the subject of the investigation was the large American transport aircraft manufactured by the Lockheed Co., the C-5a. Details of the principal mathematical model were obtained from a NASA contract report [1]. In the course of the work some adjustments to the model were made but before indicating what these were it is appropriate first to briefly explain the structure of the model involved. Only longitudinal motion was considered, and only small perturbations from the equilibrium flight path were involved. Thus, there was no coupling between longitudinal and lateral motion. Secondly, the form of the equations used to represent the motion of any flexible aircraft depends upon the displacement co-ordinates chosen to describe the motion, i.e., whether these co-ordinates are relative to some inertial axis set, or to a non-inertial set fixed in the aircraft. The results obtained finally from using either set ought to be identical. In this work the body-fixed axis system was chosen because most criteria for assessing aircraft handling and performance are expressed in terms of this set [2]. The equations of motion were represented in state variable form, i.e.:

$$\left. \begin{array}{l} \underline{\dot{x}} = A\underline{x} + B\underline{u} + D\underline{\eta} , \\ \\ \underline{y} = C\underline{x} + E\underline{u} , \end{array} \right\} \quad (1)$$

where

$\underline{x}$ is the state vector of dimension n;
$\underline{u}$ is the control vector of dimension m;
$\underline{y}$ is the output vector of dimension p;
$\underline{\eta}$ is the disturbance vector, of dimension k.

The elements of the matrices A, B, C, D and E all lie in the real field, R. Since η was simply a scalar noise variable in this work, k was 1. The importance of the form of equation (1) for the analytical work resides in the size of n, m and p.

The state vector, $\underline{x}$ was composed of variables which represented:
(i) the rigid-body motion of the aircraft;
(ii) the bending motion due to flexibility;
(iii) the motion due to unsteady aerodynamics;
(iv) the dynamics of the actuators used with the control surfaces;
(v) the dynamics of a mathematical model which allowed the effects of atmospheric turbulence to be incorporated into the problem.

Some of the models used in the work had all these components present; some did not. Table 1 shows the appropriate dimensions for the various models. Most of the results presented in this paper concern the model CLEMENTI. The state vector of this model is expressed in equation (2) in which λ_i represents the displacement of the ith bending mode, w and q represent the heave and pitch motion of the rigid-body aircraft and w_g represents the vertical velocity of atmospheric turbulence. The state variables $x_{15} - x_{20}$ represent the Küssner dynamics of the wing and the tail. The final three states represent the deflections of the control surfaces, i.e., aileron, inboard elevator section, and outboard elevator section. Only the signals to the actuators associated with the ailerons and inboard elevator section were used as controls.

$$\underline{x}' = [w, q, \dot{\lambda}_1, \dot{\lambda}_2, \dot{\lambda}_3, \dot{\lambda}_4, \dot{\lambda}_5, \dot{\lambda}_6, \lambda_1, \lambda_2, \lambda_3, \lambda_4, \lambda_5, \lambda_6, x_{15}, x_{16}, x_{17},$$
$$x_{18}, x_{19}, x_{20}, w_g, \delta_A, \delta_{e_i}, \delta_{e_o}] \quad . \tag{2}$$

Since the aim was to reduce the bending moments of the wing by active control, and since these moments were not state-variables (see equation (2)), it was necessary to define these moments as output variables $y_{21} - y_{32}$. W, q, δ_A, δ_{e_i}, $\dot{\delta}_A$ and $\dot{\delta}_{e_i}$ were selected as the final six output variables.

Table 1

Model	Vector Dimensions		
	State (n)	Control (m)	Output (p)
ARNE	79	2	56
BÀCH	42	2	56
CLEMENTI	24	2	38
FAURÉ	17	2	38
GERSHWIN	14	2	38
HANDEL	5	2	5

The model ARNE represented the most complete mathematical model and differed from CLEMENTI principally in two ways: it contained the first fifteen significant bending modes rather than the first six, and it has a set of Wagner dynamics which were used to represent the dynamics of lift growth on both the wing and the horizontal tail. However, investigation showed that these dynamics were approximated very closely by a unity transfer function, so that their contribution to the overall dynamic performance of the aircraft was likely to be insignificant. By neglecting the Wagner dynamics the model BACH resulted, and it was determined from digital simulation studies that the responses obtained from BACH did not differ significantly from those obtained from ARNE. When the upper nine bending modes were also omitted the model BACH became CLEMENTI, and the effects of the omission were again adjudged, from considering the simulation results, to be negligible. When the aircraft was considered to be flying in a steady, non-turbulent atmosphere the gust velocity and the Küssner dynamics were ineffective, and hence omitted, to obtain the model FAURÉ. The model GERSHWIN reintroduced both the gust and the Küssner dynamics but omitted the second through fifth bending moment. When the gust, the Küssner dynamics, and all the bending modes were neglected the model was referred to as HANDEL. When referring to the closed-loop responses resulting from the application of reduced-order feedback control a shorthand notation is used. Thus, a 17-state feedback means the use of that control law derived on the basis of the model FAURÉ; 14-state feedback means the use of that control law derived on the basis of GERSHWIN.

3. OPTIMAL CONTROL

The feedback control laws were obtained by solving the linear quadratic problem outlined below:

the performance index of equation (3), i.e.,

$$J = \frac{1}{2} \int_0^\infty \{\underline{y}'Q\underline{y} + \underline{u}'G\underline{u}\}dt , \qquad (3)$$

is minimized when the optimal control $\underline{u}^0$, given in equation (4), is applied to the dynamic system of equation (1), i.e.,

$$\underline{u}^0 = F\underline{x} , \qquad (4)$$

where

$$F = -\hat{G}^{-1}(E'QC + B'K) , \qquad (5)$$

$$\hat{G} = (G+E'Q\ E) , \qquad (6)$$

and K is the solution of the algebraic Riccati equation:

$$C'\hat{Q}C + K\hat{A} + \hat{A}'K = KB\hat{G}^{-1}B'K , \qquad (7)$$

in which

$$\hat{A} = (A - B\hat{G}^{-1}E'QC) ,\qquad (8)$$

and

$$\hat{Q} = (Q - QE\hat{G}^{-1}E'Q) .\qquad (9)$$

The resultant control of equation (4) depends upon the complete state vector although the performance index is expressed in terms of the output vector, $\underline{y}$. The gains for the feedback control laws were obtained by choosing appropriate weighting matrices Q and G and solving the Riccati equation (7), by means of a suitable computer programme [3].

4. SOME RESULTS

To illustrate the effectiveness of the active control, 3 test cases were considered:

	Case	Condition
	A	All control surfaces set at zero; initial heave motion of 7m/s. No pitching motion.
Uncontrolled Aircraft	B	All states zero. Elevator sections zero, but constant aileron deflection of .025 rad.
	C	All states zero. Outboard elevator section set at zero, but both inboard section and aileron set at 0.01 rad.

Corresponding to these cases, the reference command vecto for the closed-loop system was chosen to ensure that the same steady-state values of rigid-body motion were achieved so that comparisons with the responses of the uncontrolled aircraft were valid.

From Figure 1 it is evident that the short period dynamics of the rigid body were slightly improved when the structural load alleviation control was employed: the handling qualities of the aircraft were therefore unimpaired when the active control system was operating. How the bending moment at the wing root was affected by the active control may be assessed from Figure 2. (Although Case A is not shown the results were impressive. Case A was omitted for clarity since its response lies close to the abscissa.) It should be noted that both the peak and the steady-state values were substantially reduced: by as much as 50%. The oscillatory motion of the uncontrolled response was also greatly reduced by the active control which is a welcome result, for such a reduction would assist in reducing the accumulation of fatigue in the wing structure, assuming that fatigue accumulates according to Minor's hypothesis [4]. Yet, although the active control seems to be effective in achieving the aim of reducing the bending moments, some other factors gave cause for concern. From Figure 3, for example, it can be seen that in Case C the torsional moment increased when the control was applied. It is possible, however, by further numerical experiment to find another choice of Q and G which would result in an optimal control which could achieve simultaneous reduction of the bending and torsional moments at all the stations of the wing. Yet all such control laws involved all twenty-four state variables, six of which had no physical existence, and one of which is particularly difficult to measure (the gust velocity). Thus, the situation of using CLEMENTI with feedback control laws derived on the bases of FAURÉ, GERSHWIN, and HANDEL was carefully examined. The results obtained can be exemplified by the results shown in Figure 4. Curve 1 represents the uncontrolled aircraft; curve 2 represents the controlled aircraft with full-state feedback. It was hoped that the simpler control laws would result in response nearly identical to curve 2. The response curve 3 almost matches, but curve 4 is very different. Curve 3 was obtained when

the Küssner dynamics and the vertical gust were not required as
feedback variables. Curve 4 required their use, but did not depend
upon the bending modes 2, 3, 4 or 5. The response of curve 5 was
obtained using the control law derived from HANDEL, referred to
as the "safety law". However, use of the full-state feedback
caused substantial reduction in the steady-state value of bending
moment at the wing root (as well as at the other stations), whereas
with the "safety law" the reduction was significantly less, about
15%. However, the effects on the steady-state values of bending
moments of using reduced-order feedback were assessed by relatively
simple algebraic methods and some of the results are given in
Table 2.

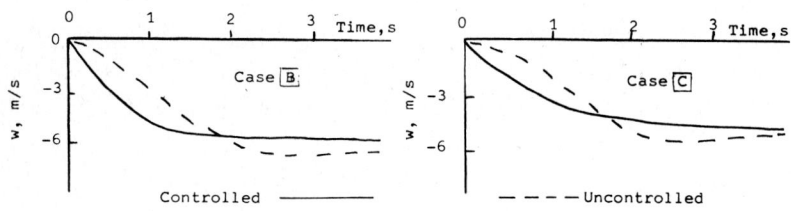

Figure 1 - Rigid Body Heave Motion

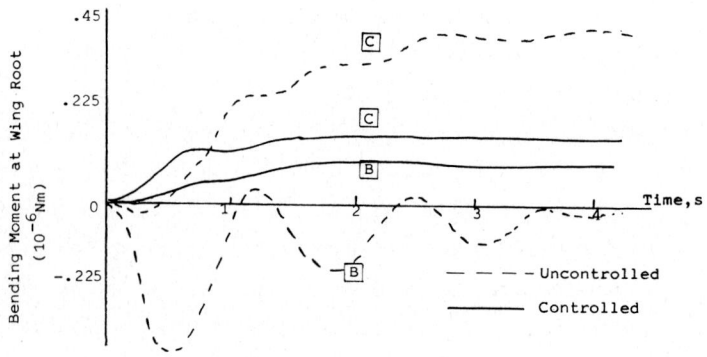

Figure 2 - Bending Moment at Wing Root

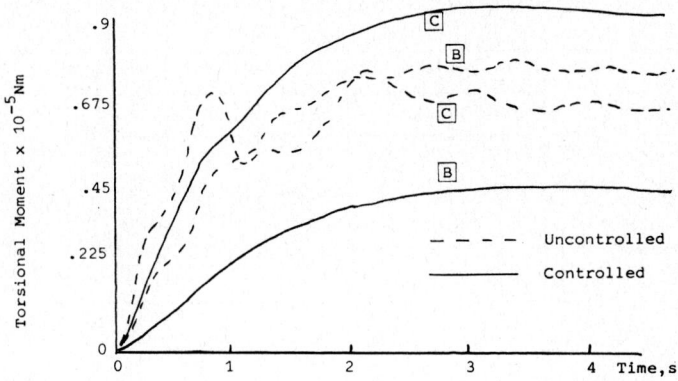

Figure 3 - Torsional Moments at Wing Root

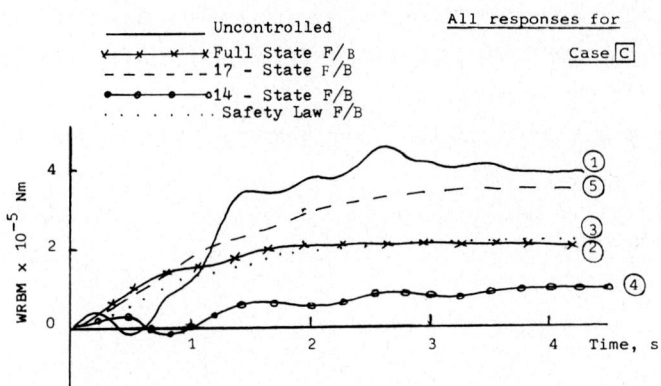

Figure 4 - Effect of Reduced State Feedback on WRBM

Table 2 - *Comparison of Steady-State Bending Moments*

Aircraft Control	WRBM		BM at WS 3		BM at WS 5	
	Case B	Case C	Case B	Case C	Case B	Case C
Uncontrolled	-.55	3.48	-1.0	.9	-.55	.1
Full State F/B	.68	1.38	-.08	-.22	-.14	-.31
17-State F/B	.827	1.43	-.02	-.05	-.12	-.28
14-State F/B	.485	.89	-.33	-.81	-.265	-.62
Safety Law F/B	1.2	3.05	.027	.08	-.14	-.35

all bending moments quoted in 10^6 Nm

It may be expected that the control surface activity required by the active control to attain the levels of load reduction being achieved will be too great for the existing arrangements. Such was not the case in this work, for both the maximum control surface deflection and the rate were considerably below the physical limits set on the aircraft. Even when the aircraft was flying in turbulence (in a digital simulation) substantial reductions were achieved in the bending moments when the active control was employed. Some representative results are presented in Table 3.

Table 3 - *RMS Values of Bending Moments Resulting from Gust*

Location	% Difference Between Controlled and Uncontrolled Aircraft
Wing Root	95.5
WS 2	98.3
WS 3	99.1
WS 4	95.3
WS 5	83.9

gust intensity 1m/s

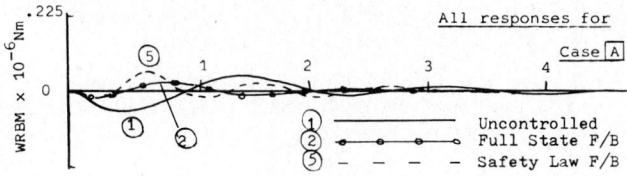

Figure 5 - Effect of Reduced State Feedback on WRBM

5. CONCLUSIONS

It is possible to obtain a useful reduction of the structural loads of a flexible aircraft by means of a feedback controller designed on the basis of linear optimal control theory. Even when the full-state feedback cannot be implemented, by reason of difficulty of measuring the necessary state variables or by virtue of using a control law derived from a less complete mathematical model, substantial reductions in the loads caused by manoeuvring or gusts can still be obtained. There is a minimum order control law, called here the "safety law", below the order of which it is unsafe to reduce the feedback as dynamic instability will result. With the safety law concept it is possible to provide the degree of flight integrity necessary to ensure that structural load alleviation by active control can be considered as a practical possibility for flight control.

REFERENCES

[1] HARVEY, C.A. and POPE, R.E., "Study of Synthesis Techniques for Insensitive Aircraft Control Systems", *NASA CR-2803*, April, 1977.

[2] SCHWANZ, R.C., "Equations of Motion of an Elastic Aircraft", *AFFDL/FGC, TM72-14*, Wright-Patterson Air Force Base, Dayton, Ohio, U.S.A., August, 1972.

[3] ANDERSON, B.D.O. and MOORE, J.B., *Linear Optimal Control*, Prentice-Hall, New Jersey, U.S.A., 1971.

[4] BURRIS, P.M. and BENDER, M.A., "Aircraft Load Alleviation and Mode Stabilisation", *AFFDL TR-68-161*, Wright-Patterson Air Force Base, Dayton, Ohio, U.S.A.

STRUCTURAL CONTROL, H.H.E. Leipholz (ed.)
North-Holland Publishing Company & SM Publications
© IUTAM, 1980

ACTIVE CONTROL OF STRUCTURES BY MODAL SYNTHESIS

L. Meirovitch and H. Öz

Department of Engineering Science and Mechanics
Virginia Polytechnic Institute and State University
Blacksburg, Virginia 24061, U.S.A.

1. INTRODUCTION

Certain structures are subjected to large transient loads acting over relatively short time durations. Examples of these are bridges and large buildings acted upon by wind gusts or loads caused by earth motion in earthquakes, off-shore drilling platforms acted upon by water waves, etc. To prevent failure, the tendency is to design for maximum loads. Sometimes the use of damping can alleviate the situation. Damping represents essentially passive control, however, so that the question arises whether active control cannot do a better job.

Active control of structures is not new. It has been used on aircraft and spacecraft structures extensively [1, 2], but applications to civil structures have only recently begun to appear [3]. One reason is that civil structures tend to be bulkier and more difficult to control. Whereas active control is not advocated for every structure, there are certain structures for which active control can permit a more efficient design.

This paper is concerned with the active control of structures by modal synthesis. Structures generally represent distributed - parameter systems. When they are discretized in space [4], the discrete models tend to have a large number of degrees of freedom. Unfortunately, modern control techniques are not suitable for high-order systems. In the case of structures with positive definite real symmetric stiffness and mass matrices, however, the system of equations of motion can be decoupled by a modal approach developed by these authors [5], thus permitting the independent control of n second-order systems instead of a coupled 2n-order system. The procedure essentially shifts the computational effort from the control problem to the eigenvalue problem for positive definite real symmetric matrices. This represents a substantial advantage, if one recognizes that the capability of the latter is at least one order of magnitude higher than that of the former. An algorithm for the optimal control of structures by modal synthesis [6] is presented.

2. THE DISTRIBUTED-PARAMETER SYSTEM EQUATION OF MOTION

Let us consider a distributed-parameter structure and assume that the kinetic and potential energies have the form

$$T = \frac{1}{2} \int_D m\dot{u}^2 dD \, , \qquad V = \frac{1}{2} [u, u] \, , \tag{1}$$

where $\dot{u}$ is the velocity of any point P in the domain of definition D of the structure, m is the mass density and $[u, u]$ is an energy inner product [4] and it contains spatial derivatives of u through order p. Moreover, letting f be the distributed force, which includes the control forces, the virtual work is

$$\delta W = \int_D f \delta u \, dD \, . \tag{2}$$

By definition, the kinetic energy is a positive definite quantity. We shall assume that the potential energy is also positive definite.

Using Hamilton's principle, we obtain the differential equation of motion

$$Lu + m \frac{\partial^2 u}{\partial t^2} = f, \qquad (3)$$

to be satisfied over the open domain D, in which L is a homogeneous self-adjoint positive definite differential operator of order $2p$, and the boundary conditions $B_i u = 0$, $(i = 1,2,\ldots,p)$ to be satisfied at every point of the boundary S of the domain D.

3. SYSTEM DISCRETIZATION

Equation (3) and the boundary conditions define a distributed control problem. The state of the art does not permit a general solution of this problem, so that the only alternative is to reduce the problem to a discrete one. Ideally, one would solve the eigenvalue problem associated with equation (3) and use the expansion theorem to reduce the partial differential equation, equation (3), to an infinite set of second-order ordinary differential equations.

More often than not, however, the above approach is not feasible because no closed-form solution of the eigenvalue problem is possible. In such cases, the only feasible alternative is discretization, which implies the reduction of the distributed-parameter system to a discrete one of finite dimension. According to the expansion theorem [4], the displacement u can be approximated by a linear combination of known space-dependent admissible functions $\Phi_i(P)$ multiplied by time-dependent generalized coordinates $q_i(t)$, or

$$u(P,t) \cong u^{(n)}(P,t) = \sum_{i=1}^{n} \Phi_i(P) q_i(t) = \underset{\sim}{\Phi}^T(P) \underset{\sim}{q}(t), \qquad (4)$$

where $\underset{\sim}{\Phi}$ and $\underset{\sim}{q}$ are n-vectors. The admissible functions Φ_i must be from a complete set, so that as $n \to \infty$ the approximate solution $u^{(n)}$ approaches the true solution u asymptotically. Although n is finite, we shall assume that it is very large. Introducing equation (4) into equations (1) and (2), the discretized kinetic energy, potential energy and virtual work become

$$T = \frac{1}{2} \dot{q}^T M \dot{q} , \quad V = \frac{1}{2} q^T K q , \quad \delta W = Q^T \delta q , \quad (5)$$

where

$$M = \int_D m \Phi \Phi^T dD , \quad K = [\Phi, \Phi] , \quad Q = \int_D f \Phi dD , \quad (6)$$

are the n × n mass matrix, the n × n stiffness matrix and the n-dimensional generalized force vector, respectively. The mass and stiffness matrices are positive definite. Using standard techniques, one can derive the Lagrange equations of motion

$$M \ddot{q} + K q = Q . \quad (7)$$

Equation (7) represents a set of n simultaneous ordinary differential equations.

Generally, the vector Q contains two classes of forces, disturbing and control forces. We shall assume that the disturbing forces produce initial disturbances, so that Q contains only control forces. The problem is to design controls to regulate the response or reduce it to a tolerable level in a reasonable amount of time.

4. STATE EQUATIONS

It is customary to treat the control problem in state form. To this end, let us introduce the 2n-dimensional state vector $x = [\dot{q}^T | q^T]^T$, as well as the associated force vector $X = [Q^T | 0^T]^T$. Then, introducing the 2n × 2n matrices

$$I = \begin{bmatrix} M & 0 \\ \hline 0 & K \end{bmatrix} , \quad G = \begin{bmatrix} 0 & K \\ \hline -K & 0 \end{bmatrix} \quad (8)$$

Structures by Modal Synthesis   509

where I is symmetric and positive definite and G is skew symmetric, equation (7) can be cast into the state form

$$I\dot{\underset{\sim}{x}} + G\underset{\sim}{x} = \underset{\sim}{X} .\qquad (9)$$

Equation (9) can be reduced to decoupled form, which requires the solution of the eigenvalue problem.

5. THE EIGENVALUE PROBLEM

Considering equations (8), the eigenvalue problem associated with equation (9) can be written in the form

$$\lambda_r \begin{bmatrix} M & 0 \\ 0 & K \end{bmatrix} \underset{\sim}{x}_r = \begin{bmatrix} 0 & -K \\ K & 0 \end{bmatrix} \underset{\sim}{x}_r .\qquad (10)$$

It is known that the eigenvalues occur in pairs of pure imaginary complex conjugates, $\lambda_r = i\omega_r$ and $\bar{\lambda}_r = -i\omega_r$, where ω_r are the "estimated" natural frequencies of the system. Moreover, the eigenvectors also occur in pairs of complex conjugates, $\underset{\sim}{x}_r$ and $\underset{\sim}{\bar{x}}_r$. Introducing the notation $\underset{\sim}{x}_r = \begin{bmatrix} \underset{\sim}{x}_{rU}^T & \underset{\sim}{x}_{rL}^T \end{bmatrix}^T$, where $\underset{\sim}{x}_{rU}$ and $\underset{\sim}{x}_{rL}$ are the upper and lower halves of the vector $\underset{\sim}{x}_r$, respectively, it is not difficult to show that $\underset{\sim}{x}_{rU} = i\omega_r \underset{\sim}{x}_{rL}$, so that the eigenvalue problem (10) can be reduced to

$$\omega_r^2 M \underset{\sim}{x}_{rL} = K \underset{\sim}{x}_{rL} \qquad (11)$$

where $\underset{\sim}{x}_{rL}$ is real.

The modal vectors are orthogonal with respect to I and G and can be arranged in the modal matrix

$$X = \begin{bmatrix} i\omega_1 \underset{\sim}{x}_{1L} & -i\omega_1 \underset{\sim}{x}_{1L} & i\omega_2 \underset{\sim}{x}_{2L} & -i\omega_2 \underset{\sim}{x}_{2L} & \cdots & i\omega_n \underset{\sim}{x}_{nL} & -i\omega_n \underset{\sim}{x}_{nL} \\ \underset{\sim}{x}_{1L} & \underset{\sim}{x}_{1L} & \underset{\sim}{x}_{2L} & \underset{\sim}{x}_{2L} & \cdots & \underset{\sim}{x}_{nL} & \underset{\sim}{x}_{nL} \end{bmatrix} .\qquad (12)$$

It will prove convenient to normalize the modal matrix so as to satisfy

$$X^H I X = 2 \times 1^{(2n)}, \quad -X^H G X = 2\Lambda, \quad (13)$$

where $X^H = \bar{X}^T$, $1^{(2n)}$ is the identity matrix of order 2n and

$$\Lambda = \text{diag}\,[i\omega_1 \ -i\omega_1 \ i\omega_2 \ -i\omega_2 \ \cdots \ i\omega_n \ -i\omega_n]. \quad (14)$$

This requires that the lower halves of the eigenvectors satisfy the orthonormality relations

$$\underset{\sim}{x}_{rL}^T K \underset{\sim}{x}_{sL} = \delta_{rs}, \quad \underset{\sim}{x}_{rL}^T M \underset{\sim}{x}_{sL} = \frac{1}{\omega_r^2} \delta_{rs}, \quad r,s = 1,2,\ldots,n, \quad (15)$$

where δ_{rs} is the Kronecker delta. The modal matrix X can be used to reduce equation (9) to canonical form.

Because the original system is distributed and not discrete, the eigensolution computed by means of the discretized system is only an approximation to the true eigensolution. Modal control schemes of system (9), however, are based on modal characteristics, so that a brief discussion of the nature of the discrete eigensolution is in order. To this end, let us consider the nth order eigenvalue problem

$$\gamma_r^{(n)} M^{(n)} \underset{\sim}{x}_r^{(n)} = K^{(n)} \underset{\sim}{x}_r^{(n)}, \quad r = 1,2,\ldots,n, \quad (16)$$

where $M^{(n)}$ and $K^{(n)}$ are n × n real symmetric positive definite mass and stiffness matrices, as given by the first two of equations (6). These matrices describe an n-degree-of-freedom discrete system resulting from the discretization of a distributed system according to equation (4). If the degree of freedom of the discretized system is increased by one, through adding one more term to the series (4), then the eigenvalue problem of order n+1 has the form

$$\gamma_r^{(n+1)} M^{(n+1)} \underset{\sim}{x}_r^{(n+1)} = K^{(n+1)} \underset{\sim}{x}_r^{(n+1)}, \quad r = 1, 2, \ldots, n+1, \quad (17)$$

where $M^{(n+1)}$ and $K^{(n+1)}$ are also real symmetric positive definite matrices and they contain $M^{(n)}$ and $K^{(n)}$ as submatrices. By the inclusion principle [4], the two sets of eigenvalues are related by

$$\gamma_1^{(n+1)} \leq \gamma_1^{(n)} \leq \gamma_2^{(n+1)} \leq \gamma_2^{(n)} \leq \cdots \leq \gamma_n^{(n+1)} \leq \gamma_n^{(n)} \leq \gamma_{n+1}^{(n+1)}. \quad (18)$$

Moreover, it can be stated that $\lim_{n \to \infty} \gamma_r^{(n)} = \gamma_r$ ($r = 1, 2, \ldots, n$), where γ_r are the true eigenvalues. In general, the estimated lowest eigenvalues converge faster to the actual eigenvalues than the higher ones and accuracy is lost as r approaches n. As a rule of thumb, for an n-degree-of-freedom discretized system only the lowest half of the eigenvalues retain sufficient accuracy.

6. CONTROL BY MODAL SYNTHESIS

From Section 5, it is clear that to obtain good estimated eigenvalues and eigenvectors (and hence eigenfunctions), it is necessary to take a sufficiently large number of terms in the series (4). Moreover, because the accuracy of the rth mode decreases as r increases, and in fact only the lower modes retain sufficient accuracy, one is advised to select the order of the discretized model appreciably larger than the number of modes to be controlled. Equivalently, one should attempt to control a number of modes 2ℓ less than half as large as the order of the discretized system, $\ell \leq n/2$.

Control by modal synthesis is defined as control based on the decoupled equations of motion. To achieve decoupling, use is made of the orthogonality of the eigenvectors. Because this is basically a real system, it is not actually necessary to work with complex quantities. Moreover, we shall control only 2ℓ modes,

$\ell \leq n/2$. Hence, let us introduce the truncated real modal matrix

$$X^{(2\ell)} = \begin{bmatrix} 0 & \omega_1 \underset{\sim}{x}_{1L} & 0 & \omega_2 \underset{\sim}{x}_{2L} & \cdots & 0 & \omega_\ell \underset{\sim}{x}_{\ell L} \\ \underset{\sim}{x}_{1L} & 0 & \underset{\sim}{x}_{2L} & 0 & \cdots & \underset{\sim}{x}_{\ell L} & 0 \end{bmatrix}. \quad (19)$$

It can be verified that the orthonormality relations (13) become

$$X^{(2\ell)^T} I\, X^{(2\ell)} = 1^{(2\ell)}, \quad -X^{(2\ell)^T} G\, X^{(2\ell)} = \tilde{\Lambda}^{(2\ell)}, \quad (20)$$

where $1^{(2\ell)}$ is the identity matrix of order 2ℓ and

$$\tilde{\Lambda}^{(2\ell)} = \text{block-diag } \tilde{\Lambda}_r, \quad (21)$$

in which

$$\tilde{\Lambda}_r = \begin{bmatrix} 0 & \omega_r \\ -\omega_r & 0 \end{bmatrix}, \quad r = 1, 2, \ldots, \ell. \quad (22)$$

Introducing the linear transformation

$$\underset{\sim}{x} \cong \underset{\sim}{x}^{(2\ell)} = X^{(2\ell)} \underset{\sim}{w}, \quad (23)$$

into equation (9), where $\underset{\sim}{w}$ is a 2ℓ-dimensional decoupled state vector, premultiplying by $X^{(2\ell)^T}$ and considering equations (20), we obtain the set of independent real second-order equations

$$\dot{\underset{\sim}{w}}_r = \tilde{\Lambda}_r \underset{\sim}{w}_r + \underset{\sim}{W}_r, \quad r = 1, 2, \ldots, \ell, \quad (24)$$

where

$$\underset{\sim}{w}_r = [\xi_r\ \eta_r]^T, \quad \underset{\sim}{W}_r = [W_{\xi r}\ W_{\eta r}]^T = [0\ \omega_r \underset{\sim}{x}_{rL}^T \underset{\sim}{Q}]^T, \quad (25)$$

in which ξ_r and η_r are conjugate generalized coordinates and $W_{\xi r}$ and $W_{\eta r}$ are associated generalized forces.

Equations (24) permit independent modal control. There are various control laws possible. We shall concentrate on proportional control and relay-type, on-off control.

7. PROPORTIONAL CONTROL

Proportional control can be regarded as artificial viscous damping. The generalized control vector is assumed to have the form

$$\underline{Q} = -2 \sum_{s=1}^{\ell} \zeta_s \omega_s^2 \eta_s(t) M \underline{x}_{sL} , \qquad (26)$$

so that, from the second of equations (15) and (25), we obtain

$$\underline{W}_r = [W_{\xi r} \ W_{\eta r}]^T = [0 \ -2\zeta_r \omega_r \eta_r(t)]^T . \qquad (27)$$

The solution of equations (24), in conjunction with equation (27), can be shown to be

$$\xi_r(t) = e^{-\zeta_r \omega_r t} \{\xi_r(0)\cos\omega_{dr}t + \frac{\omega_r}{\omega_{dr}} [\zeta_r \xi_r(0) + \eta_r(0)]\sin\omega_{dr}t\} ,$$

$$r = 1, 2, \ldots, \ell \qquad (28)$$

$$\eta_r(t) = e^{-\zeta_r \omega_r t} \{\eta_r(0)\cos\omega_{dr}t - \frac{\omega_r}{\omega_{dr}} [\zeta_r \eta_r(0) + \xi_r(0)]\sin\omega_{dr}t\} ,$$

where $\omega_{dr} = (1-\zeta_r^2)^{1/2}\omega_r$, and we note that the closed-loop eigenvalues are $-\zeta_r \omega_r \pm i\omega_{dr}$. Equations (28) indicate that the response decays with time.

8. ON-OFF CONTROL

Proportional control has the disadvantage that it must operate continuously. A scheme not suffering from this drawback is on-off control with deadband. The deadband implies that some low-amplitude oscillations can be tolerated. In this case, the generalized control vector has the form

$$Q = \sum_{s=1}^{\ell} u_s M x_{sL} \, , \qquad (29)$$

where u_s are nonlinear functions of $\eta_s(t)$ given explicitly by

$$u_s = \left\{ -k_s, \; \eta_s > d_s \; ; \; 0 \, , \; |\eta_s| \leq d_s \; ; \; k_s \, , \; \eta_s < -d_s \right\} , \qquad (30)$$

where d_s is a constant defining the deadband region. Introducing equation (29) into the second of equations (25) and recalling the second of equations (15), we obtain

$$\underset{\sim}{W}_r = [W_{\xi r} \; W_{\eta r}]^T = [0 \; u_r/\omega_r]^T . \qquad (31)$$

The solution of equations (24) can be obtained separately for each of the three intervals prescribed in equation (30), with the result

(i) For $\eta_r > d_r$,

$$\xi_r(t) = -\frac{k_r}{\omega_r^2} + \left[\xi_r(0) + \frac{k_r}{\omega_r^2}\right] \cos\omega_r t + \eta_r(0)\sin\omega_r t ,$$

$$\eta_r(t) = -\left[\xi_r(0) + \frac{k_r}{\omega_r^2}\right] \sin\omega_r t + \eta_r(0)\cos\omega_r t , \qquad r = 1,2,\ldots,\ell \, , \qquad (32a)$$

(ii) For $|\eta_r| \leq d_r$,

$$\xi_r(t) = \xi_r(0)\cos\omega_r t + \eta_r(0)\sin\omega_r t ,$$
$$\eta_r(t) = -\xi_r(0)\sin\omega_r t + \eta_r(0)\cos\omega_r t , \qquad r = 1,2,\ldots,\ell \, , \qquad (32b)$$

(iii) For $\eta_r < -d_r$ the response is given by equations (32a) but with k_r replaced by $-k_r$.

9. OPTIMAL CONTROL

The control laws presented in Sections 7 and 8 were not optimal in any sense. However, for practical reasons, it may be desirable to control the system while satisfying a certain performance measure. One such performance measure is known as quadratic performance measure. For control by modal synthesis, a quadratic modal performance measure J_r ($r = 1,\ldots,\ell$) can be defined for each mode independently of the performance measures for any other mode, leading to a total modal performance measure of the form

$$J = \sum_{r=1}^{\ell} J_r = \sum_{r=1}^{\ell} \frac{1}{2} \left\{ \underset{\sim}{w}_r^T(t_f) H_r \underset{\sim}{w}_r(t_f) + \int_{t_0}^{t_f} \left[\underset{\sim}{w}_r^T(t) Q_r \underset{\sim}{w}_r(t) + \underset{\sim}{W}_r^T R_r \underset{\sim}{W}_r \right] dt \right\}, \quad (33a)$$

where $\underset{\sim}{w}_r$ satisfies equation (24). The final time t_f is assumed to be fixed, the matrices H_r and Q_r are semidefinite and R_r is positive definite. Choosing the matrix Q_r as a 2 × 2 unit matrix, the first term in the integrand can be interpreted as the total energy of the system and the second term corresponds to the control effort. The term associated with the matrix H_r represents a penalty on the distance from the origin of the state space at the end of the control period. Because each J_r is independent of the others

$$\min J = \sum_{r=1}^{\ell} \min J_r . \qquad (33b)$$

Assuming that the controls are not bounded and that $\underset{\sim}{w}_r(t_f)$ is free, and denoting optimal quantities by asterisks, the optimal control law has the form [6]

$$\underset{\sim}{W}_r^*(t) = -R_r^{-1} K_r(t) \underset{\sim}{w}_r^*(t) = f_r^*(t) \underset{\sim}{w}_r^*, \quad r = 1,\ldots,\ell, \qquad (34)$$

where $K_r(t)$ satisfies the differential equation

$$\dot{K}_r(t) = -K_r \tilde{A}_r - \tilde{A}_r^T K_r - Q_r + K_r R_r^{-1} K_r, \quad K_r(t_f) = H_r . \qquad (35)$$

Equation (35) is a 2 × 2 matrix Riccati equation for each mode and its solution can be obtained easily. The resulting optimal gain matrix f_r^* is time dependent. Because $W_{\xi r}$ must be zero, according to equation (25), the weighting matrix R_r will be assumed to be of the form $R_r = \text{diag}[\infty \ R_{r2}]$. Then, from equations (34), it follows that the first row of the matrix f_r^* is zero. The steady-state solution of equation (35), defined by $\dot{K}_r(t) = 0$, is of special interest, because it leads to a constant gain matrix. It is shown in [6] that the steady-state solution is

$$K_{12} = K_{21} = -\omega_r R_{r2} + \sqrt{\omega_r^2 R_{r2}^2 + R_{r2}} \ ,$$

$$K_{22} = (R_{r2} - 2\omega_r^2 R_{r2}^2 + 2\omega_r R_{r2}\sqrt{\omega_r^2 R_{r2}^2 + R_{r2}})^{1/2} \quad (36)$$

$$K_{11} = [\omega_r^{-2} + 2\omega_r^{-1} R_{r2}^{-1}(\omega_r^2 R_{r2}^2 + R_{r2})^{3/2} - 2R_{r2}^2 \omega_r^2 - R_{r2}]^{1/2} \ .$$

The transient solution of equation (35) is also given in [6], but will not be presented here, for brevity. The optimal modal control law for the steady-state case has the form

$$W_{\eta r}^* = \left[\omega_r - \sqrt{\omega_r^2 + R_{r2}^{-1}} - \left[2\omega_r\left(-\omega_r + \sqrt{\omega_r^2 + R_{r2}^{-1}}\right) + R_{r2}^{-1}\right]^{1/2}\right] \begin{bmatrix} \xi_r^* \\ \eta_r^* \end{bmatrix} \ . \quad (37)$$

The advantage of the modal performance measure, as given by equation (33a), is that it leads to a set of second-order Riccati equations, independent of the order of the system. This is a significant advantage over coupled control methods, which yield Riccati equations equal in order to the order of the dynamical system. The solutions of Riccati equations of high order create problems of accuracy and computational time.

10. CONTROL IMPLEMENTATION

The proportional control, on-off control and optimal control were derived in terms of generalized control forces, which in turn were given in terms of the generalized coordinates $\xi_r(t)$, $\eta_r(t)$ ($r = 1, 2, \ldots, \ell$). The question remains as to the actual control forces.

Let us assume that the actuator forces $F_j(t)$ are applied at the points $P = P_j$ ($j = 1, 2, \ldots, \ell$). Then, the distributed force $f(P,t)$ can be described by $f(P,t) = F_j(t)\delta(P-P_j)$, where $\delta(P-P_j)$ is a spatial delta function. From the third of equations (6), however, we can write

$$\underset{\sim}{Q} = \int_D f\underset{\sim}{\Phi}dD = \int_D F_j(t)\delta(P-P_j)\underset{\sim}{\Phi}dD = \sum_{j=1}^{\ell} \underset{\sim}{\Phi}(P_j)F_j(t) . \quad (38)$$

Introducing the actual control vector $\underset{\sim}{F} = [F_1 \; F_2 \; \ldots \; F_\ell]^T$, as well as the matrix $B = [\underset{\sim}{\Phi}(P_1) \; \underset{\sim}{\Phi}(P_2) \; \ldots \; \underset{\sim}{\Phi}(P_\ell)]$, the relation between the actual and the generalized control vector is

$$\underset{\sim}{F} = B^{-1}\underset{\sim}{Q} , \quad (39)$$

where $\underset{\sim}{Q}$ is given in terms of $\eta_r(t)$, ($r = 1, 2, \ldots, \ell$) by equation (26) for proportional control and by equations (29) and (30) for on-off control. Note that the actual control $\underset{\sim}{F}$ for the latter case has a staircase form, as it is a linear combination of on-off functions.

The realization of the actual force vector is of course predicated upon the knowledge of the coordinates η_r, ($r = 1, \ldots, \ell$). We shall assume that the coordinates η_r can be inferred from the measurements of the response of the system. To this end, let us consider equation (7) and assume that the displacement u is measured at ℓ distinct points. Denoting these measurements by $u_k(t) = u(P_k, t)$ and using equation (7), we can write the output vector in the form

$$\underline{u}(t) = C \underline{q}^{(2\ell)}(t), \qquad (40)$$

where $C = [\underline{\Phi}(P_1) \; \underline{\Phi}(P_2) \; \ldots \; \underline{\Phi}(P_\ell)]^T$, in which the points $P = P_k$, $(k = 1,2,\ldots,\ell)$ are not necessarily the same points as those for the actuators. From equations (19) and (23), we can write

$$\underline{\dot{q}}^{(2\ell)} = \sum_{r=1}^{\ell} \omega_r \underline{x}_{rL}^T \eta_r, \qquad (41)$$

so that considering the second of equations (15) and equation (40), we obtain

$$\eta_r = \omega_r \underline{x}_{rL}^T MC^{-1} \underline{\dot{u}}, \qquad (42)$$

which implies that the measurement vector should consist of the velocities $\dot{u}_k = \dot{u}(P_k, t)$, $(k = 1,2,\ldots,\ell)$ instead of the displacements. Finally using equation (42) in equation (27) the modal control input has the form

$$W_{\eta r} = -2\zeta_r \omega_r^2 \underline{x}_{rL}^T MC^{-1} \underline{\dot{u}}. \qquad (43)$$

The optimal control case is different because it uses both generalized coordinates. Using equations (19), (23), (37) and (40), it can be shown that

$$W^*_{\eta r} = f^*_{1r}\xi^*_r + f^*_{2r}\eta^*_r = f^*_{1r}\omega_r^2 \underline{x}_{rL}^T M\underline{q}^{(2\ell)} + f^*_{2r}\omega_r \underline{x}_{rL}^T M\underline{\dot{q}}^{(2\ell)}$$

$$= \left[f^*_{1r}\omega_r^2 \underline{x}_{rL}^T MC^{-1} \; \vdots \; f^*_{2r}\omega_r \underline{x}_{rL}^T MC^{-1} \right] \begin{bmatrix} \underline{u} \\ \vdots \\ \underline{\dot{u}} \end{bmatrix}, \qquad (44)$$

where f^*_{1r} and f^*_{2r} are the entries in the bottom row of f^*_r. Equation (44) implies that the optimal control law requires the measurement of both displacements and velocities.

Finally, the actual control vector can be written in terms of the generalized control forces in the form

$$F = B^{-1}M \sum_{s=1}^{\ell} \omega_s x_{sL} W_{\eta s} \, . \qquad (45)$$

Hence, from equations (24) and (43), or from equation (44), we conclude that for proportional control the actual control is an output feedback control. Note that the asterisk has been omitted from equation (45), as the equation is valid for both optimal and nonoptimal control.

11. CONCLUSIONS

Active control of structures by modal synthesis is presented. The control scheme consists of independent modal control providing active damping for the controlled modes of the structure. Proportional and relay-type, on-off controls are discussed. Proportional optimal control providing active damping and stiffening of the structure is also presented. The relations between generalized controls and actual controls are derived for each of the above schemes. Control by modal synthesis is particularly attractive for high-order systems.

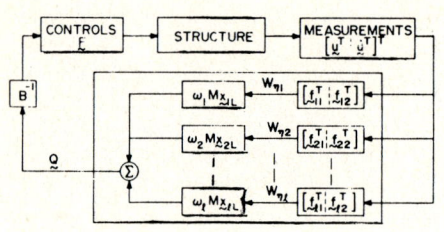

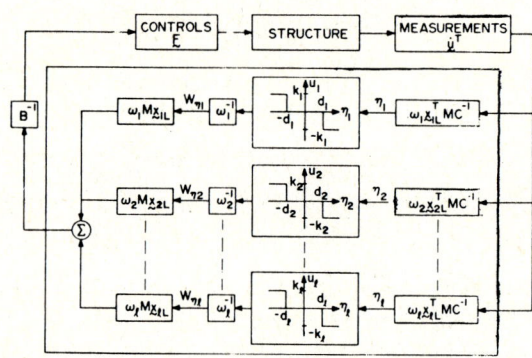

Figure 1 - Block-Diagrams for Control by Modal Synthesis
(a) Proportional Control
(b) On-Off Control

REFERENCES

[1] BALAS, M.J. and CANAVIN, J.R., "An Active Modal Control System Philosophy for a Class of Large Space Structures", *Proc. of the First VPI & SU/AIAA Sym. on Dynamics and Control of Large Flexible Spacecraft*, Blacksburg, Virginia, June 13-15, 1977, pp. 271-285.

[2] MEIROVITCH, L., VanLANDINGHAM, H.F. and ÖZ, H., "Distributed Control of Spinning Flexible Spacecraft", *Proc. of the First VPI & SU/AIAA Sym. on Dynamics and Control of Large Flexible Spacecraft*, Blacksburg, Virginia, June 13-15, 1977, pp. 249-269. Also, *J. of Guidance and Control*, Vol. 2, No. 5, 1979, pp. 407-415.

[3] YANG, J.N., "Application of Optimal Control Theory to Civil Engineering Structures", *ASCE, J. Eng. Mech. Division*, Vol. 10, No. EM6, 1975, pp. 822-838.

[4] MEIROVITCH, L., *Computational Methods in Structural Dynamics*, Sijthoff and Noordhoff International Publishers, Alphen aan den Rijn, The Netherlands, (to appear).

[5] MEIROVITCH, L. and ÖZ, H., "Modal Control of Distributed Gyroscopic Systems", presented as *Paper 78-1421 at the AIAA/AAS Astrodynamics Conference*, Palo Alto, California, August 7-9, 1978. Also, *Journal of Guidance and Control*, (to appear).

[6] ÖZ, H. and MEIROVITCH, L., "Optimal Control of Flexible Gyroscopic Systems", presented as *Paper 78-103 at the XXIXth Congr. of the Int. Astronautical Federation*, Dubrovnik, Yugoslavia, October 1-8, 1978. Also, *Journal of Guidance and Control*, (to appear).

STRUCTURAL CONTROL, H.H.E. Leipholz (ed.)
North-Holland Publishing Company & SM Publications
© IUTAM, 1980

ON OPTIMAL FORCE ACTION AND REACTION ON STRUCTURES

Zenon Mróz

Institute of Fundamental Technological Research
Polish Academy of Sciences
Swietokrzyska 21, 00-049 Warsaw, Poland

1. INTRODUCTION

In most structural synthesis problems both loading and support conditions have been assumed as specified and the solution was sought which provided some design variables representing cross-sectional areas or dimensions of load carrying members. However, in many cases the optimal design problems can be formulated in a broader context, including the position of supports and their stiffness as design variables or allowing for varying distribution of loading. Such problems were discussed within the beam or frame theory [1 - 6]. In this paper, the previous analysis will be generalized by formulating general theorems on optimal force action and reaction. Such theorems may also find application beyond the scope of optimal design theory, for instance, in active control problems when positioning of force actuators is essential in achieving effective damping of excessive deflections or stresses. In Section 2, the conditions for optimal force or couple actions will be discussed for a nonlinear elastic body whose compliance is to be minimized. In Section 3, the optimal loading distribution problem will be discussed and in Section 4 the optimal action associated with any functional of stress or displacement will be considered using the concept of adjoint systems. In Section 6, two simple examples of optimal reactions will be discussed.

2. CONDITIONS FOR OPTIMAL FORCE REACTION: MEAN COMPLIANCE DESIGN

When a structure is subjected to a prescribed set of loads which produce excessive deformation, additional forces may be applied that counteract given loads and reduce deflections or stresses. We shall use the term *optimal reaction* when we want to determine such magnitudes, positions or directions of forces which correspond to minimum of static or dynamic compliance, maximum safety factor for plastic collapse or maximum buckling loads. In particular, such reactions may be introduced by imposing deflection or rotation constraints through rigid point or line supports and their work is either negative or vanishes, $R_i u_i \leq 0$. The related problem of *optimal action* of loads whose work is positive, $P_i u_i \geq 0$ was discussed in [2], where such distribution of loading was sought which correspond to minimum of static compliance or maximum of limit load. In many technological problems, the optimal action of loads corresponds to maximum of mean or local deflections for specified upper bound on magnitudes of forces.

Consider a nonlinear elastic material for which stress and strain potentials $W(\sigma)$ and $U(\varepsilon)$ generate the stress-strain relations

$$\varepsilon = \frac{\partial W(\sigma)}{\partial \sigma} \quad , \quad \sigma = \frac{\partial U(\varepsilon)}{\partial \varepsilon} \quad , \tag{1}$$

where $U(\varepsilon) = \sigma \cdot \varepsilon - W(\sigma)$ and the dot between two symbols denotes their inner product. If $W(\sigma)$ is a homogeneous function of order n, then

$$\sigma \cdot \varepsilon = \sigma \cdot \frac{\partial W}{\partial \sigma} = nW(\sigma) = \frac{n}{n-1} U(\varepsilon) \quad . \tag{2}$$

In the following, let us assume the global complementary energy as a measure of mean structure compliance C, thus

$$C = \int W(\underset{\sim}{\sigma})dV , \qquad (3)$$

or the potential energy as a measure of structure stiffness

$$\Pi = \int U(\underset{\sim}{\varepsilon})dV - \int \underset{\sim}{t}^\circ \cdot \underset{\sim}{u}dS_t . \qquad (4)$$

Note that $C = -\Pi$. When (2) occurs, from (3) and (4) we obtain

$$C = \frac{1}{n} \int \underset{\sim}{\sigma} \cdot \underset{\sim}{\varepsilon}\, dv = \frac{1}{n} \int \underset{\sim}{t}^\circ \cdot \underset{\sim}{u}dS_t \quad , \quad \Pi = -\frac{1}{n} \int \underset{\sim}{t}^\circ \cdot \underset{\sim}{u}dS_t . \qquad (5)$$

In this work, we shall assume the convexity of stress and strain potentials, thus

$$W(\underset{\sim}{\sigma_2})-W(\underset{\sim}{\sigma_1}) - (\underset{\sim}{\sigma_2}-\underset{\sim}{\sigma_1}) \cdot \underset{\sim}{\varepsilon_1} \geq 0 , \quad U(\underset{\sim}{\varepsilon_2}) - U(\underset{\sim}{\varepsilon_1}) - (\underset{\sim}{\varepsilon_2}-\underset{\sim}{\varepsilon_1}) \cdot \underset{\sim}{\sigma_1} \geq 0 , \qquad (6)$$

for any two values of stress and strain.

Assume now that the compliance C is a function of surface tractions $\underset{\sim}{t}^\circ$ acting on S_t. Consider two systems of loads t_1° and t_2° and let the corresponding stress, strain and displacement fields be $\sigma_1, \varepsilon_1, u_1$ and $\sigma_2, \varepsilon_2, u_2$. From (3) and (6) it follows that convexity of global compliance occurs, that is

$$C(\underset{\sim}{t_2^\circ}) - C(\underset{\sim}{t_1^\circ}) - (\underset{\sim}{t_2^\circ}-\underset{\sim}{t_1^\circ}) \cdot \underset{\sim}{u_1}dS_t \geq 0 . \qquad (7)$$

2.1 Optimal Force Reaction

Consider a body shown in Figure 1(a), supported on S_u and loaded by prescribed tractions $\underset{\sim}{t}^\circ$ on S_t. The work of reactions on S_u vanishes, thus

$$\int \underset{\sim}{t} \cdot \underset{\sim}{u}dS_u = 0 . \qquad (8)$$

Moreover, let us assume that the force R is applied to the body. Its magnitude, points of application and direction should be specified in order to optimize the force reaction. Consider first the conditions for stationarity of the elastic compliance C.

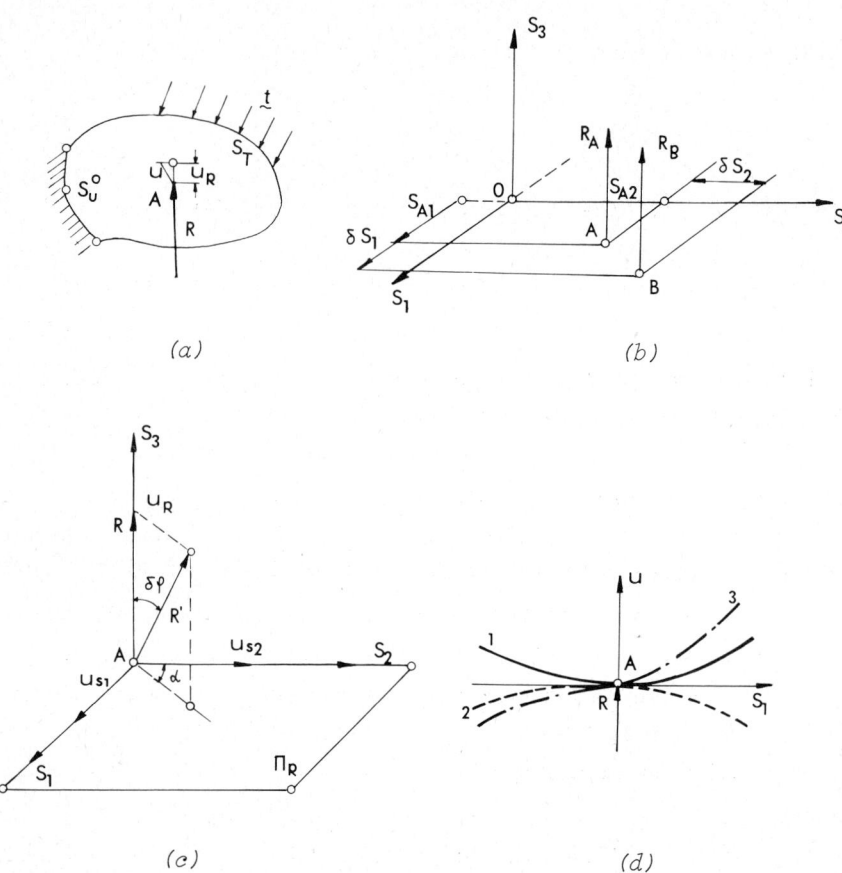

Figure 1 - (a) A Concentrated Action R is applied at A to an Elastic Body Loaded on S_T and Supported on S_U^0,
(b) Translation of Point of Application of Force R within the Plane Π
(c) Rotation of Force R
(d) Deflection Forms at the Stationarity Point A

Let the position and direction of the force R be fixed, but the magnitude of R be varied. Consider the values R and $R = R + \delta R$ and denote the corresponding stress, strain, and displacement fields by σ, ε, u and $\sigma', \varepsilon', u'$. In view of the principle of virtual work, we have

$$\int t^\circ \cdot u dS_t - Ru_R = \int \sigma \cdot \varepsilon \, dV , \qquad (9)$$

$$\int t^\circ \cdot u dS_t - \int R'u_R = \int \sigma' \cdot \varepsilon \, dV , \qquad (10)$$

and

$$-\delta Ru_R = \int \delta\sigma \cdot \varepsilon \, dV , \qquad (11)$$

where u_R denotes the displacement component along the direction of R and $\delta\sigma = \sigma' - \sigma$. In view of (11), the variation of compliance corresponding to the variation of R equals

$$\delta C = \int \frac{\partial W}{\partial \sigma} \cdot \delta\sigma \, dV = \int \varepsilon \cdot \delta\sigma \, dV = -\delta Ru_R = 0 , \qquad (12)$$

and since δR is arbitrary, the stationarity of compliance occurs when

$$u_R = 0 , \qquad (13)$$

that is the acting force should correspond to *rigid support reaction* for which the displacement component vanishes.

Assume now that the direction of R is fixed but its position may vary. Choose the rectangular cartesian coordinate system s_1, s_2, s_3 for which s_3 is parallel to R and s_1, s_2 determine the plane Π_R normal to R, Figure 1(b). Consider, for instance, the parallel translation of R along the s_1-axis. For the two positions s_{A1} and s_{B1} of R, the virtual work equation provides the equalities

$$\int t^\circ \cdot u dS_t - R_A u_{AR} = \int \sigma \cdot \varepsilon \, dV , \qquad (14)$$

$$\int t^\circ \cdot u dS_t - R_B u_{BR} = \int \sigma' \cdot \varepsilon \, dV , \qquad (15)$$

and

$$\int \delta\underset{\sim}{\sigma}\cdot\underset{\sim}{\varepsilon}\, dV = -(R_B-R_A)u_{AR} - R_B(u_{BR}-u_{AR}) \tag{16}$$

where, as previously, $\delta\underset{\sim}{\sigma} = \underset{\sim}{\sigma}'-\underset{\sim}{\sigma}$ and u_{AR}, u_{BR} denote the displacement components along the direction of R at A and B. Let us note that u_{AR} and u_{BR} denote components of the same displacement field $\underset{\sim}{u}$ corresponding to position of R at A and B. Considering infinitesimal variation of R and its infinitesimal translation, the variation of compliance C can be expressed as follows

$$\delta C = \int \frac{\partial W}{\partial \underset{\sim}{\sigma}}\cdot \delta\underset{\sim}{\sigma}\, dV = \int \underset{\sim}{\varepsilon}\cdot\delta\underset{\sim}{\sigma}\, dV = -\delta R u_{AR} - R\delta u_R , \tag{17}$$

and for a displacement field of the class C^1 there is

$$\delta u_R = \frac{\partial u_R}{\partial s_1}(A)\delta s_1 = u_{R,1}\delta s_1 . \tag{18}$$

Thus, the stationarity of C requires that

$$u_{AR} = 0 , \quad Ru_{R,1}(A) = 0 . \tag{19}$$

The first condition (19) coincides with (13) and the second condition requires either $R = 0$ when $u_{R,1} \neq 0$ or $u_{R,1} = 0$ when $R \neq 0$. When the force is allowed to translate independently in directions s_1, s_2, s_3 the stationarity conditions (19) become

$$u_{AR} = 0 , \quad Ru_{R,1}(A) = Ru_{R,2}(A) = Ru_{R,3}(A) = 0 . \tag{20}$$

Assume now that for the fixed position of R, its direction is allowed to vary. Let u_R, u_{s_1}, u_{s_2} be the displacement component along the direction of R and along the s_1, s_2-axes in the plane normal to R, Figure 1(c). Considering a rotated direction of R through the angle $\delta\phi$, the force components along s_1, s_2, and s_3 axes are respectively $R\sin\delta\phi\sin\alpha$, $R\sin\delta\phi\cos\alpha$ and $R\cos\delta\phi$, where the angles $\delta\phi$ and α are shown in Figure 1(c). Setting $\sin\delta\phi \approx \delta\phi$

and $\cos\delta\phi \approx 1$, instead of (17), the variation of compliance is now expressed as follows

$$\delta C = \int \delta\sigma \cdot \varepsilon \, dV = -\delta R u_R - R\delta\phi\cos\alpha u_{S2} - R\delta\phi\sin\alpha u_{S1} , \qquad (21)$$

and the stationarity of C requires that

$$u_R = u_{S1} = u_{S2} = 0 , \qquad (22)$$

that is all displacement components should vanish. Combining (20) and (22), we obtain the stationarity conditions for the most general case when both translation, rotation and magnitude variation of R has been allowed.

2.2 Optimal Couple Reaction

Consider now the case when there are two forces acting at two points A and B at the distance AB = 2a, Figure 2. Assume that R_A and R_B are not independent but may be varied proportionately, $R_A = R_B \beta$. Repeating the same analysis as before, we obtain instead of (20)

$$R_A u_{R,1}(A) + R_B u_{R,1}(B) = 0 , \quad R_A u_{R,2}(A) + R_B u_{R,2}(B) = 0 ,$$
$$R_A u_{R,3}(A) + R_B u_{R,3}(B) = 0 , \qquad (23)$$

and instead of (22)

$$\beta u_R(A) + u_R(B) = 0 , \quad R_A u_{S1}(A) + R_B u_{S1}(B) = 0 ,$$
$$R_A u_{S2}(A) + R_B u_{S2}(B) = 0 . \qquad (24)$$

When the two forces are equal, $R_A = R_B$, we obtain

$$u_{R,i}(A) + u_{R,i}(B) = 0, \quad i = 1,2,3,$$

$$u_R(A) + u_R(B) = 0, \quad u_{S_1}(A) + u_{S_1}(B) = 0, \quad (25)$$

$$u_{S_2}(A) + u_{S_2}(B) = 0,$$

and for opposite forces $R_A = -R_B$, we obtain

$$u_{R,i}(A) = u_{R,i}(B) = 0, \quad i = 1,2,3,$$

$$u_R(A) - u_R(B) = 0, \quad u_{S_1}(A) - u_{S_1}(B) = 0, \quad (26)$$

$$u_{S_2}(A) - u_{S_2}(B) = 0..$$

Figure 3 illustrates these conditions for the case $R_A = R_B$ and $R_A = -R_B$. Let us note that in the latter case the conditions $u_R(A) = u_R(B)$ and $u_{R,1}(A) = u_{R,1}(B)$ cannot be simultaneously satisfied for u_R monotonically increasing along AB, Figure 3(b), and u_R should have extrema within the interval AB, Figure 3(c).

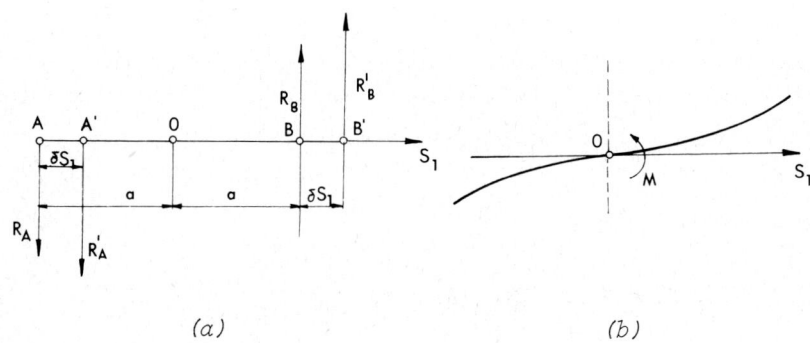

Figure 2 - (a) Translation of Two Forces Along S_1
(b) Optimal Deformation Mode for Couple Reaction

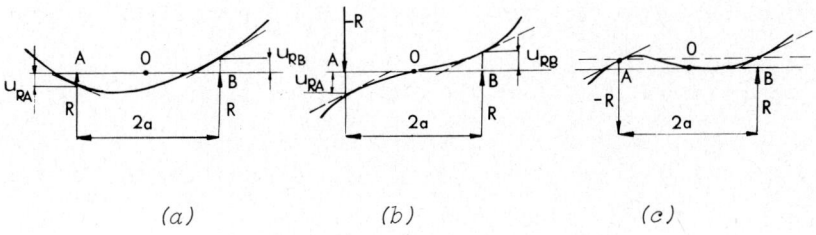

Figure 3 - Optimal Action of Two Equal (a), and Opposite Forces (b), (c)

The case of action of a concentrated couple M can be discussed by applying the optimality conditions (26) and considering the limiting case $a \to 0$, $R_A \cdot 2a \to M$. Let us express the displacement component u_R and its gradient in terms of their values at the midpoint 0. For small a, we can write

$$u_R(B) = u_R(0) + u_{R,1}(0)a + \frac{1}{2} u_{R,11}^{+}(0)a^2 + \cdots$$
$$u_R(A) = u_R(0) - u_{R,1}(0)a + \frac{1}{2} u_{R,11}^{-}(0)a^2 + \cdots \quad (27)$$

In writing (27), we assumed that $u_{R,1}(0)$ is continuous but the second derivative $u_{R,11}(0)$ may change abruptly when passing 0 from left to right along s_1. In this way, we account for types of discontinuities in curvatures of deflection fields occuring in beams or plates when acted on by concentrated or line couples. Translating AB along s_1 through the distance δs_1, the displacement u_R at A' and B' can be expressed similarly to (27), Figure 2(a).

$$u_R(B') = u_R(0) + u_{R,1}(0)(a+\delta s_1) + \frac{1}{2} u_{R,11}^{+}(0)(a+\delta s_1)^2 + \cdots$$
$$u_R(A') = u_R(0) - u_{R,1}(0)(a-\delta s_1) + \frac{1}{2} u_{R,11}^{-}(0)(a-\delta s_1)^2 + \cdots \quad (28)$$

In view of (27) and (28), the principle of virtual work

$$\int (\sigma'-\sigma') \cdot \varepsilon \, dV = -R'_B[u_R(B')-u_R(A')] + R_B[u_R(B)-u_R(A)], \qquad (29)$$

can be expressed as follows for infinitesimal displacement δs_1

$$\int (\sigma'-\sigma) \cdot \varepsilon \, dV = -(R'_B-R_B)[u_R(B)-u_R(A)] - R'_B[u_R(B')-u_R(A')-u_R(B)+u_R(A)]$$

$$= -(R'_B-R_B)\left[2u_{R,1}(0)a + \left\{\frac{1}{2}u^+_{R,11}(0) - \frac{1}{2}u^-_{R,11}(0)\right\}a^2\right] -$$

$$- R_B a \delta s_1 \left[u^+_{R,11}(0) + u^-_{R,11}(0)\right]. \qquad (30)$$

Setting $a \to 0$ and $R_B 2a \to M$, equation (30) takes the form

$$\delta C = -\delta M u_{R,1}(0) - \frac{1}{2} M \left[u^+_{R,11}(0)+u^-_{R,11}(0)\right]\delta s_1, \qquad (31)$$

and the stationarity condition $\delta C = 0$ requires that

$$u_{R,1}(0) = 0, \quad M\left[u^+_{R,11}(0) + u^-_{R,11}(0)\right] = 0. \qquad (32)$$

Thus, if M does not vanish, the second condition gives $u^+_{R,11}(0) = -u^-_{R,11} = 0$. In particular, when continuity of second derivatives occurs, we have $u_{R,11}(0) = 0$ and the optimality condition would require $u^+_{R,111}(0) = u^-_{R,111}(0)$. Thus, *the optimal couple action should correspond to vanishing angle of rotation of the s_1-line at 0 and antisymmetric deformation mode of this line with respect to 0, Figure 2(b)*. Let us note that the stationarity conditions (32) coincide with those derived in [1] for elastic beams. For translation of couple along s_2, s_3, the conditions analogous to (32) result.

So far, we have derived the stationarity conditions for the mean compliance C. However, the sufficient conditions for global minimum of C can easily be established by considering finite differences between two states corresponding to two force positions and orientations. Using the virtual work equations (16) or (21), we can write

$$C_2 - C_1 = \int W(\underset{\sim}{\sigma}_2) - W(\underset{\sim}{\sigma}_1) - (\underset{\sim}{\sigma}_2 - \underset{\sim}{\sigma}_1) \cdot \underset{\sim}{\varepsilon}_1 \, dV + R_B \Delta u_R, \qquad (33)$$

where the labels 2 and 1 correspond to two states and $\Delta u_R = u_R^1(B) - u_R^1(A)$; in (33) it is assumed that the stationarity condition $u_R^1(A) = 0$ occurs for the force position at A, and the associated states is $\sigma_1, \varepsilon_1, u_1$. As for the convex stress potential $W = W(\underset{\sim}{\sigma})$, the integrand of (33) is positive-definite, the strong inequality $C_2 > C_1$ occurs whenever $R_B \Delta u_R > 0$, that is *the reaction expends a positive work when translated or rotated from the stationarity position.* Figure 1(d) shows the deflection form 1 corresponding to the global minimum of C and the forms 2, 3 for which the global minimum cannot be assured though they satisfy the stationarity conditions. It was found, however, that for some beam structures the stationarity forms of Figure 1(d) correspond to optimal positions [3 - 5].

3. CONDITIONS FOR OPTIMAL FORCE ACTION

A closely related class of problems is associated with the optimal loading distribution. For instance, it may be assumed that the total load is specified but its position or distribution is optional and may be varied to achieve the optimal structural performance. Such cases can occur in some practical situations, for instance, in foundation design, in distributing service loads on particular floors of a frame structure or in active control problems. This class of problems was discussed in detail in [2]. Here we only generalize some of the results achieved in [2] and provide mechanical interpretation of the optimality conditions.

Similarly as in the preceding section, consider a non-linearly elastic body of volume V and the boundary surface S, supported on the portion S_u and loaded on the portion S_T of its boundary. Let the surface tractions have specified directions

$$\underline{t}^\circ = \underline{r}\,t(\underline{x}), \qquad (34)$$

where $\underline{r}$ denotes the unit vector specifying the force direction and $t(\underline{x})$ denotes its modulus. Assume that $t(\underline{x})$ is unspecified function of the position on S_T and let it satisfy the global constraint

$$\int t(\underline{x})f(\underline{x})dS_T \geq P, \qquad (35)$$

where $f(\underline{x})$ is a function of position. To find the stationarity of the global compliance C, let us consider the functional

$$C' = C - \eta^2\left[\int t(\underline{x})f(\underline{x})dS_T - P\right], \qquad (36)$$

and its first variation equals

$$\delta C' = \int \frac{\partial W}{\partial \underline{\sigma}} \cdot \delta\underline{\sigma}\,dV - \eta^2 \int \delta t f(\underline{x})dS_T - 2\eta\delta\eta\left[\int t(\underline{x})f(\underline{x})dS_T - P\right]$$
$$= \int \left[u_t - \eta^2 f(\underline{x})\right]\delta t\,dS_T - 2\eta\delta\eta\left[\int t(\underline{x})f(\underline{x})dS_T - P\right] = 0. \qquad (37)$$

Hence, the stationarity conditions are

$$u_t = \eta^2 f(\underline{x}), \qquad \int t(\underline{x})f(\underline{x}) - P = 0. \qquad (38)$$

In transforming (37), we used the virtual work principle

$$\int \frac{\partial W}{\partial \underline{\sigma}} \cdot \delta\underline{\sigma}\,dV = \int \underline{\varepsilon}\cdot\delta\underline{\sigma}\,dV = \int u_t \cdot \delta t\,dS_T, \qquad (39)$$

where u_t denotes the displacement component on S_T directed along the force, i.e., $u_t = \underline{u}\cdot\underline{r}$; and η is the Lagrange multiplier.

Consider, for instance, the plane boundary and the loading distribution of the same direction, Figure 4. When $f(x) = 1$, the condition (35) provides a lower bound on the total force and the stationarity condition (38) requires the displacement of the loaded portion to be uniform. When $f(x) = 1+\gamma x$, γ = const., the inequality (35) provides a bound on linear combination on total force and moment of loading; the corresponding displacement $u_t = \eta^2(1+\gamma x)$ is a linear function of position. In other words, when constraints are imposed on the total force or moment of surface tractions, the associated displacement field corresponds to *a rigid body of motion of S_T*. It may be conceived that the loading is transmitted through a rigid foundation block subjected to the eccentric loading, Figure 4.

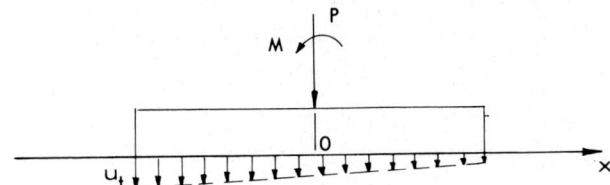

Figure 4 - *Optimal Force and Couple Action*

The other particular case of loading is a set of concentrated parallel forces $P_1, P_2, \ldots, P_n$ with the specified lower bound on their resultant, $P_1+P_2+P_3+ \cdots P_n \geq P$. The stationarity condition now is

$$u_1^P = u_2^P = u_3^P = \cdots = u_n^P = \eta^2 , \tag{40}$$

that is requiring equal displacements in the direction of forces.

The global minimum of the compliance C can easily be deduced from the convexity property (7). In fact, assuming that

t_1^0 is the optimal loading distribution satisfying the stationarity condition (38), from (7) we obtain

$$C(\underset{\sim}{t_2^0}) - C(\underset{\sim}{t_1^0}) \geq \eta^2 \int (t_2-t_1)f(x)dx \geq 0 ,\qquad(41)$$

and

$$C(\underset{\sim}{t_2^0}) \geq C(\underset{\sim}{t_1^0}) .\qquad(42)$$

4. OPTIMAL SUPPORT REACTION ASSOCIATED WITH AN ARBITRARY FUNCTIONAL

So far, we considered only the mean compliance C of a nonlinear elastic structure, measured by the total complementary energy. However, in many cases we wish to control the local displacements or stresses and strains. Then the behaviour functional need not coincide with the mean compliance C. For instance, when the excessive stresses are to be avoided, the support reaction should minimize the functional

$$G_1 = \frac{1}{n+1}\int <\phi(\underset{\sim}{\sigma}) - \sigma_0>^{n+1} dV = \int \Phi(\underset{\sim}{\sigma})dV ,\qquad(43)$$

where $\phi(\underset{\sim}{\sigma})$ is a homogeneous function of stress of order one, for instance, the effective stress based on the second stress deviator invariant, σ_0 is the admissible, non-penalized, stress level. Here G_1 can be regarded as "penalty functional" for exceeding the stress level σ_0. The symbol $<y>$ means $<y> = y$ for $y > 0$ and $<y> = 0$ for $y < 0$, thus the effective stress below σ_0 does not contribute to (43). An alternative global stress measure is the so-called p-norm

$$G_2 = \frac{1}{V}\left(\int \phi^p(\underset{\sim}{\sigma})dV\right)^{\frac{1}{p}} ,\qquad(44)$$

where p is a positive integer. When $p \to \infty$, then $G_2 \to \sup\phi(\underset{\sim}{\sigma})$ and (44) provides the measure for maximum effective stress. Similar global measures can be introduced for displacements, for instance

$$H_1(\underset{\sim}{u}) = \frac{1}{m+1}\int <\psi(\underset{\sim}{u})-u°>^{m+1} dS_c = \int \Psi(\underset{\sim}{u}) dS_c, \qquad (45)$$

where $\psi = \psi(\underset{\sim}{u})$ is a homogeneous function of order one of displacements on the portion S_c of the boundary where the excessive displacements are to be avoided; u_0 is the admissible level for the function $\psi(u)$. Analogously to (44), the p-norm could also be used as an alternative measure of the displacement of S_c.

Our analysis will be limited to *linear* elastic materials for which the Betti's reciprocity theorem is valid. Consider first the stress functional and introduce an *adjoint structure* of the same form, support conditions on S_u and the same material, with no surface tractions on S_t, but loaded by the initial strain field $\underset{\sim}{\varepsilon}^i(x)$ defined by the potential law

$$\underset{\sim}{\varepsilon}^i = \frac{\partial \Phi}{\partial \underset{\sim}{\sigma}} = <\phi(\underset{\sim}{\sigma})-\sigma_0>^{n-1} \frac{\partial \phi}{\partial \underset{\sim}{\sigma}}. \qquad (46)$$

This initial strain field induces the residual stress state $\underset{\sim}{\sigma}^r(x)$ and the associated strain $\underset{\sim}{\varepsilon}^r(x) = L\underset{\sim}{\sigma}^r$, so that

$$\underset{\sim}{\varepsilon}^a = \underset{\sim}{\varepsilon}^i + \underset{\sim}{\varepsilon}^r, \qquad \underset{\sim}{\sigma} = \underset{\sim}{\sigma}^r, \qquad (47)$$

where $\underset{\sim}{\varepsilon}^a$ and $\underset{\sim}{u}^a$ denote the resulting strain and displacement fields of the adjoint structure; L denotes the compliance matrix of the linear elastic material.

Using the concept of adjoint structure, let us derive the optimal support conditions. Considering the variation of $G_1(\underset{\sim}{\sigma})$, by virtue of (46) and (47), let us write

$$\delta G_1(\underset{\sim}{\sigma}) = \int \frac{\partial \phi}{\partial \underset{\sim}{\sigma}} \cdot \delta\underset{\sim}{\sigma} dV = \int \underset{\sim}{\varepsilon}^i \cdot \delta\underset{\sim}{\sigma} dV = \int (\underset{\sim}{\varepsilon}^a - \underset{\sim}{\varepsilon}^r) \cdot \delta\underset{\sim}{\sigma} dV = \int \underset{\sim}{\varepsilon}^a \cdot \delta\underset{\sim}{\sigma} dV \qquad (48)$$

In fact, using the reciprocity relation, we have

$$\int \underset{\sim}{\varepsilon}^r \cdot \delta\underset{\sim}{\sigma} dV = \int \underset{\sim}{\sigma}^r \cdot \delta\underset{\sim}{\varepsilon} dV = 0, \qquad (49)$$

since $\underset{\sim}{\sigma}^r$ is the residual stress field within the adjoint structure and $\delta\varepsilon$ is the compatible strain variation of the original structure. Allowing for variation, of R, translation and rotation of the support, analogously to (16) and (21), we can write

$$\delta G_1 = \int \underset{\sim}{\varepsilon}^a \cdot \delta\underset{\sim}{\sigma}dV = -u_R^a \cdot \delta R - \sum_i Ru_{R,i}^a \delta s_i - R\delta\phi\cos\alpha u_{S2}^a - R\delta\phi\sin\alpha u_{S1}^a, \quad (50)$$

and the stationarity of G_1 requires that

$$u_R^a = Ru_{R,1}^a = Ru_{R,2}^a = Ru_{R,3}^a = u_{S1}^a = u_{S2}^a = 0. \quad (51)$$

Thus the conditions for optimal support reaction are similar to (20) and (22): *the support reaction should correspond to vanishing displacements of the adjoint structure at the support points and vanishing deflection slope in the direction of support translation.*

Similar optimality conditions can be derived for the functional (45) expressed in terms of surface displacements on S_c. Let us introduce the surface tractions on S_c defined by the relations

$$\underset{\sim}{t}^a = \frac{\partial \Psi(\underset{\sim}{u})}{\partial \underset{\sim}{u}}, \quad (52)$$

and let us denote the corresponding state by $\underset{\sim}{u}^a$, $\underset{\sim}{\varepsilon}^a$, $\underset{\sim}{\sigma}^a$. Considering the variation of $H_1(\underset{\sim}{u})$, we have

$$\delta H_1(\underset{\sim}{u}) = \int \frac{\partial \Psi}{\partial \underset{\sim}{u}} \cdot \delta\underset{\sim}{u}dS_c = \int \underset{\sim}{t}^a \cdot \delta\underset{\sim}{u}dS_c = \int \underset{\sim}{\sigma}^a \cdot \delta\underset{\sim}{\varepsilon}dS_c = \int \underset{\sim}{\varepsilon}^a \cdot \delta\underset{\sim}{\sigma}dS_c, \quad (53)$$

and the formula (50) applies. Thus, the stationarity conditions (51) are valid.

The case of optimal loading distribution can be treated similarly. Following the derivation of the previous section, we find that

$$u_t^a = \eta^2 f(x), \quad (54)$$

where $u_t^a(\underset{\sim}{x})$ denotes the displacement of the adjoint structure on the loaded surface.

5. OPTIMAL FORCE REACTION WITH COST ASSIGNED

The optimality conditions can further be generalized by assuming that the force or couple reaction is associated with a cost C_s which is in general a function of force magnitude, its position and orientation. Thus, we can add to the structure compliance C the support "compliance" proportional to its cost, for instance,

$$\overline{C} = C + kC_s(R, s_i, \phi_i) , \qquad (55)$$

where k is a constant parameter. Now the stationarity conditions of C can easily be derived by following (9-22). We have

$$\delta\overline{C} = \delta C + k\left(\frac{\partial C_s}{\partial R}\delta R + \frac{\partial C_s}{\partial s_i}\delta s_i + \frac{\partial C_s}{\partial \phi_i}\partial\phi_i\right) = 0 , \qquad (56)$$

and using (17) and (21), we obtain

$$u_R = k\frac{\partial C_s}{\partial R}, \quad Ru_{R,j} = k\frac{\partial C_s}{\partial s_j}, \quad Ru_{si} = k\frac{\partial C_s}{\partial \phi_i}, \quad \begin{array}{l} i = 1,2 \\ j = 1,2,3 \end{array} \qquad (57)$$

and ϕ_i denote the force angles with the cartesian reference system. Referring to Figure 1(c), we have $\cos\phi_3 = \cos\phi$, $\cos\phi_2 = \sin\phi\cos\alpha$, $\cos\phi_1 = \sin\phi\sin\alpha$. The first equation (57) provides the magnitude of displacement u_R at the support point whereas two other conditions define the slope and the lateral displacements.

6. EXAMPLES

Extensive applications of the derived stationarity conditions were presented in [1 - 6] for beam structures with compliance, free frequency or buckling load as objective functions. Because of space limitation, we shall only discuss briefly two simple illustrative examples.

Example 1. Prestressed Beam

Consider a beam prestressed by a curvilinear cable transmitting a constant tensile force and exerting an upward pressure on the beam whereas the external loading is directed downward, Figure 5. Let us find the optimal cable action, neglecting the cost associated with prestress. The compliance due to lateral deflections is associated with the bending moments, thus

$$C = \int \frac{M^2(x)}{2s_b} dx , \qquad (58)$$

where s_b denotes the beam stiffness and $M(x)$ is the bending moment field. The total lateral deflection is the sum of deflection u_1 due to loading and deflection u_s due to prestressing action, thus

$$u = u_1 + u_s . \qquad (59)$$

Let us now regard the prestressing as supporting action on the beam. The stationarity of C now equals

$$\delta C = \int \frac{M}{s_b} \delta M dx = \int k\delta M dx = \int u\delta p_s dx , \qquad (60)$$

where p_s is the pressure of prestressing cable on the beam. The stationarity of C now requires

$$u = u_1 + u_s = 0 , \qquad (61)$$

that is vanishing of total deflection. From the beam equations it then follows that $p_s(x)$ should be equal and opposite to loading $p(x)$. In this way, we derived the well known "load balancing method" proposed earlier by Lin for prestressed beams and slabs.

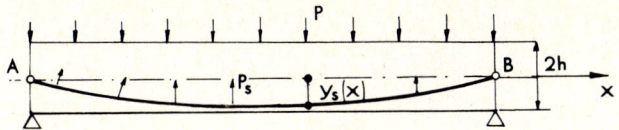

Figure 5 - Prestressed Beam: Optimal Prestress with Vanishing Lateral Deflection

Example 2. Vibrating Beam

Consider a free vibrating beam simply supported at both ends on rigid supports and two unspecified supports at A and B, Figure 6. Let us find the optimal support position and their stiffness so that the fundamental frequency of free vibrations ω_I be maximized. The relevant optimality conditions were derived in [4, 5]. Assuming the cost of support to be proportional to its stiffness, we have

$$Ru_{,x} = 0 , \quad u_A = u_B = \alpha = \text{const} , \tag{62}$$

that is the deflection at support has a constant value and the product $Ru_{,x}$ vanishes. Deleting the details of solution of [5] let us present only final results, Figure 6. Depending on the value of the relative support stiffness $\tilde{S} = S_s/S_b$, where S_s and S_b denote the support and beam stiffness, we have three different optimal solutions. For small values of $\tilde{S}$, $0 < \tilde{S} \leq \tilde{S}_1$, the two supports are located at the centre of the span and the condition $u_{,x} = 0$ is satisfied for the first eigenmode. For $\tilde{S}_1 \leq \tilde{S} \leq \tilde{S}_2$, the supports move from the centre to points A and B of zero slope of the first eigenmode, Figure 6(b). When $\tilde{S}_2 \leq \tilde{S} \leq \tilde{S}_2^1$, the second frequency may become a fundamental frequency for some values of s. Finally, when $\tilde{S}_2^1 \leq \tilde{S} \leq \tilde{S}_3$, the analytical extremum $u_{1_x} = 0$ does not correspond to optimum since the two modes occur at the same frequency. Thus instead of $u_{,x} = 0$ we have to satisfy the condition $\omega_I = \omega_{II}$, Figure 6(d). Such bimodal optimal solutions were also obtained in [7] for stability problems and their general discussion is presented in [8].

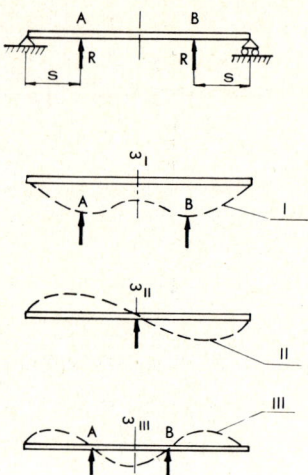

Figure 6(a) - *Optimal Support Reaction on Vibrating Beams*

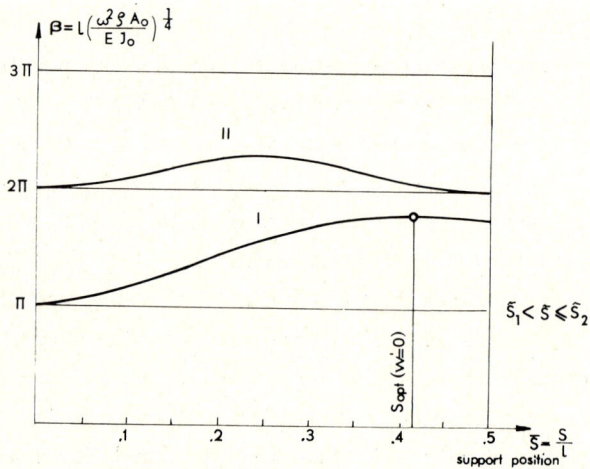

Figure 6(b) - *Optimal Support position S Depending on Relative Stiffness of Support and Beam: (b), (c) Zero-Slope position, (d) Bimodal Position*

(continued)

(continued)

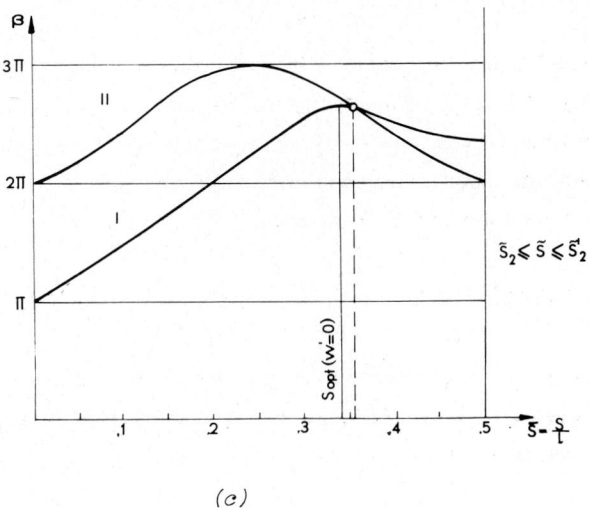

(c)

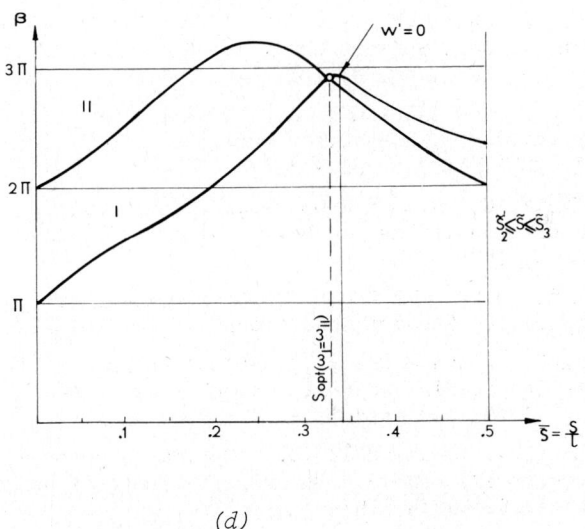

(d)

Figure 6(c),(d) - Optimal Support Position S Depending on Relative Stiffness of Support and Beam: (b),(c) Zero-Slope Position, (d) Bimodal Position

7. CONCLUDING REMARKS

The present work provides a uniform derivation of optimality conditions for reaction of supports or action of loads and generalizes the previous results [1 - 6]. In active control problems, some of these conditions can be utilized, for instance, in designing optimal force actuator locations.

ACKNOWLEDGEMENT

The present work was supported through NSF - Maria Sklodowska-Curie Fund/Grant No. INT-75-08722/established by the United States and Polish Governments.

REFERENCES

[1] MRÓZ, Z. and ROZVANY, G.I.N., "Optimal Design of Structures with Variable Support Conditions", *J. Opt. Theory and Appl.*, Vol. 15, 1975, pp. 85-101.

[2] MRÓZ, Z. and GARSTECKI, A., "Optimal Design of Structures with Unspecified Loading Distribution", *J. Opt. Theory and Appl.*, Vol. 20, 1976, pp. 359-380.

[3] ROZVANY, G.I.N. and MRÓZ, Z., "Column Design: Optimization of Support Conditions and Segmentation", *J. Struct. Mech.*, Vol. 5, 1978.

[4] SZELAG, D. and MRÓZ, Z., "Optimal Design of Elastic Beams with Unspecified Support Conditions", *ZAMM*, Vol. 58, 1978, pp. 501-5

[5] SZELAG, D. and MRÓZ, Z., "Optimal Design of Vibrating Beams with Unspecified Support Action", *Comp. Meth. in Appl. Math. Eng.*, 1979, (in print).

[6] LEKSZYCKI, T. and MRÓZ, Z., "Optimal Support Reaction in Frame Structures", *Research Report*, Institute of Fundamental Technological Research, 1979.

[7] OLHOFF, N. and RASMUSSEN, S.H., "On Single and Bimodal Optimum Buckling Loads of Clamped Columns", *Int. J. Solids Struct.*, Vol. 13, 1977, pp. 605-614.

[8] MASUR, E.F. and MRÓZ, Z., "On Non-Stationary Optimality Conditions in Structural Design", *Int. J. Solids Struct.*, 1979, (in print).

STRUCTURAL CONTROL, H.H.E. Leipholz (ed.)
North-Holland Publishing Company & SM Publications
© IUTAM, 1980

OPTIMAL CONTROL OF UNILATERAL STRUCTURAL ANALYSIS PROBLEMS

P.D. Panagiotopoulos

Institute of Steel Structures
Aristotelian University
Thessaloniki, Greece

1. INTRODUCTION

There is a variety of problems in structural analysis, for which the principle of virtual work holds in variational inequality form. This is caused by the unilateral character of the variations of some quantities of the problems, either due to physical non-linearities (plasticity, locking effects, cracks, etc.), or to inequality constraints imposed on the boundary (friction, unilateral contact, etc.) or in the body (non-negative stresses in cable elements, in rock mechanics). These problems are called "unilateral" and their study began only very recently (see e.g., [1, 2]). The numerical treatment of the most static unilateral problems is based on the fact that the solution minimizes the potential or the complementary energy, which in this case may include non-differentiable terms and the minimum may be constrained by inequalities [3]. In the dynamic case this procedure must be combined with a time integration scheme. The aim of the present paper is to formulate and to study the optimal control problems of these non-classical structural analysis problems. Just as the study of the unilateral problems requires the application of new mathematical theories, as convex analysis, the theory of maximal monotone operations [4] etc., in the same extent the respective optimal control problem is of non-classical nature since the controlled physical system is governed by a variational inequality. This is a largely unexplored field of the optimal control theory except of some recent results concerning the existence and the approximation of the optimal control of systems governed by elliptic and parabolic variational inequalities [5, 6], and the static optimal control problem for the unilateral Kirchhoff plate theory [7]. In the present paper the optimal control problem of discretized unilateral problems is formulated and studied. Existence and approximation procedures for the "construction" of the optimal control are given and the general theory is illustrated in examples.

2. VARIATIONAL PRINCIPLES IN UNILATERAL STRUCTURAL ANALYSIS

Let us consider in a cartesian system $0x_1x_2x_3$ a body Ω with a boundary Γ, which is made up by three mutually disjoint parts Γ_U, Γ_F, and Γ_S. On Γ_U (respectively Γ_F) the displacements (respectively the tractions $S_i = \sigma_{ij} n_j - n_i$ is the outward normal to Γ- are prescribed $u_i = U_i$ (respectively $S_i = F_i$). On Γ_S the subdifferential condition of the form (see Appendix)

$$-\underline{S} \epsilon \partial f(\underline{u}) \ , \tag{1}$$

hold. It may hold also in the form $-S_i \epsilon \partial f_i(u_i)$ (i = 1,2,3) or in the form $-S_N \epsilon \partial f_N(u_N)$ and $-\underline{S}_T \epsilon \partial f_T(\underline{u}_T)$, $i = 1,2$ referred to an intrinsic coordinate system, where N (respectively T) denotes the normal (respectively the tangential) direction to the boundary. Here $f:R^3 \to (-\infty, +\infty]$, $f_i \not\equiv \infty$, is a convex, lower semicontinuous (l.s.c.) function which is called the "superpotential" of the respective boundary unilateral constraint. In the framework of a nonpolar, geometric linear theory a constitutive law of the form

$$\underline{\sigma} \epsilon \partial w(\underline{\varepsilon}) \ , \tag{2}$$

is introduced. Here $\underline{\sigma}$ (respectively $\underline{\varepsilon}$) is the stress (respectively, strain) tensor and $w: R^6 \to (-\infty, +\infty]$, $w \not\equiv \infty$, is a convex, l.s.c. function, called the superpotential of the constitutive law. Functions f and w are generally non-differentiable and may become $+\infty$. By means of the conjugate functions (see equation (A.2)), equations (1) and (2) are equivalently written in the inverse form

$$\underline{u} \in \partial f^c(-\underline{S}) \,, \tag{3}$$

$$\underline{\varepsilon} \in \partial w^c(\underline{\sigma}) \,. \tag{4}$$

For the physical meaning of (1) and (2) the reader is referred to [3], [8]. Let us study first a general static unilateral B.V.P., defined by the equations of equilibrium and the compatibility conditions

$$\sigma_{ij,j} + P_i = 0 \,, \tag{5}$$

and

$$\varepsilon_{ij} = \frac{1}{2}\left(u_{i,j} + u_{j,i}\right) \text{ in } \Omega \,, \tag{6}$$

by the material law (2), and the boundary conditions on Γ_U, Γ_F and Γ_S. Further the set X (respectively Y) of the kinematically admissible displacements $\underline{v}$ and strains $\underline{\eta}$ (respectively statically admissible stresses $\underline{t}$) is introduced

$$X = \{\underline{v}, \underline{\eta} | v_i = U_i \text{ on } \Gamma_U, f(\underline{v}) < \infty \text{ on } \Gamma_S, w(\underline{\eta}) < \infty \text{ in } \Omega \,, \text{ equation (6) holds}\} \tag{7}$$

respectively,

$$Y = \{\underline{t} | T_i = F_i \text{ on } \Gamma_F, f^c(-\underline{T}) < \infty \text{ on } \Gamma_S, w^c(\underline{t}) < \infty \text{ in } \Omega \,, \text{ equation (5) holds}\} \tag{8}$$

By multiplying equation (5) (respectively, equation (6)) by $v_i - u_i$ (respectively, $t_{ij} - \sigma_{ij}$) and integrating over Ω, follows, the relation (Green-Gauss theorem, summations convention)

$$\int_\Omega \sigma_{ij}(\eta_{ij}-\varepsilon_{ij})d\Omega = \int_{\Gamma_F} F_i(v_i-u_i)d\Gamma + \int_{\Gamma_S} S_i(v_i-u_i)d\Gamma + \int_\Omega p_i(v_i-u_i)d\Omega , \qquad (9)$$

respectively

$$\int_\Omega \varepsilon_{ij}(t_{ij}-\sigma_{ij})d\Omega = \int_{\Gamma_U} U_i(T_i-S_i)d\Gamma + \int_{\Gamma_S} u_i(T_i-S_i)d\Gamma . \qquad (10)$$

By means of the subdifferential inequality (A.1) applied to the subdifferential relations (1), (2) (respectively, (3), (4)), the variational inequality

$$\int_\Omega [w(\underline{\eta})-w(\underline{\varepsilon})]d\Omega + \int_{\Gamma_S} [f(\underline{v})-f(\underline{u})]d\Gamma \geq \int_{\Gamma_F} F_i(v_i-u_i)d\Gamma +$$

$$\int_\Omega p_i(v_i-u_i)d\Omega\, \forall \underline{v},\underline{\eta}\epsilon X , \qquad (11)$$

$$\int_\Omega [w^c(\underline{t})-w^c(\underline{\sigma})]d\Omega + \int_{\Gamma_S} [f^c(-\underline{T})-f^c(-\underline{S})] \geq \int_{\Gamma_U} U_i(T_i-S_i)d\Gamma \; \forall \underline{t}\epsilon Y , \qquad (12)$$

results from equation (9) (respectively, equation (10)). These inequalities are the expressions of the principles of virtual and complementary virtual work for the considered body. Both principles hold in inequality form and this fact explains in a global sense the characterization of the respective problems as unilateral. Because the aim of the present paper is the formulation of a structural analysis theory, we consider further a discretized continuum by a finite element scheme. Only the part Γ_S of the boundary is taken now into account, since the vector $\underline{F}$ (respectively, $\underline{U}$) of the boundary tractions (respectively, of the boundary displacements) is incorporated into the load vector $\underline{p}$ (respectively, the initial strain vector $\underline{\bar{e}}$). The variational inequalities (11) and (12) become now

$$\sum_{\alpha=1}^{\nu} [w_\alpha(\underline{n}_\alpha - \overline{\underline{e}}_\alpha) - w_\alpha(\underline{e}_\alpha - \overline{\underline{e}}_\alpha)] + \sum_{\beta=1}^{\nu'} [f_\beta(\underline{v}_\beta) - f_\beta(\underline{u}_\beta)] \geq \underline{p}^T(\underline{v}-\underline{u}) \quad \forall \underline{v}, \; \underline{n} \in X_d,$$
(13)

and

$$\sum_{\alpha=1}^{\nu} [w_\alpha^c(\underline{t}_\alpha) - w_\alpha^c(\underline{s}_\alpha)] + \sum_{\beta=1}^{\nu'} [f_\beta^c(-\underline{T}_\beta) - f_\beta^c(-\underline{S}_\beta)] \geq \underline{e}^{-T}(\underline{t}-\underline{s}) \quad \forall \underline{t} \in Y_d.$$
(14)

Here ν (respectively ν') is the number of elements (respectively, the nodes of Γ_S), $\underline{e}$ (respectively $\underline{s}$) is the strain (respectively, stress) vector, $\underline{S}$ is the boundary traction vector ($\underline{S} = Q\underline{s}$),

$$X_d = \{\underline{v},\underline{n} \mid f_\beta(\underline{v}_\beta) < \infty, \; w_\alpha(\underline{n}_\alpha - \overline{\underline{e}}_\alpha) < \infty \; \forall \alpha,\beta, \; \underline{n} = \underline{G}^T\underline{v}\}$$

and

$$Y_d = \{\underline{t} \mid f_\beta^c(-\underline{T}_\beta) < \infty, \; w_\alpha^c(\underline{t}_\alpha) < \infty \; \forall \alpha,\beta, G\underline{t} = \underline{p}, \; \underline{T} = Q\underline{t}\},$$

and $\underline{G}^T$ is the compatibility matrix of the discretized continuum. We introduce the proper convex, l.s.c. functions W_1, W_2, Φ_1 and Φ_2 defined by

$$W_1(\underline{u}) = \begin{cases} \sum_{\alpha=1}^{\nu} w_\alpha(\underline{e}_\alpha - \overline{\underline{e}}_\alpha), & \text{if } \underline{e} = \underline{G}^T\underline{u} \text{ and } w_\alpha(\underline{e}_\alpha - \overline{\underline{e}}_\alpha) < \infty \; \forall \alpha, \\ \text{otherwise } \infty \end{cases},$$

$$W_2(\underline{s}) = \begin{cases} \sum_{\alpha=1}^{\nu} w_\alpha^c(\underline{s}_\alpha), & \text{if } w_\alpha^c(\underline{s}_\alpha) < \infty \; \forall \alpha, \text{ otherwise } \infty \end{cases},$$

$$\Phi_1(\underline{u}) = \begin{cases} \sum_{\beta=1}^{\nu'} f_\beta(\underline{u}_\beta), & \text{if } f_\beta(\underline{u}_\beta) < \infty \; \forall \beta, \text{ otherwise } \infty \end{cases},$$

$$\Phi_2(\underline{s}) = \begin{cases} \sum_{\beta=1}^{\nu'} f_\beta^c(-\underline{S}_\beta), & \text{if } G\underline{s} = \underline{p}, \; \underline{S} = Q\underline{s}, \text{ and } f_\beta^c(-\underline{S}_\beta) < \infty \; \forall \beta, \\ \text{otherwise } \infty \end{cases},$$

and thus the inequalities (13) and (14) become for $\underline{u} \in R^n$ and $\underline{s} \in R^{n'}$

$$W_1(\underline{v}) - W_1(\underline{u}) + \Phi_1(\underline{v}) - \Phi_1(\underline{u}) \geq \underline{p}^T(\underline{v}-\underline{u}) \quad \forall \underline{v} \in R^n , \qquad (15)$$

$$W_2(\underline{t}) - W_2(\underline{s}) + \Phi_2(\underline{t}) - \Phi_2(\underline{s}) \geq \underline{\bar{e}}^T(\underline{t}-\underline{s}) \quad \forall \underline{t} \in R^{n'} . \qquad (16)$$

It is obvious that (15) (respectively, (16)) is equivalent to the problem of the minimum of the potential (respectively, complementary) energy min $\{W_1(\underline{u}) + \Phi_1(\underline{u}) - \underline{p}^T\underline{u} | R^n\}$ (respectively, min $\{W_2(\underline{s}) + \Phi_2(\underline{s}) - \underline{\bar{e}}^T\underline{s} | R^{n'}\}$. In the case of a linear elastic body with unilateral constraints on the boundary, (15) and (16) take the form

$$\underline{u}^T \underline{K}(\underline{v}-\underline{u}) + \Phi_1(\underline{v}) - \Phi_1(\underline{u}) \geq \underline{p}^T(\underline{v}-\underline{u}) \quad \forall \underline{v} \in R^n , \qquad (17)$$

$$\underline{s}^T \underline{F}_0(\underline{t}-\underline{s}) + \Phi_2(\underline{t}) - \Phi_2(\underline{s}) \geq \underline{\bar{e}}^T(\underline{t}-\underline{s}) \quad \forall \underline{t} \in R^n , \qquad (18)$$

where $\underline{K}$ and $\underline{F}_0$ are the stiffness and the natural flexibility matrix of the structure. In the framework of an incremental theory, for example a large displacement theory, (17) holds again, but all the quantities are replaced by their increments and $\underline{K} = \underline{K}_E + \underline{K}_G$, where $\underline{K}_E = \underline{G}^T \underline{K}_0 \underline{G}$, $\underline{K}_0 = \underline{F}_0^{-1}$. Here $\underline{F}_0$ is the tangential flexibility matrix of the respective step, and $\underline{K}_G$ is the geometric stiffness matrix. Similarly (15) holds in a modified form, where besides the replacement of all quantities by their increments, the term $\underline{\dot{u}}^T \underline{K}_G (\underline{\dot{u}}-\underline{\dot{u}})$ must be added in the left hand side. In all these cases the resulting variational inequalities are equivalent to respective minimization problems. For dynamic unilateral problems a variational inequality of the form (15) results by means of the same method, with the difference, the $\underline{p}$ is replaced by $\underline{p} - \underline{M}\frac{d^2\underline{u}}{dt^2} - \underline{C}\frac{d\underline{u}}{dt}$, where $\underline{M}$ and $\underline{C}$ are the mass and damping matrices. Of course the initial conditions $\underline{u}(o) = \underline{u}_0$ and $\frac{d\underline{u}}{dt}(o) = \underline{u}_1$ must be taken into account. Finally it should be pointed out, that the variational inequalities derived include as special cases the classical variational equality principles, holding for structures with bilateral behaviour.

3. THE FORMULATION OF THE OPTIMAL CONTROL PROBLEM

To formulate the static optimal control problem we introduce a "control vector" $\underline{z} \in R^m$, the convex closed set of the admissible controls $U_{ad} \subset R^m$, and the "observation vector" $\underline{h} \in R^p$. Further two matrices $\underline{B}: R^m \to R^n$ and $\underline{M}: R^n \to R^p$ are considered. To every control $\underline{z}$ a state function $\underline{u}(\underline{z})$ is associated by means of the variational inequality

$$W_1(\underline{v}) - W_1(\underline{u}(\underline{z})) + \Phi_1(\underline{v}) - \Phi_1(\underline{u}(\underline{z})) \geq (\underline{p}+\underline{Bz})^T(\underline{v}-\underline{u}(\underline{z})) \quad \forall \underline{v} \in R^n ,$$
$$\underline{z} \in U_{ad} . \quad (19)$$

The "performance index" R adopted has the form

$$R(\underline{z}) = ||\underline{Mu}(\underline{z}) - \underline{h}||^2 + \underline{z}^T \underline{Nz} , \quad (20)$$

where $\underline{N}$ is a positive definite matrix. This performance index combines in some sense the aspects of safety and of economy and is widely used in the abstract mathematical theory [9]. The "load optimal control" problem has now the form: Find a vector $\underline{J} \in U_{ad}$ such that:

$$R(\underline{J}) = \inf\{R(\underline{z}) | \underline{z} \in U_{ad}\} . \quad (21)$$

In the case of the "initial strain optimal control problem" to every control $\underline{z}$ a state function $\underline{s}(\underline{z})$ is associated by means of the variational inequality (16), where $\underline{e}$ is replaced by $\underline{e} + \underline{Bz}$. Again the performance index has the form (20) with $\underline{u}(\underline{z})$ replaced by $\underline{s}(\underline{z})$.

For a dynamic unilateral problem we assume that $t \in (0, T)$, $T < \infty$, and we associate to every control $\underline{z}(t) \in U_{ad}$ a state vector $\underline{u}(\underline{z},t)$ by means of the variational inequality

$$\left(\underline{M}\frac{d^2\underline{u}(\underline{z},t)}{dt^2} + \underline{C}\frac{d\underline{u}(\underline{z},t)}{dt}\right)^T (\underline{v}-\underline{u}(\underline{z},t)) + W_1(\underline{v}) - W_1(\underline{u}(\underline{z},t))$$

$$+ \Phi_1(\underline{v}) - \Phi_1(\underline{u}(\underline{z},t)) \geq (\underline{p}+\underline{B}\underline{z})^T(\underline{v}-\underline{u}(\underline{z},t)) \quad \forall \underline{v} \in R^n, \quad (22)$$

which describes the motion of the system together with the initial conditions $\underline{u}(\underline{z},0) = \underline{u}_0$ and $\frac{d\underline{u}}{dt}(\underline{z},0) = \underline{u}_1$. A reasonable performance index has then the form

$$R(\underline{z}) = \int_0^T (||\underline{M}\underline{u}(\underline{z},t)-\underline{h}||^2 + \underline{z}^T\underline{N}\underline{z})dt. \quad (23)$$

Analogously the optimal control problem for an incremental unilateral problem can be formulated. In this case either a performance of the form (20) or (23) is adopted and we could speak about an incremental or a global control. For all the optimal control problems considered some algorithms will be proposed for their numerical calculation. However many questions remain still open. In the next sections a thorough study is made of the optimal control problem (20) with a variational inequality of the type (17). Physically it corresponds to the shape, cost or behaviour optimization of an elastic body with unilateral boundary conditions. Then by means of the same methods the remaining problems are studied.

4. SOME PROPOSITIONS ON THE STATIC UNILATERAL OPTIMAL CONTROL PROBLEM

Let us consider the state variational inequality

$$\underline{u}^T(\underline{z})\underline{K}(\underline{v}-\underline{u}(\underline{z})) + \Phi_1(\underline{v}) - \Phi_1(\underline{u}(\underline{z})) \geq (\underline{p}+\underline{B}\underline{z})^T(\underline{v}-\underline{u}(\underline{z})) \quad \forall \underline{v} \in R^n, \quad \underline{z} \in U_{\alpha d}. \quad (24)$$

First the case, where $\underline{K}$ is positive definite, is considered. The following proposition holds.

Proposition 1

At least one optimal control function $\underline{J}$ exists, on the assumption, that int $D_{\Phi_1} \neq \phi$ ($D_{\Phi_1} = \{\underline{v} | \underline{v} \epsilon R^n, \Phi_1(\underline{v}) < \infty\}$). Proof: Let $\{\underline{J}_n\}$ be a minimizing sequence in U_{ad}, $R(\underline{J}_n) \to \inf R(\underline{z}) = I$ for $\underline{z} \epsilon U_{ad}$. We denote by $\underline{v}_n = \underline{u}(\underline{z}_n)$ the unique [10] solution of (24). It is obvious that $-\infty \leq I < \infty$. Then $\underline{z}_n$ is bounded i.e., $||\underline{z}_n|| \leq c$ (c denotes a constant). If this is not the case, a subsequence $\{\underline{z}_{n'}\}$ exists, such that $||\underline{z}_{n'}|| \to \infty$ and then, because of the positive definiteness of $\underline{N}$, $R(\underline{z}_{n'}) \to \infty$ i.e., $I = \infty$. From (24) follows $\forall \underline{v} \epsilon$ int D_{Φ_1} the relation

$$(\underline{v}-\underline{v}_n)^T \underline{K}(\underline{v}-\underline{v}_n) \leq \Phi_1(\underline{v}) - \Phi_1(\underline{v}_n) + (\underline{p}+\underline{B}\underline{z}_n)^T(\underline{v}_n-\underline{v}) + \underline{v}^T \underline{K}(\underline{v}-\underline{v}_n) \ . \tag{25}$$

Due to the positive definiteness of $\underline{K}$, $\underline{v}^T \underline{K}\underline{v}$ is a norm on R^n equivalent to $||\underline{v}||$. Accordingly $\underline{v}^T \underline{K}\underline{v} \geq c||\underline{v}||^2$. Moreover for $\underline{v} \epsilon$ int $D\Phi_1$, $\partial \Phi(\underline{v})$ is a non-empty and bounded set ([11] th. (23.4)) and thus a vector $\underline{x}^* \epsilon R^n$ exists, such that $\Phi(\underline{v}_n) - \Phi(\underline{v}) \geq \underline{x}^{*T}(\underline{v}_n-\underline{v})$. By means of these inequalities and the inequality of Schwarz it follows immediately, that $||\underline{v}_n|| \leq c$. Since the unit ball in a finite dimensional normed space is compact and because of the boundedness of $\{\underline{z}_n\}$ and $\{\underline{v}_n\}$, we may extract subsequences $\{\underline{z}_\mu\}$ and $\{\underline{v}_\mu\}$ such that $\underline{v}_\mu \to \underline{v}_0$. Since U_{ad} is closed $\underline{z}_0 \epsilon U_{ad}$. If (24) is satisfied by $\underline{v}_\mu$ and $\underline{z}_\mu$ the limit can be taken and by means of the relation $\lim \inf \Phi_1(\underline{v}_n) \geq \Phi_1(\underline{v}_0)$ for $\underline{v}_n \to \underline{v}_0$ (Φ_1 is l.s.c.) it follows, that $\underline{v}_0$ and $\underline{z}_0$ satisfy the variational inequality (24) i.e., $\underline{v}_0 = \underline{u}(\underline{z}_0)$. Function R satisfies then the relation

$$I = \inf_{U_{ad}} R(\underline{z}) = \liminf_{n \to \infty} R(\underline{z}_n) \geq R(\underline{z}_0) \ , \tag{26}$$

since R is l.s.c. as well. But this leads to a contradiction unless $R(\underline{z}_0) = \inf R(\underline{z}) \; \forall \underline{z} \; U_{ad}$. Thus $I \neq -\infty$ and we may take in (26) $\underline{z}_0 = \underline{J}$ q.e.d. Suppose further that $\underline{K}$ is positive semidefinite, as it may happen for example in the case of an incremental theory. The proof of the existence of a solution is in this case an open problem. However, if we assume additionally that $D_{\tilde{\Phi}_1}$ is a closed bounded set and Φ_1 is strictly convex, then we can show by the same method as in Proposition 1 that at least one solution exists. Moreover an approximation proposition can be proved. By means of a symmetric positive definite matrix $\overline{K}$ and a load vector $\overline{p}$ the symmetric positive definite matrix $\underline{K}_\varepsilon = \underline{K} + \varepsilon \overline{\underline{K}}$ and the vector $\underline{p}_\varepsilon = \underline{p} + \varepsilon \overline{\underline{p}}$, $\varepsilon > 0$, are formulated. By $\underline{u}_\varepsilon$ is denoted the solution of the variational inequality

$$\underline{u}_\varepsilon^T(z)\underline{K}_\varepsilon(\underline{v}-\underline{u}_\varepsilon(z)) + \Phi_1(\underline{v}) - \Phi_1(\underline{u}_\varepsilon(z)) \geq (\underline{p}_\varepsilon + Bz)^T(\underline{v}-\underline{u}_\varepsilon(z)) \forall \underline{v} \in R^n, \; \underline{z} \in U_{ad} \; . \quad (27)$$

Due to the positive definiteness of $\underline{K}_\varepsilon$, (27) admits a unique solution $\underline{u}_\varepsilon(z)$, $\forall \underline{z} \in U_{ad}$. It is also known (see e.g., [9], Chapter I, Th. 5.1) that (24) has a convex closed set X_z of solutions for $\underline{K}$ positive semidefinite.

Proposition 2

Suppose that X_z is non-empty. Then $\underline{u}_\varepsilon(z) \to \underline{u}(z)$ in R^n for $\varepsilon \to 0$ and $\underline{z} \in U_{ad}$, where by $\underline{u}(z)$ is denoted the solution of the variational inequality (24). Proof: It is a special case of the proof given in [9] (Chapter I, Th. 5.3). Further, we denote by R_ε the functional $R_\varepsilon(\underline{z}) = ||M\underline{u}_\varepsilon(z) - \underline{h}||^2 + \underline{z}^T N\underline{z}$ and let $\underline{J}_\varepsilon$ be the solution of the problem

$$R_\varepsilon(\underline{J}_\varepsilon) = \inf \{R_\varepsilon(\underline{z}) | \underline{z} \in U_{ad}\} \; . \quad (28)$$

Because of Proposition 1, at least one solution $\underline{u}_\varepsilon$ and $\underline{J}_\varepsilon$ satisfying (27) and (28) exists. Assume moreover that at least one solution $\underline{J}$

$\underline{u}(\underline{J})$ of the positive semidefinite optimal control problem exists. Then the following approximation is possible.

Proposition 3

For $\varepsilon \to \hat{0}$, $\underline{J}_\varepsilon \to \underline{J}$ in U_{ad}, $\underline{u}_\varepsilon(\underline{J}_\varepsilon) \to \underline{u}(\underline{J})$ in R^n and $R_\varepsilon(\underline{J}_\varepsilon) \to R(\underline{J})$.
Proof: It is obvious, that $R_\varepsilon(\underline{z}) \to R(\underline{z})$ $\forall \underline{z} \in R^m$ (Proposition 2). Hence, if $I_\varepsilon = \inf R_\varepsilon(\underline{z})$, $\underline{z} \in U_{ad}$,

$$R_\varepsilon(\underline{J}_\varepsilon) \leq R_\varepsilon(\underline{J}) \to R(\underline{J}) = \inf R(\underline{z}) = I \ \forall \underline{z} \in U_{ad} \tag{29}$$

where I is finite, and thus

$$\limsup R_\varepsilon(\underline{J}_\varepsilon) \leq R(\underline{J}) . \tag{30}$$

From the positive definiteness of $\underline{N}$ and from (30) we can show that $\{\underline{J}_\varepsilon\}$ is bounded and thus we can extract a subsequence $(\underline{J}_{\varepsilon'})$ such that $\underline{J}_{\varepsilon'} \to \underline{J}_0$. Since U_{ad} is closed, $\underline{J}_0 \in U_{ad}$. Then from Proposition 2, $\underline{u}_\varepsilon(\underline{J}_\varepsilon) \to \underline{u}(\underline{J}_0)$. But $\liminf R_\varepsilon(\underline{J}_\varepsilon) \geq R(\underline{J}_0)$, which with (30) yields $\underline{J}_0 = \underline{J}$ q.e.d.

5. REGULARIZATION OF Φ_1 AND EXTENSIONS

The generally non-differentiable term Φ_1 in the variational inequality (24) does not permit the application of the usual numerical techniques of the optimal control theory. We shall approximate Φ_1 by a sequence of continuously differentiable convex functionals Φ_{1_ρ} depending on the parameter $\rho > 0$, such that $\Phi_{1_\rho}(\underline{v}) \to \Phi_1(\underline{v})$ $\forall \underline{v} \in R^n$, when $\rho \to 0$. There are many possibilities of constructing Φ_{1_ρ} (see [2], [5], [10]). In this case, we assume generally that $\Phi_1(\underline{v}) \leq \liminf \Phi_{1_\rho}(\underline{v}_\rho)$ for $\underline{v}_\rho \to \underline{v}$ and $\rho \to 0$. Then for $\underline{z} \in U_{ad}$,

$$\underline{u}_\rho^T(\underline{z})\underline{K}(\underline{v}-\underline{u}_\rho(\underline{z})) + \Phi_{1\rho}(\underline{v}) - \Phi_{1\rho}(\underline{u}_\rho(\underline{z})) \geq (\underline{p}+\underline{Bz})^T(\underline{v}-\underline{u}_\rho(\underline{z})) \quad \forall \underline{v} \in R^n ,$$
(31)

is the "regularized variational inequality" and is equivalent to the equality

$$\underline{Ku}_\rho(\underline{z}) + \frac{d\Phi_\rho}{d\underline{u}_\rho}(\underline{u}_\rho(\underline{z})) = \underline{p}+\underline{Bz} , \quad \underline{z} \in U_{ad} ,$$
(32)

which constitutes the state equation of the following problem. Find $\underline{J}_\rho \in U_{ad}$ such that

$$\text{where } R_\rho(\underline{J}_\rho) = \inf R_\rho(\underline{z}) \forall \underline{z} \in U_{ad} ,$$
(33)

$$R_\rho(\underline{z}) = ||\underline{Mu}_\rho(\underline{z})-\underline{h}||^2 + \underline{z}^T\underline{Nz} .$$
(34)

It is obvious, that at least one solution $\underline{J}_\rho$ and $\underline{u}_\rho(\underline{J}_\rho)$ of (32), (33) exists (Proposition 1).

Proposition 4

For every $\underline{z} \in U_{ad}$, and for $\rho \to 0$, $\underline{u}_\rho(\underline{z})$ and $\underline{u}(\underline{z})$ are unique and

$$\lim \underline{u}_\rho(\underline{z}) = \underline{u}(\underline{z}) .$$
(35)

Proof: The proof of this proposition is a simplfied form of the proof given in [10] (p.34) for the infinite dimensional case, q.e.d.

The following proposigion gives an approximation result which allows for the numerical calculation of the optimal control $\underline{J}$ and $\underline{u}(\underline{J})$ as the limit of the solutions $\underline{J}_\rho$ and $\underline{u}_\rho(\underline{J}_\rho)$ of a sequence of classical optimal control problems involving state equations instead of variational inequalities.

Proposition 5

For $\rho \to 0$, $\underline{J}_\rho \to \underline{J}$ in U_{ad}, $\underline{u}_\rho(\underline{J}_\rho) \to \underline{u}(\underline{J})$ in R^n and $R_\rho(\underline{J}_\rho) \to R(\underline{J})$.
Proof: We have $R_\rho(\underline{z}) \to R(\underline{z}) \forall \underline{z} \in U_{ad}$, (see Proposition 4),

$$R_\rho(\underline{J}_\rho) \leq R_\rho(\underline{J}) \to R(\underline{J}) \quad \inf\{R(\underline{z}) | \underline{z} \in U_{ad}\} . \tag{36}$$

Since $\underline{N}$ is positive definite we conclude by means of

$$\limsup R_\rho(\underline{J}_\rho) \leq R(\underline{J}) , \tag{37}$$

that $||\underline{J}_\rho|| \leq c$. Thus a subsequence $\{\underline{J}_{\rho'}\}$ of $\{\underline{J}_\rho\}$ exists, which converges to an accumulation point $\underline{J}_0$ of $\{\underline{J}_\rho\}$. But U_{ad} is closed and thus $\underline{J}_0 \in U_{ad}$. It follows easily, that

$$\underline{u}_\rho(\underline{J}_\rho) \to \underline{u}(\underline{J}_0) . \tag{38}$$

Due to the lower semicontinuity of R_ρ, $\liminf R_\rho(\underline{J}_\rho) \geq R(\underline{J}_0)$. The last relation with (37) gives $\underline{J} = \underline{J}_0$, which combined with (36) and (38) completes the proof q.e.d.

In the case of $\underline{K}$ positive semidefinite, Proposition 5 holds in a slightly different form: first function Φ_1 in (27) is regularized. The solution of the respective optimal control problem is denoted by $\underline{J}_{\epsilon\rho}$, $\underline{u}_{\epsilon\rho}(\underline{J}_{\epsilon\rho})$. This solution exists (Proposition 1) and moreover $\underline{J}_{\epsilon\rho} \to \underline{J}_\epsilon$, $\underline{u}_{\epsilon\rho}(\underline{J}_{\epsilon\rho}) \to \underline{u}_\epsilon(\underline{J}_\epsilon)$, (Proposition 3).

Proposition 6

If a solution $\underline{J}$, $\underline{u}(\underline{J})$ of the positive semidefinite optimal control problem exists, then $\underline{J}_{\epsilon\rho} \to \underline{J}$ in U_{ad}, $\underline{u}_{\epsilon\rho}(\underline{J}_{\epsilon\rho}) \to \underline{u}(\underline{J})$ in R^n and $R_{\epsilon\rho}(\underline{J}_{\epsilon\rho}) \to R(\underline{J})$. Proof: It is a immediate consequence of Proposition 2 and Proposition 5, q.e.d. The propositions 1, 3 and 5 hold in the case of the initial strain control problem, as is obvious. If $\underline{N} = \underline{0}$, then the additional assumption, U_{ad} is bounded, is needed to guarantee the boundedness of the control vector sequences. Suppose further that (15) holds, i.e., we have "internal unilateral" constraints of the form (2). In this case, the existence and the approximation of the solution $\underline{J}$, $\underline{u}(\underline{J})$ are open problems. A heuristic algorithm for the numerical calculation can be proposed. To this end Φ_1 and W_1 are regularized and the solution $\underline{J}$, $\underline{u}(\underline{J})$ is approximated by the solution of $\underline{J}_\rho$, $\underline{u}_\rho(\underline{J}_\rho)$ of the regularized optimal control problem, i.e., instead of (15), the state equation

$$\frac{dW_{1\rho}}{d\underline{u}_\rho}(\underline{u}_\rho(\underline{z})) + \frac{d\Phi_{1\rho}}{d\underline{u}_\rho}(\underline{u}_\rho(\underline{z})) = \underline{p} + \underline{Bz} , \qquad (39)$$

holds. Let us study finally the dynamic optimal control problem. In this case the regularization procedure allows again for the use of the classical optimal control numerical methods since for every ρ we have instead of the variational inequality (22) the state equation

$$\underline{M}\frac{d^2\underline{u}_\rho(\underline{z},t)}{dt^2} + \underline{C}\frac{d\underline{u}_\rho(\underline{z},t)}{dt} + \frac{dW_{1\rho}}{d\underline{u}_\rho}(\underline{u}_\rho(\underline{z},t)) + \frac{d\Phi_{1\rho}}{d\underline{u}_\rho}(\underline{u}_\rho(\underline{z},t)) = \underline{p} + \underline{Bz} . \qquad (40)$$

For this general dynamic optimal control problem, the existence of at least one solution and the convergence of the regularization procedure remains a still open problem.

6. APPLICATIONS

(a) As a first application the optimal control problem of a linear elastic body with friction boundary conditions is considered. It can be shown (compare [2]), that the solution $\underline{u}$ satisfies the variational inequality (17) with

$$\Phi_1(\underline{u}) = \left\{ \sum_{\beta=1}^{\nu'} f |C_{N\beta} \underline{u}_{T\beta}| \quad , \quad \text{if } \underline{u}_T = \underline{Pu}, \text{ otherwise } \infty \right\}.$$

Here $\underline{P}$ is a matrix relating the tangential displacement vector $\underline{u}_T$ on Γ_S with $\underline{u}$, f is the Coulomb's coefficient of friction and $\underline{C}_N$ is the prescribed value of $\underline{S}_N$ on Γ_S. Matrix $\underline{K}$ is positive definite and if we assume, that the cost function to be minimized has the form (20), the respective optimal control problem has at least one solution. The solution can be approximated numerically by means of the solution of the regularized problem, having the following form: find $\underline{J}_\rho \epsilon \underline{U}_{ad}$, minimizing (34) with the subsidiary condition

$$\underline{Ku}_\rho(\underline{z}) + \sum_{\beta=1}^{\nu'} f \frac{|C_{N\beta}|}{1+\rho} \frac{d|(P\underline{u}_\rho)_\beta|^{1+\rho}}{d\underline{u}_\rho} = \underline{p} + \underline{Bz} \quad , \quad \underline{z} \epsilon \underline{U}_{ad} \quad . \quad (41)$$

This is a usual convex optimization problem and can be numerically solved by an appropriately chosen optimization algorithm. For the respective dynamic problem the functional (23) must be minimized by taking into account the state equation

$$\underline{M} \frac{d^2 \underline{u}_\rho(\underline{z},t)}{dt^2} + \underline{C} \frac{d\underline{u}_\rho(\underline{z},t)}{dt} + \underline{Ku}_\rho(\underline{z},t) + \sum_{\beta=1}^{\nu'} f \frac{|C_{N\beta}|}{1+\rho} \frac{d|(P\underline{u}_\rho)_\beta|^{1+\rho}}{d\underline{u}_\rho} =$$

$$= \underline{p} + \underline{Bz} \quad , \quad \underline{z} \epsilon \underline{U}_{ad} \quad , \tag{42}$$

and the initial conditions $\underline{u}_\rho(\underline{z},0) = \underline{u}_0$, $\frac{d\underline{u}_\rho}{dt}(\underline{z},0) = \underline{u}_1$.

(b) The plastic analysis of structures (holonomic or incremental with large displacements and physical instabilizing effects) reduces to the solution of a linear complementarity problem (L.C.P.) of the form [12]

$$\underline{Mx} + \underline{a} \geq \underline{0} \quad , \quad \underline{x} \geq \underline{0} \quad , \quad \underline{x}^T(\underline{Mx}+\underline{a}) = 0 \quad . \tag{43}$$

Here $\underline{M}$ (respectively, $\underline{a}$) is a given matrix (respectively, vector) and $\underline{x}^T = \{\underline{u}^{T}+,\underline{u}^{T}-,\underline{\lambda}^T\}$ - or $\underline{x}^T = \{\dot{\underline{u}}^T,\dot{\underline{u}}^T-,\dot{\underline{\lambda}}^T\}$ for an incremental theory - with $u_{i+} = \sup(o,u_i)$, $u_{i-} = \sup(o,-u_i)$ and $\underline{\lambda}$ the vector of the plastic multipliers. Problem (43) is equivalent to the inclusion ([11], p. 226),

$$-(\underline{Mx}+\underline{a}) \in \partial\Phi(\underline{x}) , \qquad (44)$$

where $\Phi(\underline{x}) = \{0 \text{ if } \underline{x} \geq \underline{0}, \infty \text{ otherwise}\}$. The relation (44) is equivalent to the variational inequality

$$\Phi(\underline{x}^*) - \Phi(\underline{x}) \geq - (\underline{Mx}+\underline{a})^T (\underline{x}^*-\underline{x}) \quad \forall \underline{x}^* \in R^n . \qquad (45)$$

We consider further the respective optimal control problem resulting by setting in (20) $\underline{x}$ instead of $\underline{u}$ and in (45) $\underline{a}+\underline{Bz}$ instead of $\underline{a}$. It is worth noting here, that this optimal control problem can be formulated with respect to other holonomic or incremental problems in structural analysis leading to L.C.P. If $\underline{M}$ is positive definite the existence of an optimal control $\underline{J}$ and an optimal state $\underline{x}$ is guaranteed. The respective regularized problem results by considering the function

$$\Phi_\rho(\underline{x}_\rho) = \sum_{i=1}^{n} \Phi_{\rho i}(x_{\rho i}) ,$$

where

$$\Phi_{\rho i}(x_{\rho i}) = \begin{cases} 0 & \text{if} \quad x_i \geq 0 , \\ x_i^2/2\rho & \text{if} \quad -1 \leq x_i \leq 0 \\ -\frac{1}{\rho}\left(x_i + \frac{1}{2}\right) & \text{if} \quad x_i \leq -1 , \end{cases} \qquad (46)$$

and thus $\underline{J}_\rho \rightarrow \underline{J}$ and $\underline{x}_\rho(\underline{J}_\rho) \rightarrow \underline{x}(\underline{J})$ for $\rho \rightarrow 0$ according to Propositions 4 and 5. If $\underline{M}$ is positive semidefinite then Proposition 6 allows for the approximation of the solution.

REFERENCES

[1] FICHERA, G., "Boundary Value Problems in Elasticity with Unilateral Constraints", in *Encyclopedia of Physics*, edited by S. Flügge, Vol. VIa/2, Springer Verlag, Berlin-Heidelberg-New York, 1972.

[2] DUVAUT, G. and LIONS, J.L., *Les inequations en Mecanique et en Physique*, Dunod, Paris, 1972.

[3] PANAGIOTOPOULOS, P.D., "Convex Analysis and Unilateral Static Problems", *Ing. Archiv*, Vol. 45, 1976, pp. 55-68.

[4] PANAGIOTOPOULOS, P.D., "Ungleichungsprobleme in der Mechanik, *Habilitationsschrift*, R.W.T.H., Aachen, 1977.

[5] YVON, J.P., "Etude de quelques problèmes de contrôle pour des systems distribués", *Thèse de Doctorat d'Etat*, Université de Paris VI, 1973.

[6] MIGNOT, F., "Contrôle dans les Inéquations Variationelles Elliptiques", *J. Funct. Anal.*, Vol. 22, 1976, pp. 130-185.

[7] PANAGIOTOPOULOS, P.D., "Optimal Control in the Unilateral Thin Plate Theory", *Archives of Mechanics*, Vol. 29, 1977, pp. 25-39.

[8] LÉNÉ, F., "Sur les Matériaux élastiques à énergie de déformation non quadratique", *J. de Mécanique*, Vol. 13, 1974, pp. 499-534.

[9] LIONS, J.L., *Optimal Control of Systems Governed by Partial Differential Equations*, Springer Verlag, Berlin-Heidelberg-New York, 1971.

[10] BREZIS, H., "Problèmes Unilatéraux", *J. Math. pures et Appl.*, Vol. 51, 1972, pp. 1-168.

[11] ROCKAFELLAR, R.T., *Convex Analysis*, Princeton University Press, Princeton, 1972.

[12] MAIER, G., "Incremental Plastic Analysis in the Presence of Large Displacements and Physical Instabilizing Effects", *Int. J. Solids and Structures*, Vol. 7, 1971, pp. 345-372.

APPENDIX

A set $K \subset R^n$ is convex, if $(1-\mu)\underline{x}_1 + \mu \underline{x}_2 \in K$ for $\underline{x}_1$ and $\underline{x}_2 \in K$ and $0 < \mu < 1$. A function $f: R^n \to (-\infty, +\infty]$, $f \not\equiv \infty$, is called proper. A proper function is defined to be convex if $\forall \underline{x}_1, \underline{x}_2 \in R^n$ $f((1-\mu)\underline{x}_1 + \mu\underline{x}_2) \leq (1-\mu)f(\underline{x}_1) + \mu f(\underline{x}_2)$ and $0 < \mu < 1$. f is lower semicontinuous at $\underline{x} \in R^n$ if $f(\underline{x}) \leq \liminf f(\underline{x}_n)$ for every sequence $\{\underline{x}_n\}$ such that $\underline{x}_n \to \underline{x}$. The set of the x*'s $\underline{x}^* \in R^n$ such that

$$f(\underline{x}_1) - f(\underline{x}) \geq \underline{x}^{T*}(\underline{x}_1 - \underline{x}) \quad \forall \underline{x}_1 \in R^n , \tag{A.1}$$

is called the subdifferential of f at $\underline{x}$ and we write $\underline{x}^* \in \partial f(\underline{x})$ [12]. If f is differentiable at $\underline{x}$, $\partial f(\underline{x}) = \{\text{grad } f(\underline{x})\}$. The conjugate function f^c of f is defined by means of the relation

$$f^c(\underline{x}^*) = \sup \{\underline{x}^{*T}\underline{x} - f(\underline{x}) | \underline{x} \in R^n\} . \tag{A.2}$$

If f is proper, convex and l.s.c., f^c has the same properties and the conditions $\underline{x}^* \in \partial f(\underline{x})$ and $\underline{x} \in \partial f^c(\underline{x}^*)$ are equivalent to each other and to (A.2).

STRUCTURAL CONTROL, H.H.E. Leipholz (ed.)
North-Holland Publishing Company & SM Publications
© IUTAM, 1980

THE PRAGER-SHIELD OPTIMALITY CRITERION -
AN EFFICIENT EXTENSION TO FINITE ELEMENT PROBLEMS

G. Pape and G. Thierauf

Essen University
Federal Republic of Germany

1. INTRODUCTION

The Prager-Shield optimality criterion represents the general condition for the plastic design of structures for single load conditions and was initiated by Marcal and Prager [1] in 1964. The method of Marcal and Prager was extended to arbitrary one- and two-dimensional structures by Prager and Shield [2] in 1967. It relates the optimality of a structure to the admissibility of an elastic system by means of a general cost function.

The problems for the calculation of the design variables for more realistic structures encountered in practice according to the Prager-Shield theory are comparable to those problems found in elasticity which were overcome by introducing the Finite Element method.

After a brief review of the essentials of the Prager-Shield theory this paper discusses the extended applications to finite dimensional problems.

Proceeding from the general optimization problem for discretized structures the plastic design problem is reduced by means of the Kuhn-Tucker conditions to a problem of elastic analysis. A method for the iterative solution of the nonlinear Kuhn-Tucker conditions of the general problem is derived and some special cases are pointed out which require no iteration. The iteration is interpreted in mechanical terms and results in a change of the governing element matrices. Applications to three structural problems (pin-jointed structure, framework, plane stress problem) with quadratic or parabolic yield functions and linear objective functions are given.

2. THE PRAGER-SHIELD THEORY

Only the so called fixed layout problem is dealt with. For this problem engineering structures are described by lines or by surfaces which are considered to be fixed.

An arbitrary point of the structure is denoted by the three-dimensional vector $\underline{x}$. The local states of stress and strain in a structure are specified by s-dimensional vectors of the generalized stresses $\underline{Q}$ and the generalized strains $\underline{q}$ in such a manner that the work of the stresses $\underline{Q}$ on the strain rates $\underline{\dot{q}}$ is given by the scalar $\underline{Q}^T \underline{\dot{q}}$.*

The well-known analysis of a structure consists of finding a one-to-one relation between the distributed load $\underline{P}(\underline{x})$ and the kinematically admissible displacement field $\underline{u}(\underline{x})$. An essential requirement for the existence of the solution is the one-to-one relation between the kinematically admissible strain $\underline{g}(\underline{x})$ and the statically admissible stress $\underline{Q}(\underline{x})$. This relation is the constitutive equation.

The Prager-Shield design theory can be described as a method of deriving the constitutive equation from the functions of the design problem (Figure 1).

---

* T denotes the transpose of a vector or a matrix.

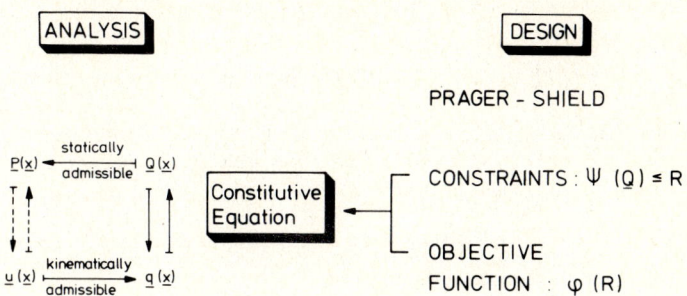

Figure 1 - Prager-Shield design theory

These functions are the objective- (cost-) function

$$\min \Phi = \min \int_V \phi(R)\, dV, \qquad (1)$$

and the constraints

$$\psi(\underline{Q}) - R \leq 0, \qquad (2)$$

$$R - R_o \geq 0, \qquad (3)$$

where

$\phi(R)$ = convex specific cost function,

R = continuously varying function of the plastic resistance,

V = volume of the structure,

$\psi(\underline{Q})$ = convex yield function,

R_o = constant minimal plastic resistance.

The mechanical behaviour of a typical element of the plastic structure is specified by the yield function $\psi(\underline{Q})$. If $R(\underline{x})$ is the plastic resistance at the point $\underline{x}$ of the structure, the components $\dot{q}_i$ ($i = 1,\ldots,s$) of the vector of the strain rate $\underline{\dot{q}}(\underline{x})$ are given by

$$\dot{q}_i = \begin{cases} 0, & \text{for } \psi(Q) < R(\underline{x}) \\ \mu(\underline{x}) \cdot \dfrac{\partial \psi}{\partial Q}, & \text{for } \psi(Q) = R(\underline{x}) \end{cases} \qquad (4)$$

where the index i indicates the i-th component of a vector and $\mu(\underline{x})$ is an unknown non-negative scalar.

An associated elastic structure with the same dimensions and the same loading and supports as the plastic structure can be defined by introducing the complementary specific energy c. To establish the connection between the plastic (A) and the elastic structure (B) it is assumed that the specific energy of B is equal to the specific cost of A:

$$c = \begin{cases} \phi(R_o) = \text{const.}, & \text{for } \psi(Q) = R < R_o \\ \phi(\psi(Q)), & \text{for } \psi(Q) = R \geq R_o \end{cases} \qquad (5)$$

According to Castigliano the partial derivative of the complementary energy with respect to the generalized stress defines the generalize strain:

$$q_i = \dfrac{\partial c}{\partial Q_i} = \dfrac{\partial \phi}{\partial Q_i} = \begin{cases} 0, & \text{for } \psi(Q) < R_o \\ \dfrac{d\phi}{dR} \cdot \dfrac{\partial \psi}{\partial Q_i}, & \text{for } \psi(Q) \geq R_o \end{cases} \qquad (6)$$

A comparison of (4) and (6) shows the proportionality of $\dot{q}$ and q, and with the solution Q^* of the elastic problem which minimizes the total complementary energy C a solution of the plastic design problem is obtained:

$$R^*(\underline{x}) = \begin{cases} R_o, & \text{for } \psi(Q^*) < R_o \\ \psi(Q^*), & \text{for } \psi(Q^*) \geq R_o \end{cases} \qquad (7)$$

The principle of minimum complementary energy for the associated elastic structure shows that there exists no statically admissible stress field with a smaller value for the total complementary energy

$$C = \int_V c(\underline{Q}^*) \, dV = \int_V \phi(R^*) \, dV , \qquad (8)$$

and hence no statically admissible stress field with a smaller value of the total cost Φ of the plastic structure can be found.

3. THE GENERAL DESIGN PROBLEM

The solution of engineering problems formulated in finite dimensional variables is usually accomplished more easily than in the infinite case. This applies as well to the Prager-Shield theory. In practice it is convenient to restrict the design problem to finite vector spaces. In the formulation presented here an approach based on Finite Elements has been adopted.

The n design variables y_i are the components of the n-dimensional vector $\underline{y}$, and the η_i basic generalized forces $\underline{F}_i$ for each of the e elements of the structure form the m-dimensional vector $\underline{F}$ where $m = \sum_{i=1}^{e} \eta_i$.

Examples of these variables are given in Figure 2 for beam elements, plane stress and shell problems. The general design problem can be stated as follows:

Minimize the cost function $Z(\underline{y})$

$$\min Z(\underline{y}) \qquad (9)$$

subject to the constraints

$$G_i(\underline{y},\underline{F}) \leq 0 \; ; \; i = 1,\ldots,k. \qquad (10)$$

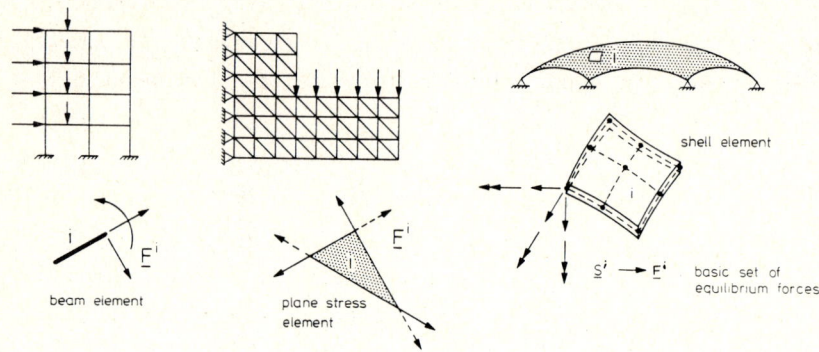

Figure 2 - Finite element Structures

It is assumed that Z admits a linear approximation and that G_i allows for a quadratic approximation. This is shown in Figure 3 by the tangential approximation of the cost-function and by the tangential approximation of the cost-function and by the elliptical approximation of G_i.

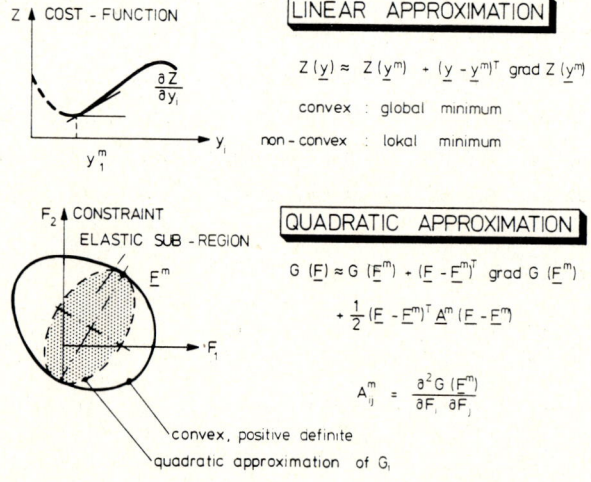

Figure 3 - Approximation of the design functions

The general optimality criterion for the design problem (9),(10) is obtained from the Lagrange-function and the corresponding Kuhn-Tucker conditions. By adding slack variables α_i^2 to the inequality constraints the Lagrangian L is defined here as

$$L = Z(\underline{y}) + \sum_{i=1}^{k} \lambda_i (G_i + \alpha_i^2) \quad , \tag{11}$$

where λ_i are the Lagrange-parameters. The corresponding Kuhn-Tucker conditions are given by the following equations:

$$\frac{\partial L}{\partial y_i} = \frac{\partial Z}{\partial y_i} + \sum_{j=1}^{k} \lambda_j \frac{\partial G_j}{\partial y_i} \equiv 0 \quad , \tag{12}$$

$$\frac{\partial L}{\partial \lambda_i} = G_i + \alpha_i^2 \equiv 0 \quad , \tag{13}$$

$$\frac{\partial L}{\partial \alpha_i} = 2 \lambda_i \alpha_i \equiv 0 \quad , \tag{14}$$

$$\frac{\partial L}{\partial \underline{F}_i} = \sum_{j=1}^{k} \lambda_j \frac{\partial G_j}{\partial \underline{F}_i} \equiv \underline{0} \quad . \tag{15}$$

The constraints with zero slack variables α_i are called active constraints. The set of active constraints is denoted by I_a.

If the active constraints can be found, the condition given by equation (12) may be presented in the form of an optimality criterion as follows:

> The gradient of the cost-function is equal to the negative of a weighted linear combination of the gradients of the active constraints.

The weighting factors are the Lagrange-parameters of the active constraints, and a geometrical interpretation is given in Figure 4.

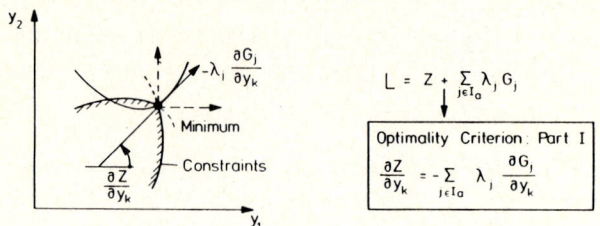

Figure 4 - *First part of the optimality criterion*

This criterion has been used successfully by Gellatly and Berke [3] for elastic design. It is only a first part of the general optimality criterion; the second part follows from (15). If the general solution for $\underline{F}$ (force-method [4,5]) is given by

$$\underline{F} = \underline{B}_o \underline{P} + \underline{B}_x \underline{X} \ , \tag{16}$$

where

$\underline{P}$ = vector of nodal laods,
$\underline{B}_o \underline{P}$ = vector of forces in a statically determined structure,
$\underline{X}$ = vector of the redundant forces,
$\underline{B}_x \underline{X}$ = vector of self-equilibrating forces,

$\frac{\partial L}{\partial \underline{F}}$ can be expressed by the differential of the redundant forces:

$$\frac{\partial L}{\partial \underline{X}} = \frac{\partial L}{\partial \underline{F}} \frac{\partial \underline{F}}{\partial \underline{X}} = \underline{B}_x^T \frac{\partial L}{\partial \underline{F}} \ . \tag{17}$$

Substituting (17) into (15) leads to the second part of the general optimality criterion, namely:

$$\underline{B}_x^T \operatorname{diag}(\underline{\lambda}) \frac{\partial \underline{G}}{\partial \underline{F}} = \underline{0} \ , \tag{18}$$

where

$$\text{diag}(\underline{\lambda}) = \text{diagonal matrix of } \lambda_i.$$

Equation (18) can be interpreted as a continuity condition for an associated elastic structure with gaps given by $\text{diag}(\underline{\lambda}) \frac{\partial G}{\partial F}$. Thus, by using only the Kuhn-Tucker conditions (12)-(15), it is possible to find an associated elastic structure which plays a central role in the Prager-Shield theory.

4. THE EXTENSION OF THE PRAGER-SHIELD THEORY

The conditions as stated here imply an extension of the Prager-Shield criterion. This can be seen by a closer inspection of the dimensions of the vectors involved. There can be more than one design variable per element and more than one constraint as well. The Prager-Shield theory is a special case with equal numbers of constraints n_c and design variables n_d (Figure 5).

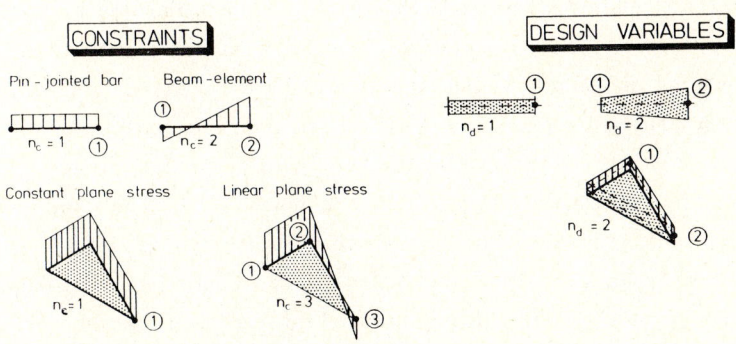

Figure 5 - *Constraints and design variables within an element*

The solution of this special case with $n_c = n_d$ is straight-forward [6], if it is assumed that Z is a linear function of $\underline{y}$ and, if G_i is given as follows (Figure 3):

$$G_i = F_i^T A_i F_i - y_i \le 0 \ . \tag{19}$$

In the special case of $n_c = n_d$ the number of unknown Lagrange-parameters λ_i is equal to the number of design variables y_i, so that $\underline{\lambda}$ can be calculated from (12). In this case the derivative,

$$\frac{\partial G}{\partial \underline{F}} = 2 \underline{A} \underline{F} = 2 \operatorname{diag}(\underline{A}_i) \underline{F} \ , \tag{20}$$

as required in the second part of the optimality criteria, is a linear function and defines the flexibility of an elastic system.

By substituting (20) into (18), the following well-known continuity condition for a linear elastic structure is obtained:

$$2 \underline{B}_X^T \operatorname{diag}(\underline{\lambda}) \underline{A} \underline{F} = \underline{0} \ . \tag{21}$$

In this special case the design of the structure does not involve iteration. It can be solved in one step by the well established Finite Element method.

The general case with more constraints than design variables per element ($n_d < n_c$) can not be solved without iteration. The number of constraints depends on the type of stress approximation within the element. One of the elementary examples is a beam-element with a linear distribution of the bending-moment. In this case two constraints ($n_c=2$) are needed. If the same element has a constant cross-section, only one design variable ($n_d=1$) is necessary. A similar situation is met when higher order elements are used. Because of the assumption of piecewise constant design variables y_i, the solution of the plastic design problem is an upper approximation of the continuously varying cross-section in the Prager-Shield theory. This is illustrated by the elementary example shown in Figure 6. In general, when more elements are used, a closer approximation of the Prager-Shield solution is obtained.

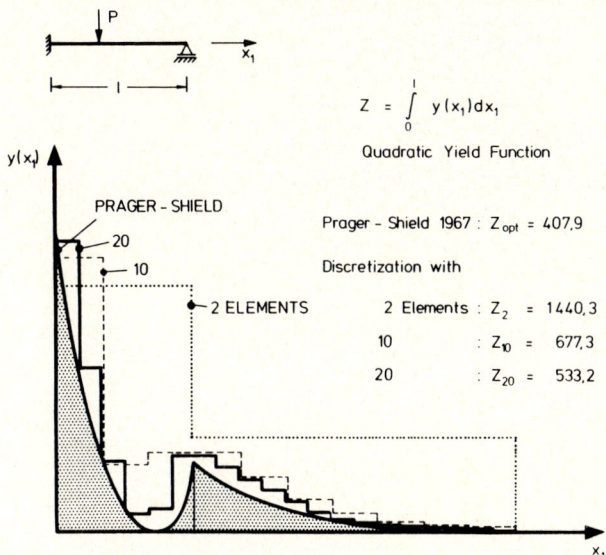

Figure 6 - *Approximation of the Prager-Shield solution*

To explain the basic iteration for the general case, the formulation of a simplified problem (22)-(25) is used as a starting point. It has one design parameter y, a linear approximation of the cost function Z, and two quadratic constraints:

$$\text{Minimize } Z = z\, y \quad \text{(linear approximation)} \tag{22}$$

subject to

$$\underline{N}\,\underline{F} - \underline{P} = \underline{0} \quad \text{(equilibrium)} \tag{23}$$

$$\underline{F}^T \underline{A}_1 \underline{F} - y \leq 0 \quad \text{(yield}\tag{24}$$

$$\underline{F}^T \underline{A}_2 \underline{F} - y \leq 0 \quad \text{conditions)}. \tag{25}$$

The Lagrangian (26) and the corresponding Kuhn-Tucker conditions (27)-(34) are:

$$L = zy + \underline{u}^T (\underline{N}\ \underline{F} - \underline{P})$$
$$+ \lambda_1 (\underline{F}^T \underline{A}_1 \underline{F} + \alpha_1^2 - y)$$
$$+ \lambda_2 (\underline{F}^T \underline{A}_2 \underline{F} + \alpha_2^2 - y) \tag{26}$$

$$\frac{\partial L}{\partial y} = z - \lambda_1 - \lambda_2 \equiv 0 , \tag{27}$$

$$\frac{\partial L}{\partial \underline{u}} = \underline{N}\ \underline{F} - \underline{P} \equiv 0 , \tag{28}$$

$$\frac{\partial L}{\partial \underline{F}} = \underline{N}^T \underline{u} + 2\lambda_1 \underline{A}_1 \underline{F} + 2\lambda_2 \underline{A}_2 \underline{F} \equiv 0 , \tag{29}$$

$$\frac{\partial L}{\partial \lambda_1} = \underline{F}^T \underline{A}_1 \underline{F} + \alpha_1^2 - y \equiv 0 , \tag{30}$$

$$\frac{\partial L}{\partial \lambda_2} = \underline{F}^T \underline{A}_2 \underline{F} + \alpha_2^2 - y \equiv 0 , \tag{31}$$

$$\frac{\partial L}{\partial \alpha_1} = \lambda_1 \alpha_1 \equiv 0 , \tag{32}$$

$$\frac{\partial L}{\partial \alpha_2} = \lambda_2 \alpha_2 \equiv 0 , \tag{33}$$

$$\lambda_1, \lambda_2 \geq 0 . \tag{34}$$

According to (27) and (34) both λ_1 and λ_2 must lie between 0 and z:

$$0 \leq \lambda_i \leq z , \text{ for } i = 1,2. \tag{35}$$

Assuming that $\lambda_1 = 0$ the problem given by (22) - (25) reduces, as before, to the elastic sub-problem described by (27), (28) and (29). From the principle of minimum complementary energy for the solution of this sub-problem it is known that both constraints (24), (25) have to be active within the element, and for one of the Lagrange-parameters the following functional is obtained [7]:

$$g(\lambda_1) = \underline{F}^T(\lambda_1) (\underline{A}_1 - \underline{A}_2) \underline{F}(\lambda_1) = 0 . \tag{36}$$

To solve the design problem it is necessary to find a zero point of this nonlinear function in the intervall given by (35).

A typical functional g and a Newton-iteration are plotted in Figure 7.

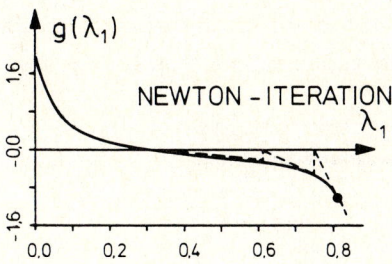

Figure 7 - *Typical functional g for one Lagrange-parameter*

The same iteration has to be performed in more dimensions. Figure 8 shows typical surfaces which define the solution for two Lagrange-parameters.

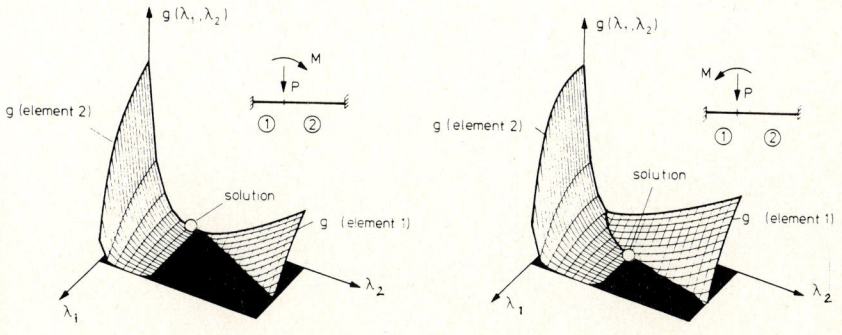

Figure 8 - *Typical functional g for two Lagrange-parameters*

The solution is obtained as their cutting point with the (λ_1, λ_2)-plane if all inequality constraints are active, otherwise it is the cutting point of one of the surfaces with one of the axes.

The solution in more than two dimensions is described by means of the flow chart in Figure 9. An outer loop performs the linear and quadratic approximation, an inner loop solves the

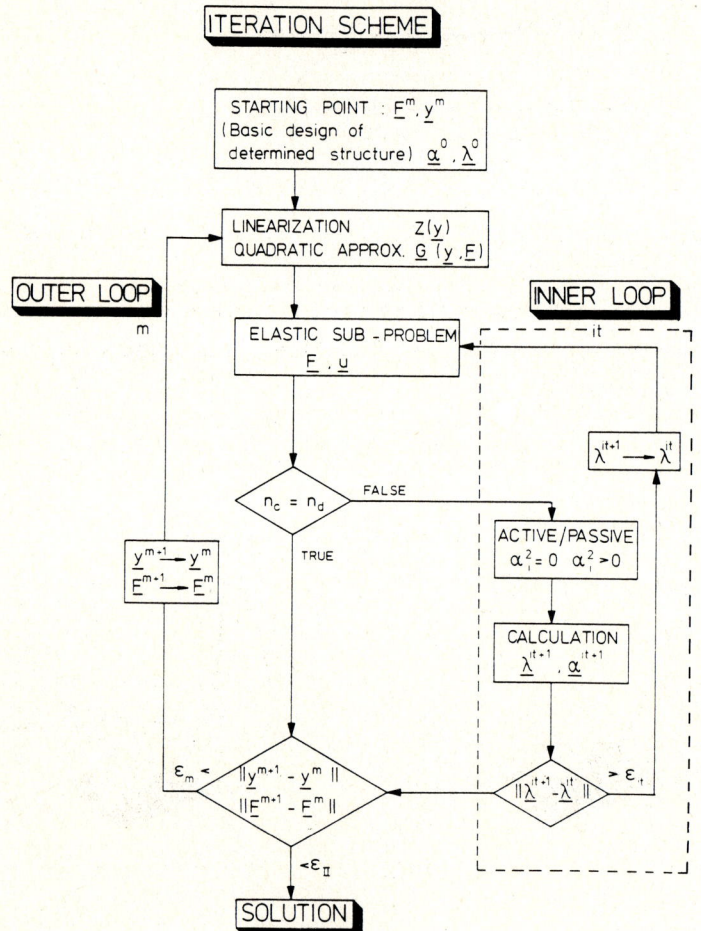

Figure 9 - Flowchart

partitioning problem into active and passive constraints. It should be mentioned that this inner loop does not exist in the direct analogue of the Prager-Shield theory. A typical iteration history for convex problems is shown in Figure 10.

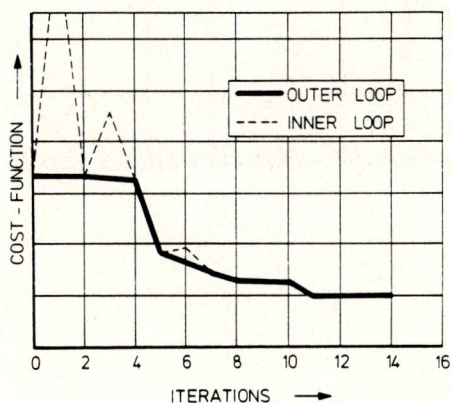

Figure 10 - Iteration history

The inner and outer iteration-loops are included. The number of iterations is usually found to be smaller than the number of design-variables involved. This is a remarkable feature of the iterative method proposed.

4. APPLICATIONS

In the following figures some typical applications are shown. The first system is one of the main girders of a railway bridge. The design obtained for one of the loading cases as shown in Figure 11 seems to be reasonable.

The distribution of material is indicated by the thickness of the lines. To obtain a linear objective function the square of the cross sectional area, i.e., A_i, was chosen as the design parameter y_i, and the yield condition for each element is

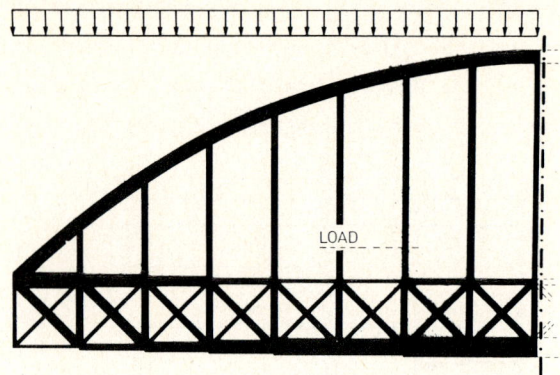

Figure 11 - Example pin-jointed bar

$$\frac{1}{\sigma_{F1}^2} F_i^2 - y_i \leq 0 \; , \tag{37}$$

where σ_{F1} is the yield stress.

The plastic design problem was solved in one "elastic" analysis of the structure ($n_c = n_d$).

The next example is a framework with linear objective and a parabolic yield condition

$$\frac{|N|}{N_{pl}} + \frac{M^2}{M_{pl}} \leq 1 \; , \tag{38}$$

where

N = axial force,

M = bending-moment,

$N_{pl} = |\sigma_{F1} A_i|$ = fully plastic axial force,

$M_{pl} = \beta N_{pl}$ = fully plastic bending-moment,

β = given scalar.

The design as indicated by the thickness of the lines in Figure 12 was obtained after 12 "elastic" iterations

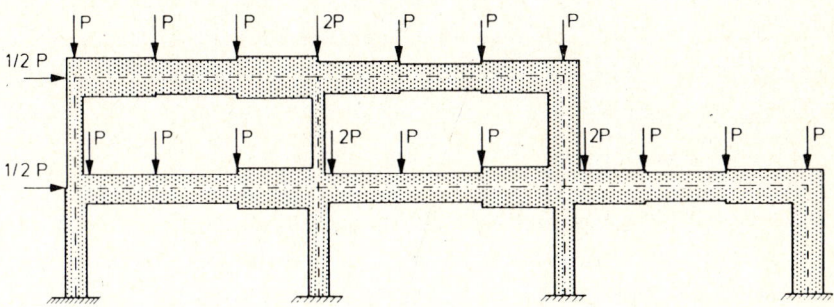

Figure 12 - Example framework

The last example is a plane stress problem with linear cost-function and quadratic constraints. The resulting design (Figure 13) was

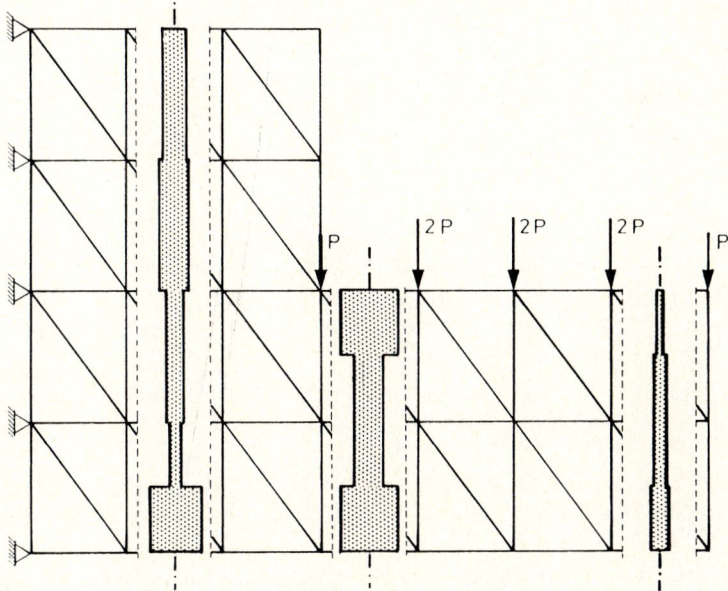

Figure 13 - Example plane stress

obtained after one analysis of the structure and again the result appears to be reasonable. The distirubtion of material coincides with the stress distribution. The few examples shown here indicate the range of possible applications. In the authors' experience, design problems with several hundreds of design variables and multiple constraints per element can be solved without undue increase in computer time.

5. CONCLUSIONS

The quadratic approximation of nonlinear constraints seems to be a powerful tool for the iterative design of structures. For convex problems the iteration described herein shows rapid convergence and can be applied without undue increase of computer time, even to problems with a large number of design parameters. For problems with non-convex objective function the same method can be used as a basic iteration for an interactive design.

REFERENCES

[1] MARCAL, P.V. and PRAGER, W., "A Method of Optimal Plastic Design," *J. de Mecanique*, Vol. 3, 1964, pp. 509-530.

[2] PRAGER, W. and SHIELD, R.T., "A General Theory of Optimal Plastic Design," *J. of Appl. Mech.*, 1967, pp. 184-186.

[3] GELLATLY, R.A. and BERKE, L., "Structural Design with Optimality-Based Algorithms," *Proc. Int. Symp. on Computers in Optimization of Structural Design*, 1972.

[4] ROBINSON, J., "Automatic Selection of Redundancies in the Matrix Force Method," *Can. Aero Space J.*, Vol. 11, 1967, pp. 9-12.

[5] THIERAUF, G. and TOPCU, A., "Structural Optimization Using the Force Method," *Proc. of Finite Element World Congress*, Bournemouth, 1975.

[6] THIERAUF, G., "A Method for Optimal Limit Design of Structures with Alternative Loads," *Comp. Meth. in Appl. Mech. and Eng.*, Vol. 16, 1978, pp. 135-149.

[7] PAPE, G., "Eine quadratische Approximation des Bemessungsproblems idealplastischer Tragwerke," Ph.D. Dissertation, Essen University, 1979.

STRUCTURAL CONTROL, H.H.E. Leipholz (ed.)
North-Holland Publishing Company & SM Publications
© IUTAM, 1980

DESIGN OF LARGE SCALE TUNED MASS DAMPERS

Niel R. Petersen
MTS Systems Corporation
Minneapolis, Minnesota

1. INTRODUCTION

The desirability of incorporating Tuned Mass Dampers (TMD's) into lightly damped structures subjected to random forces is well documented. Literature evaluates the effectiveness of these devices for motion control of both sharply resonant structures exposed to random excitations, and all types of structures subjected to steady state periodic forces. An example of such technology applied to Civil Engineering Structures is the use of two small tuned mass dampers to reduce the 2nd and 4th modes of vibration of the Canadian National Tower in Toronto in order to minimize antenna bending loads.

The application of TMD's to control the first mode bending of high rise buildings while theoretically highly desirable, presents a massive practical problem of scale. When LeMessurier Associates/SCI included TMD concepts into the structural analysis for the new Citicorp Center, Manhattan, a considerable design challenge resulted.

2. TMD SPECIFICATION SUMMARY

The Citicorp TMD is installed on the 63rd floor in the crown of the structure. At this elevation, the building could be represented by a simple modal mass of 20,000 tons (18.2 x 10^6 kg) biaxially resonant at approximately 6 1/2 second period with an inherent (self) damping factor of about 1% of critical in each axis. The system specifications as generated by LeMessurier Associates/ SCI required the moving mass to be 400 tons, (363,000 kg), biaxially resonant on the building structure with a variable operating period of 6.25 seconds $\pm 20\%$, adjustable linear damping from 8% to 14% critical, a maximum operating stroke of ± 45 inches (114.3 cm) free travel and with overtravel to ± 54 inches (137.2 cm). The system was to operate automatically and use a minimum amount of energy. Such a TMD could be expected to reduce the RMS building sway amplitude by about 50%; such a reduction by other means would require the equivalent of raising the basic structural damping to about 4%.

3. THE DESIGN PROBLEM

For scale purposes, the relative size of a TMD may be thought of as proportional to the potential energy that must be stored in the spring(s) at maximum mass deflection. When this criteria was applied to the Citicorp specification, the required TMD would be over 250 times larger than had ever been built before.

Initial concepts centered around a fully active servohydraulic drive system. The hydraulic actuator(s) used would have to be able to generate peak forces of 177 kip (788 kN) and peak velocities of 68 inches per second (1.72 M/sec) to accommodate the most severe relative TMD displacements. To accomplish a maximum level continuous sinusoidal operation in one axis coincident with 70% performance in the other axis would require about 3000 Hp (2240 KW) of hydraulic power. Although such a steady state sinusoidal criteria is obviously over-conservative, the amount of hydraulic power required would make the system prohibitively complex, and expensive to operate and maintain.

On the other hand, the system was required to operate smoothly and accurately at very low excitation. Since the TMD system is used for building occupant comfort, it was necessary that motions below a human sensitivity threshold cause appropriate mass block motion. Lateral motions less than 1 milli-g could be expected which should result in TMD action. This would require a very low friction sliding mechanism with a breakout friction coefficient less than .001. Conventional air levitation systems were capable of such low coefficients with a 400 ton mass but they required considerable space and consumed over 150 horsepower (112 KW) of energy during operation. Since most of the TMD operation would be at low levels such continuous consumption was deemed excessive for the task. Hydraulically operated hydrostatic bearings were also considered due to the lower energy consumption, but the necessary tight control of clearances could not be economically achieved with the deflections encountered in typical civil engineering structures.

4. DETAILED TMD SYSTEM DESIGN

A singular solution to the design dilemma above became apparent

    A. An approximatley linear passive spring connected between the mass block and the building in each axis would greatly reduce the necessary hydraulic actuator force and flow.

    B. A servocontrolled TMD using instrumentation to generate the TMD motion program would give accurate control of damping and natural frequency despite the non-linear system characteristics.

    C. A servohydraulic hydraulic actuator generally need only furnish limited driving forces (enough to take care of the bearing support system friction) when compared to the retarding force requirements. The use of a special servovalue design would allow the supply pressure to such a servo system to be greatly reduced, saving energy.

D. A mass block supporting system with larger inherent friction could then be used.

4.1 *Passive Spring System*

Of the above items, only the passive linear spring demanded new technology. Early concepts involved the use of torsion bars with varying length arms to change the spring rate in order to vary the natural frequency. A most critical design value of mechanical springs is the maximum stress allowable as the energy storage is proportional to the square of the maximum stress. In addition, a configuration factor was necessary in analyzing different spring techniques to account for the fact that most spring concepts do not stress the material uniformly. Problems of scale persisted however, for the torsion bar concepts. If an allowable shear design stress of 25,000 psi is used, the torsion bar would have to be an impractical 12 inches (305 mm) in diameter and 135 feet (41.1 M) long to meet the system requirements in each axis.

The passive spring problem was solved by using two opposed pneumatically precharged cylinders which are trunnion mounted with the piston rods connected as shown in Figure 1. When the mechanical assembly is deflected perpendicular to the axis of the cylinders, as shown in Figure 2, the inherent non-linearity of adiabatic gas compression in a closed vessel is almost exactly compensated by the geometric considerations. The result is a compact assembly which is very linear in its force-deflection characteristics to a substantial extension as can be seen in Figure 3. A further major advantage of the gas spring is that

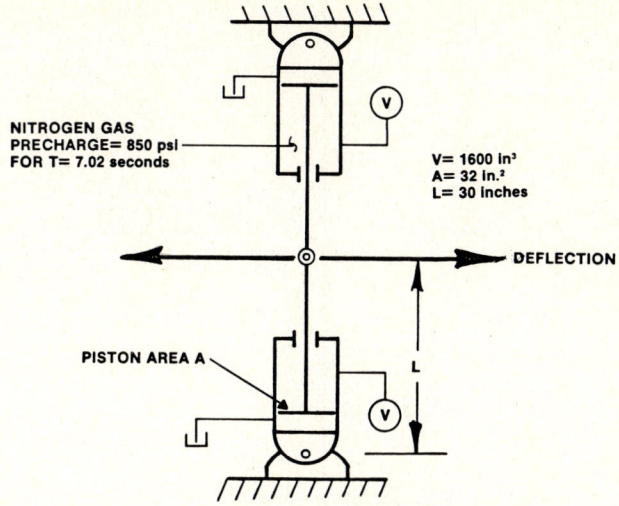

Figure 1 - Linear Gas Spring

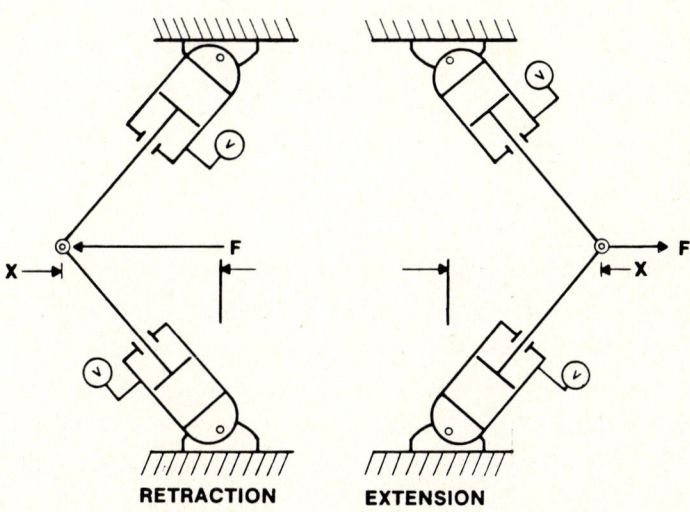

Figure 2 - Linear Spring Operation

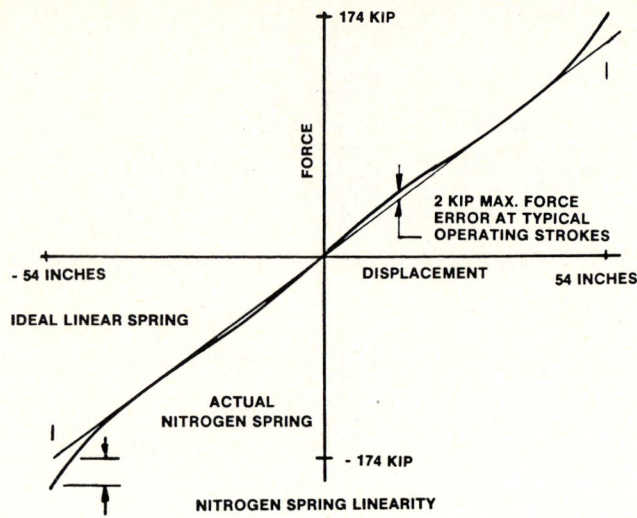

Figure 3 - Nitrogen Spring Linearity

only simple adjustment of the pre-charge pressure is necessary to vary the spring rate. For the Citicorp TMD the operating precharge is approximatley 850 psi (5865 kN/M^2) of dry nitrogen and should be maintained within 50 psi (345 kN/M^2) for optimum system operation.

4.2 *Mass Block Support Bearings*

The 400 ton (363,000 kg) mass block was supported on a series of twelve 22 inch (55.9 cm) diameter hydraulic pressure balanced bearings (PBB). A cross section of a single such bearing is shown in Figure 4. The PBB uses elastomeric seals throughout so that external leakage is minimized, even in the presence of substantial clearance. An intermediate piston shoe slides on a steel floor surface with a face seal and also is elastomerically sealed to a cylinder diameter on the bottom of the mass block. A slight difference in the diameters of the two seals causes a force offset in the piston shoe to keep the shoe against the floor surface.

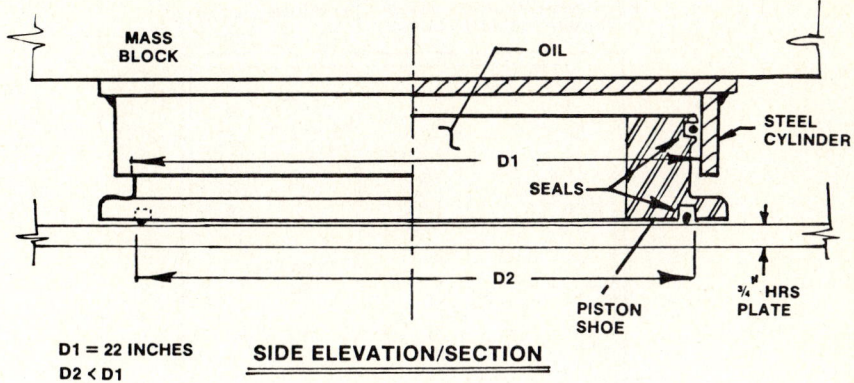

Figure 4 - Pressure Balanced Bearing

Supporting the mass block on the twelve bearings requires an internal bearing pressure of about 175 psi (1207 kN/M^2). At this pressure the elastomeric seals are capable of spanning gaps up to about 1/16 inch (1.6 mm). This requires that the floor surface need only be locally flat within 1/16 inch (1.6 mm) in any single 22 inch (55.9 cm) diameter area. This is readily accomplished by using a plate of grouted 3/4 inch (19 mm) hot rolled steel as the floor running surface. The two plates required for the twelve bearings were approximately 12 ft x 30 ft (3.6 M x 9.1 M) each and were polished on site with simple terrazzo grinding equipment.

When the TMD is not in operation, the mass block rests on a set of phenolic blocks with rubber springs to prevent undue floor load concentration.

During operation, the bearings are supplied oil from a separate 5 Hp 10 gpm (3.7 KW, .04 M^3/min) hydraulic pump which is capable of raising the mass block about 3/4 of an inch to its operating position in about 3 minutes. The leakage oil is collected by a trough around the bearing plates, and pumped through a large filter into the main hydraulic reservoir of the system. The individual bearings are hydraulically coupled together as shown in Figure 5 into three groups, with each group containing its own

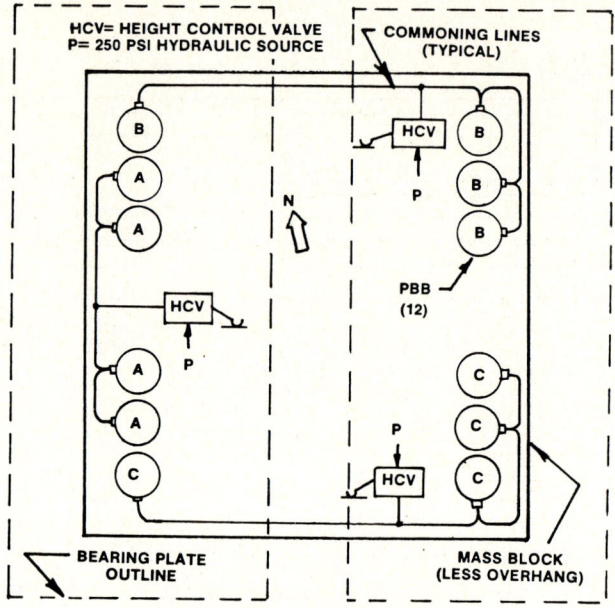

REFLECTED PLAN VIEW

Figure 5 - Pressure Balanced Bearing System

height control valve. Thus the bearings evenly distribute the floor loading despite the unevenness that may be present in the bearing plate. The overall achieved coefficient of friction for the pressure balanced bearing installation is about .003, requiring about 2400 lbs. (10,600 N) drive force to compensate for friction losses.

4.3 *Control Actuator*

The simultaneous motion of the mass block in the two directions is controlled by two double acting equal area hydraulic actuators. Oil is supplied from a fixed pressure source (700 psi (4830 kN/M^2) supply pressure) through a modified 4-way flow controlled three stage servovalue. The maximum design velocity of the actuator is 68 inches per second (1.72 M/sec) and the piston area is 13 square inches (83.9 cm^2) for a maximum damping force of 47 kips (209 kN).

The maximum driving velocity at zero actuator force is 28 inches per second (71 cm/sec). All velocities greater than 28 inches per second (71 cm/sec) require that the control actuator absorb energy from the forcing device. A summary of the capabilities of the control actuator is shown in Figure 6.

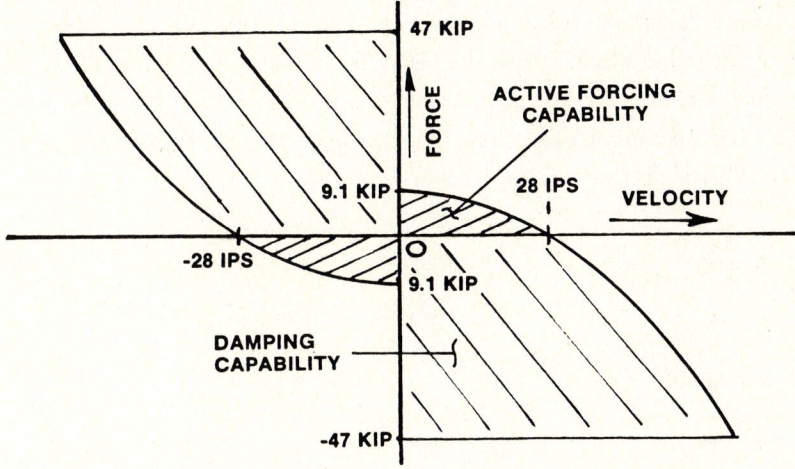

Figure 6 - *Control Actuator Force/Velocity Capabilities*

The initial detailed design of the control actuator was critical in buckling due to the comparatively small piston rods and the 108 inch (2.75 M) stroke. A reverted design was therefore used (See Figure 7) which does not have the buckling problem because the piston rod is always in tension.

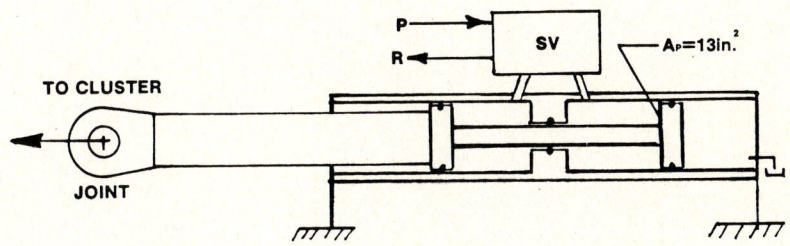

Figure 7 - *Control Actuator Configuration*

4.4 TMD System Configuration (see Figure 8)

Each control actuator and spring actuator assembly was mounted in a single linear motion assembly for each axis of the TMD. The linear motion fixture contained a cluster joint to collect the actuator forces and 40 foot long (12.2 M), 12 inch (30.5 cm) diameter steel tube was used to connect each fixture to the mass block. The tube struts were crossed underneath the mass block to clear each other, and attached to the mass block via a large swivel at the side farthest away from each respective fixture. This was done to minimize the arcing effects of the struts.

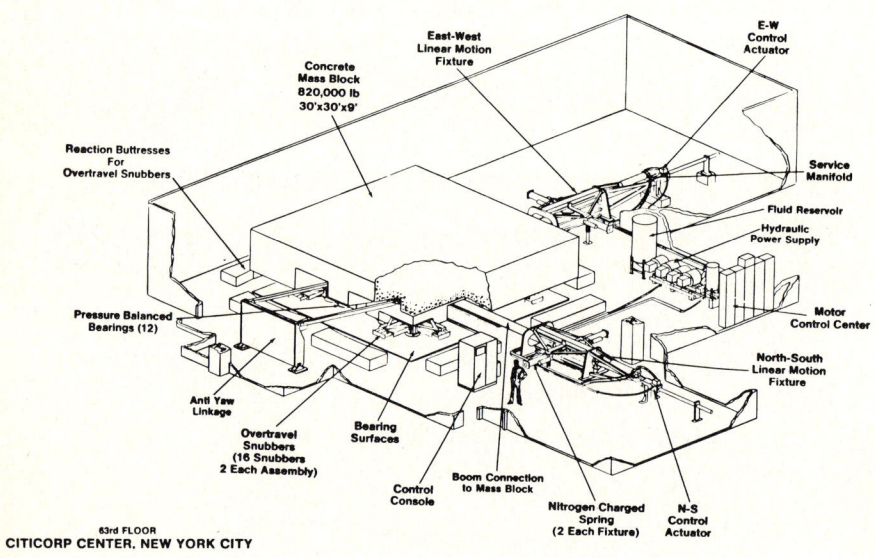

Figure 8 - MTS Tuned Mass Damper System

The concrete mass block has overall dimensions of approximately 30 feet by 30 feet by 8-1/2 feet (9.1 M x 9.1 M x 2.6 M) high. The lower portion of the mass block was restricted to 18 feet by 18 feet (5.5 M x 5.5 M) to provide appropriate bearing access and spacing. Other mechanical equipment was mounted in the lower recess including a total of 16 hydraulic over-travel snubbers. The snubbers have an operating stroke of 11 inches (28 cm). The operating space of the system allows a free motion of ± 43 inches (109.2 cm) in each axis without snubber contact. Some over-travel capability is provided although electrical limit switches cause the TMD to shut down when the stroke exceeds ± 45 inches ($\pm 1.14M$). The snubbers impact into concrete abutments formed as part of the floor and its associated reinforcement.

The mass block is constrained from rotational motion about its vertical axis (yaw) by a torsion box assembly mounted on the floor connected to the mass block by two radius rods.

4.5 *Hydraulic Power Supply*

The TMD system is supplied hydraulic oil at 700 psi (4830 kN/M^2) for operation of the two servo actuators. The pumping system includes a reservoir filled with about 200 gallons (.76 M^3) of commercial petroleum base hydraulic fluid. Four main pumps are provided which are 40 Hp (30 KW) each. Accumulators are included in pressure lines to furnish short term high flow demands. The pumps operate only as necessary on a demand basis with one pump being adequate for the majority of operation. Under the most severe conditions only three of the four units would be necessary. In the event the TMD motion demands exceed the capacity of a single pump, additional pumps are automatically started as necessary to maintain the system pressure.

The hydraulic power supply is water cooled with an oil-water heat exchanger. Under normal operating conditions, little or no water is required but under severe excitation conditions, up to about 40 gpm (.16 M^3/min) cooling water demand may be expected.

4.6 *Electronic Control*

The TMD system is electroncially controlled from a console as shown in Figure 9. The mass block position with respect to the building is programmed from accelerometers mounted on the building

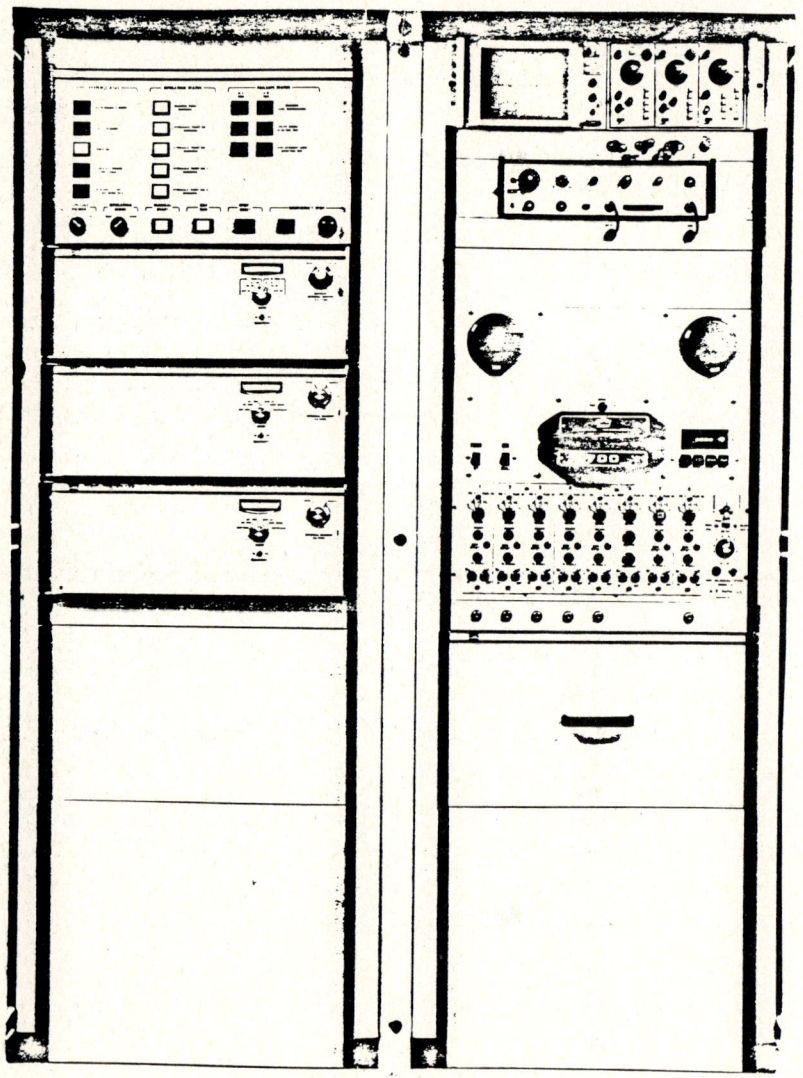

Figure 9 - CITICORP TMD Control Console

Large Scale Tuned Mass Dampers

which measure the horizontal sway in each axis. The TMD hydraulic system is activated automatically whenever the horizontal motion exceeds 3 milli-g's for two cycles, initiating the start sequence. After 4 minutes, the mass block is allowed to move in response to the electronically generated commands from the system instrumentation. The system controls the mass block motion from the output of an accurate analog computer into which is set the desired natural frequency and the desired damping (see Figure 10). Thus, the system is forced by the servo actuator to be repeatable for these

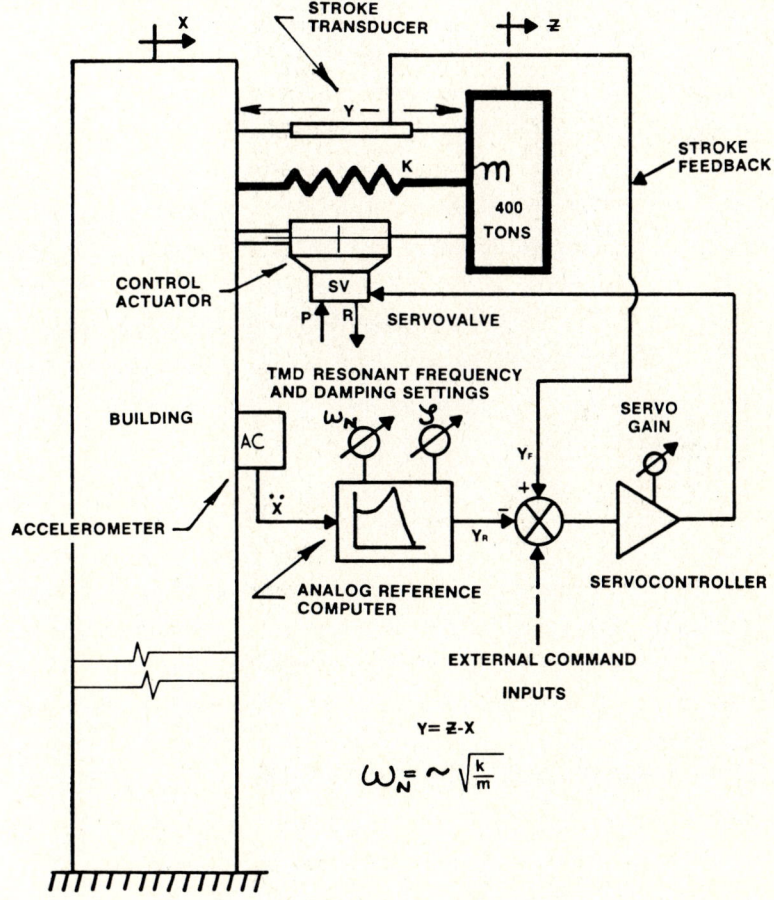

Figure 10 - TMD Control System

critical parameters. The non-linearities of the mechanical system are compensated for by the servocontrol hardware and the active force capability of the servohydraulic actuator. For example, small changes in the spring rate of the gas springs are fully compensated by the control system. Attempts were made to determine the long term repeatability of the natural frequency of the control system reference signal and it was found to be well within 0.1% over a period of many weeks. In a similar manner the damping values of the reference system showed such stability that changes were also undetectable.

A primary feature of the electronic reference system is the ability to easily vary the critical TMD parameter values electronically. Although the damping of this system was finally set to 80% (it is electronically increased to 70% immediately before shut down) other TMD systems have been built in which the damping automatically changes to optimize the TMD performance while staying within the operating limits of the system. The system contains a full range of interlocks to automatically protect and reset itself. The TMD will automatically shut itself down when 30 minutes have elapsed in which the building sway has not exceeded .75 milli-g's in either axis.

4.7 *System Installation and Checkout*

The primary force generating elements of the TMD system are connected directly to the building steel structure. These include the linear motion assemblies, the anti-yaw assembly, and the snubber impact abutments. An extra heavy concrete floor was poured under the mass block area. The bearing plates were lifted onto the floor in four sections, welded together, leveled and grouted to make the two bearing plates. The plates were then ground to finish the top surfaces. The concrete mass block was formed from four steel weldments in the base, with conventional wood forming being used for the larger 30 foot by 30 foot (9.1 M x 9.1 M) upper section.

The mass was poured and the necessary attachments post tensioned 28 days later.

The TMD machinery was installed, connected and when the area was closed and heated, the hydraulic system was flushed. The control system and its associated instrumentation was subjected to an analog computer simulation at MTS-Minneapolis to verify the correctness of the control concepts used. A view of the completed facility is shown in Figure 8.

One of the significant features of an electronically programmed TMD is the ability to electronically introduce external programs or disturbances into the system. This allows the mass block to be positioned dynamically within the force and velocity limits of the control actuator as modified by the passive spring system. This feature was used extensively to perform a series of tests on the TMD and to determine the dynamic characteristics of the building.

For example, after initial calibration, the mass block was deliberately impacted into the snubbers to check their performance and the building response to such an impulse. In another series of tests, the mass was cycled back and forth sinusoidally at as high velocity as could be attained with the nitrogen spring pressure being reduced to give the maximum resonant excitation possible. The nitrogen spring pressure was adjusted to permit the TMD to be used to excite the building at its natural frequency. Since the excitation and response could be studied in a steady state operating condition, it was possible to very accurately determine the building period and the inherent structural damping.

Test results at 2 milli-g building excitation levels indicated a N-S 1st mode natural period of 6.89 seconds and an E-W 1st mode natural period of 7.22 seconds. In each direction the period increased about .15 seconds when the excitation level was increased to the maximum achievable with the facility (10 milli-g's). The existing structural damping present in each mode was found to be about .009 of critical with no discernible shift with amplitude.

As a final test the TMD was used to excite the structure to a 5 milli-g level. The excitation signal was then removed and the resulting actual TMD motion decay vs. the theoretical (i.e., reference system) motion decay was recorded. Error analyses were performed on the relative velocity signal with checks being done on the zero crossing accuracy (i.e., phase lag) and on the achieved amplitude accuracy. The peak velocity present in the test was about 9 inches per second (23 cm/sec). The TMD system was able to respond to the simulated transient with less than 0.2 inch per second (0.5 cm/sec) error and with undetectable phase shift error.

CONCLUSION

The TMD system built and described herein shows a practical means to create very large scale TMD's with established technology. The scale of this particular installation was readily achieved and substantially larger systems would seem practical.

The technique of using an electronic reference program to a servocontrolled TMD eliminates the accuracy, dynamic range, and flexibility problems of purely passive systems.

REFERENCES

[1] AMYOT, J.R., VAN BLOKLUND, G., WARDLAW, R.L., STANDEN, N.M. and COOPER, K.R., *Computer Studies of a Vibration Damper for Wind-Induced Motion of a Tall Building*, National Research Council of Canada, D.M.E. LTR-AN-8.

[2] CRANDALL, S.H. and MARK, W.D., *Random Vibrations in Mechanical Systems*, Wiley and Sons, Inc., 1968.

[3] DEN HARTOG, J.P., *Mechanical Vibrations*, McGraw-Hill, 4th edition, 1956.

[4] McNAMARA, R.J., "Tuned Mass Dampers for Buildings " *ASCE Journal of Structural Division*, Vol. 103, Sept. 1977.

[5] WIESNER, K.B., "Tuned Mass Dampers to Reduce Building Wind Motion," presented at the ASCE Boston Convention, April 1979.

STRUCTURAL CONTROL, H.H.E. Leipholz (ed.)
North-Holland Publishing Company & SM Publications
© IUTAM, 1980

STRUCTURAL OPTIMIZATION BASED ON THE WORKHARDENING ADAPTATION CONCEPT

C. Polizzotto and C. Mazzarella

Facoltà di Architettura, Università di Palermo
Via Maqueda 175, I-90133 Palermo, Italy

1. INTRODUCTION AND FUNDAMENTALS

Workhardening adaptation is a concept due to Prager [1, 2] and studied by others [3 - 7]. We say that a rigid-plastic structure is able to adapt to loads which vary arbitrarily within a given domain, when, after an initial plastic phase, it eventually reaches a (deformed) rigid state in which no further deformation occurs.

The workhardening adaptation concept, assumed as design criterion [8, 9], leads to optimal structures which are more economic than those obtainable by the usual Limit Design [9]. A completely correct application of the above criterion should also introduce suitable limitations to deformation produced during the adaptation process; but such limitations have not been considered so far.

The present paper is devoted to structural optimization for workhardening adaptation with limitations to deformation. The structure is conceived as a discrete model, each constituent being characterized by a piecewise linear yield surface and a piecewise linear workhardening law. This leads to an optimization problem having the format of a mathematical programming problem which in general is nonlinear and nonconvex.

The load is described by the nodal forces $\underline{f}$ which vary with time $t \geq 0$ according to an unknown time history but remaining always within a given bounded domain. Denoting the nodal displacements, the element stresses and strains by $\underline{u}$, $\underline{\sigma}$ and $\underline{\varepsilon}$ respectively, we write the equilibrium equations as

$$\underline{C}^T \underline{\sigma} + \underline{M}\,\underline{\ddot{u}} = \underline{f}(t), \qquad \forall t \geq 0, \tag{1}$$

where $\underline{C}^T$ is the equilibrium matrix and $\underline{M}$ is the mass matrix, and the compatibility equations as

$$\underline{\varepsilon} = \underline{C}\,\underline{u}, \qquad \forall t \geq 0, \tag{2a}$$

$$\underline{\dot{u}} = \underline{\dot{u}}_0, \qquad \text{at } t = 0, \tag{2b}$$

where $\underline{\dot{u}}_0$ is the initial velocity vector.

The plastic behaviour of the structural elements is characterized by the inequalities

$$\underline{N}^T \underline{\sigma} - \underline{H}\,\underline{\lambda} - \underline{k} \leq \underline{0}, \qquad \forall t \geq 0, \tag{3}$$

which define a polyhedron whose faces have unit external normals specified by the (constant) matrix $\underline{N}$ and distances from the origin specified by the vector $\underline{k} + \underline{H}\,\underline{\lambda}$, $\underline{\lambda}$ being a non-negative vector describing plastic strain. At $t = 0$, $\underline{\lambda}(0) = \underline{0}$, so that in the virgin state these distances are specified by the vector $\underline{k} > \underline{0}$ (*plastic resistances*).

Introducing the plastic potential vector $\underline{\phi}$ defined by

$$\underline{\phi} = \underline{N}^T \underline{\sigma} - \underline{H}\,\underline{\lambda} - \underline{k}, \qquad \forall t \geq 0, \tag{4}$$

we apply the usual flow-rule to obtain the plastic strain rates, i.e.,

$$\underline{\dot{\varepsilon}} = \frac{\partial \underline{\phi}}{\partial \underline{\sigma}}\,\underline{\dot{\lambda}} = \underline{N}\,\underline{\dot{\lambda}}, \qquad \forall t \geq 0, \tag{5}$$

to which the following conditions must be added:

$$\underline{\phi} \leq \underline{0}, \quad \underline{\dot{\lambda}} \geq \underline{0}, \quad \underline{\phi}^T \underline{\dot{\lambda}} = \underline{\dot{\phi}}^T \underline{\dot{\lambda}} = 0, \qquad (6a\text{-}d)$$

all these equations being equivalent to the usual concepts of associated plasticity theory [10]. A number of workhardening laws, such as for instance isotropic and kinematic workhardening laws, can be taken into account by suitably defining the workhardening matrix $\underline{H}$ which by hypothesis is positive semidefinite.

As *adaptation criterion* we use the boundedness of the non-negative and nondecreasing vector-valued function

$$\underline{\lambda} = \int_0^{t_1} \underline{\dot{\lambda}} \, dt, \qquad (7)$$

where t_1 is any subsequent time. Theorems for workhardening adaptation were given in [3 - 7], in particular in [6] is stated that *a structure able to adapt to some given quasi-static loads is also able to adapt to the same loads supposed to act dynamically*. Bounding theorems for deformation produced during the adaptation process were formulated in [4, 6, 7] and a new theorem is given in the next Section.

2. A BOUNDING THEOREM

Let the loading domain be a hyperpolyhedron whose vertices are specified by n *relevant loading conditions*, say $\underline{f}_i$, $(i = 1, 2, \ldots, n)$, and let a scalar time-dependent *deformation parameter*, say $\eta = \eta(t)$, be associated with the real loading history and expressed as

$$\eta = \underline{d}^T \underline{\lambda}, \qquad (8)$$

where $\underline{\lambda} = \underline{\lambda}(t)$ is the plastic strain intensity vector given by equation (7) while the constant vector $\underline{d}$, called *perturbation vector*, is assigned.

Given a *load multiplier* $\gamma > 0$ and a *perturbation multiplier* $\omega > 0$, if a time-independent plastic strain intensity vector $\underline{\mu} \geq \underline{0}$ and n time-independent stress vectors $\underline{\rho}_i$ can be found such as to satisfy

$$\underline{C}^T \underline{\rho}_i - \underline{f}_i \gamma = \underline{0} , \tag{9a}$$

$$\underline{N}^T \underline{\rho}_i - \underline{H}\,\underline{\mu} - \underline{k} + \underline{d}\,\omega \leq \underline{0} , \tag{9b}$$

$$\forall i \in (1,2,\ldots,n) , \tag{9c}$$

then the following inequality is true:

$$\eta \leq \frac{1}{\omega}\left(\frac{1}{2}\underline{\mu}^T \underline{H}\,\underline{\mu} + K_0\right) , \qquad \forall t \geq 0 , \tag{10}$$

where $K_0 = \frac{1}{2}\underline{\dot{u}}_0^T \underline{M}\,\underline{\dot{u}}_0$ (initial kinetic energy), *whatever the real (static or dynamic) loading history inside the polyhedral domain of vertices* $\underline{f}_i \gamma$, $(i = 1,2,\ldots,n)$.

Proof. Let us observe that the load $\underline{f}(t)$ can be represented by:

$$\underline{f}(t) = \gamma \sum_{i=1}^{n} \underline{f}_i \tau_i , \qquad \forall t \geq 0 , \tag{11}$$

where the coefficients $\tau_i = \tau_i(t)$ satisfy

$$\tau_i \geq 0 , \qquad \forall i \in (1,2,\ldots,n) , \tag{12a}$$

$$\sum_{i=1}^{n} \tau_i = 1 . \tag{12b}$$

Then, introducing the stress vector

$$\underline{\rho} = \sum_{i=1}^{n} \underline{\rho}_i \tau_i , \tag{13}$$

multiplying the two relations (9a, b) by τ_i and summing up with respect to $i \in (1,2,\ldots,n)$ yields respectively

$$\left. \begin{array}{l} \underline{C}^T \underline{\rho} - \underline{f}(t) = \underline{0} \\ \underline{N}^T \underline{\rho} - \underline{H} \underline{\mu} - \underline{k} + \underline{d}\omega \leq \underline{0} \end{array} \right\} \quad \forall t \geq 0 . \quad (14a,b)$$

Now multiplying inequality (14b) by $\underline{\dot{\lambda}}$ and subtracting from the resulting inequality the orthogonality condition (6c) written in explicit form gives the inequality

$$-(\underline{\sigma} - \underline{\rho})^T \underline{\dot{\varepsilon}} + (\underline{\lambda} - \underline{\mu})^T \underline{H} \underline{\dot{\lambda}} + \omega \underline{d}^T \underline{\dot{\lambda}} \leq 0 , \quad \forall t \geq 0 . \quad (15)$$

Since by the virtual work principle

$$(\underline{\sigma} - \underline{\rho})^T \underline{\dot{\varepsilon}} + \underline{\dot{u}}^T \underline{M} \underline{\ddot{u}} = 0 , \quad (16)$$

we can rewrite inequality (15) in the form

$$\omega \underline{d}^T \underline{\dot{\lambda}} \leq -(\underline{\lambda} - \underline{\mu})^T \underline{H} \underline{\dot{\lambda}} - \underline{\dot{u}}^T \underline{M} \underline{\ddot{u}} , \quad (17)$$

whose right-hand member is the negative of the positive semidefinite quadratic function

$$B = \frac{1}{2} (\underline{\lambda} - \underline{\mu})^T \underline{H} (\underline{\lambda} - \underline{\mu}) + \frac{1}{2} \underline{\dot{u}}^T \underline{M} \underline{\dot{u}} . \quad (18)$$

An integration of equation (17) over the time interval $(0, t_1)$, t_1 being any subsequent time, finally gives

$$\omega \underline{d}^T [\underline{\lambda}(t_1) - \underline{\lambda}(0)] \leq B(0) - B(t_1) \leq B(0) , \quad (19)$$

and the latter, since $\underline{\lambda}(0) = \underline{0}$, coincides with equation (10). So, the theorem is proven.

The above theorem is a unified formulation of several theorems of the workhardening adaptation theory. For example, if $\underline{d} = \underline{k}$ and ω is suitably selected within the interval $(0, 1)$, it identifies with Prager's adaptation theorem in the generalized form given in [3, 6]. Bounding theorems for plastic strain intensities, for plastic strains and for displacements [6, 7] are also included as is easy to show by choosing $\underline{d}$ as in the following:

(i) If all the components of $\underline{d}$ are zero except the hth, which is one, then η coincides with λ_h.

(ii) If $\underline{d}$ is given the form $\underline{d} = \underline{N}^T \underline{s}$, $\underline{s}$ being an arbitrary stress vector, then by equation (5) we have $\eta = \underline{s}^T \underline{\varepsilon}$, which is a measure of strain.

(iii) If the stress vector $\underline{s}$ is in equilibrium with some arbitrary load $\underline{g}$, i.e., if $\underline{C}^T \underline{s} = \underline{g}$, then by the virtual work principle we have $\eta = \underline{g}^T \underline{u}$, which is a measure of displacements.

The consideration of a further parameter with the meaning of plastic work dissipated within a part of the structure would require introducing a rather different perturbation ingredient and for simplicity we ignore it.

3. THE DESIGN PROBLEM

As a design criterion we use the workhardening adaptation concept meant in a narrower sense than usual [8, 9]. In fact, we require that the structure, in addition to being able to adapt to the given loads, also satisfies some safety requirements implying that certain deformation parameters (such as strains, strain intensities and displacements) prove to be always not greater than given limits whatever the real loading history may be inside the given domain. A *safe design* is one which satisfies such a criterion.

In order to simplify the problem of specifying the most economic design among the set of safe ones, we consider the design layout as fixed and make the hypothesis that the cost function G as well as the plastic resistances $\underline{k}$ are linearly dependent on $\underline{x}$, the latter being the (independent) design variable vector ($\underline{x} \geq \underline{0}$), i.e.,

$$G = \underline{c}^T \underline{x} , \qquad (20)$$

$$\underline{k} = \underline{D}^T \underline{x} + \underline{k}_0 , \qquad (21)$$

where $\underline{k}_0$ refers to the weakest design $\underline{x} = \underline{0}$, $\underline{c}$ is the cost gradient vector and $\underline{D}$ is a matrix depending on the types of available elements. Moreover we suppose that the matrix $\underline{N}$ is independent of $\underline{x}$, which implies that the faces of the yield polyhedron can only translate without rotation as design variables change [11, 12]. The matrix $\underline{H}$ is a function of $\underline{x}$ and this circumstance cannot be ignored now as was in [9].

In order to formulate the design problem, let us introduce a set of m deformation parameters, say η_p, ($p = 1,2,\ldots,m$), along with the corresponding upper limits, say b_p. Then, let us consider the following minimization problem

$$\text{minimize } G = \underline{c}^T \underline{x} , \qquad (22a)$$

subject, for every $i \in (1,2,\ldots,n)$ and for every $p \in (1,2,\ldots,m)$, to $\omega_p > 0$ and

$$\underline{c}^T \underline{\rho}_{ip} - \underline{f}_i \gamma_p = \underline{0} , \qquad (22b)$$

$$\underline{N}^T \underline{\rho}_{ip} - \underline{H}(\underline{x})\underline{\mu}_p - \underline{D}\,\underline{x} - \underline{k}_0 + \underline{d}_p \omega_p \leq \underline{0} , \qquad (22c)$$

$$\tfrac{1}{2} \underline{\mu}_p^T \underline{H}(\underline{x})\underline{\mu}_p - b_p \omega_p + K_0 \leq 0 , \qquad (22d)$$

$$\underline{\mu}_p \geq \underline{0} , \quad \underline{x} \geq \underline{0} , \qquad (22e)$$

where the load multipliers $\gamma_p > 0$ are assigned.

Let us observe that any admissible design, i.e., satisfying the constraints (12b-e), is a safe design with respect to the set of m deformation parameters η_p, which in fact satisfy the inequalities $\eta_p \leq b_p$ for every $p \in (1,2,\ldots,m)$. Since the latter inequalities may be connected with different limit states (working or failure states) and hence with different safety factors, the load multipliers γ_p have been introduced into equation (22b). For $m = 1$, $\underline{d}_1 = 0$, $\omega_1 = 0$ and b_1 very big, we obtain the design formulation with no limitations to deformation, as in [8, 9].

In spite of the simplifications introduced, the minimization problem (22a-e) is nonlinear and in general nonconvex, and an *ad hoc* computer code will be provided at a later time. The Kuhn-Tucker conditions, through a procedure and transformations which we disregard for simplicity, prove to be as in the following:

$$\underline{C}^T \underline{\rho}_{ip} - \underline{f}_i \gamma_p = \underline{0}, \qquad \forall i \in I_n \text{ and } \forall p \in I_m, \qquad (23)$$

$$\left. \begin{array}{l} \underline{\psi}_{ip} = \underline{N}^T \underline{\rho}_{ip} - \underline{H}\,\underline{\mu}_p - \underline{D}\,\underline{x} - \underline{k}_0 + \underline{d}_p \omega_p \\[4pt] \underline{\psi}_{ip} \leq \underline{0}, \quad \underline{\ell}_{ip} \geq \underline{0}, \quad \underline{\psi}_{ip}^T \underline{\ell}_{ip} = 0 \end{array} \right\} \forall i \in I_n \text{ and } \forall p \in I_m, \quad (24\text{a-d})$$

$$\underline{N}\,\underline{\ell}_{ip} - \underline{C}\,\underline{a}_{ip} = \underline{0}, \qquad \forall i \in I_n \text{ and } \forall p \in I_m, \qquad (25)$$

$$\underline{\ell}_{Rp} = \sum_{i=1}^{n} \underline{\ell}_{ip}, \quad \underline{a}_{Rp} = \sum_{i=1}^{n} \underline{a}_{ip}, \qquad \forall p \in I_m, \qquad (26)$$

$$\underline{H}\,\underline{\ell}_{Rp} = \underline{H}\,\underline{\mu}_p \beta_p, \quad \underline{\mu}_p \geq \underline{0}, \qquad \forall p \in I_m, \qquad (27)$$

$$\underline{d}^T \underline{\ell}_{Rp} = b_p \beta_p, \quad (\omega_p > 0), \qquad \forall p \in I_m, \qquad (28)$$

$$\left. \begin{array}{l} \vartheta_p = \tfrac{1}{2} \underline{\mu}_p^T \underline{H}\,\underline{\mu}_p - b_p \omega_p + k_0 \\[4pt] \vartheta_p \leq 0, \quad \beta_p \geq 0, \quad \vartheta_p \beta_p = 0 \end{array} \right\} \forall p \in I_n, \qquad (29\text{a-d})$$

$$\left.\begin{array}{l} \chi_j = \sum\limits_{p=1}^{m} \left[\underline{D}_j + \dfrac{\partial H}{\partial x_j} \underline{\mu}_p \right]^T \underline{\ell}_{Rp} - \dfrac{1}{2} \sum\limits_{p=1}^{m} \beta_p \underline{\mu}_p^T \dfrac{\partial H}{\partial x_j} \underline{\mu}_p - \alpha c_j \\[2mm] \chi_j \leq 0 , \quad x_j \geq 0 , \quad \chi_j x_j = 0 , \quad \forall j \in I_r , \end{array}\right\} \quad (30\text{a-d})$$

where $I_q = (1,2,\ldots,q)$, $(q=n,m,r)$, r is the number of design variables, $\underline{D}_j$ is the jth column of the matrix $\underline{D}$ and α is a positive constant. In the above relations the new symbols are Lagrangian variables of quite obvious meanings. The following comments may be useful:

(a) If, for a $p \in I_m$, the Lagrangian variable β_p is positive, then from equations (29) it follows that $\vartheta_p = 0$ and thus the relevant deformation constraint is activated. It would be possible to show that the bound $\eta_p \leq b_p$ proves to be the most stringent. On the contrary, if $\beta_p = 0$, ϑ_p is likely to be negative and thus the deformation limitation is not active, i.e., $\eta_p < b_p$.

(b) The group of equations (30a-d) can be considered as the extension to the present context of the *uniform energy dissipation principle* by Drucker and Shield [13], and in view of this they have sometimes been called *uniformity conditions* and given a geometrical and economical interpretation [11, 12, 9].

4. EXAMPLE

As an illustrative example we consider the simple bar of Figure 1(a), of length L and cross area A, with a mass M at the free end. A force $F = F_0 \tau$, $(0 \leq \tau \leq s)$, acts upon it, F_0 being some reference force. The workhardening is kinematic (Figure 1(b)) and characterized by the coefficient $E_h A/L$, E_h being the workhardening modulus of the material. We want to determine the minimum-volume bar whish is able to adapt and also satisfies a single deformation constraint of the form $\eta \leq b$. For this the optimality conditions of Section 3 are applied.

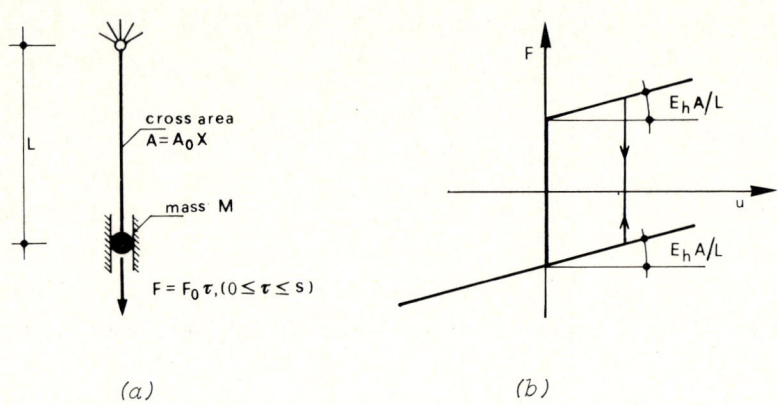

Figure 1 - Optimum Design of a Simple Bar.
    (a) Geometric and Loading Schemes
    (b) Force-Displacement Diagram

Denoting by $X = A/A_0$ the design variable, A_0 being a reference area, we expect that the optimal area X satisfies the continued inequality

$$X" \leq X \leq X' \, , \tag{31}$$

where X' is the optimal plastic design and $X"$ is the optimal design for unrestricted workhardening adaptation, i.e.,

$$X' = \frac{sF_0}{\sigma_0 A_0} \, , \qquad X" = \frac{1}{2} X' \, . \tag{32}$$

σ_0 being the yield stress of the material. The optimal design X depends on the upper limit b assigned ($X = X'$ for $b = 0$).

We consider the following four cases:

(a) The deformation parameter is:

$$\eta = \lambda^+ + \lambda^- = \int_0^t ||\dot{\varepsilon}|| dt \, , \tag{33}$$

which is obtained by the choice $d^+ = d^- = 1$.

(b) The deformation parameter is

$$\eta = \lambda^+ = \int_0^t ||z\dot{\varepsilon}||dt, \quad \begin{cases} z = 1 \text{ for } \dot{\varepsilon} \geq 0 \\ z = 0 \text{ for } \dot{\varepsilon} < 0 \end{cases} \tag{34}$$

obtained when $d^+ = 1$, $d^- = 0$.

(c) The deformation parameter is

$$\eta = \lambda^- = \int_0^t ||z\dot{\varepsilon}||dt, \quad \begin{cases} z = 1 \text{ for } \dot{\varepsilon} < 0 \\ z = 0 \text{ for } \dot{\varepsilon} \geq 0 \end{cases} \tag{35}$$

obtained when $d^+ = 0$, $d^- = 1$.

(d) The deformation parameter is the displacement u of the mass M, i.e.,

$$\eta = \lambda^+ - \lambda^- = u(t), \tag{36}$$

which requires $d^+ = -d^- = 1$.

For the sake of simplicity we disregard the computations and show only the results obtained in Figure 2, where the optimal ratio X/X' is plotted as a function of $(E_h/\sigma_0)b$. We see that the limitations on deformation produce optimal designs which are greater than X'' but rather smaller than X'.

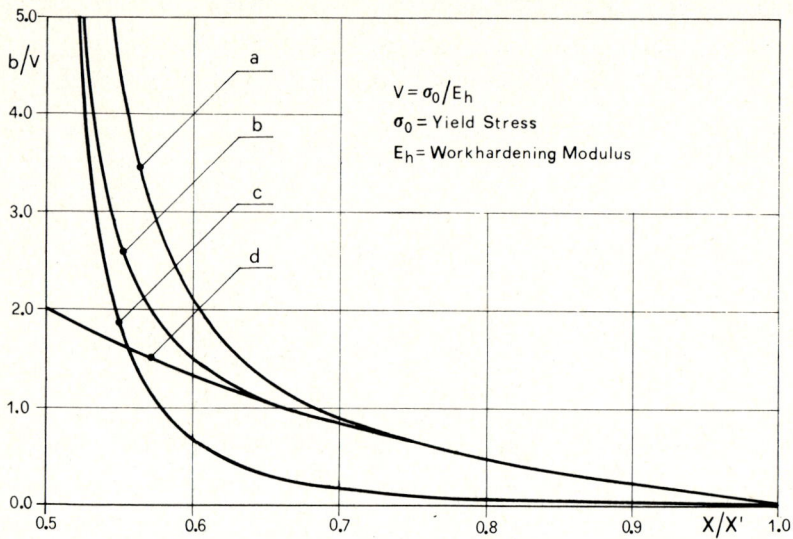

Figure 2 - Optimum Design of the Bar of Figure 1. Ratio X/X' as a Function of the Upper Limit b for the Four Cases (a,b,c,d)

5. CONCLUSION

In the present paper we have formulated the problem of optimum plastic design assuming the workhardening adaptation concept as design criterion and imposing limitations on deformation parameters, such as displacements, plastic strain and plastic strain intensity. The structure is conceived as a discrete model formed up by rigid-plastic elements all characterized by a piecewise linear yield surface and a piecewise linear workhardening law. The load acts statically or even dynamically, along an unknown path which develops inside a given loading domain.

The above limitations on deformation are stated by introducing suitable upper limits suggested by practice or other safety criteria (such as serviceability and fatigue, for instance) and then applying these limits to some time-independent functions each of which is an upper bound to a real deformation parameter. Such bounding functions are constructed according to a perturbation method [6, 7] which requires modifying the yield surface of one or more elements in a suitable manner depending on what kind of deformation parameter is to be controlled.

We have formulated a bounding theorem which is a unified version of several theorems of workhardening adaptation theory. It indicates how the bounding functions can be constructed and the associated perturbations specified. In formulating the optimization problem we have introduced factorized loads, each load multiplier having to be specified in connection with the relevant deformation parameter to be controlled and the associated limit state (working or failure state).

In spite of the simplifying hypothesis introduced and linearizations operated, the optimization problem proves to be a nonlinear and in general also a nonconvex one. The Kuhn-Tucker conditions are written out and utilized for solving the design problem of a simple bar. Computational aspects have been disregarded.

Also on the basis of the example, it seems reasonable to consider the optimum design for workhardening adaptation to be able to produce fairly more economic optimal structures than the classical limit design can do, also in the presence of imposed limitations on deformation. The problem is far from being completely clarified even from the theoretical point of view and further concern is hoped for. In the author's opinion, two main questions seem to deserve a deeper consideration: one is a better assessment of the workhardening adaptation design reliability in engineering practice, the other is providing an efficient computational procedure.

ACKNOWLEDGEMENTS

The results presented in this paper were obtained in the course of a research project sponsored by the National (Italian) Research Council, C.N.R., PAdIS Committee.

REFERENCES

[1] PRAGER, W., "Bauschinger Adaptation of Rigid, Workhardening Trusses", *Mech. Res. Comm.*, Vol. 1, 1974, pp. 253-256.

[2] PRAGER, W., "Adaptation Bauschinger d'un solide plastique à écrouissage cinématique", *C.R. Acad. Sc.*, Paris, t.280, Serie B, 1975, pp. 585-587.

[3] POLIZZOTTO, C., "Workhardening Adaptation of Rigid-Plastic Structures", *Meccanica*, Vol. 10, No. 4, 1975, pp. 280-288.

[4] KÖNIG, J.A. and MAIER, G., "Adaptation of Rigid-Workhardening Discrete Structures Subjected to Load and Temperature Cycles and Second-Order Geometric Effects", *Computer Meth. Appl. Mech. Eng.*, Vol. 8, 1976, pp. 37-50.

[5] POLIZZOTTO, C., "Adaptation of Rigid-Plastic Continua under Dynamic Loadings", *J. Struc. Mech.*, Vol. 6, No. 3, 1978, pp. 319-329.

[6] POLIZZOTTO, C., "On Workhardening Adaptation of Discrete Structures under Dynamic Loadings", *Tech. Rep. SISTAR-78-OMS-6*, Facoltà di Architettura di Palermo, Palermo, July, 1978, and *Archives of Mechanics* (accepted for publication).

[7] MAZZARELLA, C. and PANZECA, T., "Adaptation of Rigid-Plastic Discrete Structures Subjected to Dynamic Loadings and Second-Order Geometric Effects", *J. Struc. Mech.*, (submitted t

[8] PRAGER, W., "Optimal Plastic Design of Trusses for Bauschinger Adaptation", from *Omaggio a Carlo Ferrari*, Libreria Editrice Universitaria Levrotto and Bella, Torino, 1974.

[9] POLIZZOTTO, C., MAZZARELLA, C. and PANZECA, T., "Optimum Design for Workhardening Adaptation", *Computer Meth. Appl. Mech. Eng.*, Vol. 12, 1977, pp. 129-144.

[10] MAIER, G., "A Matrix Structural Theory of Piecewise Linear Elastoplasticity with Interacting Yield Planes", *Meccanica*, Vol. 5, 1970, pp. 54-66.

[11] POLIZZOTTO, C., "Optimum Plastic Design of Structures under Combined Stresses", *Int. J. Solids and Structures*, Vol. 11, 1975, pp. 539-553.

[12] POLIZZOTTO, C., "Optimum Plastic Design for Multiple Sets of Loads", *Meccanica*, Vol. 9, No. 3, 1974, pp. 206-213.

[13] DRUCKER, D.C. and SHIELD, R.T., "Design for Minimum Weight", *Proc. Int. Congr. Appl. Mech.*, Book 5, Brussels, 1956, pp. 212-222.

[14] MANGASARIAN, O.L., *Nonlinear Programming*, McGraw-Hill, New York, 1969.

[15] BRYSON, A.E., Jr. and HO, Y.C., *Applied Optimal Control*, John Wiley and Sons, New York, 1975.

STRUCTURAL CONTROL, H.H.E. Leipholz (ed.)
North-Holland Publishing Company & SM Publications
© IUTAM, 1980

AN ACTIVE SYSTEM FOR NEUTRALIZATION OF VIBRATORY
FORCE INTERACTIONS IN STRUCTURES

J.W. Roberts and M.C. Borgohain

School of Engineering
University of Edinburgh
Edinburgh EH9 3JL, Scotland

1. INTRODUCTION

Dynamic forces arising from unbalance and similar effects in structure borne machinery may cause vibration of the primary structure and additional problems with radiated noise. The traditional solution is to make use of anti-vibration mountings for the internal equipment. However, alignment difficulties and other problems arise from the excessive flexibility of machinery on anti-vibration mountings and impose limits on the extent to which passive anti-vibration techniques may be used.

The present paper discusses an exploratory application of active feedback control to achieve some degree of neutralization of vibratory interactive forces. This may be envisaged as an alternative to the A/V mounting approach or as a supplement to obtain additional attenuation over a specific bandwidth. Active control methods have been proposed as a means of reducing structural vibration by increasing the damping of structures [1 - 7]. The force neutralization approach reported here is novel, and has the advantage of being effective at both resonant and non-resonant frequencies of excitation. The basic scheme is illustrated in

Figure 1. Primary dynamic forces transmitted between a mounted machine and structure are detected by a force gauge and used to derive a cancelling force to be applied at an adjacent point. The cancelling force generator is necessarily required to be 'structure-borne', i.e., the cancelling force reaction is sustained by the body of the actuator unit flexibly suspended from the structure itself as indicated in Figure 2. Consequently, even in the case of ideal performance there will be a residual transmitted force through the suspension point. This may be made arbitrarily small by design. It is seen that in this approach the problems of excessive flexibility and static deflection are transferred from the machine to the sacrificial cancelling force actuator unit. (For convenience we use the abbreviations AFC for the active force cancelling system and SBU for the structure-borne actuator unit.)

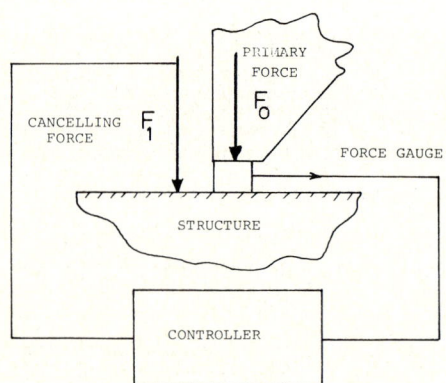

Figure 1 - Basic AFC Scheme

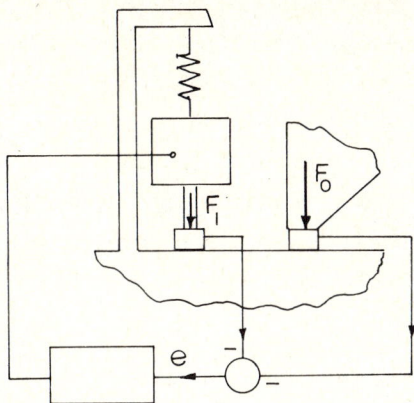

Figure 2 - AFC with Feedback Control and Structure-Borne Force Cancelling Unit

Figure 2 shows the arrangement implemented to operate with a wide band periodic or random primary excitation force F_0. A second force gauge is used to detect the cancelling force F_1 in a feedback configuration. In the exploratory work we have considered a single mounting point in isolation. Here we present an analysis of the AFC system and its interaction with the dynamic characteristics of the structure.

2. AFC SYSTEM ON A HIGH IMPEDANCE STRUCTURE

It is convenient to first assume negligible motion of the structure at the machine mounting point. The AFC system is assumed to be based upon an electromechanical force actuator driven by a power amplifier with (variable) voltage gain g from a force error signal e. The various parameters of the SBU are defined in Figure 3(a), F_m being the electromagnetic force due to the coil current I and F_1 the cancelling force detected by the force gauge. Figure 3(b) shows the equivalent output circuit of the power amplifier and load having a total electrical impedance $z_E \equiv z_O + z_C$. A transfer function may be derived between F_1 and the force error e in the form:

$$F_1(s) = G_1(s)e(s), \qquad (1)$$

where

$$G_1(s) = \frac{\frac{\mu g}{Z_E}\left[(K_A+sC_A)(Ms^2+C_1s+K_1)-(K_v+sC_v)(m_c s^2+C_A s+K_A)\right]}{\left[(Ms^2+C_1s+K_1)(m_c s^2+C_A s+K_A)\right] + \frac{\mu\eta s}{Z_E}\left[(M+m_c)s^2+(C_1+C_A)s+(K_1+K_A)\right]} \qquad (2)$$

In (2), μ is the electromagnetic force per unit current of the actuator, η is the back e.m.f. voltage per unit relative velocity of the armature and s is the complex argument of the Laplace transforms. Also $K_1 \equiv K+K_v$; $C_1 \equiv C+C_v$. Equation (2) defines a complex frequency response function between F_1 and e with the following approximations. At low frequencies in the neighbourhood of the SBU suspension frequency $\sqrt{K_1/M}$

$$G_1(j\omega) \approx \frac{\mu g}{Z_E}\left[\frac{K-M\omega^2+j\omega C}{K_1-M\omega^2+j\omega\left(C_1+\frac{\mu\eta}{Z_E}\right)}\right]. \qquad (3)$$

At high frequencies

$$G_1(j\omega) \approx \frac{\mu g}{Z_E}\left[\frac{K_A+j\omega C_A}{K_A-m_c\omega^2+j\omega\left(C_A+\frac{\mu\eta}{Z_E}\right)}\right]. \qquad (4)$$

At intermediate frequencies

$$G_1(j\omega) \approx \frac{\mu g}{Z_E}. \qquad (5)$$

Typical frequency response behaviour is characterized by (a) a low frequency resonance and anti-resonance in the vicinity of the SBU suspension frequency. This may be set at a low value by choice of

M and K. (b) a high frequency resonance determined by the mass and stiffness distribution in the actuator armature which imposes an upper limit to the operating bandwidth of the SBU. Figure 4 shows a typical SBU frequency response plot which confirms this behaviour. The SBU may be incorporated into the envisaged AFC feedback system provided additional compensation filtering is added to achieve stable operation with an adequate amount of open loop gain. It is therefore assumed that a compensation filter with transfer function $G_2(s)$ is added before the power amplifier. $G_2(s)$ will depend on the actual resonant frequencies and the desired AFC system bandwidth in any application. Now writing

$$F_1 = G_1(s)G_2(s)e \equiv k\,G(s)e\;. \tag{6}$$

$G(s)$ is the combined transfer function of the SBU with associated gain multiplier k. In the standard feedback system configuration of Figure 2, then,

$$e = -(F_0+F_1)\;, \tag{7}$$

and

$$F_1 = \frac{kG}{1+kG}(-F_0)\;. \tag{8}$$

If $F_R \equiv (F_0+F_1)$ = Resultant Structure Force, it is convenient to define a Force Reduction Ratio $R(j\omega)$ as a frequency domain performance index,

$$R(j\omega) \equiv |F_R/F_0| = \left|\frac{1}{1+kG(j\omega)}\right|\;. \tag{9}$$

Good performance at any frequency requires high open loop gain, $|kG(j\omega)| \gg 1$, the limit being determined by stability considerations. Figure 5 illustrates a typical AFC reduction ratio plot with a system operating near its stability limit.

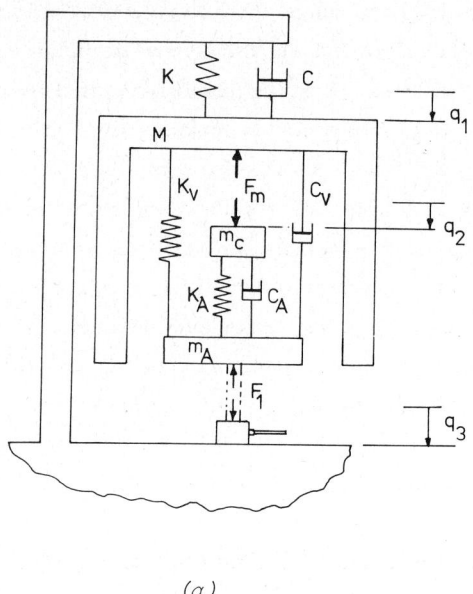

(a)

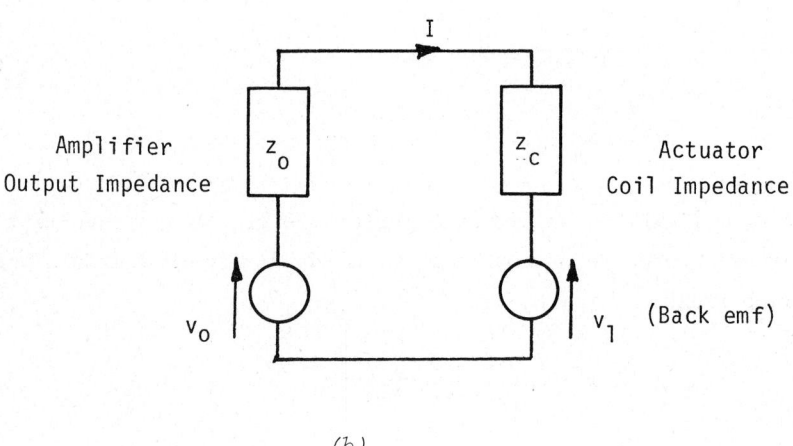

(b)

Figure 3 - (a) Dynamic Model of SBU
(b) SBU Electrical Circuit

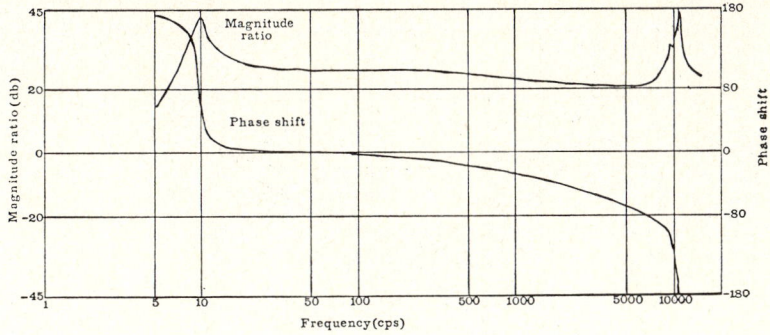

Figure 4 - SBU on Infinite Impedance: Experimental Transfer Characteristics without Filters

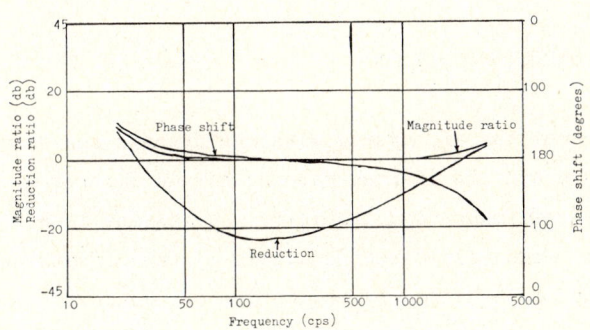

Figure 5(a) - SBU on Infinite Impedance: Closed loop performance at maximum stable gain with 2-pole high mass filter cut-off 112 cps and 1-pole Low pass filter cut-off 160 cps.

(continued)

(continued)

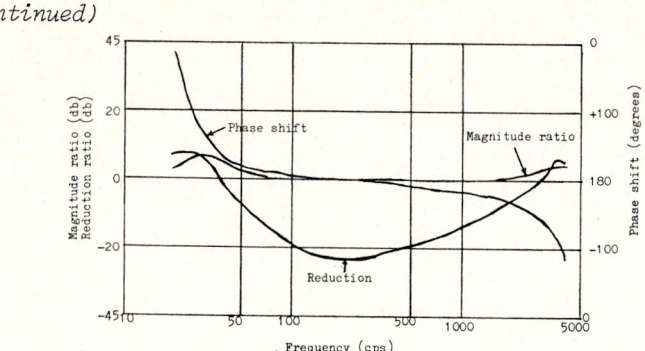

Figure 5(b) - *SBU on Infinite Impedance: Closed loop performance at maximum stable gain with 2-pole high pass filter cut-off 112 cps and 1-pole low pass filter cut-off 500 cps.*

3. EFFECT OF STRUCTURE MOTION

The interaction of the AFC system with the dynamic characteristics of the structure and the mounted machine has an important effect on the system performance and stability. We assume linear structure behaviour. The most direct analysis of the effects of the finite structure is obtained by making use of mechanical impedance functions. We assume that at the mounting point the structure has an absolutely integrable impulse response, implying a well behaved transfer function between force and displacement. With the substitution $j\omega \equiv s$ the corresponding steady state frequency response function is obtained which also yields the steady state frequency response function between force and velocity at the point, the latter being the reciprocal of the mechanical impedance function. It is convenient to write the differential equations directly in terms of the Laplace Transforms of corresponding quantities using mechanical impedances generalized as functions of s. This is implicit in the following. Introduce coordinate q_3 as indicated in Figure 3(a) to represent structure motion. Let f_0 and f_1 now represent the primary and cancelling forces transmitted by the force gauges for $q_3 \neq 0$. Then reconstruction of the system equations leads to

$$f_1 = kGe - \left[ms^2 + \left(C_v + \frac{\mu n}{z_E} \right) s + K_v \right] q_3 , \qquad (10)$$

where $m \equiv m_A + m_C$. In arriving at (10) a number of terms have been considered to have negligible effect within the bandwidth of the FC system determined by G. Equation (10) has the form

$$f_1 = F_1 - Z_1 \, sq_3 , \qquad (11)$$

where Z_1 is the equivalent source mechanical impedance of the SBU and F_1 is the SBU output force in the case of infinite structure impedance. In the case of transmitted force detected by the primary force gauge, the former is now an integral component in a complex dynamical system. We assume linearity in the force generating mechanism and consider the primary force as a 'blocked' force F_0, together with an associated source mechanical impedance Z_0. F_0 is then the force developed at the force gauge for the case of an infinite impedance structure, while the actual force transmitted has the form

$$f_0 = F_0 - Z_0 \, sq_3 , \qquad (12)$$

F_0 depends on the structure response and cannot be regarded as an arbitrary system input. With Z_3 the structure mechanical impedance at the mounting point, then

$$q_3 = \frac{(f_0 + f_1)}{s \, Z_3} + q_3^0 . \qquad (13)$$

q_3^0 is included as an additive structure motion arising from other excitation sources. System equations may be formulated in terms of F_0 and $f_R \equiv (f_0 + f_1)$. We obtain

$$f_R = \frac{1}{\left[1 + \frac{Z_0}{Z_3}\right]\left[1 + \frac{kG+Z_1/Z_3}{1+Z_0/Z_3}\right]} [F_0 - s(Z_0+Z_1)q_3^0] . \quad (14)$$

F_0 is seen to be attenuated by a combination of a passive impedance factor and an active factor from the AFC system. Consider the steady-state cancelling performance with $q_3^0 = 0$. Then the reduction ratio $R(j\omega) \equiv f_R/f_R$ (no AFC) is given by:

$$R(j\omega) = \frac{1}{1 + \frac{kG+Z_1/Z_3}{1+Z_0/Z_3}} . \quad (15)$$

This shows clearly how the various mechanical impedances interact with the active system to affect performance. We expect $Z_1 \ll Z_3$ so that the ratio Z_0/Z_3 is the significant impedance effect in (15). For good performance it is required that

$$\left|\frac{kG \, Z_3}{Z_0+Z_3}\right| \gg 1 . \quad (16)$$

For any given AFC system function kG, it is seen that R is improved at frequencies for which (Z_0+Z_3) has minima, i.e., resonant frequencies of the loaded structure, but will be degraded at frequencies corresponding to minima of Z_3, i.e., resonant frequencies of the unloaded structure. Generally performance will deteriorate with increasing magnitude of the ratio Z_0/Z_3.

System stability will also be influenced by the finite structure impedance, and this may imply a reduction in the operating gain k of the active system with a corresponding degradation in cancelling performance in addition to that due to the impedance function in (15). The characteristic equation is given by

$$\left[1 + \frac{Z_0}{Z_3}\right]\left[1 + \frac{kG+Z_1/Z_3}{1 + \frac{Z_0}{Z_3}}\right] = 0 . \quad (17)$$

The first factor in (17) implies stable characteristic roots. In the second the characteristic roots are functions of the mechanical impedances and the AFC system transfer function. The effect of the mechanical impedances on system stability may be determined from root locus plots of these characteristic roots in the s-plane. A simpler alternative is to interpret f_R in (14) as the error function of a feedback system with input given by $F_0/[1 + \frac{Z_0}{Z_3}]$. The degree of stability of the system may then be inferred from the behaviour of the equivalent open loop frequency response function given by

$$\left[\frac{kG + Z_1/Z_3}{1 + \frac{Z_0}{Z_3}} \right],$$

in the complex plane, using simple Nyquist stability considerations. This form of stability analysis is particularly amenable to applications where the relevant mechanical impedances are available as measured frequency response data. An even simpler analysis is possible if the ratio Z_1/Z_3 is negligibly small, in which case the AFC gain multiplier k operates on a modified system transfer function $[\frac{G Z_3}{Z_0 + Z_3}]$, and complete system stability and closed loop performance data may be determined from a single frequency response plot of the latter. If Z_0 is an appreciable fraction of Z_3 then it is clear that strong destabilizing effects will arise from fluctuations in Z_3 due to structure resonant frequencies within the system passband. These will be particularly severe if the structure is lightly damped. It may also be deduced that the AFC system will give a better performance as a supplementary system for a machine on an A/V mounting than for the directly mounted machine, since the primary source mechanical impedance will invariably be lower in the former case.

Finally, we consider the effect of externally generated motion at the mounting point on the system behaviour. With $F_0 = 0$

and $q_3^0 \neq 0$ in (14) it is seen that the AFC system is activated by such motional inputs. As a result of the nonzero impedance loading on the force gauges, forces are induced by the motional input and the system reacts by trying to neutralize them. This will generally be an undesirable feature of the AFC system. The dynamic response of the structure to external forces will be altered by the presence of AFC systems, and a danger will exist of any AFC system being saturated by motionally induced loadings and thus being prevented from operating against internal forces.

4. MOTIONAL COMPENSATION TECHNIQUE

A possibility exists for improving the AFC system performance in the presence of the mechanical impedance effects discussed in the previous section. The structure motion at the mounting point may be detected using for example an accelerometer and a filtered motion signal may be used to compensate the AFC error signal. The system is shown schematically in Figure 6. Let the modified error signal be

$$\bar{e} = e - h(s) \, s^2 q_3 , \qquad (18)$$

with filter characteristic chosen to be

$$h(s) = \left[\frac{Z_0 + Z_1}{s} \right] . \qquad (19)$$

Then,

$$F_1 = kG\bar{e} = -kG[F_0 + F_1] , \qquad (20)$$

and the resultant structure interactive force f_R may be derived as

$$f_R = \frac{1}{\left[1 + \frac{Z_0 + Z_1}{Z_3} \right]} \left[\frac{F_0}{1 + kG} - (Z_0 + Z_1) s q_3^0 \right] . \qquad (21)$$

This expression for f_R with motional compensation shows a considerable advantage. The Reduction Ratio for steady state cancelling performance is seen to be

$$R(j\omega) = \frac{1}{1+kG(j\omega)}, \qquad (22)$$

and the system characteristic equation is given by

$$1 + kG(s) = 0 . \qquad (23)$$

Both of these expressions correspond to the case of the AFC system on an infinite impedance structure. Finally, it is noted that with $F_0 = 0$ in (21) it is seen that the force resultant arising from externally generated motion at the mounting point is independent of kG and involves only passive mechanical impedances. The active system no longer reacts to the forces induced by impedance loading of the force gauge. The problem of saturation of the AFC system by externally generated motion is diminished, but not removed entirely since the force gauges and their circuits will still respond to these inputs. In practical situations difficulty may exist in realising the form of compensation filter h(s). Generally, Z_1 in (19) will be negligible compared with Z_0. Z_0 will have a particularly simple form if the source impedance may be represented by a mass rigidly coupled to the force gauge, or a mass coupled through an A/V mounting.

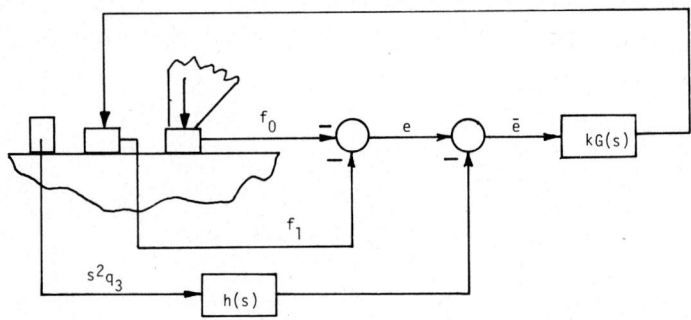

Figure 6 - Accelerometer Compensation

5. CONCLUSION

The paper has described how active control may be applied to the problem of vibration reduction of structures under the action of internal dynamic forces. An analysis of the basic scheme has been presented and the interaction of the active system with the dynamic characteristics of the structure has been analyzed from the point of view of stability and steady state performance in terms of mechanical impedance functions.

Experimental work has been carried out on a number of pilot AFC systems and results obtained are encouraging. These will be presented in a further publication in due course.

REFERENCES

[1] ROCKWELL, T.H. and LOWTHER, J.M., "Theoretical and Experimental Results on Active Vibration Dampers", *J. Acoust. Soc. America*, Vol. 36, 1964, pp. 1507-1515.

[2] LOWTHER, J.M. and ROCKWELL, T.H., "Compensation Technique for Active Damping Improvement", *J. Acoust. Soc. America*, Vol. 38, 1965, pp. 481-482.

[3] ROCKWELL, T.H., "Investigation of Structure Borne Active Vibration Dampers", *J. Acoust. Soc. America*, Vol. 38, 1965, pp. 623-628.

[4] ABU-AKEEL, A.K., "The Electrodynamic Vibration Absorber as a Passive or Active Device", *Trans. ASME, J. Eng. for Ind.*, 1967, pp. 741-753.

[5] OLSON, H.F., "Electronic Control of Noise, Vibration and Reverberation", *J. Acoust. Soc. America*, Vol. 28, 1956.

[6] KRYAZEV, A.S. and TARTAKOVSKII, B.D., "Application of Electromechanical Feedback for the Damping of Flexural Vibrations in Rods", *Soviet Physics-Acoustics*, Vol. 17, 1965, pp. 150-154.

[7] WALKER, L.A. and YANESKE, P.P., "Characteristics of an Active Feedback System for the Control of Plate Vibrations", *J. Sound and Vibration*, Vol. 46, 1976, pp. 157-176.

STRUCTURAL CONTROL, H.H.E. Leipholz (ed.)
North-Holland Publishing Company & SM Publications
© IUTAM, 1980

EXPERIMENTS IN FEEDBACK CONTROL OF STRUCTURES

John Roorda

Department of Civil Engineering
University of Waterloo
Waterloo, Ontario, Canada

1. INTRODUCTION

Researchers in the relatively new field of automatic vibration control of structures must have their feet planted firmly in two worlds - the real, everyday working world and the abstract, scientifically oriented world. Problems with excessive deflections, vibrations or accelerations originate in the real world and should be defined on the basis of information gathered in the real world. Such problems are then analyzed with the help of various mechanics principles of the abstract world to produce idealized representation of the real problems. These idealized abstractions are then manipulated in an effort to simulate reality, and hopefully a prediction or expectation will emerge from this activity. Expected behaviours must then be verified in the real world by evaluation and testing.

While considerable progress is currently being made on the theoretical side [1, 2, 4, 5, 8 - 10, 12 - 18], very little experimental work [3, 6, 7, 11] is being pursued in the feedback vibration control field. Yet, if future designers are to accept such new concepts they must be thoroughly convinced of the benefits and advantages of feedback structural control. Hence researchers must do experiments. These could take the form of large scale tests on complete buildings or other structures, or small scale laboratory set-ups that are easily manageable and relatively inexpensive. Tests of either type can give good insights and generate new ideas if they are done with logic and care. Tests will also expose dangerous pitfalls and uncover phenomena not hither to encountered or explained.

The purpose of this paper is to present the results of a number of small scale experiments designed and performed under the author's supervision during the last six or seven years. They range from a very simple deflection control scheme that utilizes a human controller to more complex electro-hydraulic and electro-magnetic control devices. In view of space limitations the description of each experiment is necessarily brief, touching only upon the basic structural and control components. The results, some of which are preliminary in nature, are presented in condensed graphical form with a minimum of discussion. Comparisons with theoretical work is here and there incomplete and in some instances non-existant. The main thrust in this presentation is to acquaint the reader with the main features of the tests and to give him an appreciation of how structural control works and what the inherent complexities are that beset experiments in this field.

2. MANUAL DEFLECTION CONTROL OF A KING POST TRUSS

The first experiment demonstrates in a simple way the essential ingredients of an active feedback control system. It involves the control of the midspan deflection of a King post truss by selectively lengthening or shortening the underslung cable in a controlled way. Figure 1 shows the experimental set-up. The standard steel C_4 x 5.4 channel beam has length $L = 72$ in., is simply supported at its ends and receives additional support at midspan from a vertical post which is fixed to the beam. The post, in turn, is supported by cables connected to the ends of the beam at an angle θ = 10°. At one end the cable is tied to a threaded eyebolt and nut, allowing the tension in the cable to be changed. The cable is free to slide over the bottom end of the post. Two dial gauges are used in this demonstration experiment, one to register the midspan deflection (Δ); the other to register the movement of the eyebolt (u). A concentrated load W = 185 lbs. is applied at a variable distance ξ from the left end of the structure.

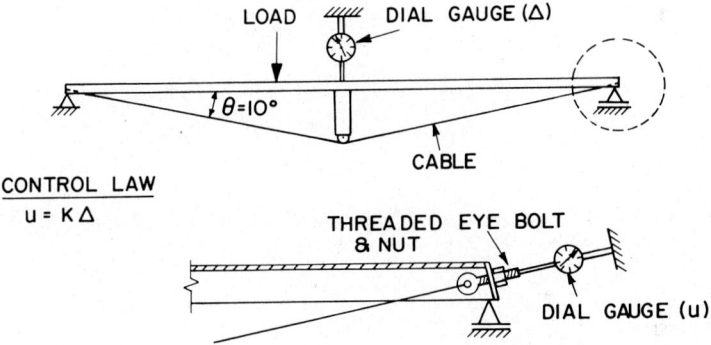

Figure 1 - *Components of the Manual Deflection Control Equipment*

If the linear control equation

$$u = K\Delta \; , \tag{1}$$

is introduced, then the structural properties of the system can be conveniently drawn in the form of the block diagram shown in Figure 2. Here EI is the bending stiffness of the beam and A_c and E_c are the effective cross-sectional area and the effective

Young's modulus of the cable. It should be noted that stranded, galvanized steel cable of diameter 1/8 in. and with a cross-section made up of 7 bundles of 19 strands each was used. This cable had the peculiar nonlinear load versus elongation property shown in Figure 3. The operation of the experiment took place with the cable tension in the linear range.

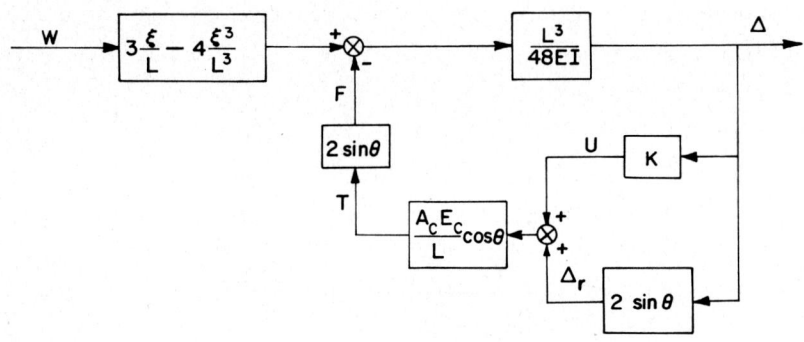

Figure 2 - Block Diagram Showing the Feedback Effects of the Control and the Post-Cable Substructure

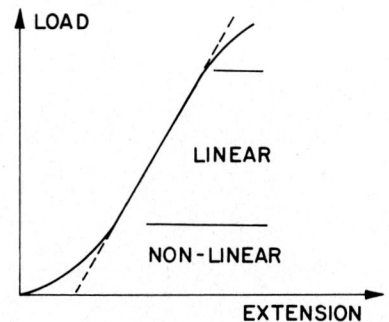

Figure 3 - Cable Properties

With this experimental set-up the essential features of a control process are easily demonstrated. The graphs in Figure 4 illustrate the behaviour of the midspan deflection (Δ) and the control variable (u) for a "step" load (W). In these graphs an artificial time interval (T) is introduced during which the human controller (the experimenter) takes certain necessary actions. During the first interval the load is applied, and a central deflection registers on the dial gauge. At time T the controller calculates the required u based on equation (1). During the second time interval the controller turns the nut on the eyebolt to achieve the required u in the structure. A new midspan deflection will result, at which point a new calculation is made for u and the same applied to the structure during the third time interval, and so on. The process is an iterative one that very quickly settles down to the final controlled deflection for the given load.

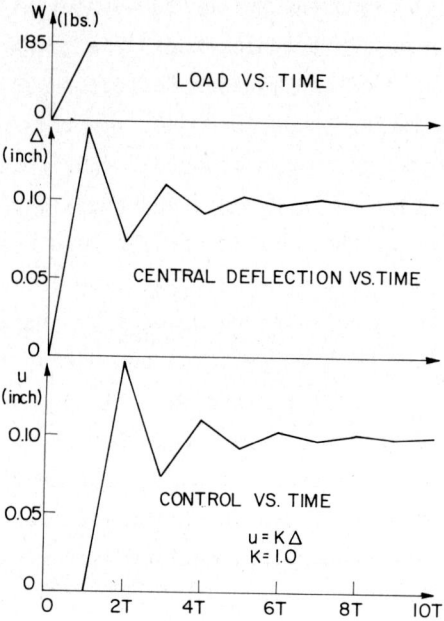

Figure 4 - *Typical Pseudo-Dynamical Behaviour of Manual Control Experiment*

The deflection response in this pseudo-dynamic experiment is reminiscent of the response of a damped spring-mass system to a step excitation. The damping effect emerges as the result of the delayed reaction on the part of the controller. Some experimental influence lines for central deflection are shown in Figure 5 for (a) the beam without cables, (b) the beam with cables installed at 375 lbs. pretension (K=0), (c) for non-zero positive values of K. In this plot q is a convenient dimensionless mid-span deflection given by

$$q = \frac{48EI}{WL^3} \Delta \ . \qquad (2)$$

A theoretical expression for q can readily be derived from the information in Figure 2 and is given by

$$q = \frac{3y-4y^3}{1+2\beta(2\sin\theta+K)\sin\theta\cos\theta}, \qquad (3)$$

where $y = \xi/L$ and β is the dimensionless ratio of cable to beam stiffness given by

$$\beta = \frac{A_c E_c L^2}{48EI} = 1.406 . \qquad (4)$$

The theoretical values lie very close to the experimentally obtained curves.

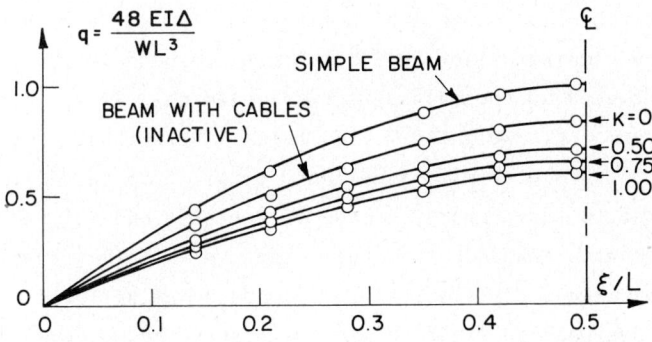

Figure 5 - *Experimental Influence Lines for a Range of Control Constants*

These results indicate that the central deflection can be reduced substantially even in such a simple control scheme. More importantly, all the basic ingredients of feedback structural control are demonstrated in a simple way. These include the *controller*, which constitutes the brain and brute-force portion of the system (in this experiment a human controller), the *control "force"* (a cable contraction), the controlled system or *plant* (i.e., the loaded beam plus its substructure) and the output or *controlled variable* (the midspan deflection). The human operator determines the proper control action to be applied based upon actual deflection information received via a deflection sensor. This experiment lends itself to further development and refinement for purposes of classroom demonstration.

In the next experiment the human operator is replaced by an electro-hydraulic servo mechanism. The resulting system therefore becomes an automatic control system.

3. AUTOMATIC VIBRATION CONTROL OF A KING POST TRUSS [1, 15]

Figure 6 shows a schematic diagram of the experimental set-up for feedback vibration control of a King post truss. For this experiment an aluminum alloy 6063-T5 channel beam of length L = 20 ft. is used as the main structural element. (I = 0.4740 in.^{4}, E = 10^7 lbs/in., unit weight = 0.175 lbs/in.) The tendons are made of 1/8 in. diameter mild steel and stretch from the 24 in. long King post to the supporting base on one side and to the hydraulic actuator shaft on the other side. With this arrangement the beam experiences only lateral forces and no axial force. The tendons are designed to yield before the forces they transmit can damage the beam. The beam is supported through a bearing system that allows free rotation at one end and free horizontal translation plus rotation at the opposite end.

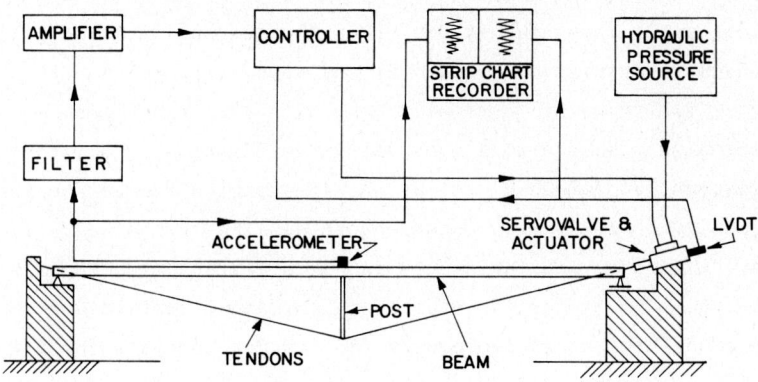

Figure 6 - Schematic Diagram of the Components used for Electro-Hydraulic Control of Beam Deflection

An accelerometer (Q-FLEX QA-116-15) fixed to the beam at midspan is used to sense the motion. The accelerometer signal is filtered, amplified and then fed into an MTS 406-11 controller which governs the activity of the servo-valve (MOOG 73-232) - actuator (MILLER 7004-2119) combination. A COLLINS LVDT monitors the actuator motion continuously. This motion, along with the accelerometer signal, is recorded on a 2 channel strip chart recorder as indicated on the schematic. The transfer characteristics of the low-pass filter coupled with the electro-hydraulic control device has the net effect of creating a phase shift, or delay, between the accelerometer signal and the actuator motion, thus giving an excellent damping effect.

Two adjustments can be made in the control loop of this experiment, these being the *gain* and the *span* adjustments. The gain control governs the speed with which the actuator responds to an incoming signal. A high gain setting means a quick response. The span setting governs the magnitude of the actuator displacement in response to a given signal and therefore determines the degree of control applied to the structure. The actuator displacement is a linear function of the input signal with the proportionality constant determined by the span setting.

Two types of tests were done with this experiment - (a) vibration response to a non-periodic excitation, and (b) vibration response to harmonic excitation. The excitation in the first test consisted simply of a manual pulse applied to the beam at, or near, midspan. This pulse was in essence a sudden push downward on the beam, immediately after which manual contact with the beam was broken and subsequent free vibration was allowed to take place. For three different gain levels (2, 4 and 6) the span setting was increased in finite steps from zero until an inherent dynamic instability was observed.

Some typical responses are shown in Figure 7 for gain setting 4 and span settings 0 (no control) 100, 300, and 500. The accelerometer trace of the uncontrolled structure responds in a gradually diminishing free vibration mode at a frequency of 7.3 Hz with a damping factor at a low 0.63% of critical. The actuator trace shows little or no movement at span zero. A span setting of 100 has a considerable effect on the response. After about 5 cycles the first mode of vibration practically disappears. The damping ratio is increased more than ten-fold to approximately 8% of critical. In these graphs the midspan beam acceleration is positive in the *upward* direction and the actuator motion is positive when it causes *increasing* tension in the cables. The actuator movement lags behind the acceleration by about 250°. As the span setting is increased through 300 to 500 the fundamental mode damps out progressively more quickly at the expense of greater actuator motion. At the 500 setting the fundamental mode has all but disappeared after 2 cycles of oscillation. At span setting exceeding 500 the system experiences instabilities in higher frequency modes which tend to build up very quickly. This implies a net input of energy into these modes beyond a certain control level.

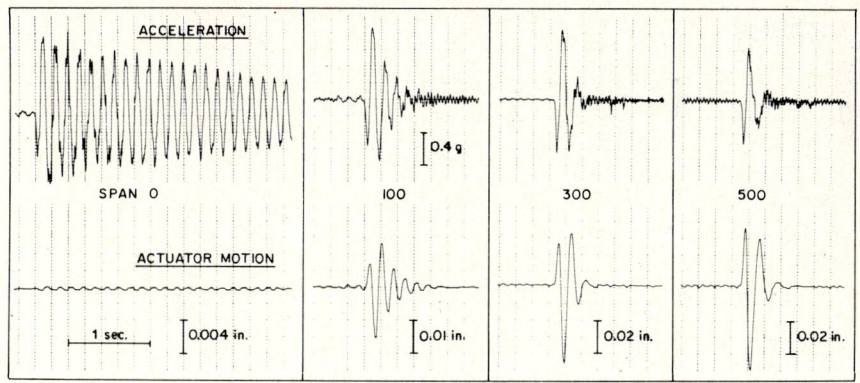

Figure 7 - Response to a Manually Applied Pulse, at Gain = 4 and Control Levels Ranging from 0 to 500

Typical traces of the acceleration and actuator responses of the controlled beam to a harmonic excitation are indicated in Figure 8 for a gain setting of 2. The excitation is supplied by a rotating unbalanced mass mounted on a small electric motor fixed to the beam near midspan. The motor speed was adjusted to 7 Hz to coincide approximately with the natural vibration frequency of the assembled structure. In Figure 8 the span settings of 0, 50, 100, 300 and 600 yield progressively lower acceleration responses with gradually increasing actuator movement. The actuator motion lags behing the accelerometer signal by approximately 300°. Beyond the 600 span setting an instability in a high frequency mode occurs requiring the experiment to be shut down.

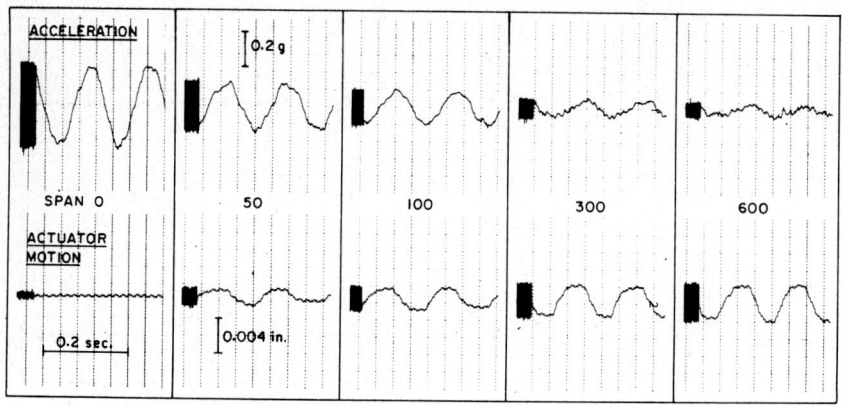

Figure 8 - Steady State Response to Sinusoidal Excitation, at Gain = 2 and Control Levels Ranging from 0 to 600

In Figure 9 the measured experimental results for steady state acceleration and actuator response are summarized for the full range of gain and span settings. From this figure it is clear that this control system is somewhat more effective at higher gain levels. For higher gains, however, the system becomes unstable at lower span levels. For example at gain 6, a span setting of 200+ yields instabilities, for gain 4 this occurs at span 500+ and for gain 2 only at 600+.

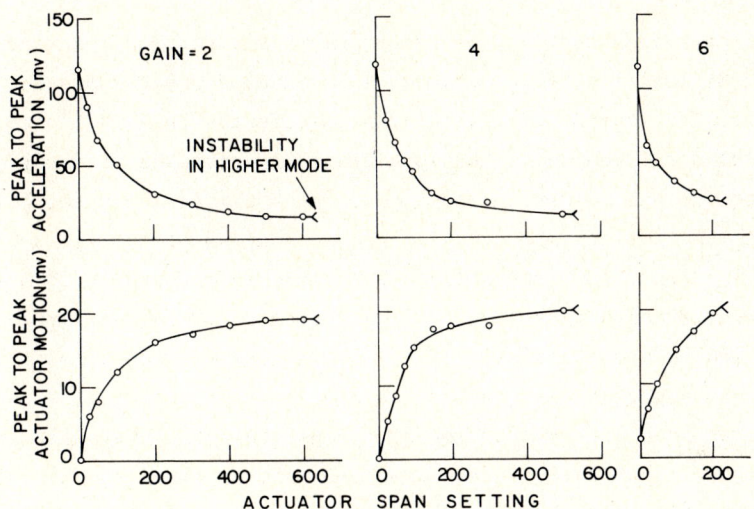

Figure 9 - *Summary of Peak Acceleration and Actuator Responses with Sinusoidal Excitation for the Full Range of Gain and Span Settings*

These results are generally speaking very encouraging. They demonstrate that an automatic control scheme of the type used here can be highly beneficial. In the extreme situation the fundamental mode response to harmonic excitation can be reduced to a low 15% of its uncontrolled value.

In order to eliminate the instabilities experienced in this experiment, additional sensors and perhaps a more rapidly responding electro-hydraulic servo-device must be designed into the system. A single accelerometer placed at midspan cannot monitor anti-symmetric modes or torsional modes of vibration. Nor can a single vertical control force applied to the beam at midspan hope to control such higher modes. In future work a more complex control scheme must be developed to cover a wider range of possible vibration modes.

4. TENDON CONTROL OF A FREE STANDING COLUMN [9, 14]

The primary component in this experiment is a vertical cantilever column of length 16 ft. and made of a standard steel cross-section S6 x 17.25. Control of this column is effected through a pair of vertical steel tendons fixed to a cross-arm attached at a variable distance up the column and to a yoke which pivots about the column centre line near the base. A schematic diagram of the set-up is shown in Figure 10. The solid steel tendons (diameter 0.1875 in.) are eccentrically attached, top and bottom at 12 in. and are pre-tensioned to about one half of the yield stress using a modified pre-stressing chuck of the type used in concrete work. The yoke is driven in a see-saw fashion by a single electro-hydraulic servo-ram, whose input derives from an accelerometer fixed to the column at a variable height. The accelerometer output is first conditioned through a low-pass filter and amplified to provide an adequate signal for the controller. Except for the filter, the ancillary equipment used to control this structure is essentially the same as that used in the previous beam experiment.

For the current control tests the gain level was fixed at 6. Two types of tests were again performed. In the first series the column was manually excited in the first mode up to a convenient amplitude while the control was active, then released and allowed to vibrate freely with the control loop functioning continuously. In the second test series a harmonic force was applied by way of an unbalanced mass rotating on the shaft of an elastic motor mounted at the top. The steady-state response to this force was monitored on a strip chart recorder.

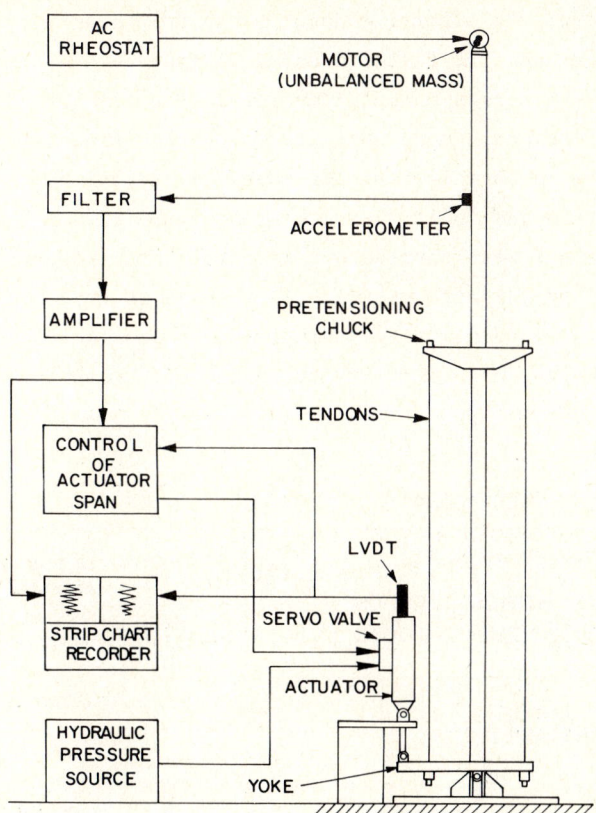

Figure 10 - Schematic Diagram of the Components used for Electro-Hydraulic Control of a Cantilever Beam

The free vibration tests were performed with the cross-arm and accelerometer both mounted at a height of 10 ft. The natural frequency of vibration for this configuration was approximately 3 Hz. At span setting zero (no control) a relatively high damping - 1.35% of critical - was observed, which may be attributed to unavoidable intercomponent friction losses.

Figure 11 shows the increase in logarithmic decrement as the control is increased through a range of span settings up to 100, at which point the effective damping has been nearly doubled. Typical traces of the filtered and amplified accelerometer output along with the corresponding actuator motion are also shown in Figure 11. Beyond a span setting of 100 the structure lost stability in an uncontrolled higher mode. A typical trace of the growth of this instability is given in Figure 12. Further free vibration tests were carried out with the accelerometer at several heights (11 ft., 12 ft., 13 ft., 14 ft. and 15 ft.), while the cross-arm remained at 10 ft. Similar behaviour was observed in each case, with an instability creeping in at some level of control.

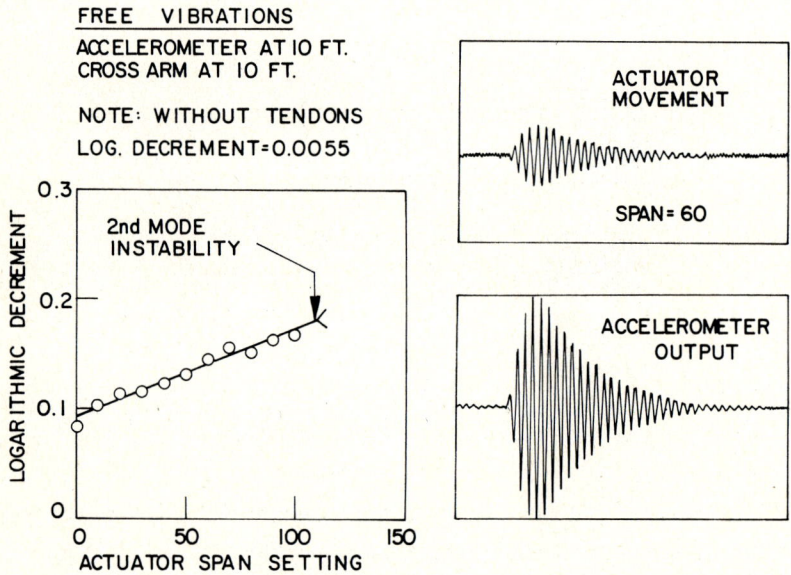

Figure 11 - The Effect of Control on Logarithmic Decrement. Typical Accelerometer and Actuator Outputs

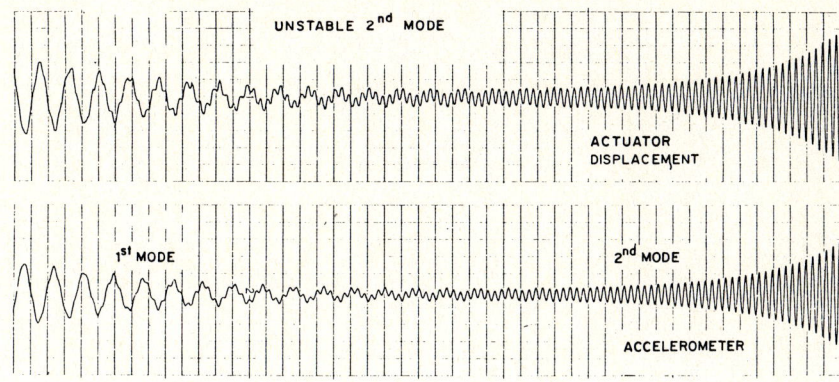

Figure 12 - Typical Trace of the Growth of an Instability

The steady-state tests were run with the cross-arm at two different heights - 10 ft. and 13 ft. - and a range of accelerometer heights. The tests were done by setting the controller span initially to zero (no control), running the unbalanced tip mass at first mode resonant speed, and then increasing the span setting in a stepwise fashion until a point of instability was reached.

Figure 13 shows the measured change in response for the 10 ft. cross-arm location and Figure 14 for the 13 ft. location. Results are plotted for a number of accelerometer locations. The trend of the curves is fairly constant in all cases. The level of control at which higher mode instabilities are experienced varies widely. With the cross-arm at 10 ft., the most effective accelerometer position is the 15 ft. level, where a maximum response attenuation of 75% is achieved at span setting 550. With the cross-arm at 13 ft. (near the node of the second mode of vibration) and the accelerometer at the same level, the greatest attentuation is achieved. An 85% reduction in response resulted at a span setting of 700.

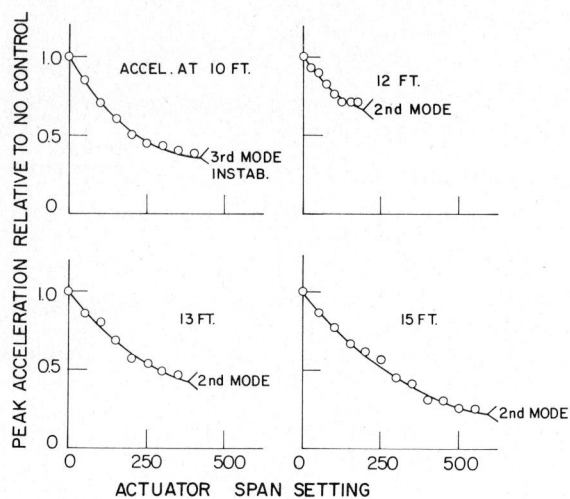

Figure 13 - Summary of Peak Acceleration Responses for Cross-Arm Location at 10 ft. and Various Accelerometer Locations

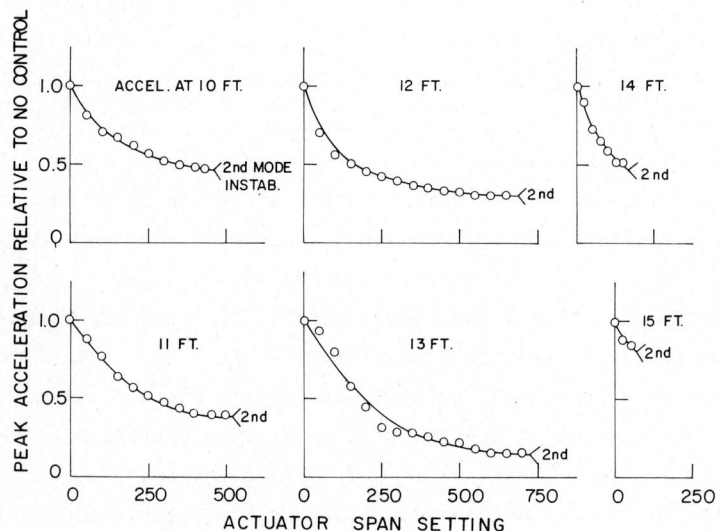

Figure 14 - Summary of Peak Acceleration Responses for Cross-Arm Location at 13 ft. and Various Accelerometer Locations

It is evident from these results that the troublesome instabilities at higher control levels arise very much as a function of the relative cross-arm and accelerometer locations, with some combinations being preferable to others. Figure 15 shows a typical trace of how the higher mode gradually builds up in a test of this kind. To pinpoint the underlying reason for this behaviour and to find a cure will require much further work, both theoretical and experimental.

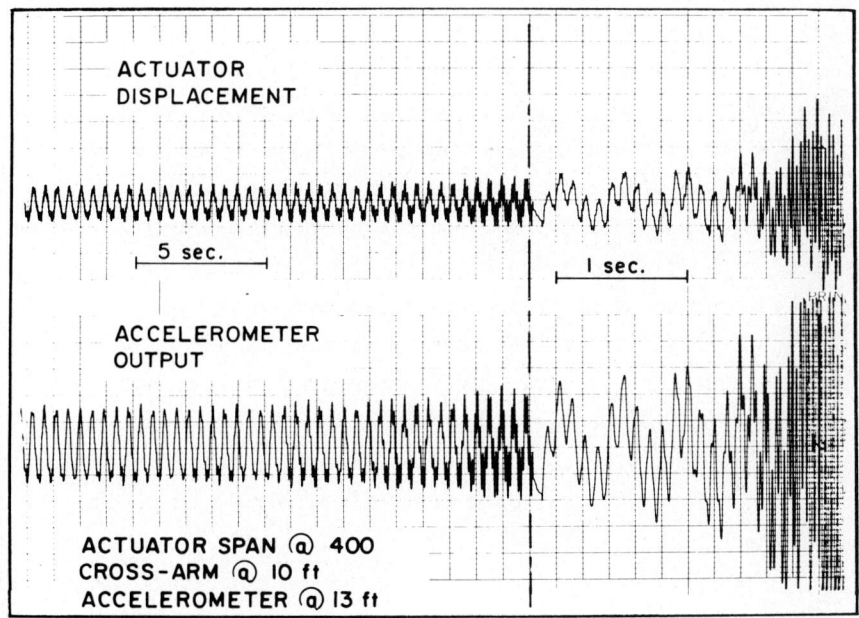

Figure 15 - *Typical Trace of the Growth of Unstable Modes in a Harmonic Excitation Test*

The results at this stage are promising. It appears that significant vibration control can be achieved in free standing columns using the tendon method, particularly in situations where only the fundamental mode is of importance.

5. AUTOMATIC CONTROL OF A CANTILEVER BEAM USING AN ELECTRO-MAGNETIC DEVICE [7]

The electro-hydraulic control schemes used in the beam and column experiments incorporated a mechanical connection in the form of cables or tendons to transmit a control force from the control device to the structure. These schemes involve a number of moving or rotating parts as well as fluid couplings connected in various ways. These ancillary mechanical components invariably introduce considerable passive damping into the system which, although of benefit in a real situation, tends to mask to a certain degree the change in the vibration properties that the experimenter is attempting to measure. Furthermore the addition of control cables tends to stiffen the structure considerably, even when the control is zero, thus giving rise to higher natural frequencies.

In the present small scale experiment a control force is applied, without an intervening mechanical connection, by electro-magnetic means to an aluminum beam of length 42 in. and cross-section 1 in. by 1/8 in. The beam is firmly clamped at one end to a frame, with the length lying in the horizontal plane and the 1 in. dimension in the vertical plane so that it vibrates in the horizontal plane. The control force, as well as the driving force is based on the force developed between a wire carrying a current in a magnetic field and a set of permanent magnets providing this field. The forcing arrangement can be seen in the schematic of Figure 16 and photographically in Figure 17. Six flat permanent magnets are appropriately bonded to a U shaped steel shield which is fastened to the free end of the beam. This arrangement provides a constant field between the magnets along the length of the U shield. The force is created by means of an independently supported coil composed of 150 turns of 28 GA magnet wire. The section of the coil between the magnets is vertical and can be thought of as a bundle of wires carrying current in the same direction in a constant magnetic field. This coil-magnet arrangement ensures that the current to force transfer function is constant for all foreseeable beam displacements. There are two identical coils side by side, one for the excitation force driven by a function generator and one for the control force. A Q-FLEX accelerometer (Model QA-116-15), attached to the beam near its free end, is used as a sensor for the control loop. Figure 16 shows the ancillary components used in this feedback control scheme.

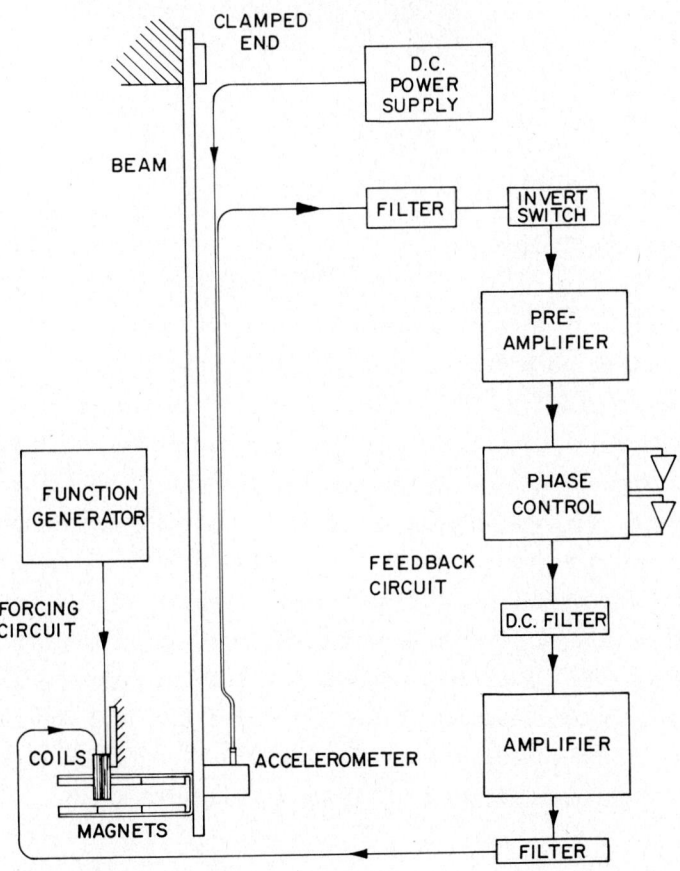

Figure 16 - *Schematic Diagram of the Components used for Electro-Magnetic Control of a Cantilever Beam*

Figure 17 - Details of the Forcing Device at the Free End of the Cantilever

The system was designed to allow variation of two independent control parameters - gain (F) and phase (ϕ). The gain control is provided by the multiplier effect of the amplifiers. A phase control is introduced in this experiment to study its effect on the efficiency of the control equipment. Continuous phase control through 360° is achieved by suitably altering the impedances into and across two operational amplifiers in series with an inverting switch. Viewed from the point of view of mechanics, and with reference to the phasor diagram for the forces and motion at the cantilever tip (Figure 18), the phase ϕ is the angle by which the feedback force lags acceleration. The gain F determines the magnitude of the feedback force. This force will contain a component of velocity as well as acceleration.

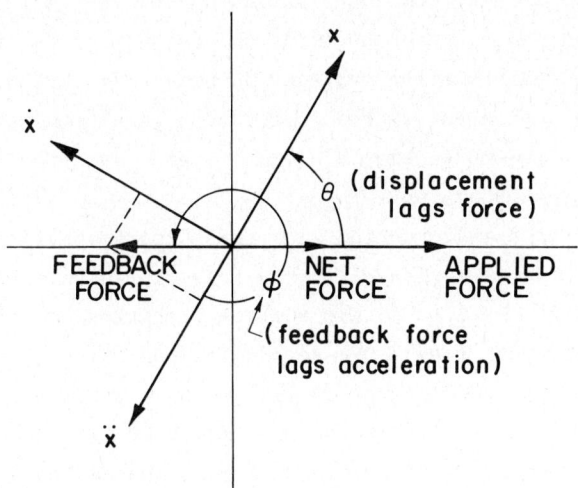

Figure 18 - *Phasor Diagram for the Forces and Motion at the Cantilever Tip*

For theoretical purposes the cantilever beam is considered as a single-degree-of-freedom system oscillating in the first mode only. First mode structural properties can be experimentally obtained with these results - beam stiffness = 18.6 N/m; equivalent mass = 0.361 Kg; damping factor = 0.139% of critical damping; natural frequency = 1.138 Hz. The constants for the feedback circuit are such that the total control force, expressed as the sum of an acceleration and a velocity component, amounts to $(2.26 \times 10^{-5} \text{ F})(\ddot{x}\cos\phi + \dot{x}\omega\sin\phi)$, where ω is the frequency of oscillation with the feedback loop in operation. The complete equation of motion governing the controlled fundamental mode is

$$(0.361 - 2.26 \times 10^{-5} F\cos\phi)\ddot{x} + (7.22 \times 10^{-3} - 2.26 \times 10^{-5} F\omega\sin\phi)\dot{x} +$$

$$+ 18.6x = 1.04 \times 10^{-4} I_0 \cos\omega t . \qquad (5)$$

The harmonic forcing term on the right hand side includes the current amplitude I_o and the transfer constant of the forcing coil. Note that both the mass and damping terms are modified by the control action and depend on feedback gain F and phase angle ϕ. It is therefore convenient to speak of these as *effective* (or *active*) mass and damping.

The tests conducted with this experiment included both free and forced vibrations. In the free vibration tests the phase ϕ and gain F were fixed, and the vibration was started by releasing the beam from a pre-set deflection using a special jig. The accelerometer output was registered on a storage oscilloscope.

Figure 19 illustrates the free vibration decay with the feedback circuit switched on after 20 seconds to demonstrate the sudden change in effective damping produced by feedback. Under certain feedback conditions negative damping can occur, as shown in Figure 20, with exponentially increasing amplitudes of vibration. The complete test results in free vibration are summarized in Figure 21. Here the damping ratio is plotted versus the complete range of ϕ for F values of 50,200 and 1,000. Relatively good correspondence is seen to exist between theory and experiment. The optimum phase angle appears to be near 270°, giving the largest positive damping ratio. Unstable vibrations occur for a wide range of phase angles around 90°, the width of the range depending on the feedback gain. These damping curves appear to be skew-symmetric about the inherent structural (F = 0) damping ratio of 1.39×10^{-3}.

Feedback Control of Structures

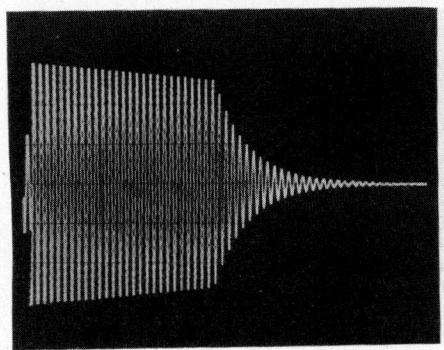
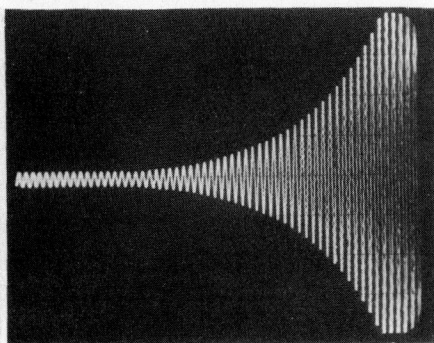

Figure 19 - Free Vibration with Feedback of Gain = 1000 and Phase = 270° Switched on After 20 Seconds (Stable)

Figure 20 - Free Vibrations with Feedback of Gain = 500 and Phase = 90° Switched on After 15 Seconds (Unstable)

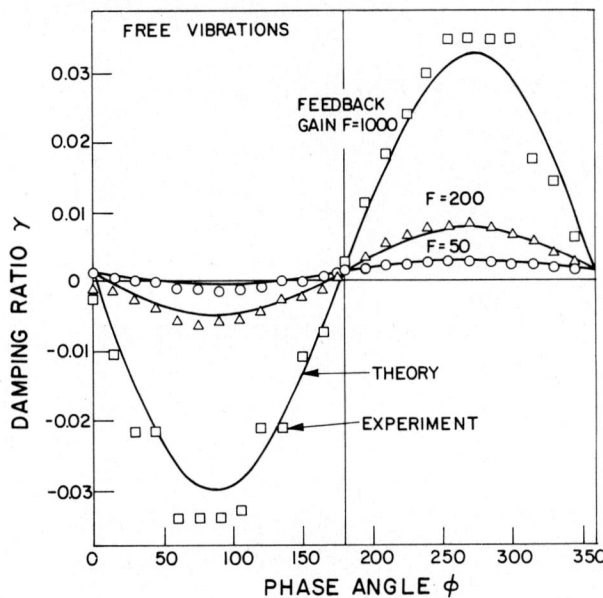

Figure 21 - The Variation of Damping Ratio with Phase Angle: Theory and Experiment, at Gains of 50, 200 and 1,000

Theoretical curves of the response of the system to harmonic excitation are shown in Figure 22(a) for the optimum ϕ value of 270° and various gain. At this phase angle the resonant frequency remains constant at 7.18 radians/sec. for all gain values. The feedback damper appears to be highly effective. Figure 22(b) shows the shift in resonance when non-optimal phase angles are used, accompanied by a decreased damping efficiency. These phenomena are directly attributable to the concepts of effective mass and effective damping expressed in equation (5).

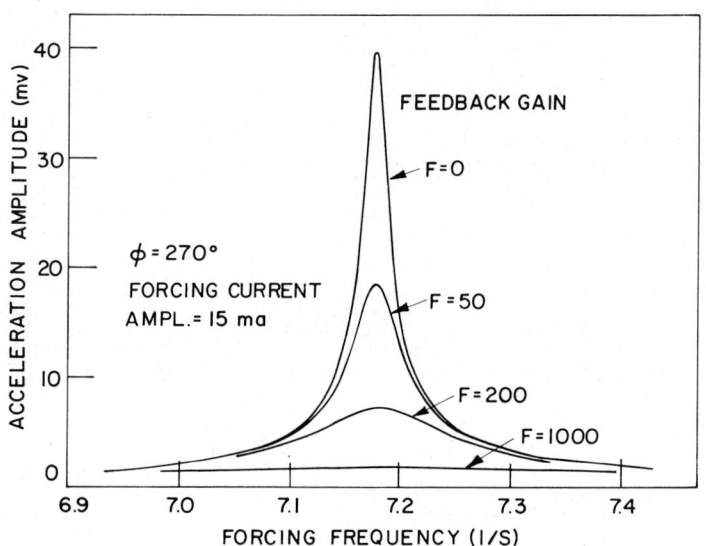

(a)

Figure 22 - *Theoretical Acceleration Response Curves*
    (a) Phase Constant at 270°; Gain Setting 0, 50, 200 and 1,000
    (b) Gain Constant at 200; Phase Settings 200°, 270° and 340°

(continued)

(continued)

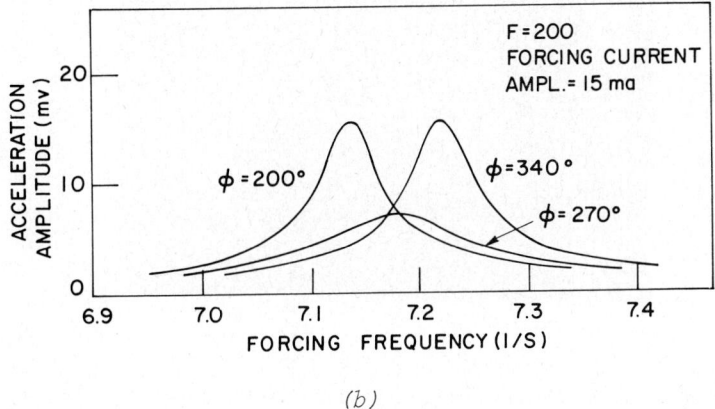

(b)

Figure 22 - Theoretical Acceleration Response Curves
  (a) Phase Constant at 270°; Gain Setting 0, 50, 200 and 1,000
  (b) Gain Constant at 200; Phase Settings 200°, 270° and 340°

The peak response at resonance was measured experimentally over the stable range of ϕ values for F settings at 50, 200, 1,000 and 5,000. A typical trace of forced vibration with the feedback circuit switched on after 10 seconds is shown in Figure 23 to demonstrate the transient behaviour and the effectiveness of the system. Summarized results are plotted in Figure 24. Minimum responses for all gain settings occur near ϕ = 270° where, for F = 50, the response is reduced by about 54%, for F = 200 by 83% and for F = 1,000 by 97%. For non-optimum phase angles the feedback device becomes increasingly less effective. In view of the functional dependence of the effective mass upon both F and ϕ, the change in resonant frequency can be appreciable, particularly at large gain settings. This is shown in Figure 25 which compares theoretical and experimental values of frequency at resonance.

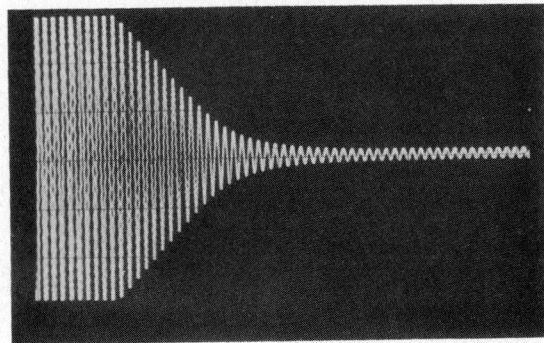

Figure 23 - Harmonically Forced Vibration with Feedback of Gain = 1,000 and Phase = 270° Switched on After 10 Seconds (Stable)

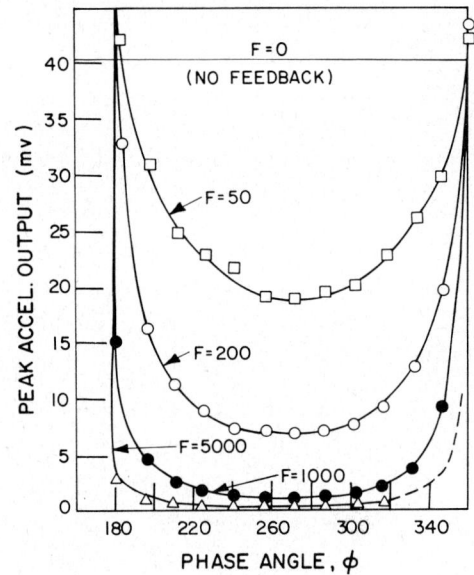

Figure 24 - Summary of Experimental Peak Acceleration Values for the Full Stable Range of Phase Angles and Gain Settings

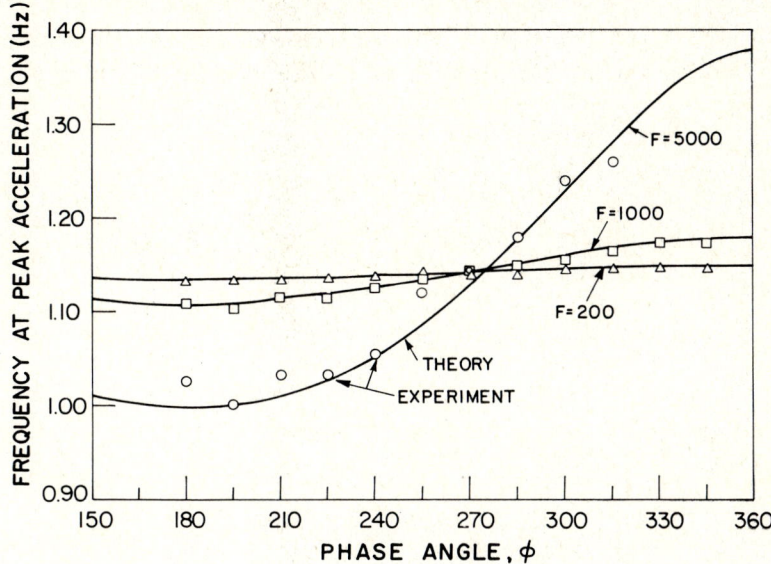

Figure 25 - Comparison of Experimental and Theoretical Values of Resonant Frequency for the Stable Range of Phase Angles and Gain Settings at 200, 1,000 and 5,000

A directly applied control force of the kind used in this electro-magnetic experiment can be very effective. The best results from the tests performed yielded a reduction to 0.2% of the uncontrolled response. This oscillation was barely visible to the naked eye, while the uncontrolled vibration had a resonant amplitude of about 7 cm. There are, however, several drawbacks to the practical use of this kind of system for the control of structures. Firstly, the direct application of force in the best position is likely to be impractical. For instance, a forcer with its own fixed base cannot be conveniently located at the top of a tall tower. Secondly, the magnitude of the force required approaches that of the driving force as gain is increased and exceeds it at non-optimal phase angles. This is shown clearly in Figure 26 where the measured values of the feedback force, taken in ratio with the applied force, are plotted against the phase angle. Such large forces seem impractical at first instinct.

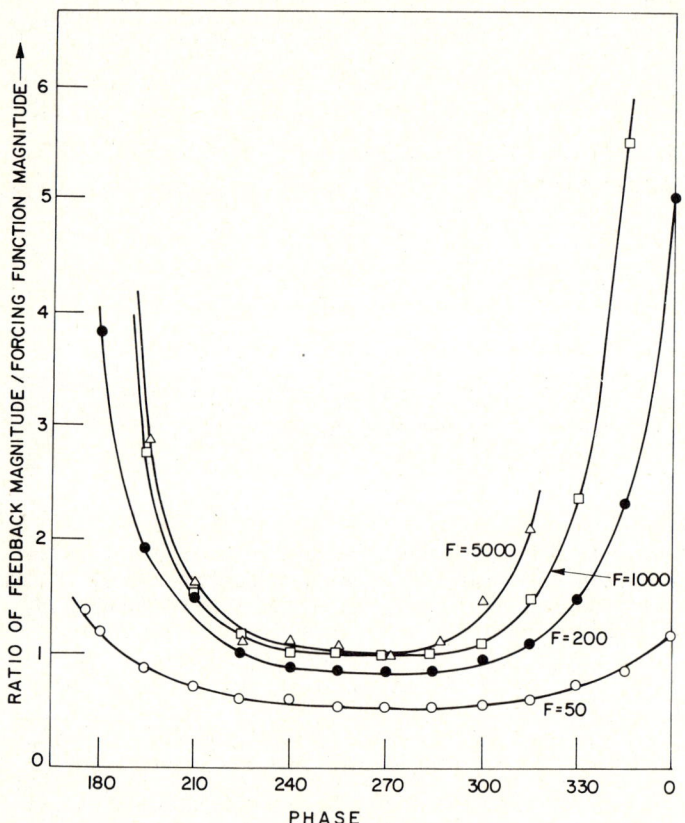

Figure 26 - *The Change in Feedback Force/Applied Force Ratio near the Optimum Phase Angle*

The system as described is effective in damping all modes provided that the sensor and forcer are placed at the same location which is not a nodal point for a particular mode.

6. CONCLUDING REMARKS

The experiments presented in this paper demonstrate that the concept of active feedback control of structures may indeed become a feasible solution for vibration problems in future years. It has been shown that vibration levels can be reduced to a degree by the introduction of suitable controlled forces. Of the four experiments presented here, the ones utilizing electro-hydraulic devices presented the most difficulties. The inherent complexities of such systems render them very sensitive to extraneous disturbance. Troublesome high frequency noise in the signals, instabilities in uncontrolled modes and excessive frictional losses which tended to mask the phenomenon under study were the major problems here. From a practical point of view, control schemes using cables and hydraulic actuators would appear to be the most useful and hence deserve much more study. For example the problem of the instabilities is one area that needs to be carefully investigated, both experimentally and from a theoretical viewpoint. The reliability of the equipment and of the power supply must be absolute for such systems to become acceptable.

The electro-magnetic device, although in a sense academic in nature, avoided many of the above mentioned difficulties because of the simplicity of its action. In the absence of a mechanical link between the structure and its controller, it was possible in this experiment to study the structural behaviour in its pure form. This experiment shows very clearly that the phase relationship between the control force and the structural motion is of paramount importance.

7. ACKNOWLEDGEMENTS

The author gratefully acknowledges the financial support of the Natural Sciences and Engineering Research Council of Canada under

Grant No. A3728. Many thanks also to George Burnham, Sherif Osman, Brian Stirling, Andy MacPhee and Frank Morison. This paper represents a joint effort in which the collective contributions of all these students constitutes the main ingredient.

REFERENCES

[1] ABDEL-ROHMAN, M. and LEIPHOLZ, H.H.E., "Active Control of Flexible Structures", *ASCE, J. of the Structural Div.*, Vol. 104, No. ST8, 1978, pp. 1251-1266.

[2] ABDEL-ROHMAN, M. and LEIPHOLZ, H.H.E., "Structural Control by the Pole Assignment Method", *ASCE, J. of the Eng. Mech. Div.*, Vol. 104, No. EM5, 1978, pp. 1159-1175.

[3] HIRSCH, G., "Critical Comparison Between Active and Passive Control of Wind Induced Vibrations of Structures by Means of Mechanical Devices", *Proc. IUTAM Symposium on "Structural Control"*, held at the University of Waterloo, Waterloo, Ontario, Canada, June 4-7, 1979.

[4] LUND, R.A., "Active Damping of Large Structures in Winds", *Proc. IUTAM Symposium on "Structural Control"*, held at the University of Waterloo, Waterloo, Ontario, Canada, June 4-7, 1979.

[5] MARTIN, C.R. and SOONG, T.T., "Modal Control of Multistory Structures", *ASCE, J. of the Eng. Mech. Div.*, Vol. 102, No. EM4, 1976, pp. 613-623.

[6] PETERSON, N.R., "Design of Large Scale Tuned Mass Dampers", *Proc. IUTAM Symposium on "Structural Control"*, held at the University of Waterloo, Waterloo, Ontario, Canada, June 4-7, 1979.

[7] ROCKWELL, T.H. and LAWTHER, J.M., "Theoretical and Experimental Results on Active Vibration Dampers", *J. of the Acoustical Society of America*, Vol. 36, No. 8, 1964, pp. 1507-1515.

[8] ROORDA, J., "Active Damping in Structures", *Report Aero No. 8*, Cranfield Institute of Technology, Cranfield, Bedfordshire, United Kingdom, 1971.

[9] ROORDA, J., "Tendon Control in Tall Structures", *ASCE, J. of the Structural Div.*, Vol. 101, No. ST3, 1975, pp. 505-521.

[10] ROORDA, J., "A Comparative Study of Vibration Damping Devices", *Proc. Fourth Canadian Congress of Applied Mechanics*, held in Montreal, Quebec, Canada, May 28-June 1, 1973, p. 553.

[11] SENSBURG, O., BECKER, J. and HÖNLINGER, H., "Active Control of Flutter and Vibration of an Aircraft", *Proc. IUTAM Symposium on "Structural Control"*, held at the University of Waterloo, Waterloo, Ontario, Canada, June 4-7, 1979.

[12] SMANCHAI, S.U. and YAO, J.T.P., "Active Control of Building Structures", *CE-STR-76-1*, School of Civil Engineering, Purdue University, Lafayette, Indiana, U.S.A., 1976.

[13] YANG, J.N., "Application of Optimal Control Theory to Civil Engineering Structures", *ASCE, J. of the Eng. Mech. Div.*, Vol. 101, No. EM6, 1975, pp. 819-838.

[14] YANG, J.N. and GIANNOPOULOS, F., "Active Tendon Control of Structures", *ASCE, J. of the Eng. Mech. Div.*, Vol. 104, No. EM3, 1978, pp. 551-568.

[15] YANG, J.N. and GIANNOPOULOS, F., "Active Control and Stochastic Response of Cable-Stayed Bridge", *ASCE Convention and Exposition*, Boston, Massachusetts, U.S.A., April 2-6, 1979, Preprint 3468.

[16] YAO, J.T.P., "Concept of Structural Control", *ASCE, J. of the Structural Div.*, Vol. 98, No. ST7, 1972, pp. 1567-1574.

[17] YAO, J.T.P. and TANG, J.P., "Active Control of Civil Engineering Structures", *CE-STR-73-1*, School of Civil Engineering, Purdue University, Lafayette, Indiana, U.S.A., 1973.

[18] ZUK, W., "Kinetic Structures", *ASCE, Civil Engineering*, Vol. 39, No. 12, 1968, pp. 62-64.

STRUCTURAL CONTROL, H.H.E. Leipholz (ed.)
North-Holland Publishing Company & SM Publications
© IUTAM, 1980

ACTIVE CONTROL OF LARGE BUILDING STRUCTURES

J.N. Juang

Jet Propulsion Laboratory
Pasadena, California, U.S.A.

S. Sae-Ung and J.N. Yang

Department of Civil, Mechanical and Environmental Engineering
The George Washington University
Washington, D.C. 20052, U.S.A.

1. INTRODUCTION

Recently, the application of active control theories to civil engineering structures under environmental loads, such as earthquakes and wind gusts, has attracted considerable interest and relevant literature is given in this proceedings [1 - 10]. The purpose of this study is to examine the feasibility of a direct application of the pole assignment approach (or modal control) to civil engineering structures. Although the method of modal control (or pole assignment) has been applied to the active control of structures [4, 10], various aspects relevant to this methodology have not been seriously considered and will be explored herein.

For a structure implemented by a set of sensors and controllers, the poles (or eigenvalues) of such a structure can be shifted to desirable locations by use of a linear state feedback control [11], if the controlled structure is controllable. Thus, when the controlled structure is controllable, its natural frequencies and dampings, which represent the imaginary and the real parts, respectively, of the poles, can be relocated to desirable locations. Since, however, the state feedback control requires the information of all the states (displacements and velocities) of the structure, an application of the state feedback control may not be practical for most of the large civil engineering structures which involve very large degrees of freedom. From the practical standpoint, it is desirable to have as few sensors and controllers as possible because of physical and economical constraints.

For a structure implemented by only a few sensors and controllers, i.e., with incomplete information of the states, the whole states can be estimated by an asymptotic state estimator [11], if the controlled structure is observable. As a result, the controllability and observability of the controlled structure should first be established. In this paper a method of determining the controllability and observability of the structure is suggested taking advantage of the method of modal control (pole assignment) in which the eigenvalues and the eigenvectors have been determined. Furthermore, the stability condition can always be maintained by the method of pole assignment such that the real part of each pole of the controlled structure and the state estimator is negative.

Most of the civil engineering structures are very complex with many degrees of freedom. Thus it is very difficult to control all the vibrational modes because of practical limitations as well as computational difficulties involved. On the other hand, a significant portion of the power of the environmental loads, such as earthquakes and wind gusts, are concentrated in certain region of frequency. Therefore, it is more effective to control the vibrational modes within such a frequency region, referred to as the critical modes. The rest of the vibrational modes which are not controlled are referred to as the residual modes.

Although the critical mode control has its attractive features, the residual modes may be excited and amplified thus resulting in the adverse effect referred to as the observation and the control spillovers [13, 14]. The observation spillover is eliminated herein by a prefilter [13], while the effect of the control spillover has been estimated such that the controllers can be designed to reduce the adverse effect.

Numerical results indicate that for building structures the shift of the critical poles along the real axis (i.e., increasing the active damping) requires less state feedback gain than the shift of the critical poles along the imaginary axis (i.e., moving the natural frequencies). Finally, an example is given to illustrate the design of state estimator with a velocity sensor placed at the top floor of the building.

2. EQUATIONS OF MOTION

A lumped mass n-story building with active control devices is shown in Figure 1. The particular control devices considered herein consist of tendons attached to hydraulic servomechanisms which can be installed in the partitions of the building. The equations of motion can be written in the form:

$$m\ddot{\underline{y}} + c\dot{\underline{y}} + k\underline{y} = h\underline{u} + \underline{p} - \underline{e}, \qquad (1)$$

in which $\underline{y} = [y_1, y_2, \ldots, y_n]' = [x_1-x_0, x_2-x_1, \ldots, x_n-x_{n-1}]'$ is the relative displacement vector, where the prime denotes the transpose of a vector (or matrix); $\underline{p}$ is the external excitation vector due to wind loads and $\underline{e} = \ddot{x}_0 [m_1, m_2, \ldots, m_n]'$ is the excitation vector due to earthquakes where $\ddot{x}_0$ denotes the ground acceleration; $\underline{u}$ is the control vector with dimension r, where $r\epsilon\{1,2,\ldots,n\}$. Each element of $\underline{u}$ represents either the pushing or pulling force generated by a control device located between any two adjacent floors (see Figure 1); h is a n×r matrix whose elements depend on the arrangement of controllers as shown in Figure 1. In equation

(1), m, c, and k are the mass matrix, the damping matrix and the stiffness matrix respectively.

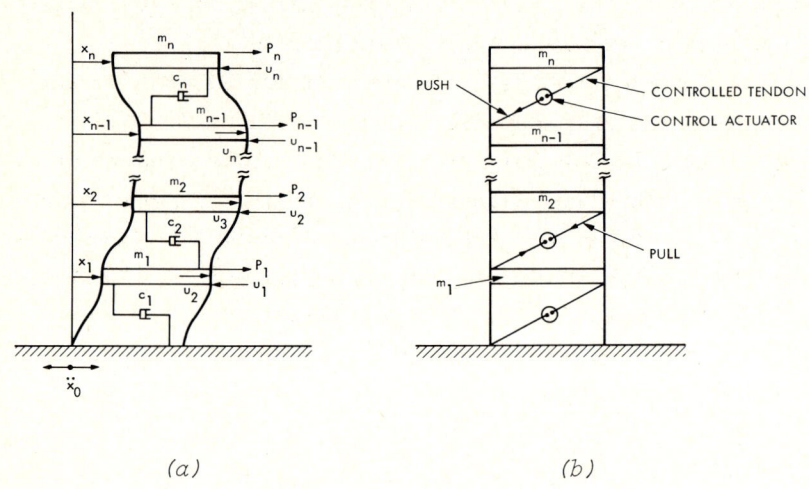

(a) (b)

Figure 1 - The Building Structure and Control Device

Without loss of generality for the subsequent discussions, the external excitations can be disregarded and hence equation (1) becomes

$$m\underline{\ddot{y}} + c\underline{\dot{y}} + k\underline{y} = h\underline{u} .\qquad(2)$$

Equation (2) can be converted into a set of 2n first order differential equations as follows

$$\underline{\dot{z}} = \tilde{K}\underline{z} + \underline{w} ,\qquad(3)$$

in which $\underline{z} = [\underline{y}',\underline{\dot{y}}']'$, $\tilde{K} = M^{-1}K^*$, $\underline{w} = M^{-1}H\underline{u}$, and

$$M = \begin{bmatrix} I & 0 \\ 0 & m \end{bmatrix}, \quad K^* = \begin{bmatrix} 0 & I \\ -k & -c \end{bmatrix}, \quad H = \begin{bmatrix} 0 \\ h \end{bmatrix},\qquad(4)$$

where the dimension of H is 2n×r.

3. CONTROLLABILITY AND OBSERVABILITY

Let $\mu_k \pm i\nu_k$ ($k = 1,2,\ldots,q$) and μ_ℓ ($\ell = 2q+1, 2q+2,\ldots,N$) be the eigenvalues of the matrix $\tilde{K}$ with the corresponding eigenvectors $a_k \pm ib_k$ and a_ℓ, respectively, where $N = 2n$. If $\tilde{K}$ is a real matrix of simple structure [12], then there exists a real transformation matrix T with bases a_1, b_1, a_2, b_2, $\ldots$, a_q, b_q, a_{2q+1}, a_{2q+2}, $\ldots$, a_N such that the $A = T^{-1}\tilde{K}T$ is a canonical matrix with the diagonal blocks $\{A_0, A_1, \ldots, A_L, A_{L+1}, \ldots, A_E\}$ in which A_0 is a zero matrix of order γ (γ = multiplicity of zero eigenvalues), and A_j, $j = 1,2,\ldots,L$ is the direct sum of the real quasi-diagonal matrix with nonzero diagonal block

$$\begin{bmatrix} \mu_j & \nu_j \\ -\nu_j & \mu_j \end{bmatrix}.$$

The order of A_j is $2\alpha_j$ with α_j being the multiplicity of the eigenvalues $\mu_j \pm \nu_j$. The matrix A_k for $k = L+1, L+2, \ldots, E$ is a diagonal matrix with nonzero elements, μ_k, and order β_k that is the multiplicity of the real nonzero eigenvalue μ_k of $\tilde{K}$.

Let $\underline{v} = T^{-1}\underline{z}$, then equation (3) can be expressed as

$$\underline{\dot{v}} = A\underline{v} + B\underline{u}, \tag{5}$$

in which

$$B = T^{-1}M^{-1}H = [B_0', B_1', \ldots, B_E']', \tag{6}$$

where B_j corresponds to the block A_j of A for $j = 0,1,\ldots,E$.

Consider the sensor equations

$$\underline{s} = \bar{C}\underline{z}, \tag{7}$$

in which $\underline{s}$ is a sensor vector with dimension p and $\overline{C}$ is the influence matrix for the sensors. Since $\underline{v} = T^{-1}\underline{z}$, equation (7) can be written as

$$\underline{s} = C\underline{v} , \qquad (8)$$

in which

$$C = \overline{C}T = [C_0, C_1, \ldots, C_E] , \qquad (9)$$

where C_j corresponds to the block A_j of A for $j = 0, 1, \ldots, E$.

The system is controllable if and only if (1) the number of controllers, $r \geq \max(\gamma, \alpha_1, \ldots, \alpha_L, \beta_{L+1}, \ldots, \beta_E)$, (2) rank $B_0 \geq \gamma$, (3) rank $[B_j, A_j B_j] \geq 2\alpha_j$ for $j = 1, 2, \ldots, L$, and (4) rank $[B_j] \geq \beta_j$ for $j = L+1, L+2, \ldots, E$.

Similarly, the system is observable if and only if (1) the number of sensors, $p \geq \max(\gamma, \alpha_1, \ldots, \alpha_L, \beta_{L+1}, \ldots, \beta_E)$, (2) rank $C_0 \geq \gamma$, (3) rank $[C_j', A_j'C_j']' \geq 2\alpha_j$ for $j = 1, 2, \ldots, L$, (4) rank $[C_j] \geq \beta_j$ for $j = L+1, L+2, \ldots, E$.

The smallest r and p such that all the conditions of controllability and observability are satisfied, respectively, are the smallest numbers of controllers and sensors required for the system to be controllable and observable.

The procedures for determining the observability and controllability of structures described above are simpler than the classical ones [4, 10]. The present approach takes advantage of the fact that in the critical mode control (or pole assignment) the eigenvalues and eigenvectors are available.

4. CRITICAL MODE CONTROL AND CONTROL SPILLOVERS

Let $\overline{v}_c$ be the critical modes (or controlled modes) to be controlled. The remaining modes denoted by $\overline{v}_r$ are referred to as the residual modes.

With the following partitions of A, B and C

$$A = \begin{bmatrix} A_c & 0 \\ 0 & A_r \end{bmatrix}, \quad B = \begin{bmatrix} B_c \\ B_r \end{bmatrix}, \quad C = [C_c, C_r].$$

Equations (5) and (7) can be written as

$$\dot{\bar{v}}_c = A_c \bar{v}_c + B_c \underline{u}, \tag{10}$$

$$\dot{\bar{v}}_r = A_r \bar{v}_r + B_r \underline{u}, \tag{11}$$

$$\underline{s} = C_c \bar{v}_c + C_r \bar{v}_r. \tag{12}$$

The term $B_r \underline{u}$ in equation (11) is referred to as the controll spillover since it excites the residual modes, while the term $C_r \bar{v}_r$ in equation (12) is referred to as the observation spillover for it is the contribution to the sensor output from the residual modes. The observation spillover can be eliminated by introducing a prefilter [13] between the sensor output and state estimator and hence $C_r = 0$ as shown in Figure 2.

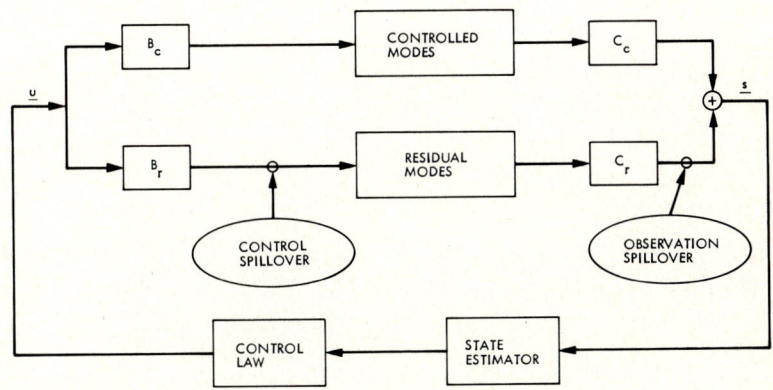

Figure 2 - The Closed Loop System

If e_c denotes the estimator error and $\hat{v}_c$ indicates the estimate of state $\bar{v}_c$, then

$$e_c = \hat{v}_c - \bar{v}_c . \tag{13}$$

The control force $\underline{u}$ is assumed to be of the form

$$\underline{u} = G_c \hat{v}_c , \tag{14}$$

in which G_c is the feedback gain matrix. For the asymptotic estimator it can be shown that [11],

$$\dot{e}_c = (A_c - K_c C_c) e_c , \tag{15}$$

in which K_c is a constant gain matrix. If $\bar{\zeta}_c$ is the state $\bar{\zeta}_c = [\bar{v}_c', e_c']'$, then it follows from equations (10) and (15) that

$$\dot{\bar{\zeta}}_c = \hat{A}_c \bar{\zeta}_c , \tag{16}$$

in which

$$\hat{A}_c = \begin{bmatrix} A_c + B_c G_c & B_c G_c \\ 0 & A_c - K_c C_c \end{bmatrix} .$$

Thus, the residual modes becomes

$$\dot{\bar{v}}_r = A_r \bar{v}_r + B_r G_c^* \bar{\zeta}_c , \tag{17}$$

where $G_c^* = [G_c, G_c]$. Finally, the spillover effects are presented in a block diagram as shown in Figure 2.

The gains G_c and K_c can be found such that the system can be stabilized to any desired margin and the state estimate $\hat{v}_c$ can be made to converge to $\bar{v}_c$ at any desired rate [11] when the system is controllable and observable. However, a large relocation of the poles might cause the system to become nonlinear or susceptible to noise.

Let N_c and N_r be, respectively, the orders of $\hat{A}_c$ and A_r. If $\hat{A}_c$ has N_c linearly independent eigenvectors corresponding to its eigenvalues $\lambda_1, \lambda_2, \ldots, \lambda_{N_c}$, then there exists a nonsingular matrix P_c such that $\hat{A}_c P_c = P_c \lambda$ where λ is a diagonal matrix with the diagonal elements $\lambda_1, \lambda_2, \ldots, \lambda_{N_c}$. Consequently, the solution of equation (16) can be obtained as

$$\overline{\zeta}_c(t) = P_c e^{t\lambda} P_c^{-1} \overline{\zeta}_c(0) . \qquad (18)$$

Similarly, if A_r has N_r linearly independent eigenvectors corresponding to its eigenvalues $\beta_1, \beta_2, \ldots, \beta_{N_r}$, then there exists a nonsingular matrix P_r such that $A_r P_r = P_r \beta$ where β is a diagonal matrix with the diagonal elements $\beta_1, \beta_2, \ldots, \beta_{N_r}$.

The solution of equation (17) can be shown as

$$\overline{v}_r = P_r e^{\beta t} P_r^{-1} \overline{v}_r(0) + P_r \hat{G} P_c^{-1} \overline{\zeta}_c(0) , \qquad (19)$$

in which the element $\hat{g}_{j\ell}$ ($j = 1, 2, \ldots, N_r$; $\ell = 1, 2, \ldots, N_c$) of the matrix $\hat{G}$ is given by

$$\hat{g}_{j\ell} = \left(e^{\lambda_\ell t} - e^{\beta_j t} \right) g_{j\ell} / (\lambda_\ell - \beta_j) , \qquad (20)$$

where $g_{j\ell}$ is the element of the matrix G given by $G = P_r^{-1} B_r G^* P_c$. The second term of equation (19) is due to the controll spillover.

Finally, for the controlled system to be stable, the real parts of every λ_ℓ and β_j must be negative. Note that $\beta_1, \beta_2, \ldots, \beta_{N_r}$ represent the poles of the residual modes; $\lambda_1, \lambda_2, \ldots, \lambda_{M_c}$ represent the poles of the controlled modes where $M_c = N_c/2$; and $\lambda_{M_c+1}, \lambda_{M_c+2}, \ldots, \lambda_{N_c}$ represent the poles of the state estimator. The gain G_c corresponding to any chosen values of $\{\lambda_1, \lambda_2, \ldots, \lambda_{M_c}\}$ and the gain K_c corresponding to any chosen values of

$\{\lambda_{M_C+1}, \lambda_{M_C+2}, \ldots, \lambda_{N_C}\}$ can be computed by the algorithm given in Reference [11].

5. NUMERICAL EXAMPLE

Consider a 10 story building, as shown in Figure 1, in which the mass of each floor is identical, i.e., m_j = 345.6 tons for j = 1,2,...,10. The stiffnesses of the columns of each story in x-direction are k_1 = 3.87 x 10^5, k_2 = 3.20 x 10^5, k_3 = 3.72 x 10^5, k_4 = 3.61 x 10^5, k_5 = 3.40 x 10^5, k_6 = 3.10 x 10^5, k_7 = 2.69 x 10^5, k_8 = 2.17 x 10^5, k_9 = 1.55 x 10^5, and k_{10} = 8.26 x 10^4 kN/m. The damping coefficient c_j = 50 kN-sec/m (for j = 1,2,...,10) is used.

The eigenvalues (poles) of the uncontrolled building structure are obtained as -1.79 x 10^{-3} $\pm$ i4.64, -1.99 x 10^{-2} $\pm$ i11.63, -5.54 x 10^{-2} $\pm$ i18.41, -9.38 x 10^{-2} $\pm$ i25.12, -0.13 $\pm$ i31.76, -0.16 $\pm$ i38.35, -0.19 $\pm$ i44.86, -0.22 $\pm$ i51.23, -0.24 $\pm$ i57.37, and -0.27 $\pm$ i62.99.

Only one control actuator is placed between the top two floors. Thus, H = [0',h']' is a 20 x 1 matrix with h = [0,0,..., 1,-1]' being a 10 x 1 matrix. From the conditions given in Section 3, the structure is shown to be controllable.

If a displacement sensor is placed at the top floor, the matrix $\overline{C}$ becomes a row vector with every element of $\overline{C}$ being zero except the 10th element that is equal to unity. From the conditions given in Section 3, the system is shown to be observable. Similarly, with only a velocity sensor being placed at the top floor, the system is also observable. It is further shown that with only one controller placed between any two adjacent floors and one displacement or velocity sensor placed at any floor, the building structure is controllable and observable.

Under the earthquake excitations where the spectrum of the ground acceleration is concentrated in the frequency region between 6 rad./sec. and 25 rad./sec., it may be desirable to move the three natural frequencies (11.63, 18.41 and 25.12 rad./sec.) out of this frequency range. If these three natural frequencies are moved to 26, 28 and 30 rad./sec., respectively, (i.e., the poles $-1.99 \times 10^{-2} \pm i11.63$, $-5.54 \times 10^{-2} \pm i18.41$ and $-9.38 \times 10^{-2} \pm i25.12$ are shifted to the locations $-1.99 \times 10^{-2} \pm i26$, $-5.54 \times 10^{-2} \pm i28$ and $-9.38 \times 10^{-2} \pm i30$, respectively, along the imaginary axis), then the gain G_c is computed by the algorithm given in [11] as [3.75×10^8, 2.54×10^6, -4.19×10^7, 4.47×10^5, 6.39×10^4, 5.65×10^3]. Such a gain results in an unrealistically large control force.

An alternate approach to reduce the structural response under earthquakes is to increase the active dampings by the state feedback. Supposing that the poles $-1.99 \times 10^{-2} \pm i11.63$, $-5.54 \times 10^{-2} \pm i18.41$ and $-9.38 \times 10^{-2} \pm i25.12$ are moved along the real axis to $-5.98 \times 10^{-2} \pm i11.63$, $-1.66 \times 10^{-1} \pm i18.41$ and $-2.81 \times 10^{-1} \pm i25.12$, respectively, the gain G_c associated with such a relocation of the poles (increasing the active damping) is computed as [5.86, -1.73×10^2, 4.32, -1.38×10^2, -25.94, 1.47×10^2]. The gain, G_c, under such a situation is 6 orders of magnitude smaller than that of the previous case and it is realistic for practical applications.

For the example considered above, it is assumed that all the state variables are available, either from sensors or asymptotic estimators. Because of physical and economical limitations, it is desirable to install as few sensors as possible. Consequently, the state estimator should be designed appropriately with a minimum number of required sensors. From the analysis presented in Section 3, the minimum number of sensors is one. For illustrative purposes, a velocity sensor is installed at the top floor and a state estimator is designed with the poles at $-3.99 \times 10^{-2} \pm i11.63$, $-1.16 \times 10^{-1} \pm i18.41$ and $-1.88 \times 10^{-1} \pm i25.12$ to estimate the state variables of the 2nd, 3rd, and 4th modes. The sensor gain K_c is found to be [-2.49×10^{-4}, -5.43×10^{-2}, 6.86×10^{-4}, -0.133,

-4.96×10^{-2}, $0.222]'$. Such a magnitude of sensor gain is also practical.

From the example given above, it appears that the increase of active damping (i.e., moving the poles along the real axis) requires much smaller control forces than shifting the natural frequencies outside the critical region (i.e., moving the poles along the imaginary axis).

6. CONCLUSION

A technique of modal control (or pole assignment has been applied to control the critical modes of large building structures. Relevant problems associated with the critical mode control, such as the observability and controllability, the design of state feedback and state estimator, the control spillover, the observation spillover, etc., have been addressed and described.

Numerical results obtained from a straightforward application of the method of pole assignment indicate that the increase of active dampings (i.e., a shift of poles along the real axis) requires much smaller control forces than the shift of natural frequencies out of the critical frequency region (i.e., moving the poles along the imaginary axis). However, the optimal relocation of poles and a meaningful trade-off among various control parameters can only be made by the consideration of external loads as well as the reduction in structural responses. Since environmental loads to civil engineering structures, such as earthquakes ground accelerations and wind gusts, are stochastic in nature, a random vibration approach described in [6 - 8] should be carried out in conjunction with the present approach of critical mode control. Then, the optimal modal strategy can be established by considered the reduction of responses in terms of structural safety and resident comfort, the magnitude of control forces, the external energy and power required for the control devices [5, 8], etc. This subject will be described in a next report.

ACKNOWLEDGEMENT

This work is partially supported by the U.S. National Science Foundation through Grant No. ENG 78-24279 and Jet Propulsion Laboratory under NASA Contract No. NAS7-100.

REFERENCES

[1] YAO, J.T.P., "Concept of Structural Control", *ASCE, J. of the Structural Div.*, Vol. 98, No. ST7, Proc. Paper 9048, July, 1972, pp. 1567-1574.

[2] ROORDA, J., "Tendon Control in Tall Structures", *ASCE, J. of the Structural Div.*, Vol. 101, No. ST3, Proc. Paper 11168, March, 1975, pp. 505-521.

[3] YANG, J.N., "Application of Optimal Control Theory to Civil Engineering Structures", *ASCE, J. of the Eng. Mech. Div.*, Vol. 101, No. EM6, Proc. Paper 11812, December, 1975, pp. 819-838.

[4] MARTIN, C.R. and SOONG, T.T., "Modal Control of Multistory Structures", *ASCE, J. of the Eng. Mech. Div.*, Vol. 102, No. EM4, Proc. Paper 12321, August, 1976, pp. 613-623.

[5] SAE-UNG, S. and YAO, J.T.P., "Active Control of Building Structures", *ASCE, J. of the Eng. Mech. Div.*, Vol. 104, No. EM2, April, 1978, pp. 335-350.

[6] YANG, J.N. and GIANNOPOULOS, F., "Active Tendon Control of Structures", *ASCE, J. of the Eng. Mech. Div.*, Vol. 104, No. EM3, June, 1978, pp. 551-568.

[7] YANG, J.N. and GIANNOPOULOS, F., "Active Control and Stability of Cable-Stayed Bridge", *ASCE, J. of the Eng. Mech. Div.*, Vol. 105, No. EM4, August, 1979, pp. 677-694.

[8] YANG, J.N. and GIANNOPOULOS, F., "Active Control of Two Cable-Stayed Bridge", *ASCE, J. of the Eng. Mech. Div.*, Vol. 105, No. EM5, October, 1979, pp. 795-811.

[9] ABDEL-ROHMAN, M. and LEIPHOLZ, H.H.E., "Active Control of Flexible Structures", *ASCE, J. of the Structural Div.*, Vol. 104, No. ST8, August, 1978, pp. 1251-1266.

[10] ABDEL-ROHMAN, M. and LEIPHOLZ, H.H.E., "Structural Control by Pole Assignment Method", *ASCE, J. of the Eng. Mech. Div.*, Vol. 104, No. EM5, October, 1978, pp. 1159-1175.

[11] CHEN, C.T., *Introduction to Linear System Theory*, Holt, Rinehart and Winston, Inc., 1970.

[12] GANTMACHER, F.R., *The Theory of Matrices*, Vols. I and II, Chelsea Publishing Co., New York, New York, 1959.

[13] BALAS, M., "Modal Control of Certain Flexible System", *SIAM, Journal of Control and Optimal*, May, 1978.

[14] JUANG, J.N. and BALAS, M., "Dynamics and Control of Large Spinning Spacecraft", to appear in *AAS Journal of Astronautical Sciences*.

STRUCTURAL CONTROL, H.H.E. Leipholz (ed.)
North-Holland Publishing Company & SM Publications
© IUTAM, 1980

HIERARCHICAL CONTROL OF DESIGN ORGANIZATION
FOR STRUCTURAL OPTIMIZATION

A.S. Sambura

Structural Mechanics Division of Civil Engineering Committee
Polish Academy of Sciences
ul. Konstytucji 11
44-101 Gliwice, Poland

1. INTRODUCTION

What most of the designers design today, should be seen as systems and not simply as objects or as structures. These systems consist of structures, services, technology processes and other elements which are strongly interconnected between themselves and with the environment.

When attempting to perform a structural optimization one has to realize that one cannot optimize the structure alone without considering the full context of the design task and situation. One has to set the global goal first, and then proceed to determine the local goals for every designed subsystem (the structure between them). It is clear that these goals should be in harmony with the global goal.

More or less this kind of procedure in solving complex design tasks is practiced today. However in most cases this process is an intuitive one and based mainly on experience.

It is evident that the quick changes in technology and growing complexity of the design tasks require more than merely intuition and experience in solving properly all the tasks involved.

Furthermore the design process (DP) takes place inside the design organization (DO) and thus the design methods and tools are determined by the DO itself. Obviously these methods and tools have to be based on a solid mathematical framework which allows a holistic approach when linking the modelling DO with the design activities (DA), and which further allows the DA coordination for the purpose of achieving a better overall design in general, and of optimal structures in detail.

After analyzing the specific features of DP and DO, the author proposes to treat the DO as an adaptive control system. This approach seems to be a most rewarding one and creates a possibility to build an operational model of DO and DP based on the theory of multilevel hierarchical systems and the mathematical coordination theory by Mesarovič et al [2].

2. MODELLING DESIGN ORGANIZATION AND DESIGN PROCESS

Design organization (DO) is obviously hierarchical by its very nature. This type of organization, as shown in Figure 1, consists of inter-related decision-making units which are arranged hierarchically.

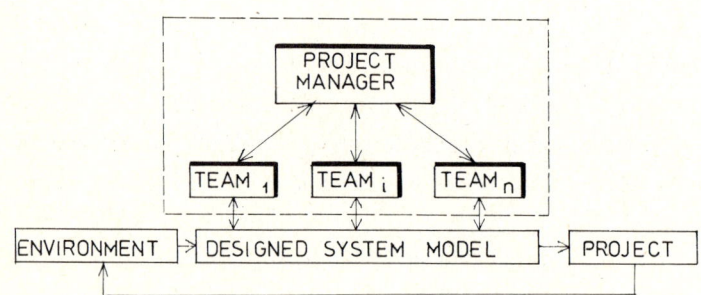

Figure 1 - Model of Design Organization and Designed System

The participants in the modelled DO (both the project manager and the design teams) can be viewed as decision-making (or goal-seeking) units in the sense of motivational approach. It means that they could be influenced or controlled in some way. Consequently, the DO (designing system) as a whole can be treated as a decision-making system and the DP can be treated as a decision process.

For further considerations, the decision system will be regarded as a system in which input specifies its decision problem and output satisfies that problem. In the case of the designing system - the decision problem is represented by the design task and the project (the result of design activities) is the solution of this task satisfying at the same time the decision problem.

To formalize mathematically the issues, the design task will be defined as a general decision problem under the condition of uncertainty that is known as a general satisfaction problem, usually specified by a quadruple:

$$(g, t, X^a, \Omega).$$

The meaning of these denotations is as follows:
- *the objective function 'g'* represents a set of design criteria. In most design situations there exists several objective functions g_i and usually some of them are contradictory;
- *the set 'X'* represents all design decisions and $x \in X$ represents a decision that fixes the values of design variables for a particular design solution. In other words, a given set X defines design solution space, and element x determines one solution;
- *the set 'X^a'* is a subset: $X^a \subseteq X$, that consists of allowable design decisions and hence represents the set of allowable (or feasible) solutions;
- *the set 'Ω'* represents the environment influence on a designed system. At the same time it represents a set of uncer-

tainties connected with designing, that is the set of all possible effects which can influence the outcome of a given design decision x;

the examples of $\omega \in \Omega$ in the built design environment can be structural, material or technological uncertainties and environmental conditions like loads or foundations;

- the *objective function* '*g*' often is specified by two functions:

$$P: X \times \Omega \to Y,$$

which represents the *designed system model*, and

$$G: X \times \Omega \times Y \to V,$$

which is the so called *performance function* and describes a behaviour or state of design model. Further G will be referred to as *state function* where Y is the set of possible results of DP and V is the set of values of state and objective functions,

$$g(x,\omega) = G(x,\omega,P(x,\omega)).$$

Normally, the state of the solution is described by several state functions, so the set V of values of state and objective functions is represented by the cartesian product of m components:

$$V = V_1 \times \cdots \times V_m.$$

The ith component of the state function can be described as:

$$G_i: X \times \Omega \times Y \to V_i.$$

Each one of those components can be treated as an objective function and described as:

$$g_i(x,\omega) = G_i(x,\omega,P(x,\omega)) \; ;$$

- *the constraint function* 't' defined as:

$$t: \Omega \to V$$

specifies an 'upper limit' of the acceptable or allowable states of the designed system. The t function can be described as the constraint function imposed on the design model P and usually consists of m functions t_i,
- *the value set* 'V' can be described as a space of design model states. If so, the t_i functions cut one or more subspaces of allowable states of the design solution.
Now the satisfaction problem in designing can be defined as:

given a subset $X^a \subseteq X$,
find $\bar{x} \in X^a$ such that,
for all $\omega \in \Omega$
$g(\bar{x},\omega) \leq t(\omega)$.

A design decision $\bar{x}$ can be considered satisfactory if it yields a design state (or performance) which does not exceed the specified constraints t for any outcome of the uncertainty.

The concepts of the theory of Mesarović [2] are used here as a base of modelling the hierarchical structure of DO. That these concepts are relevant to the DO and DP becomes quite clear when comparing an organization controlling an industrial process with the DO coordinating design process. The model of an industrial organization shown in Figure 2 bears such a resemblance to the model of DO shown in Figure 1 that nearly a total analogy can be drawn between their system's activities.

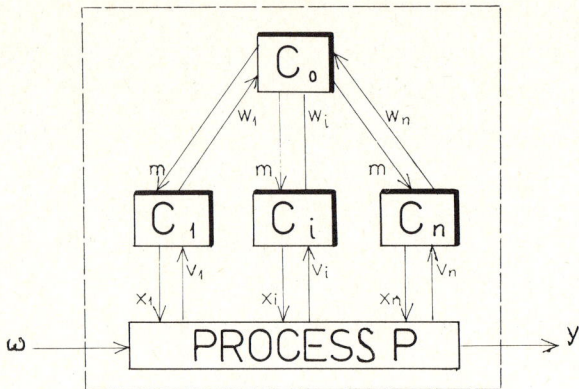

Figure 2 - Organization as a Two-Level System

In all subsequent analyses, the model of a two-level system shown in Figure 2 will be considered as the model of DO as well.

The modelled DO (designing system) has n+2 subsystems: the supremal control unit C_o representing the project manager (further called 'coordinator'), the n infimal control units C_i representing the design teams (further called 'design units') and the controlled process representing the designed system (further called 'design model').

The coordinator C_o has only one input and one output. The input is the feedback information w from the n design units, concerning the behaviour of these units. The output m represents a coordination input which serves as a means of influencing the design units. Hence a mapping:

$$C_o : W \rightarrow M ,$$

where M is the set of feedback information inputs from the units C_i, and M is the set of coordination inputs influencing the units C_i.

The design unit C_i has two inputs: the coordination input m provided by the coordinator C_o and defining its decision problem (design task), and the feedback information input v_i coming from the design model P and representing the design state (performance). Due to the existence of n design units, the design decision set X is represented by the cartesian product of n components:

$$X = X_1 \times \cdots \times X_n ,$$

so that the design unit C_i affects the design model P selecting the component x_i from X_i. The design unit C_i is then assumed to be a mapping:

$$C_i : M \times V_i \to X_i ,$$

where V_i is the set of design state values v_i, and M, X_i are as explained before.

The design model P has also two inputs: the control input x representing the decision fixing the values of design variables, and the input ω from the environment. The output y represents the results of DP. Hence a mapping:

$$P : X \times \Omega \to Y ,$$

(which are the same as described earlier). The design model P has to be decomposed into n subsystems, each of which is subject to design decision x_i of a specific design unit C_i. The decomposed model P is shown in Figure 3.

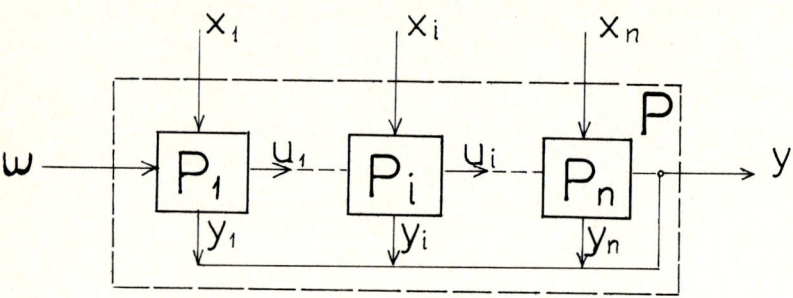

Figure 3 - Decomposition of the Design Model P

Each of the subsystems can be assumed as a mapping:

$$P_i : X_i \times \Omega \times U_i \to Y_i ,$$

where U_i is the set of inputs u_i through which the subsystem P_i is coupled or interfaced with other designed subsystems, and X_i, Ω, Y_i are the same as described earlier.

The inputs u_i are called 'interface inputs' and the design units consider the interface inputs between design model subsystems as signals, the rest of the design model is treated as an environment and no causal relationship is assumed between the decisions of the design units and the response of the whole system. The interface inputs $U = U_1, \ldots, U_n$ can be expressed as a function of the design decision x and the environment input ω. It means that us is given by a mapping:

$$K : X \times \Omega \to U.$$

This mapping represents the interaction function K which embodies the overall design model P.

Sometimes it is convenient to consider K as the specific subsystem which generates the interactions of the designed subsystems P_i. This is shown on Figure 4.

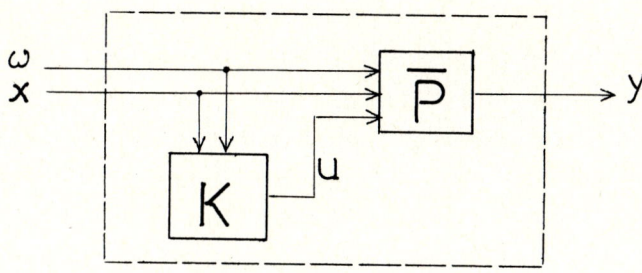

Figure 4 - The Relationship Between the Interaction Function K and the Uncoupled Subsystems Represented by $\bar{P}$

In structural design, for example, particular elements or superelements can be treated as designed subsystems. Internal forces or displacements on the element's boundaries can be considered as the interface inputs. Consequently, the finite element method (or any one analytical method) represents the K subsystem generating these interface inputs.

3. HIERARCHICAL CONTROL OF DESIGN ORGANIZATION FOR DESIGN OPTIMIZATION

In the previous section, the design task was defined as a general decision problem under condition of uncertainty, and was called a satisfaction problem. Any design solution that satisfies all the design criteria and does not violate constraints, represents the satisfaction problem solution. The existence of contradictory design criteria (or goals represented by the objective functions) in real design situations creates the need for finding several satisfying but compromising solutions and then choosing the one that fulfils in the best possible way the global design goal.

This kind of procedure in solving a design task is treated here as a design optimization process. The same meaning has the notion of structural optimization.

It is obvious that in the case of several design units participating in solving the design task, their activities have to be controlled in some way to achieve the optimal solution. When the design units are themselves goal-seeking units and have some freedom of action, the process of controlling their activities should become rather the process of coordinating these activities.

To coordinate DO is to influence the design units to act harmoniously in aiming to achieve the overall goal described in terms of design solution quality.

As was found by March and Simon [3], a participant in an organization can be influenced through: (1) factors related to his goals; (2) factors related to the expectations of the consequences of his decisions; and (3) factors related to the set of alternative actions available at the moment of decision.

Mesarovič [2] calls these modes of coordination respectively: goal intervention, model intervention and constraint intervention. These modes differ in the ways in which the interface inputs between designed subsystems are handled.

To formalize the concept of coordination, achieving a goal is represented by the solution of a decision problem. As can be shown [6], further consideration based on the above mentioned modes of coordination leads to the three formal coordination principles which not only determine a strategy of the coordinator but define his decision problem as well. From this moment on the coordination task can be treated mathematically.

Solving the coordination problem for DO, being a large nonlinear system, represents an exceedingly complex and difficult task.

However, some particular properties of the designed system structure will bring the needed simplification. The most useful properties have systems consisting of serially connected subsystems.

After an analysis of the design process of large, complex systems where several design teams have to be involved, it becomes evident that most of these systems can be treated as serial systems. There exists a strict ordering of subsystems designing. Furthermore, the results of consecutive subsystem designing create the base for the next subsystem design problem definition and at the same time they define constraints resulting from the subsystem compatibility.

Heuristic approach and methods derived by Singh [7] for serially connected dynamical subsystems provide the starting point for further consideration.

The design model P after decomposition consists of the n subsystems P_i, which can be treated as serially connected to each other by the interface inputs $U = K(x)$. This is schematically shown in Figure 5 together with the two-level coordination structure of a design organization.

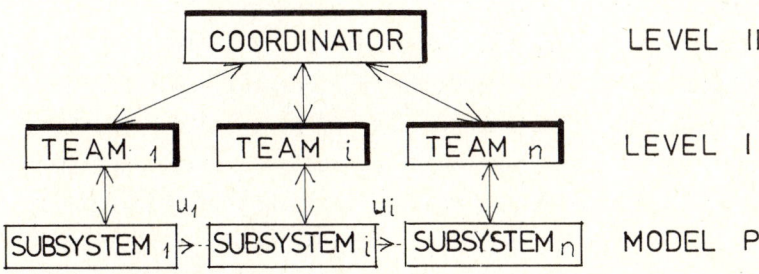

Figure 5 - Coordination Structure for Serial Design Model

To facilitate presentation of the coordination procedures for serial systems, the ideas of 'design decision problem tree' and 'design solution tree', as shown on Figure 6 are introduced.

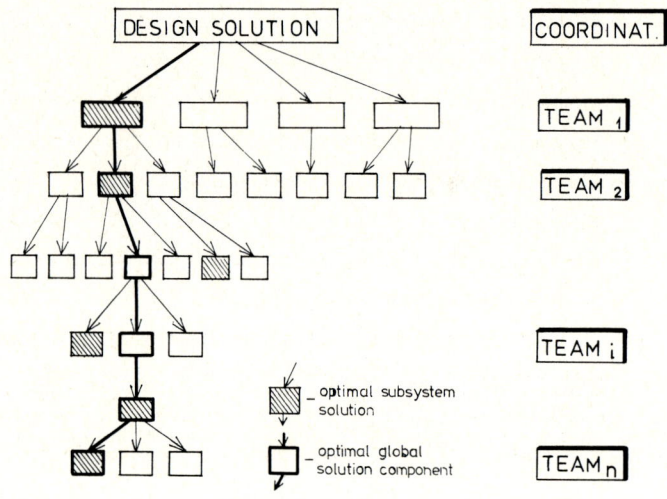

Figure 6 - *The Tree of Permissible Discrete Design Solutions*

The solution tree represents the result of solving the decision problem tree of the identical topology. The design solutions on each level being the solutions of the decision problem family of each design unit represent the optimal and suboptimal subsystem solutions. They are created by proper discretization of the subsystem design solution space and then searching through that discrete design space by means of suitable optimization methods. The very promising interactive graphic optimization routine results from the author's research and investigations [4, 5].

Several runs 'down' the solution tree represent consecutive cycles of the iterative DP and each cycle covers the analysis of fewer but more detailed solutions.

The coordination procedure of the DP begins with defining the parameters (g_1, t_1, X_1^a, Ω_1) of the C_1 design unit decision problem. Which of these parameters will be defined by the coordinator and which are left for the design unit to fix, depends mainly on the coordinator intervention mode. After finding the discrete subsystem solutions by the C_1 unit, the coordinator defines the decision problem parameters for the next design unit. The coordination procedure ends on choosing these subsystem solutions which create the most satisfying global system solution.

One of the main issues in efficient coordination is the problem of communication within the DO.

Introducing the concept of multi-stage design data base, e.g., [1], as an integral part of the designed system model, creates, in the author's opinion the best possibility to overcome all the problems of communication within the DO. The computerized data base covering all design project information has to replace traditional cumbersome mediums of communication between the participants in DO by providing direct access to the relevant information for all parties involved.

After creating the design project data base, the communication between the coordinator and the design unit goes through two channels created by: (1) interlevel coordination inputs, and (2) design data base.

Most of the information flows through the second channel and the control of this flow as well as the classification of the data to proper data base stages have to become a part of the coordinator responsibilities in order to make the coordination a successful one. It has to be remembered, however, that the communication through the data base cannot replace completely the direct contacts between the participants in DO which is the technical, yet social organization and consists mainly of human beings.

4. FINAL REMARKS

It is worthwhile to underline that when using the two-level system model as a basic block, one can easily: (1) model much more complex system structures, and (2) apply the same coordination procedure to the subsystem optimization process (e.g., structural optimization). In the second case, the application of interactive optimization techniques which leaves all the critical decisions to the designer is essential for the success of such an optimization and the success of the global design coordination as well.

In the author's opinion further research could develop the concept of hierarchical coordination of design units and provide practical methods of coordinating DO and optimizing structures. Furthermore, it is equally important that the concept could be extend into the new concept of hierarchical coordination of control units in order to achieve optimal structural responses. The design data base provides all the necessary information on the structure being designed, and the structure model and coordination methods used during structural optimization could be easily used during structural control.

REFERENCES

[1] EASTMAN, Ch.M., "Databases for Physical System Design: A Survey of US Efforts", *Proc. of CAD-76 Conference*, IPC Press, London, England, March, 1976.

[2] MESAROVIČ, M.D., MACKO, D. and TAKAHARA, Y., *Theory of Hierarchical, Multilevel Systems*, Academic Press, 1970.

[3] MARCH, I.G. and SIMON, H.A., *Organizations*, Wiley, 1958.

[4] SAMBURA, A., "Computer Graphic Techniques in Structural Optimization", *Proc. of CAD-76 Conference*, IPC Press, London, England, March, 1976.

[5] SAMBURA, A., "Interactive Design Optimization Methods", *XXIII PZITB Conference*, Krynica, September, 1977.

[6] SAMBURA, A., "Mathematical Modelling of Hierarchical Structures in Designing", *Proc. of PArC' 79 Conference*, AMK Berlin, Online, Berlin, June, 1979.

[7] SINGH, M.G., *Dynamical Hierarchical Control*, North-Holland, 1977.

STRUCTURAL CONTROL, H.H.E. Leipholz (ed.)
North-Holland Publishing Company & SM Publications
© IUTAM, 1980

ACTIVE CONTROL OF FLUTTER AND VIBRATION OF AN AIRCRAFT

O. Sensburg, J. Becker and H. Hönlinger

Messerschmitt-Bölkow-Blohm GmbH
Airplane Division
Postfach 801160, D-8012 Ottobrunn, West Germany

1. INTRODUCTION

The design of combat aircraft is influenced by performance requirements of prescribed tasks like the high maneuvrability in the air to air or air to ground mission.

While the successful operation of a the rigid aircraft is often achieved in the wanted flight envelope by suitable aerodynamic design, there will remain structural dynamic problems which may lead to restrictions in the total useability of the flight envelope due to flutter and vibration and dynamic loading of the aircraft and its substructures. Though optimal dynamic structural design methods are always used in addition to the aerodynamical and flight mechanical design concepts, they often give undesirable high disadvantages due to stiffness or mass increments which may reduce the flight operation significantly.

However active control of flutter or vibrations of the aircraft may re-establish the original design properties.

2. ACTIVE FLUTTER SUPPRESSION (AFS)*

During the life time of high performance combat aircraft a large number of new external stores will be developed which must be carried for tactical reasons. Most of these new stores will be accommodated within the already flutter cleared range of weights and radii of gyration of existing stores. Some of these wing mounted stores or store combinations may cause dangerous flutter instabilities within the operational flight envelope. This results in extensive and costly store clearance programmes.

The classical means of preventing flutter such as stiffness increase or mass balance are very expensive during service life and also penalize the aircraft's performance. Because of the wide range of different stores and store combinations mounted on the wing pylon, the only suitable means of preventing flutter is limiting the speed. This in turn could mean an increase in the vulnerability of the aircraft during a ground attack mission.

Many contemporary fighter attack aircraft have flutter restrictions in the range of Ma = 0.7 to Ma = 1.0 at sea level. For large and unconventional stores flutter occurs even at Ma = 0.4 S.L.

In some cases (especially asymmetric combinations) flutter was not actually encountered; low damping, however, limited the performance of the aircraft.

The inventory of advanced fighter-attack aircraft of the future will most likely be limited in system type and quality because of soaring development costs. This trend will dictate a high performance combat aircraft and the use of many external store configurations. Based on the growth patterns of aircraft/weapon systems of the past, store flutter limits on these advanced aircraft appear inevitable and within the same speed range as they are found today unless improved flutter suppression techniques

---

*Chapters 2, 3 and 4 of this paper are part of References [14, 15]

are defined. It is also believed that the increased use of composite materials will not solve all these wing store flutter problems.

Active flutter suppression is therefore a possible and promising solution. Sometimes also flutter modes occur with large contributions of fuselage motions which are very difficult to use especially when aerodynamic interferences between wing and tail occur (13). These problems could be solved with AFS.

Several attempts have been made and reported to actively suppress flutter on fighter airplanes either "on paper" or accompanied by wind tunnel tests [1], [2], [3], [4].

As an example research done at MBB, sponsored by the German Ministry of Defence, is reported here.

2.1 *Active Flutter Suppression on Wings with External Stores*

A system which is capable of suppressing wing store flutter was developed and tested on a full flying subsonic wind tunnel model in the flutter tunnel of the Eidgenössisches Flugzeugwerk in Emmen [5].

The model is shown in Figure 1, the diagram of the vane control system is shown in Figure 2.

Figure 1 - Subsonic Flutter Model

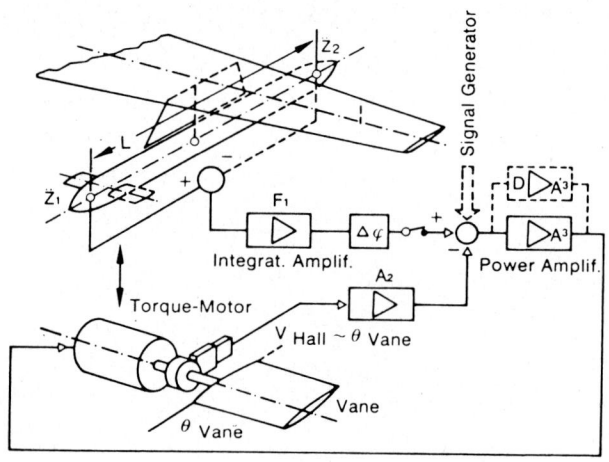

Figure 2 - Block Diagram of the Vane Control System

The control system drives a vane attached to a store controlled by a feedback signal in such a way that it counteracts the store motion. The developed mechanism can also be utilized for conventional flight flutter testing excitation techniques such as frequency sweep or harmonic signals together with a method for a quick finding of the frequency and damping of the flutter mode. Another application of the system is the reduction of external store amplitudes created by buffet or air turbulence, thus increasing the fatigue life of wing pylon attachments, improving the target aiming of weapons and sharpness of pictures shot by reconnaissance cameras in wing mounted pods. Figure 3 shows the time histories for speeds above the flutter speed. These graphs were produced in the wind tunnel by switching the active flutter suppression system (AFSS) on and off. Since the system is unstable, it flutters when the suppression is off and it decays when it is on.

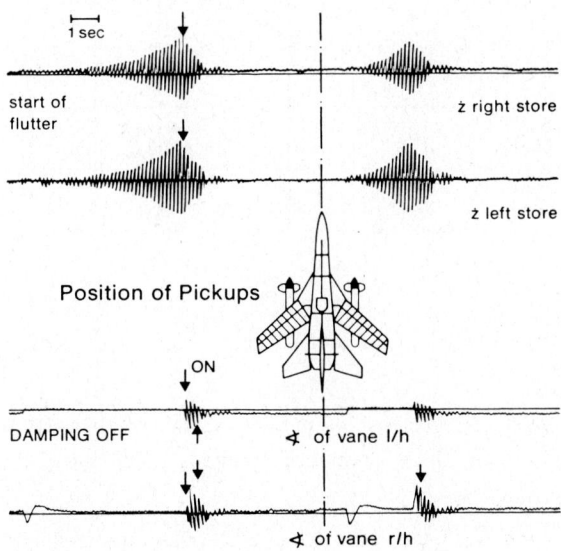

Figure 3 - Times Histories of Various Signals

In Figure 4, a comparison of measured and calculated damping values for a store configuration is given. One percent of structural damping must be added to mode 2 damping, because the drag force is creating additional damping when the model was supported on its rod in the tunnel. Considering the complicated flutter mechanism, correlation of test and analysis is very good. The analysis underestimates the tunnel flutter speed only by about ten percent (AFSS off) and gives the same damping trend (AFSS off and on).

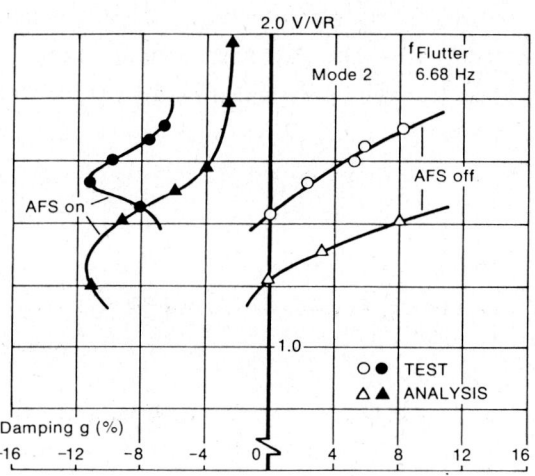

Figure 4 - Comparison of Measured and Calculated Velocity

Because the model is free flying in the tunnel a high angle of attack could be simulated that caused wing stall and a high noise environment.

Figure 5 shows that the AFSS reduces the response of the store considerably. The wing response is not attenuated as much. This is due to the fact that the wing mainly responds in its bending mode. Very little damping force can be introduced into this mode at the wing pylon station.

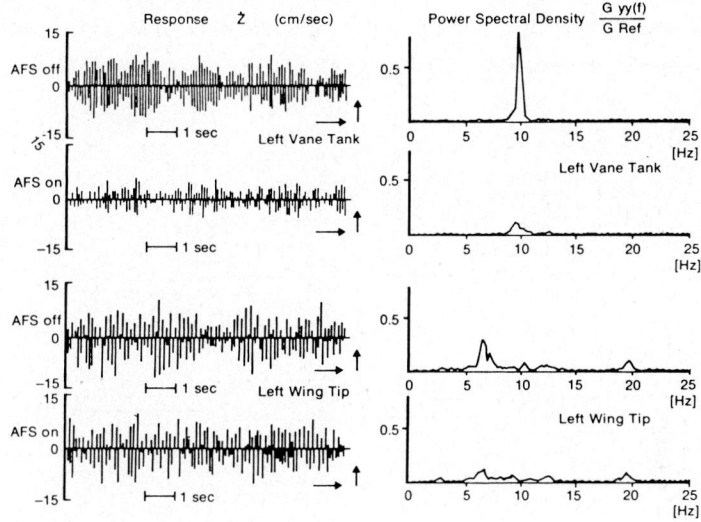

Figure 5 - Mode Buffet Response

2.2 Active Control of Empennage Flutter

After a successful application of the active flutter control technology to a wing store flutter problem, an extension of this technology to am empennage flutter problem was considered to be rewarding [6].

A control system was developed which actively suppresses a total airplane model flutter problem by counteracting with a hydraulically driven rudder. Such a system could be easily implemented into modern fighter designs. The flutter phenomenon suppression is characterized by a large contribution of fuselage torsional movement, therefore producing high inertial forces in comparison with unsteady aerodynamic forces. These properties lead to a mild onset of flutter. Such a mode is very apt to active control technology.

For the wind tunnel tests, the aforementioned subsonic flutter model and the control system have been modified. Figure 6 shows a block diagram of this flutter suppression system.

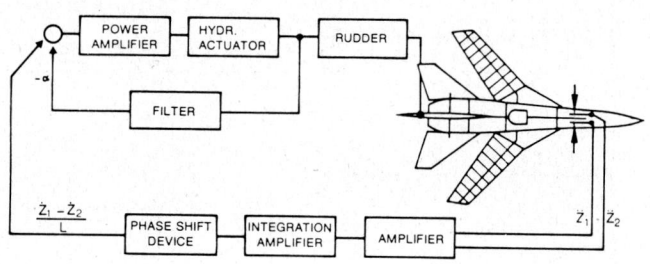

Figure 6 - System Block Diagram

It can be said that the aim to control an empennage flutter mode was achieved. Figure 7 shows the flutter speed versus damping for the stabilized and unstabilized system as well as a comparison with the analytical prediction for the unstabilized system.

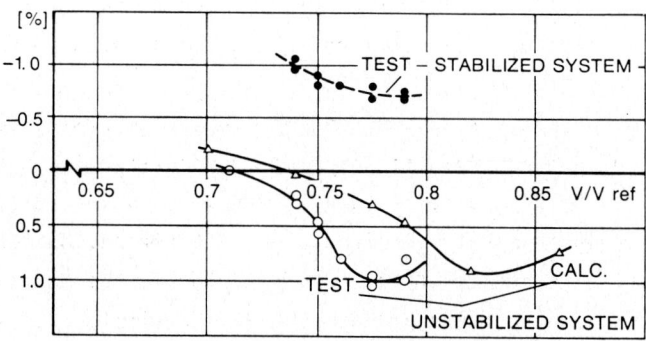

Figure 7 - Flutterspeed versus Damping for Stabilized and Unstabilized Systems

2.3 Flight Test Programme of a Wing Store Flutter Suppression on FIAT G 91

A follow-up effort was focused on the flight test of a wing/store flutter suppression system with store mounted vanes [7]. This programme, started in 1975, with the design of the system and the instrumentation of a FIAT G 91 as flying test bed.

Test objectives of this study were to:
- provide first flight experience with a flutter suppression system on external stores;
- evaluate new methods of flight flutter testing of external stores;
- investigate the system's behaviour in turbulent air.

The aircraft with stores and AFSS is shown in Figure 8.

Figure 8 - FIAT G 91 with Automatic Flutter Suppression System

The aim of flutter suppression could be demonstrated in flight and the automatic mode excitation was successfully applied. Time histories of the stable and artificially unstable system are shown in Figure 9. The efficiency of the system is presented in Figure 10.

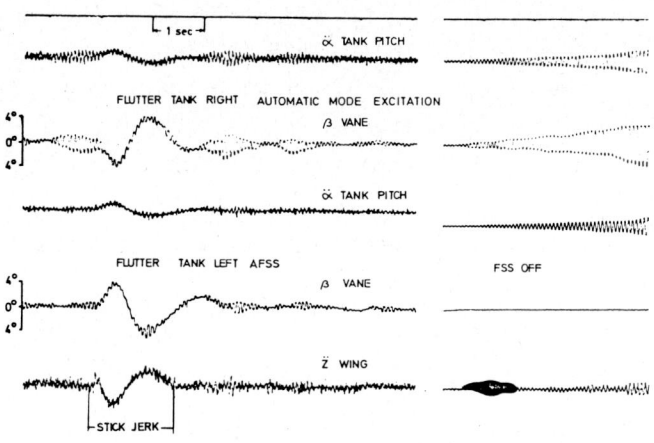

Figure 9 - Time Histories of Stable and Artificial Unstable System

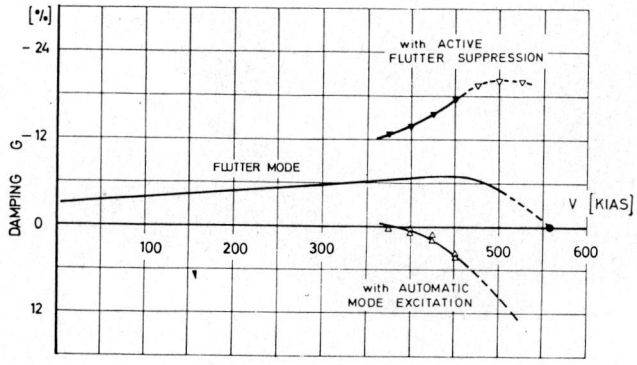

Figure 10 - Efficiency of AFSS

2.4 Active Flutter Suppression on a F4 Phantom Aircraft

A flight test programme is now performed to study flutter suppression using already existing control surfaces (ailerons) and sensors on the wing of the F-4F. This programme will be conducted in co-operation with the AFFDL (Air Force Flight Dynamics Laboratory). The objective of this programme is to develop and flight test an active flutter suppression system (AFSS) which can become a possible candidate for an operational flutter suppression system on any aircraft. A flutter calculation for the proposed system is shown in Figure 11.

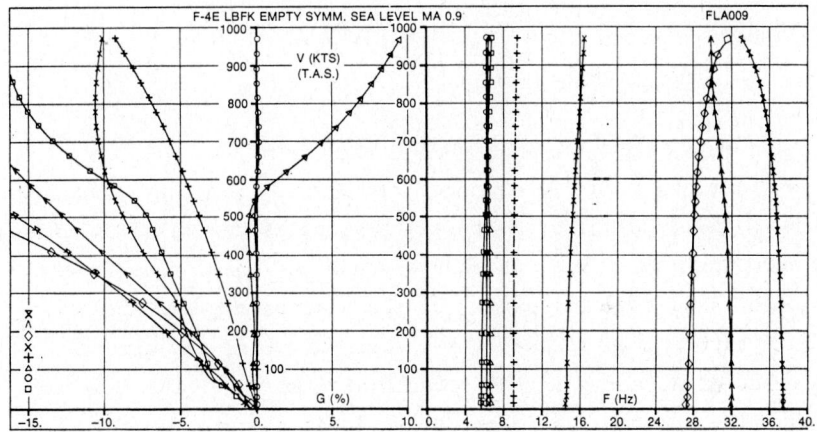

Figure 11 - *Damping and Frequency versus Speed for F-4F with Dummy Store*

3. USE OF OPTIMAL CONTROL LAWS FOR FLUTTER SUPPRESSION

Theoretical applications of optimal control laws for control of an elastic aircraft have been reported in [8] and [9].

The equations of motion for the forced dynamic response of an aeroelastic system can be written in matrix differential equation form:

$$m_r b_r^2 \begin{bmatrix} M_{qq} & M_{q\beta_0} \\ M_{\beta_0 q} & M_{\beta_0 \beta_0} \end{bmatrix} \begin{Bmatrix} \ddot{q} \\ \ddot{\beta}_0 \end{Bmatrix} + \frac{s_R}{kv} \begin{Bmatrix} \omega_r^2 m_r b_r^2 \end{Bmatrix} \begin{vmatrix} gK_{qq} & 0 \\ 0 & K''_{\beta_0 \beta_0} \end{vmatrix} +$$

$$+ \frac{\rho}{2} v^2 F s_R \frac{b_r^2}{s_R^2} \begin{vmatrix} c''_{qq} & c''_{q\beta_0} \\ c''_{\beta_0 q} & c''_{\beta_0 \beta_0} \end{vmatrix} \begin{Bmatrix} \dot{q} \\ \dot{\beta}_0 \end{Bmatrix} + \begin{Bmatrix} \omega_r^2 m_r b_r^2 \end{Bmatrix} \begin{vmatrix} K_{qq} & 0 \\ 0 & K'_{\beta_0 \beta_0} \end{vmatrix} +$$

$$+ \frac{\rho}{2} v^2 F s_R \begin{vmatrix} c'_{qq} & c'_{q\beta_0} \\ c'_{\beta_0 q} & c'_{\beta_0 \beta_0} \end{vmatrix} \begin{Bmatrix} q \\ \beta_0 \end{Bmatrix} = \begin{Bmatrix} Q(t) \end{Bmatrix}, \qquad (1)$$

where m, br and ωr are chosen reference mass, length and frequency and M, K and C are referred to as the generalized mass, stiffness and aerodynamic matrices which are nondimensional. The true airspeed v and the semispan s_R of the reference plane are used to form the reduced frequency $k = (\omega s_R)/v$. F is the area of reference plane and g is the structural damping of the elastic modes. The generalized forces Q are equal to zero for the conventional flutter problem. The generalized coordinate q describes the amplitude of the rigid body modes and the elastic airplane modes including elastic control surface modes for a system with actuators assumed to be rigid, whereas β_0 denotes the rotation of the rigid control surface according to the complex actuator stiffness represented by the impedance function of equation (2),

$$K_{\beta_0 \beta_0} = K'_{\beta_0 \beta_0} + iK''_{\beta_0 \beta_0} \ . \qquad (2)$$

For the controlled aircraft the servo-induced control deflection $\Delta\beta$ has to be introduced as an additional degree of freedom for each control surface. The generalized forces Q generated by the servo-induced control deflections $\Delta\beta$ can be described as the right-hand term of equation (1) by

$$\{Q(t)\} = -m_r b_r^2 \begin{Bmatrix} M_{q\Delta\beta} \\ M_{\beta_0\Delta\beta} \end{Bmatrix} \Delta\ddot{\beta} - \frac{\rho}{2} v^2 Fs_R \frac{b_r^2}{s_R^2} \frac{s_R}{k \cdot v} \begin{Bmatrix} c''_{q\Delta\beta} \\ c''_{\beta_0\Delta\beta} \end{Bmatrix} * \Delta\dot{\beta}$$

$$- \frac{\rho}{2} v^2 Fs_R \frac{b_r^2}{s_R^2} \begin{Bmatrix} c'_{q\Delta\beta} \\ c'_{\beta_0\Delta\beta} \end{Bmatrix} \Delta\beta \ . \tag{3}$$

Assuming normalized rigid control surface modes β_0 and $\Delta\beta$, the rotation of each control surface can be superimposed by

$$\beta = \beta_0 + \Delta\beta \ . \tag{4}$$

For the case described here, β is the aileron rotation.

Dividing equation (1) by $m_r \cdot b_r^2 \cdot \omega_r^2$, approximating the unsteady aerodynamic forces with a polynomial in $s = i\omega$ for the reduced frequency k at the flutter point

$$(C' + iC'') = a_0 + a_1 s + a_2 s^2 \ , \tag{5}$$

and introducing the actuator function as

$$\frac{\Delta\beta}{x_i} = F_{ACT} = \frac{1}{1 + b_1 s + b_2 s^2} \ , \tag{6}$$

which provides the necessary condition for the added control degree of freedom $\Delta\beta$ then the state-space-description of (1) is as follows:

$$\{\dot{x}\} = [A]\{x\} + \{B\}x_i \ , \tag{7}$$

where

$$\{x\} = \begin{Bmatrix} q \\ \beta_0 \\ \Delta\beta \\ \dot{q} \\ \dot{\beta}_0 \\ \dot{\Delta\beta} \end{Bmatrix} = \text{state vector} \quad , \tag{8}$$

$$[A] = \begin{vmatrix} 0 & 1 \\ [s_1]^{-1} \cdot [s_2] & [s_1]^{-1} \cdot [s_3] \end{vmatrix} \quad , \tag{9}$$

$$[s_1] = \left[-\frac{1}{\omega_r^2} \begin{vmatrix} M_{qq} & M_{q\beta_0} & M_{q\Delta\beta} \\ M_{\beta_0 q} & M_{\beta_0\beta_0} & M_{\beta_0\Delta\beta} \\ 0 & 0 & \omega_r^2 b_2 \end{vmatrix} + \frac{\rho}{2} v^2 \frac{F}{s_R} \frac{1}{m_r \omega_r^2} * \begin{vmatrix} a_{2qq} & a_{2q\beta_0} & a_{2q\Delta\beta} \\ a_{2\beta_0 q} & a_{2\beta_0\beta_0} & a_{2\beta_0\Delta\beta} \\ 0 & 0 & 0 \end{vmatrix} \right] \quad , \tag{10}$$

$$[s_2] = \left[\begin{vmatrix} K_{qq} & 0 & 0 \\ 0 & K'_{\beta_0\beta_0} & 0 \\ 0 & 0 & 1 \end{vmatrix} + \frac{\rho}{2} v^2 \frac{F}{s_R} \frac{1}{m_r \omega_r^2} \begin{vmatrix} a_{0qq} & a_{0q\beta_0} & a_{0q\Delta\beta} \\ a_{0\beta_0 q} & a_{0\beta_0\beta_0} & a_{0\beta_0\Delta\beta} \\ 0 & 0 & 0 \end{vmatrix} \right] \quad , \tag{11}$$

$$[s_3] = \left[\frac{s_R}{vk} \left(\begin{vmatrix} gK_{qq} & 0 & 0 \\ 0 & K''_{\beta_0\beta_0} & 0 \\ 0 & 0 & \frac{vk}{s_R} b_1 \end{vmatrix} + \frac{\rho}{2} v^2 \frac{F}{s_R} \frac{1}{m_r \omega_r^2} * \begin{vmatrix} a_{1qq} & a_{1q\beta_0} & a_{1q\Delta\beta} \\ a_{1\beta_0 q} & a_{1\beta_0\beta_0} & a_{1\beta_0\Delta\beta} \\ 0 & 0 & 0 \end{vmatrix} \right) \right] \quad , \tag{12}$$

and

$$\{B\} = \begin{Bmatrix} 0 \\ 0 \\ 0 \\ [S_1]^{-1} \cdot \begin{Bmatrix} 0 \\ 0 \\ 1 \end{Bmatrix} \end{Bmatrix}$$

To get the optimal control law K_{opt} the quadratic performance criterion is minimized

$$J = \int_0^\infty (\{x\}^T [Q]\{x\} + x_i R x_i) dt . \qquad (14)$$

Q is a weighting matrix found by trial and error using a screen together with the computer. R is a scalar, to be selected, because there is only one control surface.

Minimizing (14) leads to the optimal control law

$$x_i = \{K_{opt}\}^T \cdot \{x\} , \qquad (15)$$

with

$$K_{opt}^T = -R^{-1} \{B\}^T [P] , \qquad (16)$$

where P is the steady state solution of the Matrix Riccati equation

$$[-P] = [P][A] + [A]^T[P] - [P]\{B\}R^{-1}\{B\}^T[P] + [Q] . \quad (17)$$

In our case the complete state can be measured and fed back: the flutter system can be described by two modes g,
  (1) first wing bending,
  (2) first wing torsion and store pitch.
These modes can be measured by two accelerometers on the wing, the signals of which will be added and subtracted.

The aileron motion can be described by:

$$\frac{m_\beta}{k_\beta} = \beta_0 , \qquad (18)$$

where m_β = hinge moment measured by strain gauge,
k_β = actuator impedance function.

4. IMPACT OF ACTIVE CONTROL ON STRUCTURES

This chapter deals with the methods of analysis and experimental procedures which have to be used when implementing ACT into an elastic aircraft. These methods can be used for the layout of the system and to assure against detrimental coupling of the structure with the control system.

Three newly designed fighter or combat airplanes have been reported to have exhibited some kind of the above mentioned interaction [10], [11], [12]. We shall restrict ourselves just to describe the investigation procedures used on the Tornado (being similar to the ones used in the other references) and want to point out, that the normally used concept of implementing notch filters into the control system is no more applicable when ACT-functions requiring the higher regime such as vibration mode control flutter mode control are installed.

Several input data to form mathematical models must be determined or predictions must be checked and if necessary corrected by tests. These tests are: Ground Resonance Test to check the structure elastic behaviour, transfer and impedance functions for the hydraulic actuators for several parameters (because of non-linearities).

Open loop tests on the complete system - aircraft-control-system - must be done and stability must be assured before the electrical loops can be closed. Niquist diagrams for the Tornado - being the first operational aircraft the features a triplex analogue fly-by-wire system and automatic stabilization - are shown in Figure 12. This figure shows an unstable structural mode because the point -1 is encircled clockwise. After a notch filter was implemented the system was stable.

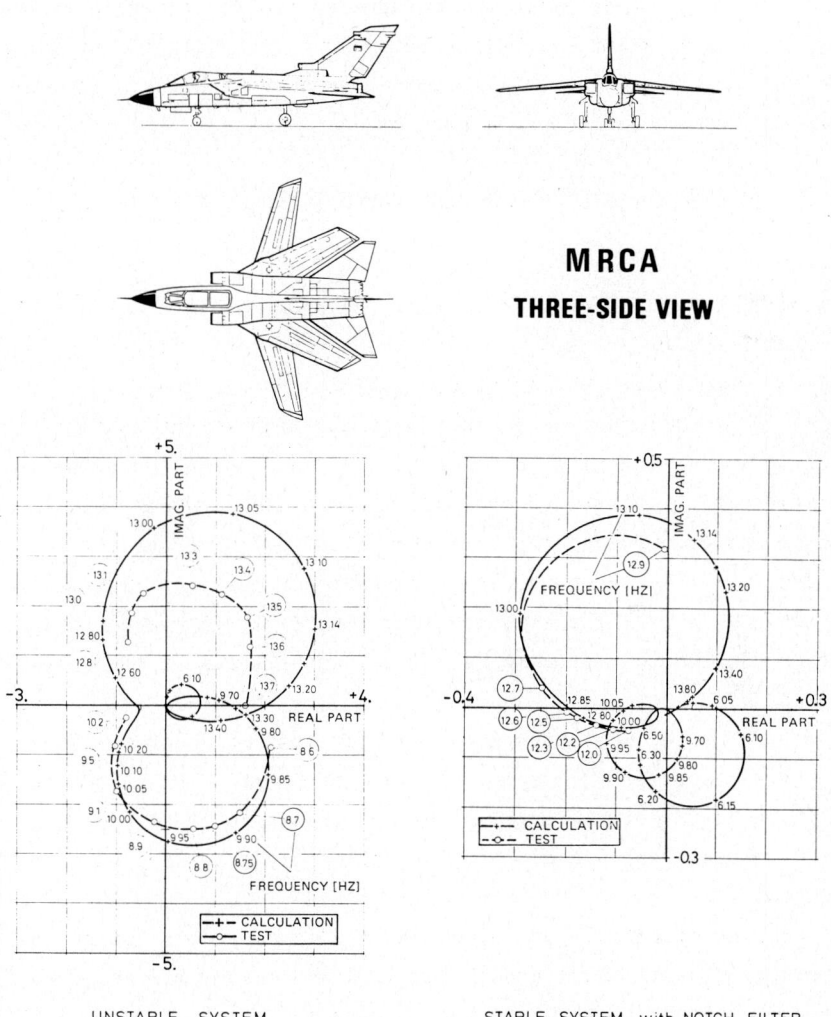

Figure 12 - Niquist Diagram to Check CSAS - Structural Mode Coupling

If other ACT-functions than damping rigid body modes would have been applied (requiring the higher frequency regime) then the stabilization of this structural mode would have required other means like relocation of rate gyros.

Flight tests with the engaged control system have to be done as well to assure stability throughout the whole flight regime. All these investigations will become more complex with an increasing number of interconnected ACT-functions.

5. ACTIVE CONTROL SYSTEM FOR RIDE COMFORT IMPROVEMENT OF THE ELASTIC AIRCRAFT

An active control concept to improve the ride qualities of a combat aircraft has been investigated for the normal level flight in the wind 68° swept back configuration at high subsonic speeds.

At different sweep angles of the wing flight test results indicate increasing pilot discomfort with increasing speed due to vibration in the first symmetrical fuselage bending mode which leads to pilot nodding at 9 Hz. The pilot feels angular motions but does not feel the vertical acceleration because he is sitting on the node line of the elastic mode. The first fuselage bending mode is excited by rear fuselage flow instabilities in combination with gusts.

For the investigation of a ride comfort control system for the alleviation of a specific elastic vibration behaviour an analytical study of the total elastic aircraft response and open loop frequency response measurements from flight testing have been combined.

An effective control of the fuselage bending motion was believed to be achieved through acceleration measurement of the mode and feedback to the taileron, Figure 13.

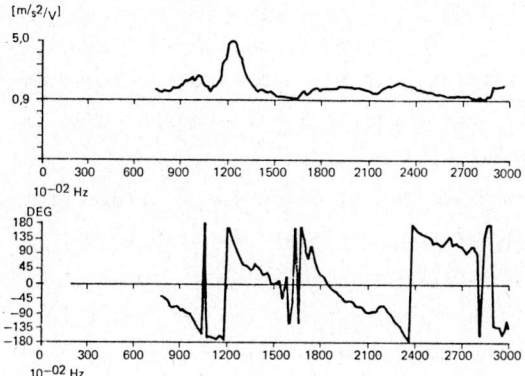

(a)

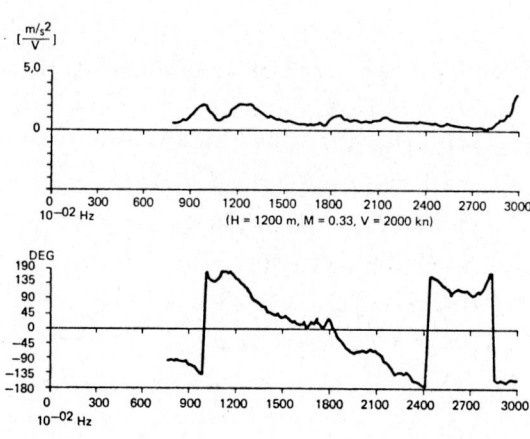

(b)

Figure 13 - (a) Vertical Acceleration at Front Fuselage Station Frequence Response in Ground Test
(b) Vertical Acceleration Front Fuselage Station Frequency Response in Flight Test

5.1 Flight Test Results

The aim of flight testing at 68° wing sweep using symmetrical excitation of the horizontal tail with engaged CSAS (stability augmentation system) was to establish frequency response characteristics of accelerometers at different fuselage stations. The order of magnitude of the accelerations at 9 Hz will give the answer whether an active control system is possible. The results indicate that accelerations are excited at 9 Hz which have similar amplitudes as in the normal level flight case.

5.2 Analytical Design of the Control Law

5.2.1 Description of the Theoretical Model

Twelve elastic eigenmodes are used to model the aircraft in the symmetrical 68° wing sweep configuration. The eigenfrequencies of these modes are listed below:

- (1)   6.08 Hz (wing bending)
- (2)   9.55 Hz (first fuselage bending)
- (3)   10.05 Hz (wing for and aft)
- (4)   11.81 Hz (first taileron mode)
- (5)   17.12 Hz (second taileron mode)
- (6)   17.66 Hz
- (7)   21.88 Hz
- (8)   24.28 Hz
- (9)   25.25 Hz
- (10) 25.95 Hz
- (11) 62.48 Hz
- (12) 191.62 Hz

The generalized masses and generalized aerodynamic forces are the same as used for the analytical flutter clearance for the aircraft. The accuracy of these input data for the analytical dynamic response model is therefore believed to be high.

The analytical model of the aircraft is the same as presented in paragraph 3. The introduction of individual structural damping for the different vibration modes derived from ground resonance tests was necessary to predict the dynamic response result.

In addition the tail actuator transfer function and the CSAS transfer function are introduced into the calculation (Figures 14, 15 and 16). The CSAS together with the tail actuator transfer function are necessary for the calculation, since the elastic aircraft response may be influenced through the elastic components in the rate gyro signal and their feedback through the CSAS feedback.

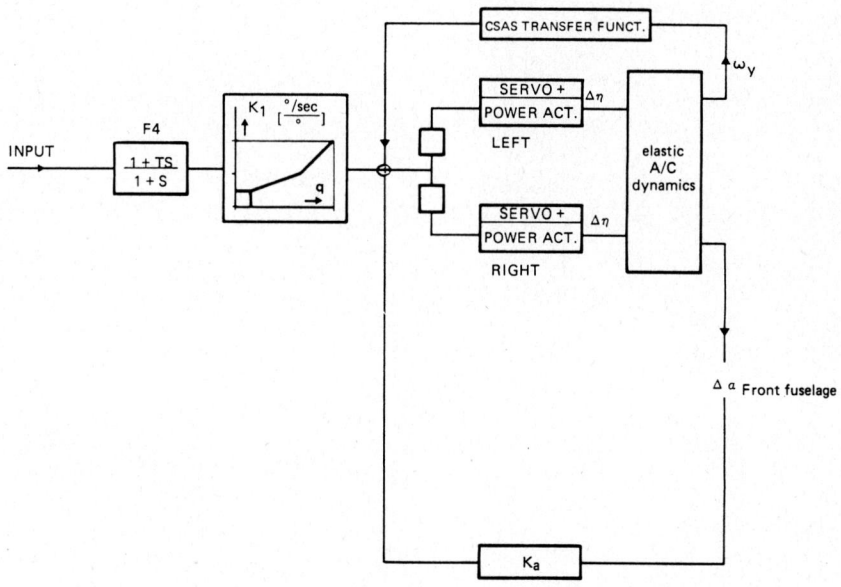

Figure 14 - *Scheme of Ride Improvement System Frig Box Excitation in CSAS-Mode*

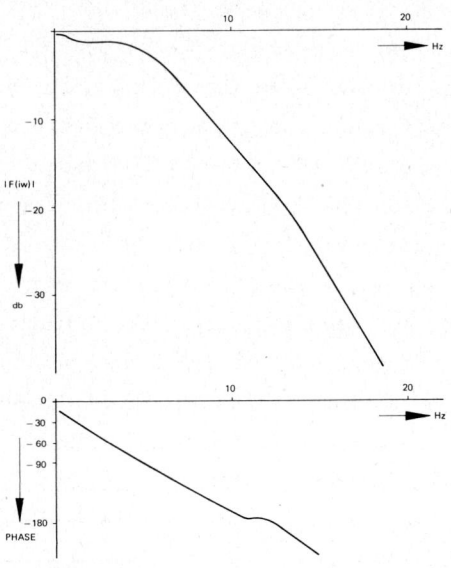

Figure 15 - Tail Actuator Frequency Response Used in Analysis

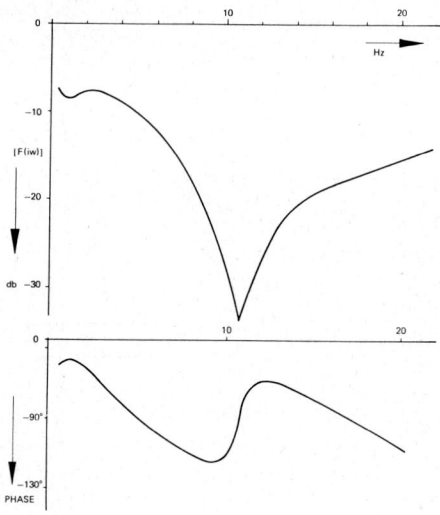

Figure 16 - CSAS Frequency Response (without Actuator) used in Analysis

Using the ride improvement acceleration feedback concept the influence of amplitude dependent nonlinearities of the actuator transfer function will limit the practical application of such a system. However here nonlinear effects are small in the frequency range less than 10 Hz, the range of interest in the considered system.

5.3 Test of the Analytical Model

The following steps have been performed to check the results of the open loop analytical model.

- Recalculation of flutter characteristics with closed CSAS and comparison with flutter analysis results without CSAS. The influence of the elastic mode feedback of the CSAS does not change the stability behaviour of the elastic modes significantly. This was expected since the structure filter of the CSAS reduces the influence of elastic mode feedback.

- Calculation of the open loop transfer function of the front fuselage accelerometer signal using the excitation of the aircraft with the installed frequency generator and comparison with flight test results (flight condition M = 0.33, altitude 1200 m, CSAS on), (Figures 17 and 18). Reasonable correlation with flight test results could be predicted especially in the frequency region of the first fuselage bending. Differences in the natural frequencies of the modes and the amplitude of the vibration are strong at higher frequencies. The amplitude of the acceleration is influenced through the introduction of structural damping.

The differences may be caused through the effects on frequency and phase shift of the frequency sweep excitation technique used in flight test.

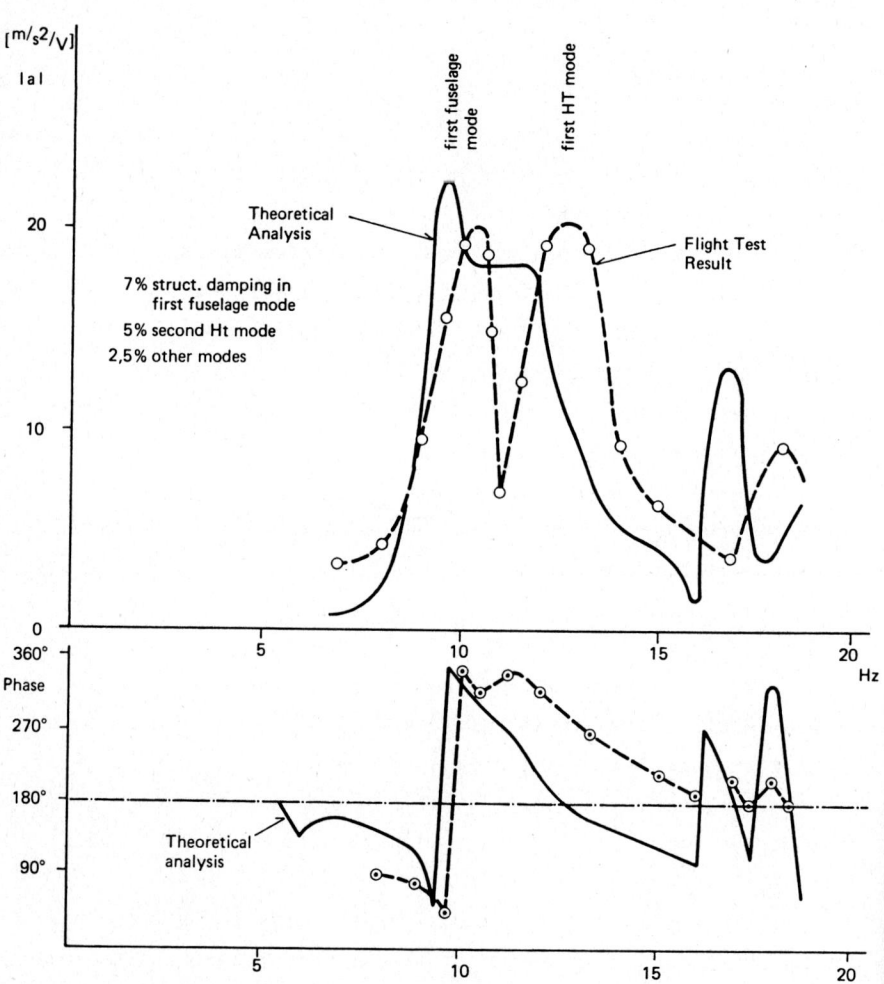

Figure 17 - Comparison Flight Test Result and Analysis open Loop Front Fuselage Acceleration

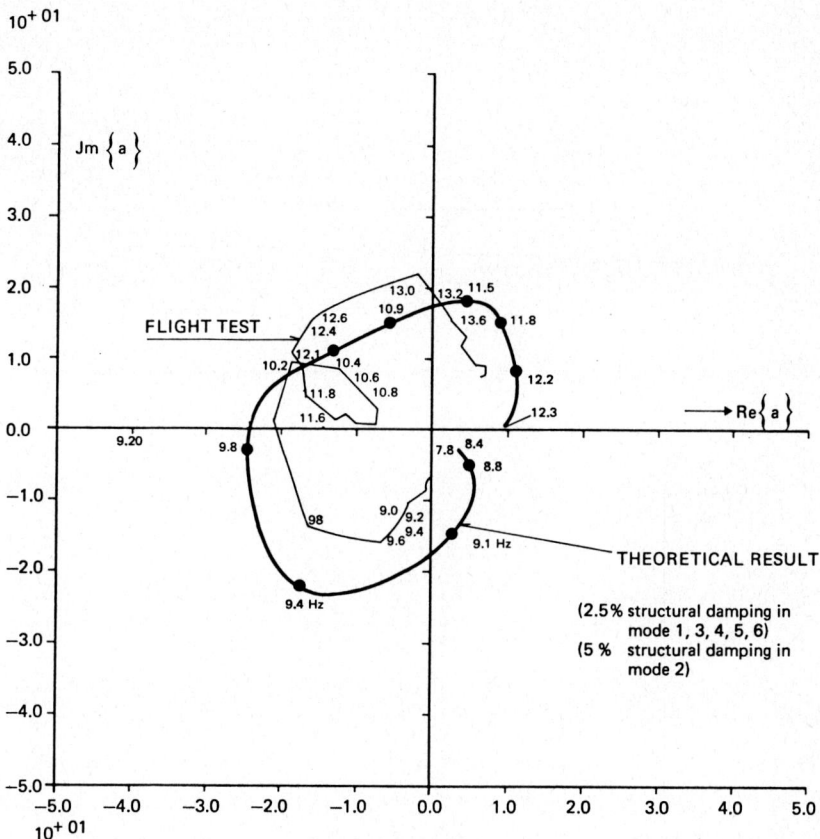

Figure 18 - Open Loop Frequency Response, $M = 0.3$, $H = 1200$ m of Front Fuselage Acceleration

5.4 Stability Investigation of the Ride Improvement Control System

A gain root locus investigation for the closed loop system is presented in Figure 19 for the design flight condition $M = 0.9$, altitude 0 m.

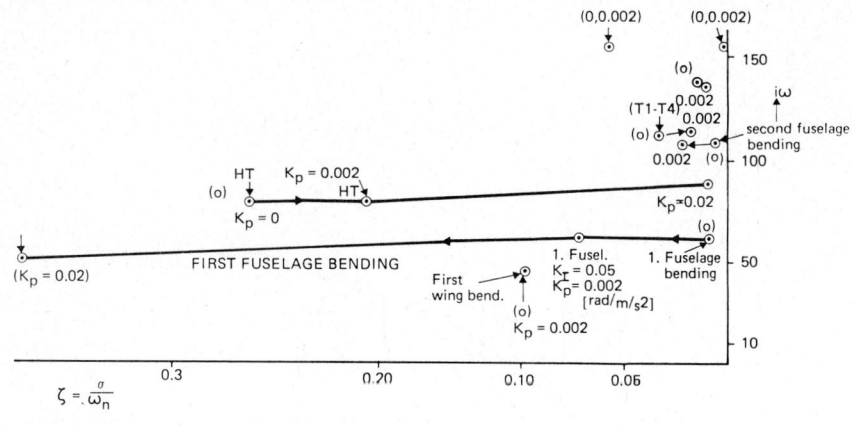

Figure 19 - *Buffet Alleviation 68°, Sensor Location 1, M = 0.9, H = 0, Gain Root Locus, CSAS on (Zero Structural Damping)*

A sufficient feedback gain of the front fuselage acceleration is reached with $K_a = 0.002$ rad/m/s². The damping of the first fuselage bending mode is increased by 6% through the feedback. Caused by the sensor location (front fuselage) the damping of the first tail mode is reduced, however this mode remains still stable at a gain of $K_a = 0.02$ rad/m/s².

5.5 *Dynamic Response*

The transfer function of the front fuselage acceleration is presented in Figure 20 both for the aircraft with and without acceleration feedback. The amplitude at 9 Hz of the signal is reduced to 1/3 of the value without feedback. Amplitudes at higher frequencies are visible. However these are produced by the impulse input. The excitation through gust and rear fuselage buffet will show strong decrease of the response amplitude with frequency because these excitation spectra have only small energy content at high frequencies.

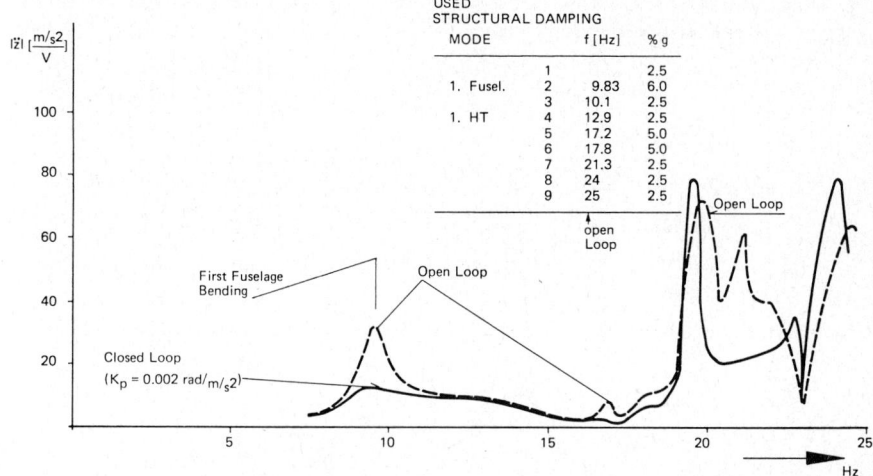

Figure 20 - *Theoretical Result Open Loop with Closed Loop (Buffet Alleviation) M = 0.9, H = 0 (600 kn)*
 $|\ddot{Z}|$ *Due to Harmonical Excitation with Frigbox in CSAS Mode*

5.6 Sensorlocation

In addition to the front fuselage sensor location the signal feedback of the sensor location near the C.G. has been studied. The result (Figure 21) shows excellent alleviation of acceleration at 9Hz but destabilization of a higher frequency mode.

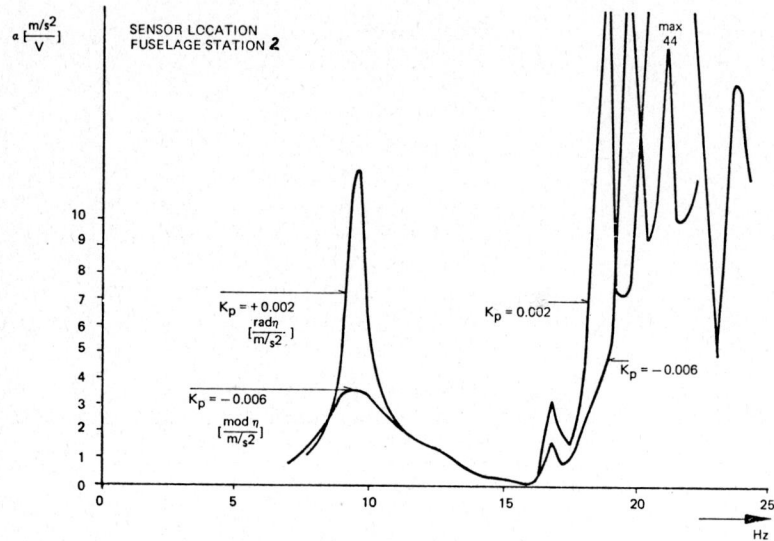

Figure 21 - Theoretical Result Closed Loop (Sensor Location Fuselage Station 2) $\ddot{z}$ due to Harmonical Excitation with Frig Box in CSAS Mode. Feed Back of c.g. Acceleration, M = 0.9, H = 0, CSAS on

6. SUMMARY

Active control concepts are summarized to give a survey on activities of MBB/UF concerning flutter prevention and vibration alleviation due to external excitation forces caused by gusts and buffet. Special examples are presented for the practical important case of store flutter prevention and the reduction of elastic fuselage and external store vibrations due to gusts and separated flow. An analytical procedure for the derivation of aeroelastic control laws is described. Examples of windtunnel testing on dynamically scaled models with active control devices are mentioned together with flight test results.

REFERENCES

[1] PERISHO, C.H., TRIPLET, W.E. and MYKYTOW, W.J., "Design Considerations for an Active Suppression System for Fighter Wing/Store Flutter", *AGARD Conference Proceedings, Specialists Meeting on Flutter Suppression and Structural Load Alleviation*, No. 175, AGARD-CP-175.

[2] DESTUYNDER, R., "Essai en Soufflerie d'un Suppresseur de Flottement sur une Aile droite", *AGARD Conference Proceedings, Specialists Meeting on Flutter Suppression and Structural Load Alleviation*, No. 175, AGARD-CP-175.

[3] DESTUYNDER, R., "Etude en Soufflerie d'un Suppresseur de Flottement", *La Recherche Aerospatiale*, No. 1976-4.

[4] ABEL, I. and SANDFORD, M.C., "Status of Two Studies on Active Control of Aeroelastic Response at NASA Langley Research Center", *AGARD Conference Proceedings, Specialists Meeting on Flutter Suppression and Structural Load Alleviation*, No. 175; AGARD-CP-175.

[5] HAIDL, G., LOTZE, A. and SENSBURG, O., "Active Flutter Suppression of Wings with External Stores", *AGARDograph*, No. 175.

[6] SENSBURG, O. and HÖNLINGER, H., "Active Control of Empennage Flutter", *AGARD Conference Proceedings, Specialists Meeting on Flutter Suppression and Structural Load Alleviation*, No. 175, AGARD-CP-175.

[7] HÖNLINGER, H., "Active Flutter Suppression on an Airplane with Wing Mounted External Stores", paper presented at the *44th Meeting of SMP*, April 17-22, 1977, Lisbon, Portugal.

[8] DRESSLER, W., "Control of an Elastic Aircraft Using Optimal Control Laws", *AGARD Conference Proceedings on Impact of Active Control Technology on Airplane Design*, No. 157, AGARD-CP-157.

[9] TURNER, M., "Active Flutter Suppression", *AGARD Conference Proceedings, Specialists Meeting on Flugger Suppression and Structural Load Alleviation*, No. 175, AGARD-CP-175.

[10] LOTZE, A., SENSBURG, O. and KÜHN, M., "Flutter Investigation of a Combat Aircraft with a Command and Stability Augmentation System", *Journal of Aircraft*, Vol. 14, No. 4, April, 1977.

[11] ARTHUR, T.D. and GALLAGHER, J.T., "Interaction Between Control Augmentation System and Airframe Dynamics on the YF 17", *AIAA*, Paper No. 75-747.

[12] PELOUBET, R.P., "YF 16 Active-Control-System/Structural Dynamics Interaction Instability", *AIAA*, Paper No. 75-824.

[13] SENSBURG, O. and LASCHKA, B., "Flutter Induced by Aerodynamic Interference Between Wing and Tail", *Journal of Aircraft*, Vol. 7, No. 4, July-August, 1970.

[14] HÖNLINGER, H. and DESTUYNDER, R., "External Store Flutter Suppression with Active Controls", *VKI Lecture Series*, 1979-1, December, 1978.

[15] SENSBURG, O. and ZIMMERMANN, H., "Impact of Active Control on Structures Design", *Fighter Aircraft Design*, AGARD-CP-241, October, 1977.

STRUCTURAL CONTROL, H.H.E. Leipholz (ed.)
North-Holland Publishing Company & SM Publications
© IUTAM, 1980

ON OPTIMAL CONTROL CONFIGURATION IN THEORY OF MODAL CONTROL

T.T. Soong

Department of Civil Engineering
State University of New York at Buffalo
Buffalo, New York 14214, U.S.A.

M.I.J. Chang

Goodyear Aerospace Corporation
Akron, Ohio 44222, U.S.A.

1. INTRODUCTION

Active feedback control of large systems such as space structures [1, 2], tall buildings [3, 4], and chemical processes [5] introduces a number of difficult problems in the applications of modern control theory. Because of their large dimensions, standard control design offers only limited help in implementation of control to these systems. One of the important problems is that of optimal control configuration, that is, the determination of appropriate locations of controllers when, due to practical and economic considerations, only a limited number of them are available.

Consider the standard state-space systems equation

$$\dot{x} = Ax + Bu, \qquad (1)$$

where x is the state vector, A is the system matrix, B is the control matrix and u is the control vector. The aforementioned problem is then one of determining the optimal form of B given a restricted number of controllers, optimal in the sense that some performance index is optimized. This problem is considered in this paper within the framework of modal control theory. Specifically, given the control objective, an optimal B is to be found among all feasible choices. The performance index considered here is an energy criterion which takes the form

$$E = \int_0^{T_o} u^T Q u \, dt, \qquad (2)$$

where Q is a positive definite weighting matrix.

The systems considered in this paper are assumed to possess distinct complex eigenvalues. Both single-input and multi-input systems are considered. Numerical results are presented by means of examples in structural control of a multistory structure.

2. SINGLE-INPUT SYSTEMS

In this section, a method of determining an optimal control matrix for single-input systems is developed. In this case, the system is described by

$$\dot{x} = Ax + bu, \qquad (3)$$

where b is an n x 1 vector and u is a scalar. For simplicity, modal control involving alteration of only one pair of conjugate complex eigenvalues is considered first.

Let $\lambda_{1,2} = \xi_1 \pm j \xi_2$ be the eigenvalues to be replaced by the desired values $\rho_{1,2} = \eta_1 \pm j \eta_2$. The design procedure given in [6] shows that the linear feedback control u is given by

$$u = (\alpha_1 D_1^1 + \alpha_2 D_1^2)^T x, \qquad (4)$$

where $D_1 = D_1^1 + j\, D_1^2$ is the eigenvector corresponding to eigenvalue λ_1 associated with A^T, the transpose of the system matrix A. The gain coefficients α_1 and α_2 are given by [6]

$$\left.\begin{array}{l} \alpha_1 = \dfrac{e_{11}v_{12}+e_{12}v_{11}}{(v_{11}^2+v_{12}^2)\,\xi_2} \; , \\[1em] \alpha_2 = \dfrac{e_{12}v_{12}-e_{11}v_{11}}{(v_{11}^2+v_{12}^2)\,\xi_2} \; , \end{array}\right\} \qquad (5)$$

where

$$\left.\begin{array}{l} e_{11} = (\eta_1-\xi_1)^2 + \eta_2^2 - \xi_2^2 \; , \\[0.5em] e_{12} = 2(\eta_1-\xi_1)\xi_2 , \\[0.5em] v_{11} = b^T\, D_1^1 \; , \\[0.5em] v_{12} = b^T\, D_1^2 \; . \end{array}\right\} \qquad (6)$$

In order to express $u(t)$ as a function of the control vector b explicitly, equation (4) can be written in the form

$$u(t) = \operatorname{Re}\left\{(\alpha_1-j\alpha_2)[\beta_1(t)/(e_{11}+je_{12})+\beta_2(t)/(e_{11}-je_{12})]\right\}, \qquad (7)$$

where Re ( ) denotes the real part and $\beta_1(t)$ and $\beta_2(t)$ are functions of system dynamics and possible external excitations but not functions of b. In equation (7), only α_1 and α_2 are functions of b as given by equations (5) and (6).

In the single-input case, it is assumed that $b^T b = 1$ without loss of generality. The substitution of equation (7) into equation (2) (with Q = 1) shows that E is a quadratic function in α_1 and α_2. In order to minimize the energy required, it is clear that a simple criterion can be established for choosing b by requiring that $\alpha_1^2 + \alpha_2^2$ is minimized subject to the controllability condition and to the constraint $b^T b = 1$.

It is seen from equation (5) that, since

$$\alpha_1^2 + \alpha_2^2 = (e_{11}^2 + e_{12}^2)/(v_{11}^2 + v_{12}^2)\xi_2^2 , \tag{8}$$

an equivalent criterion is that $v_{11}^2 + v_{12}^2$ be maximized.

The result given above can now be stated below as a theorem. Let us first give the following definition.

Definition

The quantity $w_1 = v_{11}^2 + v_{12}^2$ as defined by equation (6) is called the energy coefficient for system defined by equation (3) for altering a pair of conjugate complex eigenvalues of the matrix A. It is noted that v_{11} and v_{12} are projections of the control vector b onto the real and imaginary components of D_1, the eigenvector associated with A^T corresponding to the eigenvalue pair undergoing alteration.

Theorem 1

Under the control law given in equation (4), a sufficient condition for optimal design of control vector b which minimizes energy E is that the energy coefficient w_1 be maximized, subject to the constraint that $b^T b = 1$ and that the matrix pair (A,b) satisfy the controllability condition.

The extension of the above result to the case in which k pairs of eigenvalues are altered can be carried out following the sequential procedure developed in [6]. Consider the case where two pairs of conjugate complex eigenvalues, $\lambda_{1,2}$ and $\lambda_{3,4}$, are to be altered to $\rho_{1,2}$ and $\rho_{3,4}$. The control is given by [6]

$$u = (\alpha_1 D_1^1 + \alpha_2 D_1^2 + \alpha_3 D_{13}^1 + \alpha_4 D_{13}^2)^T x , \tag{9}$$

where D_{13}^1 and D_{13}^2 are real and imaginary parts of the eigenvector D_{13} corresponding to λ_3 of the matrix A_1^T where A_1 is defined by

$$A_1 = A + b(\alpha_1 \, D_1^1 + \alpha_2 \, D_1^2)^T . \tag{10}$$

As a function of b, the control u is again expressible as a quadratic function of α_1, α_2, α_3, and α_4, where α_1 and α_2 are given in equation (5) and

$$\left. \begin{array}{l} \alpha_3 = \dfrac{e_{23}v_{24}+e_{24}v_{23}}{(v_{23}^2+v_{24}^2)\xi_4} , \\[6pt] \alpha_4 = \dfrac{e_{24}v_{24}-e_{23}v_{23}}{(v_{23}^2+v_{24}^2)\xi_4} , \end{array} \right\} \tag{11}$$

where

$$\left. \begin{array}{l} e_{23} = (\eta_3-\xi_3)^2 + \eta_4^2 - \xi_4^2 , \\ e_{24} = 2(\eta_3-\xi_3)\xi_4 , \\ v_{23} = b^T D_{13}^1 , \\ v_{24} = b^T D_{13}^2 , \end{array} \right\} \tag{12}$$

where it can be shown that D_{13} can be expressed as a linear function of eigenvectors associated with A^T, A being the original system matrix.

It is now easy to deduce control result when more than two pairs of eigenvalues are to be altered. Following the same procedure as one used previously, we arrive at the following general result for determining the optimal control vector b.

Definition

For altering k pairs of eigenvalues, the energy coefficient for the system given by equation (3) is defined by

$$w_k = v_{11}^2+v_{12}^2+v_{13}^2+v_{14}^2+ \cdots + v_{1(2k-1)}^2+v_{1(2k)}^2 , \tag{13}$$

where

$$v_{1(2k-1)} = b^T D_{2k-1}^1, \quad v_{1(2k)} = b^T D_{2k-1}^2. \quad (14)$$

In other words, $v_{1(2k-1)}$ and $v_{1(2k)}$ are projections of the control vector b onto the real and imaginary components of D_{2k-1}, the eigenvector associated with A^T corresponding to the eigenvalue λ_{2k-1} undergoing alteration.

Theorem 2

For modal control of k pairs of eigenvalues in single-input systems, a sufficient condition for optimal design of control vector b which minimizes energy E is that the energy coefficient w_k be maximized subject to the constraint that $b^T b = 1$ and that the matrix pair (A, b) satisfy the controllability condition.

3. MULTI-INPUT SYSTEMS

The optimal design of control matrix for multi-input systems is now developed. Consider first the design of an n x 2 matrix B given by

$$B = [b_1 \ b_2]. \quad (15)$$

For control purposes, controllability conditions must be satisfied in the design process. However, as is shown in [6], the matrix B may be designed in such a way that each column b_j of the matrix B is determined by single-mode control. Assuming that two pairs of eigenvalues are to be altered, then b_1 may be designed to alter only the first pair and b_2 the second. Let

$$u = \begin{bmatrix} u_1 \\ u_2 \end{bmatrix}. \quad (16)$$

The controls u_1 and u_2 are then given by [6]

$$u_1 = (\alpha_1 D_1^1 + \alpha_2 D_1^2)^T x ,$$
$$u_2 = (\alpha_3 D_{13}^1 + \alpha_4 D_{13}^2)^T x .$$
(17)

Where α_1 and α_2 are defined by equations (5) and (6) with b_1 replacing b and α_3 and α_4 defined by equations (11) and (12) with b_2 replacing b.

The procedure for obtaining criterion for optimal control matrix design now follows that used for single-input systems. Generalizing to the case of altering k pairs of conjugate complex eigenvalues and again noting that D_{13} can be expressed as a linear function of eigenvectors associated with the transpose of matrix A, we have the following result.

Theorem 3

For modal control of k pairs of eigenvalues with $B = [b_1 \; b_2 \; \ldots \; b_k]$, a sufficient condition for optimal design of control matrix B which minimizes energy E is that the energy coefficient w_k be maximized subject to the constraints that $b_j^T b_j = 1$, $j = 1,\ldots,k$, and that the matrix pair (A,B) satisfy the controllability condition.

In the above, the energy E is defined by

$$E = \int_0^{T_o} u^T u \, dt ,$$
(18)

with

$$u = [u_1 \; u_2 \; \ldots \; u_k]^T ,$$
(19)

and w_k is defined by equation (13) with

$$v_{k(2k-1)} = b_{2k-1}^T D_{2k-1}^1 \; , \quad v_{k(2k)} = b_{2k-1}^T D_{2k-1}^2 .$$
(20)

4. NUMERICAL EXAMPLES

In this section, numerical examples are presented for the case of structural control of a three-story structure. Let the state vector be written in the form

$$x = [x_1 \ x_2 \ x_3 \ \dot{x}_1 \ \dot{x}_2 \ \dot{x}_3]^T, \qquad (21)$$

where x_j and $\dot{x}_j$ denote, respectively, displacement and velocity of the jth floor. An approximate equation of motion for the structure is [4]

$$\dot{x} = Ax + Bu + w, \qquad (22)$$

where u is the control vector and w is external excitation. To approximate a lightly damped structure, A is given by

$$A = \begin{bmatrix} 0 & 0 & 0 & 1 & 0 & 0 \\ 0 & 0 & 0 & 0 & 1 & 0 \\ 0 & 0 & 0 & 0 & 0 & 1 \\ -.5 & .25 & 0 & -.02 & .01 & 0 \\ .25 & -.5 & .25 & .01 & -.02 & .01 \\ 0 & .25 & -.25 & 0 & .01 & -.01 \end{bmatrix} \qquad (23)$$

and

$$w = f(t)[0 \ 0 \ 0 \ 1 \ 1 \ 1]^T. \qquad (24)$$

The eigenvalues of the matrix A are

$$\left.\begin{aligned} \lambda_{1,2} &= -0.001 \pm j0.2225, \\ \lambda_{3,4} &= -0.0078 \pm j0.6234, \\ \lambda_{5,6} &= -0.0162 \pm j0.9008, \end{aligned}\right\} \qquad (25)$$

and the corresponding normalized eigenvectors are

$$C_{1,2} = C_1^1 \pm jC_1^2 ,$$
$$C_{3,4} = C_2^1 \pm jC_2^2 , \qquad (26)$$
$$C_{5,6} = C_3^1 \pm jC_3^2 ,$$

where

$$C_1^1 = \begin{bmatrix} .2420 \\ .4361 \\ .5438 \\ .0464 \\ .0836 \\ .1043 \end{bmatrix}, \quad C_1^2 = \begin{bmatrix} -.2096 \\ -.3777 \\ -.4710 \\ .0541 \\ .0974 \\ .1215 \end{bmatrix},$$

$$C_2^1 = \begin{bmatrix} .6218 \\ .2707 \\ -.4986 \\ -.0468 \\ -.0208 \\ .0375 \end{bmatrix}, \quad C_2^2 = \begin{bmatrix} .0673 \\ .2299 \\ -.0539 \\ .3871 \\ .1723 \\ -.3104 \end{bmatrix}, \qquad (27)$$

$$C_3^1 = \begin{bmatrix} .4311 \\ -.5376 \\ .2393 \\ -.0820 \\ .1022 \\ -.0455 \end{bmatrix}, \quad C_3^2 = \begin{bmatrix} .0832 \\ -.1038 \\ .0462 \\ .3870 \\ -.4826 \\ .2148 \end{bmatrix}.$$

The quantities λ_j, $j = 1, \ldots, 6$, are also eigenvalues of A^T but the corresponding eigenvectors of A^T are

$$D_{1,2} = D_1^1 \pm jD_1^2 ,$$
$$D_{3,4} = D_2^1 \pm jD_2^2 , \qquad (28)$$
$$D_{5,6} = D_3^1 \pm jD_3^2 ,$$

where

$$D_1^1 = \begin{bmatrix} -.0687 \\ -.1238 \\ -.1544 \\ -.0858 \\ -.1545 \\ -.1907 \end{bmatrix} , \quad D_1^2 = \begin{bmatrix} -.0188 \\ -.0338 \\ -.0422 \\ .3085 \\ .5558 \\ .6931 \end{bmatrix} ,$$

$$D_2^1 = \begin{bmatrix} -.3494 \\ -.1555 \\ .2802 \\ .2707 \\ .1205 \\ -.2171 \end{bmatrix} , \quad D_2^2 = \begin{bmatrix} .1731 \\ .0771 \\ -.1388 \\ .5638 \\ .2510 \\ -.4521 \end{bmatrix} , \qquad (29)$$

$$D_3^1 = \begin{bmatrix} .3841 \\ -.4790 \\ .2132 \\ .1128 \\ -.1407 \\ .0626 \end{bmatrix} , \quad D_3^2 = \begin{bmatrix} .0947 \\ -.1181 \\ .0526 \\ -.4244 \\ .5292 \\ -.2359 \end{bmatrix} .$$

Let us assume that

$$f(t) = 3 \sin 0.25t + 5 \sin 0.35t + 7 \sin 1.25t + 4 \sin 1.5t , \qquad (30)$$

whose first frequency is close to that of the system. The control objective is to reduce excessive displacements by applying control so that the system matrix of the resulting system has eigenvalues $\rho_{1,2}$, $\rho_{3,4}$, and $\lambda_{5,6}$ where

$$\rho_{1,2} = -0.099 \pm j4 ,$$
$$\rho_{3,4} = -0.0778 \pm j0.6234 .$$
(31)

Example 1

Consider the case of placing two controllers on two of the three floors. In the single-input case, Theorem 2 immediately determines the best locations to be the second and third floors with

$$b_{2,3} = [0 \ 0 \ 0 \ 0 \ .4766 \ .8791]^T .$$
(32)

The values of its associated energy coefficient w_2 and energy E are compared with two other possible control configurations, $b_{1,2}$ and $b_{1,3}$, and are given in Table 1. Figure 1 also gives the displacements of the top floor under all control schemes. It is seen that the optimal control design also gives optimal performance in terms of displacement reduction. Similar results are also obtained for floor acceleration reductions. Hence, from the standpoint of safety and comfort control [3] as well as energy consideration, the choice $b_{2,3}$ is uniformly superior in the single-input case.

Table 1 - Energy Coefficient and Energy in Example 1

Control Vector b	Energy Coefficient w_2	Energy E ($T_0 = 50$)
$b_{1,3}$	0.79	56,000
$b_{1,2}$	0.85	42,000
$b_{2,3}$	0.92	33,000

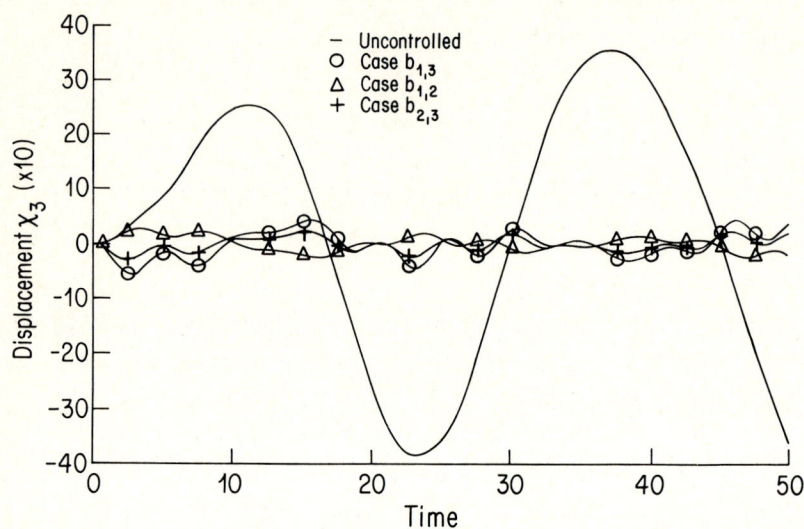

Figure 1 - Displacement x_3 in Example 1

Example 2

The same problem is now considered using a two-loop control scheme. Let us determine b_1 and b_2 of the matrix B such that b_1 is designed to alter only the first pair of complex conjugate eigenvalues, $\lambda_{1,2}$, and b_2 is designed to alter the second pair, $\lambda_{3,4}$.

Under these assumptions, the following designs of control matrix are of practical interest.

Case A: Different control forces applied to first and third floors:

$$B_{1,3} = \begin{bmatrix} 0 & 0 & 0 & .6256 & 0 & .7801 \\ 0 & 0 & 0 & .9136 & 0 & -.4060 \end{bmatrix}^T$$

Case B: Different control forces applied to first and second floors:

$$B_{1,2} = \begin{bmatrix} 0 & 0 & 0 & -.4066 & .9130 & 0 \\ 0 & 0 & 0 & .8744 & -.4852 & 0 \end{bmatrix}^T$$

Case C: Different control forces applied to second and third floors:

$$B_{2,3} = \begin{bmatrix} 0 & 0 & 0 & 0 & .8744 & .4852 \\ 0 & 0 & 0 & 0 & .7801 & -.6256 \end{bmatrix}^T$$

All three designs satisfy the controllability conditions.

As predicted by Theorem 3 and verified by Table 2, Case C produces the optimal control configuration and, as shown in Figure 2, it also is the most effective in reduction of displacement. Calculations also show that Case C is best in reduction of acceleration as well.

Table 2 - *Energy Coefficient and Energy in Example 2*

Control Matrix B	Energy Coefficient w_2	Energy E ($T_0 = 50$)
Case A	1.18	25,000
Case B	0.33	91,000
Case C	1.29	20,000

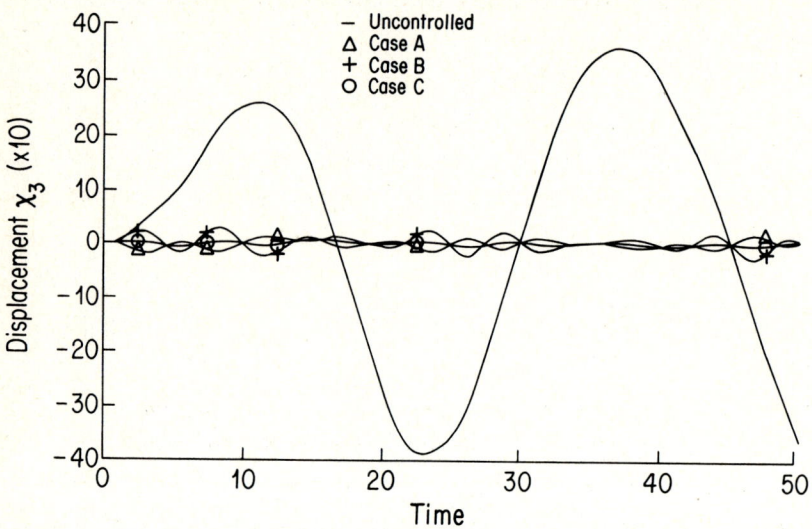

Figure 2 - Displacement x_3 in Example 2

Furthermore, a comparison of values in Tables 1 and 2 shows that Case C in two-loop control scheme is best overall under energy criterion. The energy coefficient in this case is maximum and it implies that the energy coefficient can be used as an effectiveness measure crossing single-input and multi-input lines.

CONCLUDING REMARKS

Within the framework of modal control, a simple criterion for optimal design of the control matrix has been established under energy constraint. In control of large systems, the question of where to place a limited number of controllers is of practical importance and this result provides a simple guide for making this determination when energy required is to be minimized. Although only complex eigenvalues are considered, it is easily applied to cases involving real eigenvalues.

It has also been shown through examples that minimum-energy control also leads to most effective control in these cases, a desirable by-product of the analysis. Also it is seen that the energy coefficients for both single-input and multi-input cases take the same form and thus can be used to determine overall optimal control configuration.

It is worth pointing out that, under energy criterion, the analysis presented herein provides some guide in determining the desired locations of the altered eigenvalues. Consider, for example, the case where the eigenvalue λ_1 is to be altered to ρ_1 using the single-input scheme. In general, some freedom exists in the choice of ρ_1 beyond requirements such as $|\rho_1-\lambda_1| \geq \varepsilon$.

Equation (8) gives

$$\alpha_1^2 + \alpha_2^2 = (e_{11}^2+e_{12}^2)/(v_{11}^2+v_{12}^2)\xi_2^2 . \qquad (33)$$

Having determined the control vector b from Theorem 1, the denominator of equation (33) is fixed and the numerator can be expressed by:

$$e_{11}^2+e_{12}^2 = (\eta_1-\xi_1)^4 + (\eta_1-\xi_1)^2 [4-2(\eta_2^2-\xi_2^2)] + (\eta_2^2-\xi_2^2)^2 . \qquad (34)$$

Since $\alpha_1^2+\alpha_2^2$ is to be minimized for minimum energy, the most energy-efficient locations of η_1 and η_2, the real and imaginary parts of ρ_1, can be determined by minimizing equation (34) with respect to η_1 and η_2. This is to be carried out, of course, under constraints such as $|\rho_1-\lambda_1| \geq \varepsilon$.

ACKNOWLEDGEMENTS

This work was supported by the National Science Foundation under Grant No. ENG 76-82226 and is gratefully acknowledged.

REFERENCES

[1] BALAS, M.J., "Modal Control of Certain Flexible Dynamic Systems", *SIAM J. Control*, Vol. 16, 1978, pp. 450-462.

[2] JUANG, J.N. and BALAS, M.J., "Dynamics and Control of Large Spinning Spacecrafts", *Proceedings of the AIAA/AAS Astrodynamics Conference*, held in Palo Alto, California, 1978.

[3] SAE-UNG, S. and YAO, J.T.P., "Active Control of Building Structures", *ASCE, J. of the Eng. Mech. Div.*, Vol. 104, 1978, pp. 335-350.

[4] MARTIN, C.R. and SOONG, T.T., "Modal Control of Multistory Structures", *ASCE, J. of the Eng. Mech. Div.*, Vol. 102, 1976, pp. 613-623.

[5] ROSENBROCK, H.H., "Distinctive Problems of Process Control", *Chem. Engr. Progress*, Vol. 58, 1962, pp. 43-50.

[6] CHANG, M.I.J., "On Optimal Control Design in Modal Control of Complex Systems", *Ph. D. Dissertation*, State University of New York at Buffalo, New York, 1978.

STRUCTURAL CONTROL, H.H.E. Leipholz (ed.)
North-Holland Publishing Company & SM Publications
© IUTAM, 1980

A UNIFIED APPROACH TO OPTIMAL CONTROL AND MECHANICS
OF CONSTRAINED CONTINUA

G. Szefer

Institute of Structural Mechanics
Technical University of Cracow
ul. Warszawska 24, 31-155 Cracow, Poland

1. INTRODUCTION

In the paper, the mechanics of bodies which configuration, state of deformations and state of stresses subjected to additional constraints will be presented. From the physical point of view, any restriction of motion is treated as a constraint. The very well known boundary conditions in mechanics belong to this class of restrictions, too. Constraints defined on the boundary of the body are called "external constraints". Recently a series of nonclassical boundary-value problems - unilateral problems, Signorini-Fichera constraints, obstacle problems, problems with friction - were thoroughly investigated by several authors [1 - 4].

If restrictions of motion are defined on points belonging to the interior of the body, than such restrictions will be called "internal constraints". An example of this kind is given by incompressible elastic materials described in the nonlinear case by Truesdell [5]. But there are many other problems which lead to the mechanics of continua with internal constraints. Structures like: bodies with internal rigid rods or discs, inextensible cords or surfaces, reinforced soil, reinforced earth wall, pile-soil systems, bodies with bounded stresses, volume or energy - as measure of compliance - are examples of this type. A general theory of continua with internal constraints was presented by Woźniak [6]. In the present paper, some generalization of the constrained problems will be discussed. Using the convex analysis notions [7 - 9] a unified variational promulation of the problem will be given. A strong connection with the control of processes and structures is easily seen. Particularly, some optimal control problems for elastic bodies will be formulated. Examples of important classes of constraints and applications in engineering - reinforced bridge bearings made of rubber-like materials, consolidation processes in soild, optimal control with constraint - show the possibilities and the potential of the theory.

2. MECHANICS OF CONSTRAINED MEDIA

2.1 *Statement of the Problem*

Let us consider a deformable body B loaded by external body forces b_i and surface traction p_i. Large deformations, displacements and nonlinear properties of the material are permitted. Let furthermore additional constraints, independent of forces, mass distribution and physical features be superposed on configuration - (state of strain) - of the body. Then, denoted by

T_{iK} - first Piola-Kirchhoff stress tensor;

F^i_K - gradient of deformation;

x^i - current position's vector of the particle $X \in B$;

r^i - reaction body force in continuum evoked by the internal constraints;

s_i - reaction boundary traction evoked by the external constraints on the boundary;

the system of relations which described the state of strain and stress of the body has the form:

(A) Definitional constraint relations:

$$r_i \in A_i(x) \quad \text{in} \quad B$$
$$s_i \in Q_i(x_0) \quad \text{on} \quad \partial B \quad \quad (1a,b)$$

(B) Field equations:

(i) $T_{iK,K} + \rho_R b_i + r_i = 0$

$T_{iK} F_K^j = T_{jK} F_K^i$

(ii) $F_K^i = x^i,_K \quad \text{in} \quad B$ \quad (2)

(iii) $T_{iK} \in F_{iK}(X,F)$

(iv) $T_{iK} n_K = p_i + s_i \quad \text{on} \quad \partial B$.

The problem which we will solve is the following: find the state variables (x^i, F_K^i, T_{iK}) and reactions (r_i, s_i), so that relations (1) and (2) are fulfilled.

In contrast to the classical formulation of continuum mechanics, the field equations given above contain additional unknowns r_i and s_i which constitute the response of the medium to the restriction of motion. It is worth while to note that the formulation presented here is more general than results given up to now (see [6]). First of all, our description of constraints admits unilateral constraints, dry friction (operators A_i and Q_i can be multivalued), etc. Moreover, the constitutive relations (iii) are also of a general type and permit multivalued mappings. Owing to this fact it is possible to consider materials with locking-effect, too. It is understood that the multivalued operators F_{iK} must be consistent with all physical material-postulates, like the principle of objectivity, principle of determinism etc. Furthermore we assume that operators A_i, Q_i, F_{ik} are maximal monotone (see [7 - 9]). Hence it follows that there exists such subdifferentiable functions like $W(F)$, $\phi_i(x)$, $\psi_i(x_0)$, (called superpotentials) whose subdifferentials coincide with the image of operators A_i, Q_i and F_{ik}. It means

$$A_i(x) = \partial \phi_i(x) ,$$

$$Q_i(x_0) = \partial \psi_i(x_0) , \qquad (3a\text{-}c)$$

$$F_{ik}(F) = \partial W(F) .$$

Usually it is

$$\partial W = \partial W_0 + \partial W_1 , \qquad (4)$$

where $W_0(F)$ is differentiable (∂W_0 contains one element only) and $W_1(F)$ subdifferentiable only (the subdifferential ∂W_1 represents that part of stresses in (iii) which follow from the restrictions in the physical law, e.g., incompressible material, ideal plasticity, material with locking-effect, etc.)

2.2 Variational Formulation

Let V be the kinematically admissible field (in the usual sense), $x \in V$. Multiplying (2(i)) by $\delta x^i = \hat{x}^i - x^i$, $\hat{x}^i \in V$ and performing integration over B we obtain

$$\int_B T_{iK,K} \delta x^i dB + \int_B \rho_R b_i \delta x^i dB + \int_B r_i \delta x^i dB = 0$$

$$\int_B (T_{iK} \delta x^i)_{,K} dB - \int_B T_{iK} \delta x^i_{,K} dB + \int_B \rho_R b_i \delta x^i dB + \int_B r_i \delta x^i dB = 0 . \qquad (5)$$

Using the Green-Gaus theorem, one obtains

$$\int_{\partial B} T_{iK} n_K \delta x^i dS - \int_B T_{iK} \delta F^i_K dB + \int_B \rho_R b_i \delta x^i dB + \int_B r_i \delta x^i dB = 0 .$$

Taking into account relation (2(iv)) and assumptions (3), one obtains furthermore

$$\int_{\partial B} (p_i + s_i) \delta x^i dS - \int_B T_{iK} \delta F^i_K dB + \int_B \rho_R b_i \delta x^i dB + \int_B r_i \delta x^i dB = 0 . \quad (6)$$

Because of inequalities (which follow from (2(iii)) and (3c))

$$\int_B [W(\hat{F}) - W(F)] dB \geq \int_B T_{iK} \delta F^i_K dB ,$$

$$\sum_{i=1}^{3} \int_B [\phi_i(\hat{x}) - \phi_i(x)] dB \geq \int_B r_i \delta x^i dB , \quad \sum_{i=1}^{3} \int_{\partial B} [\psi_i(\hat{x}_0) - \psi_i(x_0)] dS \geq$$

$$\geq \int_{\partial B} s_i \delta x^i dS , \quad (7)$$

and notation

$$L(\delta x^i) = \int_B \rho_R b_i \delta x^i dB + \int_{\partial B} p_i \delta x^i dS , \quad (8)$$

it is true that

$$\int_B [W(\hat{F}) - W(F)] dB + \sum_{i=1}^{3} \int_B [\phi_i(\hat{x}) - \phi_i(x)] dB + \sum_{i=1}^{3} \int_{\partial B} [\psi_i(\hat{x}_0) - \psi_i(x_0)] dS -$$

$$- L(\hat{x} - x) \geq 0 \quad \forall x \in V . \quad (9)$$

Hence, there exists a functional

$$T(x) = T_0(x) + Б(F) + \sum_{i=1}^{3} \Phi_i(x) + \sum_{i=1}^{3} \Psi_i(x_0) =$$

$$= \int_B W_0(F)dB - L(x) + Б(F) + \sum_{i=1}^{3} \Phi_i(x) + \sum_{i=1}^{3} \Psi_i(x_0) , \quad (10)$$

the minimum of which represents the solution of (2).

Here the following notations

$$\Phi_i(x) = \int_B \phi_i(x)dB ,$$

$$\Psi_i(x_0) = \int_{\partial B} \psi_i(x_0)dS ,$$

$$Б(F) = \int_B W_1(F)dB ,$$

$$L(x) = \int_B \rho_R b_i x^i dB + \int_{\partial B} p_i x^i dS ,$$

$$T_0(x) = \int_B W_0(F)dB - L(x) ,$$

are used. Inversely, the critical point of the functional

$$T(x) = \int_B W_0(F)dB - [\int_B \rho_R b_i x^i dB + \int_{\partial B} p_i x^i dS] + \int_B W_1(F)dB +$$

$$+ \sum_{i=1}^{3} \Phi_i(x) + \sum_{i=1}^{3} \Psi_i(x_0) ,$$

leads to the relation

$$0 \in \partial T(x) \to \partial T = \partial T_0 + \partial(W_1 + \Sigma\Phi_i + \Sigma\Psi_i) = \delta T_0(x,\delta x) + Б(\hat{x}) - Б(x) +$$

$$+ \sum_{i=1}^{3} [\Phi_i(\hat{x}) - \Phi_i(x)] + \sum_{i=1}^{3} [\Psi_i(\hat{x}_0) - \Psi_i(x_0)] \geq 0 \quad \forall \hat{x} \in V .$$

It means

$$\int_B \frac{\partial W_0}{\partial F_K^i} (\hat{F}_K^i - F_K^i) dB + \int_B [W_1(\hat{F}) - W_1(F)] dB + \sum_{i=1}^{3} \int_B [\phi_i(\hat{x}) - \phi_i(x)] dB +$$

$$+ \sum_{i=1}^{3} \int_{\partial B} [\psi_i(\hat{x}_0) - \psi_i(x_0)] dS \geq \int_B \rho_R b_i (\hat{x}^i - x^i) dB + \int_{\partial B} p_i (\hat{x}^i - x^i) dS ,$$

$$\forall \hat{x}^i \in V . \quad (11)$$

The above inequality is a generalization of the very well known principle of virtual work.

2.3 A Certain Case of Constraints

It is frequently the case in applications, that constraints (1a) are defined by a set. Two important cases of such a situation are discussed below.

2.3.1 Constraints Described by a Set of Equations and Inequalities

Consider the set

$$K = \{x: \ h(X, x, F(x)) \leq 0 \}. \quad (12)$$

In such a case it is convenient to introduce the indicator [9]

$$I_K(x) = \begin{matrix} 0 & x \in K \\ \infty & x \notin K \end{matrix} , \quad (13)$$

and consider the functional

$$T(x) = T_0(x) + I_K(x) \to \min_{x \in V} . \quad (14)$$

Including the optimality criterion

$$\partial T = \partial T_0 + \partial I_K \ni 0 , \quad \partial I_K = \{r: \ I_K(\hat{x}) - I_K(x) \geq \int_B r_i \delta x^i dB\} ,$$

we obtain after simple transformations the field equations

$$T_{iK,K} + \rho_R b_i + \lambda \frac{\partial h}{\partial x^i} + \left(\lambda \frac{\partial h}{\partial F_K^i}\right)_{,K} = 0 \quad \text{in} \quad B$$

$$T_{iK} n_K = p_i + \left(-\lambda \frac{\partial h}{\partial F_k^i}\right) n_K \quad \text{on} \quad S \,, \tag{15}$$

and the formulas for the reactions

$$r_i = \lambda \frac{\partial h}{\partial x^i} + \left(\lambda \frac{\partial h}{\partial F_k^i}\right)_{,K}$$

$$s_i = -\lambda \frac{\partial h}{\partial F_k^i} n_K \,. \tag{16}$$

2.3.2 Programmable Constraints

Let us set

$$K = \{x: \int_B k(X,x,F)dB \leq c_0\} \quad c_0 \in R \,. \tag{17}$$

Similarly, as above, we introduce a functional

$$T(x) = T_0(x) + I_K(x) \to \min_{x \in V} \,,$$

$$\partial I_K = \left\{r: \quad 0 \geq \int_B r_i(\hat{x}^i - x^i)dB, \; \forall \hat{x} \in K\right\} \to r_i = \lambda \frac{\partial f}{\partial x^i} \,,$$

$$\lambda \in R$$

where

$$f(x) = \int_B k(X,x,F)dB - c_0 \,.$$

Finally, there is

$$T_{iK,K} + \rho_R b_i - \lambda\left[\frac{\partial k}{\partial x^i} + \left(\frac{\partial k}{\partial F^i_K}\right)_{,K}\right] = 0 \quad \text{in} \quad B,$$

$$T_{iK} n_K = p_i + \left(-\lambda \frac{\partial k}{\partial F^i_K}\right) n_K \quad \text{on} \quad S,$$

$$r_i = \lambda\left[\frac{\partial k}{\partial x^i} + \left(\frac{\partial k}{\partial F^i_K}\right)_{,K}\right],$$

$$s_i = -\lambda \frac{\partial k}{\partial F^i_k} n_K.$$

(18)

3. CONNECTION WITH OPTIMAL CONTROL AND VARIATIONAL THEORY

It is easy to see, that the theory of constrained media presented previously, has a strong connection with optimal control problems and general variational theory. This is so because of two reasons: (1) the formal structure of optimal control and variational problems is of this same type, (2) optimal control problems for constrained continua, especially control with constraints, can be formulated. To show this, consider a typical abstract formulation of the optimal control problem [10]. It has the form:

- state equation $\quad S(x,u) = 0 \quad x \in V, \; u \in U,$
- constraints $\quad u \in U_{ad} \subset U,$ (19a-c)
- cost function $\quad T(u) \to \min_{u \in U_{ad}}.$

Where the following notations denote

V - space of states,
U - space of controls,

U_{ad} - set of admissible controls,
x - state functions,
u - control,
S - state operator.

In mechanics, operator S described the equations of motion [11] (equilibrium) or principle of virtual work:

$$S(x,u) = B(x(u),z) - L(z) = 0, \quad \forall z \in V \text{ where } B(x,z)$$

is the work of the deformation and $L(z)$ denotes the work of external forces. From the general variational point of view there are three kinds of variational problems [10]:

- unconditional extremem

$$T(x) \underset{x \in V}{\to \min} \to B(x,z) = L(z) \quad \forall z \in V,$$

- conditional extremum

$$T(x) \underset{x \in V_{ad}}{\to \min} \to B(x,z-x) \geq L(z-x) \; \forall z \in V_{ad}, \qquad (20a\text{-}c)$$

- conditional extremum on a cone

$$V_{ad} = K, \quad K \text{ - cone in } V$$

$$T(x) \underset{x \in K}{\to \min} \quad \begin{matrix} B(x,z) \geq L(z) & z \in K \\ B(x,x) = L(x) & \end{matrix}.$$

We see immediately, that the optimal control problem (19) belongs to the class of (20b), where the restrictional conditions contain relations (19a,b). But questions of mechanics of constrained media lead to the same form of problems as is shown in (11). So, the common features of problems (19), (20) and (1-2) are fully apparent.

4. APPLICATIONS

Two examples of application of the presented theory are given below.

4.1 Deformation of Expansion Bearings of Rubber-Like Material

As an example of a body with internal constraints of type (12) we consider an elastic bearing in the plane state of strain. The internal constraints constitute reinforcement in a form of rigid plates. These plates allow shift and rotation only. Hence it follows that constraints defined in points occupied by these plates must satisfy the relations:

$$\frac{\partial v}{\partial x} = 0 \quad \text{for symmetrical loading},$$

$$\frac{\partial^2 v}{\partial x^2} = 0 \quad \text{otherwise, } v \text{ - vertical displacement.}$$
(21)

The material of the bearing is linear elastic (neoprene) but exhibits large deformations. So, the nonlinear theory of elasticity must be applied. To solve this difficult nonlinear boundary-value problem, the incremental variational formulation and total Lagrangian approach [12] was used. Numerical calculations* by means of the finite element technique were performed. Figures 1 and 2 show the finite element net in the reference state and in the deformation state, i.e., in the current configuration. For the sake of comparison, the bearing with and without reinforcement is presented.

---

*Computer programmes and all calculations were done on Computer CYBER 70 and were performed by Mr. Mikolajek from the Institute of Structural Mechanics, Technical University of Cracow.

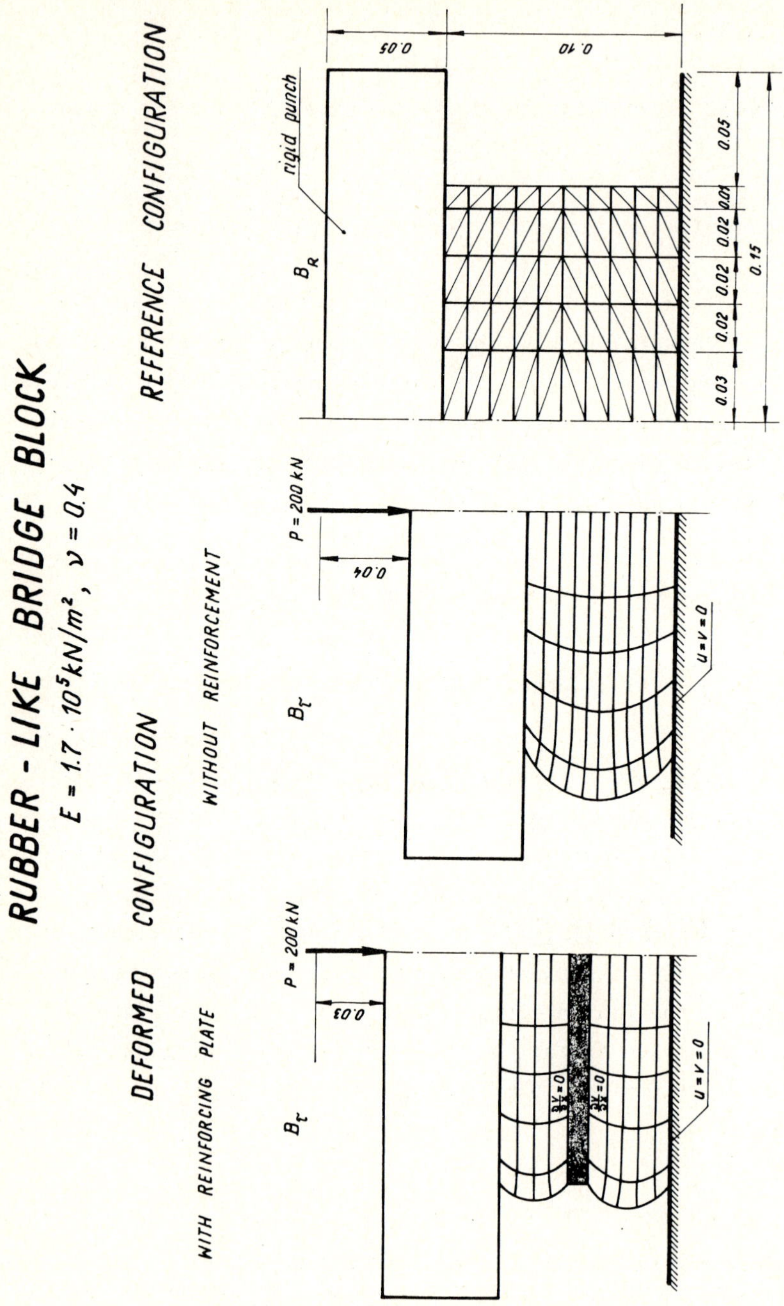
RUBBER - LIKE BRIDGE BLOCK
E = 1.7 · 10^5 kN/m², ν = 0.4
DEFORMED CONFIGURATION
REFERENCE CONFIGURATION
WITH REINFORCING PLATE
WITHOUT REINFORCEMENT
rigid punch
B_R
B_τ
B_τ
P = 200 kN
P = 200 kN
0.05
0.10
0.04
0.03
0.05
0.01
0.02
0.02
0.02
0.03
0.15
u = v = 0
u = v = 0

Figure 1

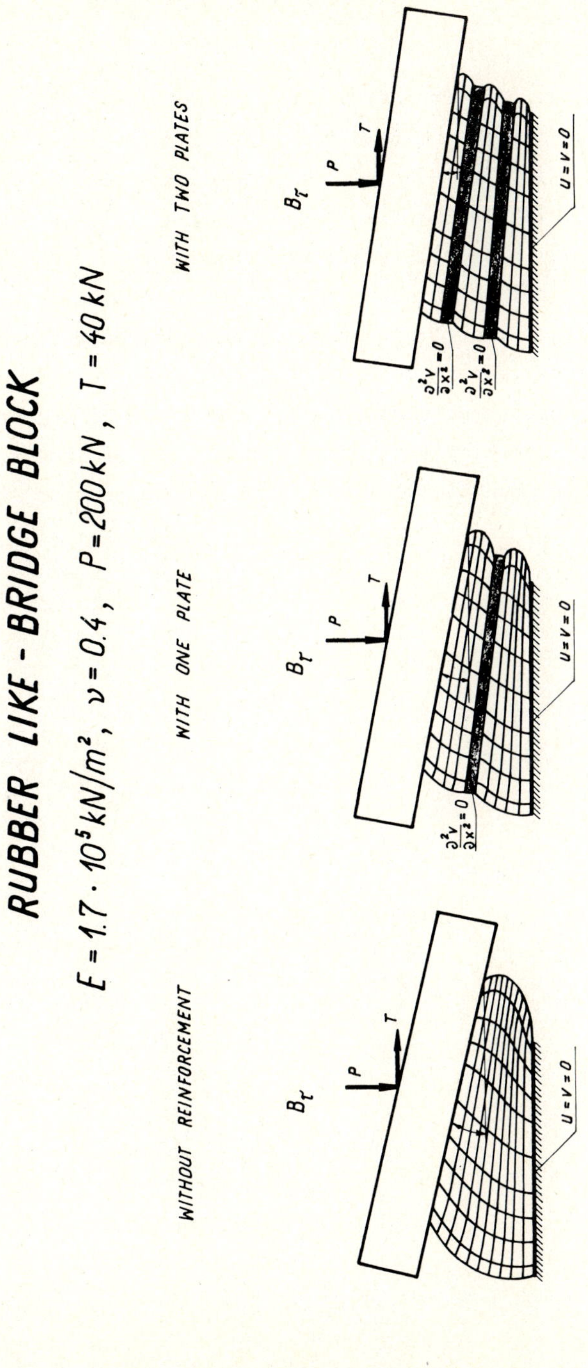
RUBBER LIKE - BRIDGE BLOCK
$E = 1.7 \cdot 10^5 kN/m^2$, $\nu = 0.4$, $P = 200 kN$, $T = 40 kN$
WITHOUT REINFORCEMENT
WITH ONE PLATE
WITH TWO PLATES
B_τ
P
T
U=V=0
$\frac{\partial^2 V}{\partial x^2} = 0$

Figure 2

4.2 *Consolidation of a Porous Layer with Initial Hydraulic Gradient*

The second example concerns an important problem of soil mechanics namely the consolidation of porous subsoil. It is very well known, that if the value of the gradient of the fluid pressure in the pores $\sigma_{,i}$ is smaller than a given characteristic quantity i_0 (called initial hydraulic gradient), then there is no flow of fluid through the porous skeleton. So, the problem of determining the state of consolidation under the influence of the initial hydraulic gradient leads in a natural way to mechanics with internal, unilateral constraints. The system of equations (on the basis of the Biot's theory) with a graph of subdifferential $\partial\Psi$ and the form of functional I, which must be minimized, is given on Figure 3.

CONSOLIDATION OF A POROUS LAYER WITH HYDRAULIC GRADIENT'S LAW

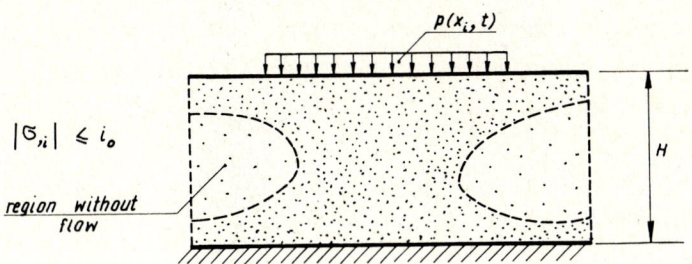

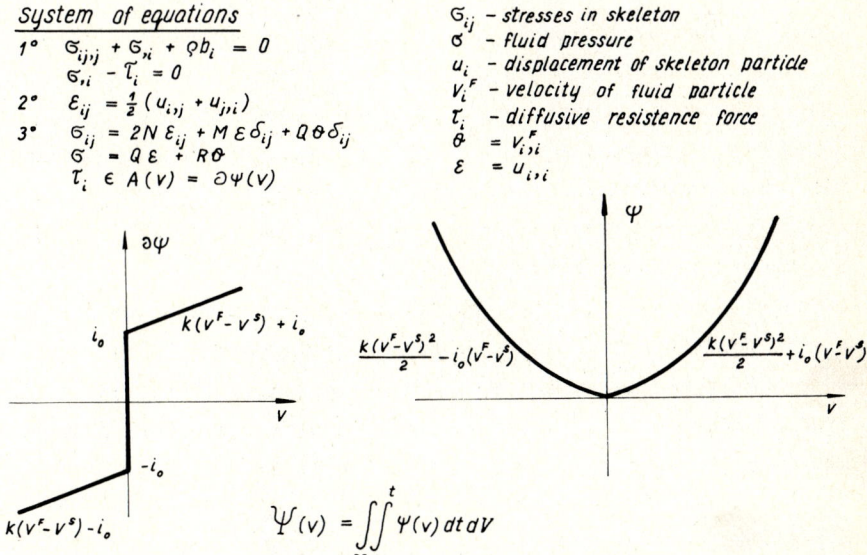

Figure 3

5. CONCLUSIONS

In this paper a new, general, convex analysis formulation of continua with internal and external constraints was given. The presented approach permits to discuss the mechanics of constrained media, optimal control theory and general variational principles in a unified manner. Three basic final corollaries can be formulated:

(i) A broad class of new nonclassical boundary value problems for bodies with internal constraints can be analyzed in the frame of a unified theory.

(ii) Several nonmaterial restrictions like: imaginary, apparent, and programmable constraints can be taken into account.

(iii) Control with constraints is possible. This fact allows the treatment of constraints as control parameters and to consider a new class of optimal control problems in mechanics (e.g., control with classes of materials, control with unilateral supports, control with deformation and stress processes by means of suitable programmable constraints and so on).

Certain immediate practical applications of the given theory are easily found. For example, problems of reinforced soils, reinforced earth walls, reinforced bridge bearings, vibration insulation washers, pile soil systems etc., can be adequately described and analyzed as continua with constraints.

REFERENCES

[1] SIGNORINI, A., "Sopra alcune questioni di elastostatica", Atti Soc. Ital. per progresso delle Science, 1933.

[2] FICHERA, G., "Boundary Value Problems in Elasticity with Unilateral Constraints", in *Handbuch der Physik*, Band VIa/2, Festkorpermechanik II, pp. 391-424, Springer-Verlag, Berlin-Heidelberg-New York, 1972.

[3] MOREAU, I.I., "La notion de sur-potential et les liaisons unilaterales en elastostatique", *C.R. Academy Sci.*, Paris, Vol. 267A, 1968, pp. 954-957.

[4] PANAGIOTOPOULOS, P., "Ungleichungen in der Mechanik", von der Mathematisch-Naturwissenschaftlichen Fakultat der Rheinisch, Westfalischen Technischen Hochschule Aachen Genehmigte Habilitationsschrift, Aachen, 1977.

[5] TREUSDELL, C. and NOLL, W., "The Nonlinear Field Theories of Mechanics", *Handbuch der Physik*, Vol. III/3, Springer-Verlag, Berlin-Heidelberg-New York, 1965.

[6] WOŻNIAK, C., "Constrained Continuous Media I, II, III", *Bull. Acad. Polon. Sci.*, Sci. Techn. Vol. 3-4, No. 21, 1973.

[7] BRESIS, H., "Problemes Unilateraux", *J. Math. Pures et. Appl.*, Vol. 51, 1972, pp. 1-168.

[8] ROCKAFELLAR, R., *Convex Analysis*, Princeton, 1970.

[9] TEMAM, R. and EKELAND, I., *Convex Analysis and Variational Problems*, North-Holland Publ. Comp., Amsterdam, 1976.

[10] LIONS, I.L., *Optimal Control of Systems Governed by Partial Differential Equations*, Springer-Verlag, Berlin-Heidelberg-New York, 1971.

[11] SZEFER, G., "Deformable Material Continuum as a Control System with Spatially Distributed Parameters", *Archives of Mechanics*, Vol. 23, No. 6, 1971, pp. 927-952.

[12] WASHIZU, K., *Variational Methods in Elasticity and Plasticity*, Second Edition, Pergamon Press, 1975.

IDENTIFICATION AND CONTROL OF STRUCTURAL DAMAGE

James T. P. Yao

Purdue University
W. Layfayette, Indiana

1. INTRODUCTION

When I was a college student, I learned to always choose a larger section and thus a more conservative alternative in structural design whenever such an option existed. The reasons were that, for civil engineering structures, (a) the weight of the structure was not a critical constraint, and (b) the cost of construction material was not a major consideration. Moreover, design codes are traditionally conservative. It is well known that most civil engineering structures to-date are relatively massive, stiff, and stable. Nevertheless, over the years, a few of these conservatively designed structures have been damaged due to unusually servere natural hazards, defective materials, as well as human errors in design and construction. Recently, more flexible structures are being designed and built because (a) more sophisticated methods of analysis are available, (b) cost of material is becoming a more significant design factor than ever before, and (c) taller (or longer) and thus more flexible structures are being attempted for architectural and other considerations.

In an ideal situation where the disturbance to be encountered and the resistance of the structure are completely known, it is relatively simple to design a comfortable and safe structure. In reality, there always exist uncertainties in predicting future loading conditions as well as in estimating structural resistance. Moreover, there exist discrepancies between the actual structural behavior and its corresponding mathematical representations used in the process of structural analysis and design. To-date, several motion-controlling devices have been used or proposed for comfort and/or safety considerations. The development of structural control has been reviewed recently [1], and the state of the art is presented in detail by various experts during this Symposium. Nevertheless, it is still difficult to design and construct a structure even with the effective use of control systems, which will completely avoid the possibility of being damaged during its intended lifetime.

The objective of this paper is to stimulate discussion of possible topics in structural control for further research and development. The literature on structural control is briefly summarized from a structural engineer's viewpont. The general applicability of structural control under various conditions is reviewed. Finally, the identification of human and/or structural response state in existing structures and its effect on structural control are discussed.

2. LITERATURE REVIEW

The history of control theory relating to structural applications was reviewed by Zuk [2]. An attempt is made herein to briefly summarize the recent development of structural control.

About ten years ago, I thought that the application of active control could solve most difficult problems in structural engineering. Ideally, flexible structures such as extremely tall buildings or long bridges can be designed to resist only the operational gravity loads and the active control system can take care of any side-sway motions resulting from lateral load effects.

Such being the case, we can bypass such problems as statistical uncertainty in predicting future loading conditions and complicated structural analysis. Of course, I learned quickly that not only the control theory is a well-established and difficult subject in itself, but also its application to structural engineering requires further investigations. Instead of providing a simple overall solution, the application of control theory has helped to create many new challenging problems in structural engineering.

In 1968, Zuk [3] discussed the concept of kinetic structures, an example of which was the application of tendon control as proposed independently by Freyssinet in 1960 and by Zetlin in 1965. Meanwhile, Wright [4] and Nordell [5] suggested the use of active systems, which can be used to resist any exceptionally high overloading of a given structure. Later, the use of initially slack cables in forming bilinear hardening structures which are subjected to earthquake loads was studied [6,7]. The concept of structural control was presented to the structural engineering profession in 1971 [8]. As an example, the use of thruster engines to generate impulsive control forces was mentioned. Meanwhile, Gaus [9] suggested that it is desirable to search for an optimum combination of passive and active control devices.

In 1970, Wirsching [10,11] studied the use of passive motion-reducing devices and suggested several means for the improvement of structural safety under earthquake loading conditions. The displacement response of one-, five-, and ten-story building structures to strong-motion earthquakes was simulated with the use of an analog computer. The Gumbel Type I distribution of maxima was used in the statistical description of these peak response data. Results of this study showed that the isolator system was the most effective one among the five passive control systems thus studied.

In 1967, Masri [12] studied the possible application of "two-particle" impact dampers. Gupta and Chandrasekaran [13] investigated the use of an absorber system for the reduction of earthquake effects. It was also mentioned that gyroscopes were being studied for use in the torsional stabilization of suspension bridges in Japan [14]. Nevertheless, most studies in this direction in recent years are concentrated on the practical implementation of isolator systems, including the use of energy-absorbing devices (through plastic deformation), and tuned mass dampers.

Green [15] suggested in 1935 to construct buildings with a flexible first story to obtain favorable response to earthquake excitations. Fintel and Kahn [16] reported that buildings without shear walls on the first floor suffered less damage than those with shear walls during the 1963 Skopje and the 1964 Caracas earthquakes. Therefore, it was suggested that buildings can be designed with a first story which is stiff enough to resist wind loads but flexible enough to isolate the upper floors from seismic effects.

In 1969, Matsushita and Izumi [17] proposed the use of non-circular rollers which would cause the building to rise when it is displaced laterally. The weight of the building would then act as a restoring force, which depends on the shape of these rollers.

During the 1970 Gediz, Turkey earthquake, it was reported by Penzien and Hanson [18] that stretching of anchor bolts at column bases of several buildings prevented the occurrence of more serious structural damage than those actually occurred. In 1973, Kelly, Skinner, and Heine [19] tested three types of energy-absorbing devices, which undergo plastic torsion and thus absorb the kinetic energy in the structure due to earthquake motions. The use of such devices as isolators in siesmic structures was explored by Skinner, Beck and Bycroft [20]. Recently, the behavior of two types of mild steel energy-absorbing devices was given by Kelly and Tsztoo [21], who showed that these devices have substantial hysteretic energy absorbing capacity. Kelly and

Tsztoo [22] also presented results of earthquake simulation tests of model frames with such energy-absorbing devices. These results indicated the feasibility of using such devices for aseismic design [23].

In 1976, Skinner, Bycroft, and McVerry [24] studied the combined use of the energy-absorbing devices and laminated rubber bearings which possess adequate horizontal flexibility for the isolation of nuclear power plants during earthquakes. They concluded that a high reliability can be achieved for base-isolator components on the basis of extensive laboratory tests. Tyler [25] reported on results of dynamic shear tests on such laminated rubber bearings, and concluded that these bearings are suitable for use as base isolators. Robinson and Tucker [26] studied a lead-rubber isolator consisting of a steel-reinforced elastomeric bearing with a lead insert fitted in its center. They also recommended for its use in base-isolation systems for the protection of structures during earthquakes. Eidinger and Kelly [27] demonstrated experimentally the possibility of using such bearings as isolators. Jolivet and Richli [28] reported on the application of similar reinforced-elastomer/friction-plate bearing systems in the foundation design of nuclear power plant in South Africa. Recently, a massive research program to study the use of steel energy absorbing restrainers and their incorporation into nuclear power plants was described in a summary report [29]. In a companion volume, the current uses of energy absorbing devices were reviewed [30]. The optimal design of an earthquake isolation system was also investigated [31].

Crandell, et al [32,33] studies the slip of friction-controlled mass under earthquake loading conditions. Recently, Nemat-Nasser [34] is making an analytical study of the vibration of a continuous viscoelastic slab resting on viscoelastic support for such practical applications.

Klein, et al [35] studied the use of shutter-like appendages to stabilize wind induced vibrations in tall buildings. This concept is being extended and experimental studies are now in progress [36,37].

A tuned mass damper was installed on the 59th floor of the 914-foot Citicorp building to minimize the discomfort experienced by occupants on windy days. The device weighs 400 tons with two spring damping mechanisms and a control system which is used to collect data and controls the motion of this mass [38,39]. Although additional steel plates were welded to bolted connections later [40], the retrofit was said to be unrelated to the effectiveness of the tuned mass damper which is working extremely well. This interesting topic was discussed by Petersen [41] recently.

In a technical report, Yao and Tang [42] discussed the application of an active control system using impulsive control forces. For a single-degree-of-freedom system, the control force was chosen as follows:

$$F = - \sum_{i=1}^{\ell} a_i \, H(|X| - \xi_i) \, \text{sgn}(X) \, H(X \cdot \dot{X})$$

where a_i denotes the force magnitude increment at the ith control level;

ξ_i denotes the specified displacement of the ith control level;

$H(.)$ denotes the Heaviside unit-step function;

sgn(.) denotes the signum function.

Similar control forces were given for a two-story building structure Results of numerical examples indicated that such impulsive control forces can be effective in reducing the displacement responses of one- and two-story structural frames to the 1940 El Centro earthquake excitation. For the purpose of illustration, the effectiveness of such control laws in reducing the displacement response of a two-story building structure is shown in Figure 1.

Control of Structural Damage

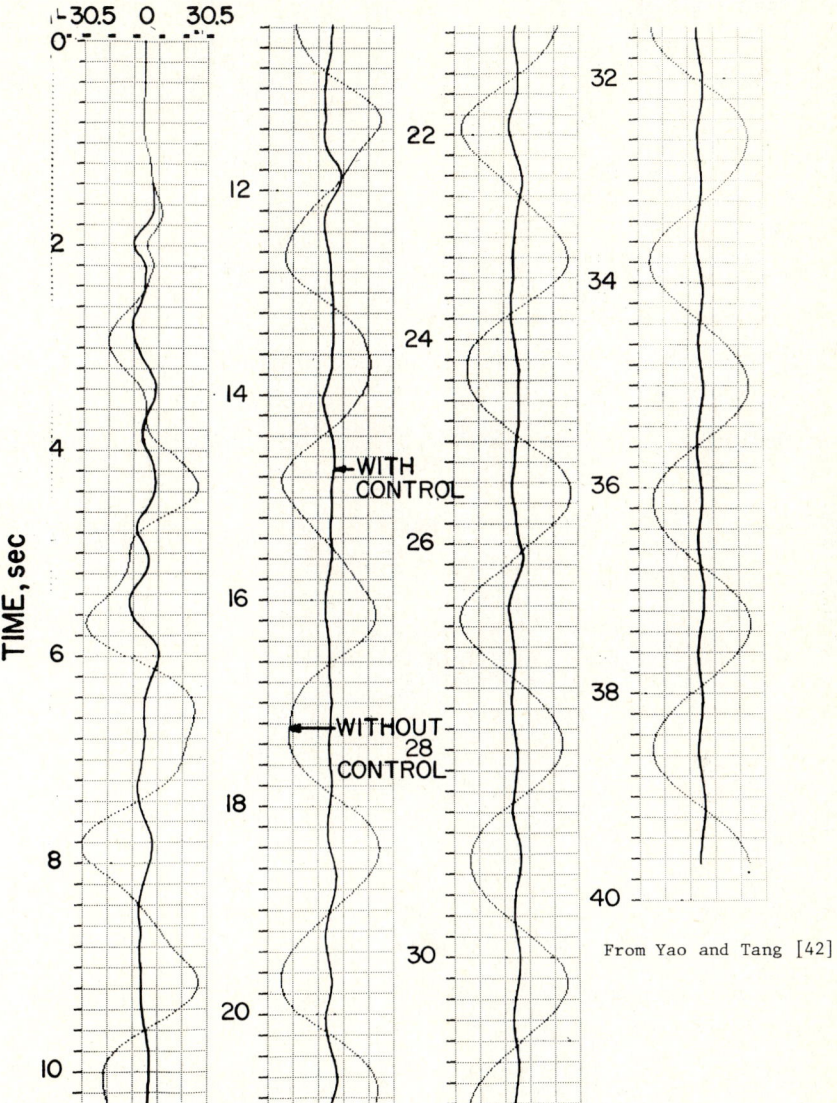

Figure 1(a) - Displacement of two-story structure - first floor relative to ground

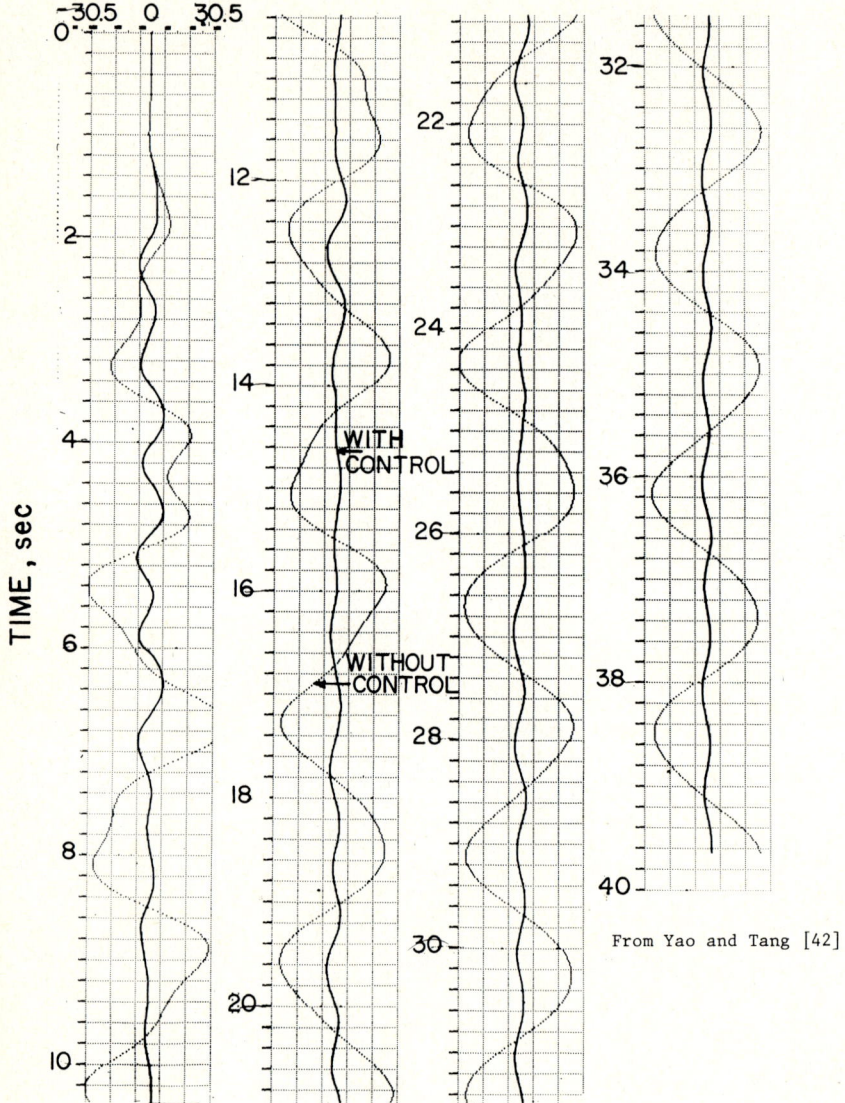

Figure 1(b) - Displacement response of two-story structure - second floor relative to first floor

Control of Structural Damage 765

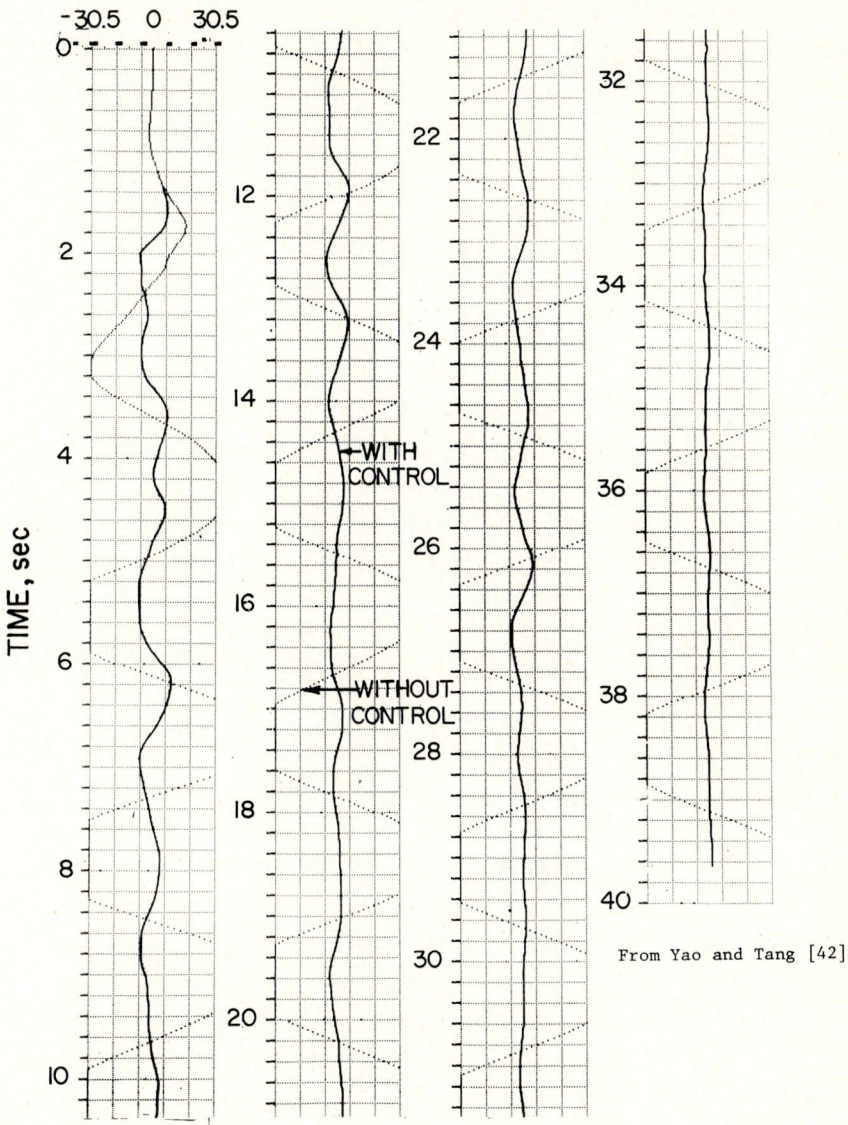

From Yao and Tang [42]

Figure 1(c) - Displacement response of two-story structure - second floor relative to ground

Yang and Yao [43] explored the application of the classical stochastic control theory to civil engineering structures. Yang [44] showed that significant reduction in covariances of structural responses to stationary wind forces as well as nonstationary earthquake effects can be obtained with the use of optimal control. It is realized, however, that this type of controller cannot be used to avoid any given peak responses.

Sae-Ung, et al [45,46] used the Monte-Carlo method to study structural control for comfort purposes. A comfort control law using impulsive control forces was applied to a 40-story building structure, and these simulation results indicated that such a control law is feasible in terms of energy requirements.

Martin and Soong [47] applied modal control in changing specific dynamic modes and system stiffness directly. A design procedure for modal control has also been developed by Chang and Soong [48]. Recently, various types of tendon control have been studied by Roorda [49], Schorn [50], Yang and Giannopoulos [51], and Abdel-Rohman and Leipholz [52,53]. In a different type of application, it was proposed to apply the control theory in the development of design codes [54].

3. APPLICABILITY OF STRUCTURAL CONTROL

Recently, structural control for the purpose of maintaining human comfort has been successfully implemented in practice [39-41]. Meanwhile, many questions on practicality remain in the application of structural control for safety purposes. Some even say that practicing engineers will never accept the idea of using active control systems to ensure structural safety. In this section, an attempt will be made to assess the applicability of structural control under various circumstances.

It may be desirable to review the following several facts. First, it is difficult to predict future loading conditions with limited amount of past records. Secondly, it is not economically feasible to continue the tradition of designing and building massive and stiff civil engineering structures. Last but not least, uncertainties exist in material resistance, mathematical representations, methods of structural analysis, and various human factors. In the case of structures with control systems, the reliability of various components of the control system cannot be overlooked. Depending on the failure consequence, a structure can be designed to possess a certain level of safety. Nevertheless, it is not practical to design and build a structure even with control systems which can completely avoid any damage during its intended lifetime.

For the sake of discussion, consider a given structure with and without structural control as shown schematically in Figure 2. Let $X = X(t)$ be a vector of random processes denoting various loading conditions. Furthermore, let S_o denote structural responses without the control system and S_i denote responses of the ith design of the structure with control system, $i=1,2,\ldots,n$.

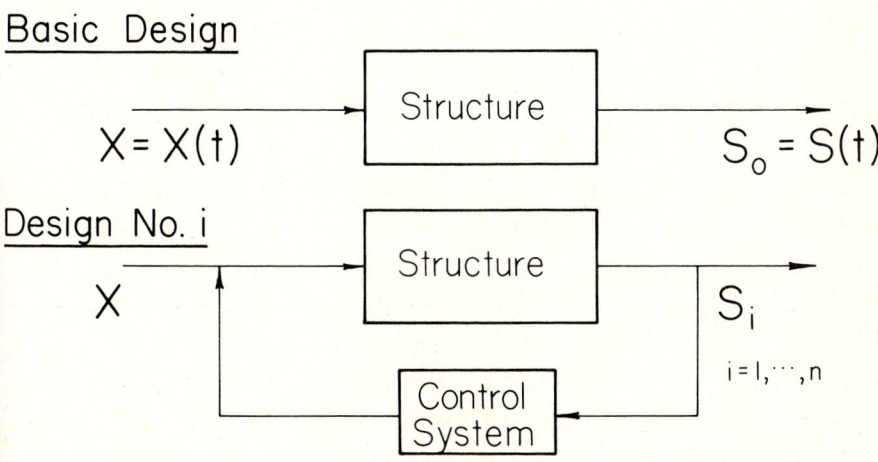

Figure 2 - Schematic diagrams of a structure with and without control system

Define the following damage states:

D_{oi} = "little or no" damage for ith design
   = $(S_i < \ell_e)$

D_{1i} = "Tolerable" damage for ith design
   = $(\ell_e \leq S_i < \ell_t)$

D_{2i} = "Repairable" damage for ith design
   = $(\ell_t \leq S_i < \ell_r)$

D_{3i} = "Severe" damage for ith design
   = $(S_i \geq \ell_r)$

where ℓ_e, ℓ_t, and ℓ_r denotes elastic limits, tolerable damage limits, and repairable damage limits, respectively. If the cost for all these designs are comparable, the design to be chosen can be the one with max $P(D_{oi})$, i=0,1,...,n. The computation of $P(D_{oi})$ can be complicated enough even for structures without control systems. Although the reliability problem related to structural control was explored in 1972 [55], practical and complete solutions are not available to-date.

A point of view concerning the applicability of structural control is summarized in Figure 3. Consider two types of control system, namely passive and active systems. The passive system such as dampers and isolators are always available and operational. On the other hand, consider three types of active control systems, say one each involving small, moderate, and large control forces. The active control system with small control forces seems to be suitable for comfort control and for keeping the structure within the null-damage state D_o. The small control force can also be associated with those loads which exceed the design values but still are lower than the exceptional ones. The active control system with moderate control forces may be appropriate for maintaining the structure within tolerable or repairable damage limits. The moderat

Purpose	Structural Response	Type of Control: Passive — Usage: Always Available	Active — Frequent Use — Control Force: Small	Active — Occasional Use — Control Force: Moderate	Active — Almost Never — Control Force: Large	Damage State
Comfort		✓	✓			
	Warning Limit(s)					l_w
Safety		✓	✓			D_0
	Elastic Limit(s)					l_e
Safety		✓		✓		D_1
	Tolerable Damage					l_t
Safety		✓		✓		D_2
	Repairable Damage					l_r
Safety		✓			✓	D_3
	Near-Collapse		x_d	x_e	x_a	
	Load / Load State	E_0	E_1	E_2	E_3	

Figure 3 - Applicability of Structural Control

control force can also be associated with those loads which range between exceptional and abnormal loads. The large control force should not be used until the structure is near collapse when it is subjected to abnormal loading conditions. The adjective "almost never" implies a very small or zero probability of occurrence.

The argument against structural control for safety considerations will certainly continue. With this brief discussion, it is hoped to call attention to the fact that (a) it is difficult to predict with great certainty the extraordinary or abnormal loading conditions which may occur during the intended lifetime of the structure, and (b) different magnitudes and types of control force may be used under various loading conditions. If and when the failure consequence of a particular structure is extremely grave, the use of one or more levels of active control systems may be warranted.

4. STRUCTURAL IDENTIFICATION

To effectively control the motions of a structure, it is necessary to be able to describe the characteristics of the particular structure. Currently available mathematical representations result from generalizations of existing knowledge in the structural engineering profession. Following the completion of the construction process, each civil engineering structure possess its own characteristics, the precise description of which is difficult to obtain with the use of any general mathematical model. In recent years, techniques of system identification [56,57] have been applied to structural engineering [58]. At present, the study of structural identification includes mathematical modelling, damage assessment, and reliability of existing structures on the basis of field observations and test data [59]. A comprehensive summary and discussion of the subject matter was presented recently [60].

The interrelationship between structural control and structural identification is shown schematically in Figure 4.

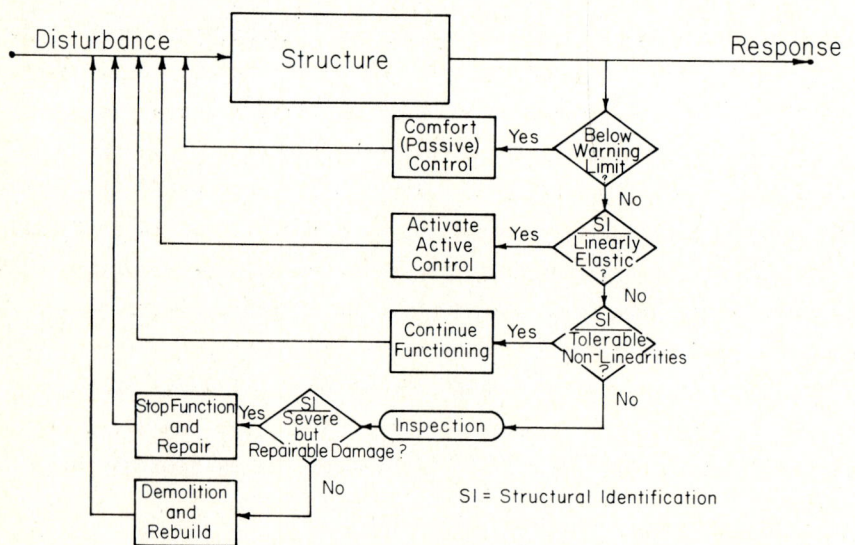

Figure 4 - *Interrelationship between structural control and structural identification*

Basically, a set of warning limits can be established such that the active control devices are activated whenever one or more of these warning limits are exceeded. Structural identification techniques can be applied to detect any damage and to decide whether such damage is tolerable. If and when permanent and moderate damage occur, the structure must undergo a detailed inspection. If it is found necessary, the structure must stop functioning and major repairs should be implemented. When the damage is found to be severe, the structure must then be demolished and rebuilt.

It is noted that such terms as "tolerable" and "repairable" damage are not clearly defined for an existing structure, which is usually a very complex system. Therefore, it is difficult to identify such limit states in reality. As a possible approach to the solution of this problem, the application of pattern recognition is being explored [61].

5. CONCLUDING REMARKS

It is encouraging to note that there has been an increasing interest in research activities concerning structural control during this past decade. With the cooperation of experts from various disciplines including structural engineering, theoretical and applied mechanics, and control theory, more significant contributions to this subject area can be expected in the near future.

One of the most challenging problems seems to be the one relating mathematical representations to the structural behavior in the real world. I believe that experimental results such as those reported by Roorda [62] will be most promising in bridging this gap. Nevertheless, there remains the problem of obtaining precise yet significant solutions of the behavior of complex systems. As it was stated by Zadeh [63], our ability of making precise and yet still significant statements concerning the system behavior diminishes with increasing complexity of the system. Consequently, the closer one looks at a complex real-world problem, the fuzzier its solution becomes.

ACKNOWLEDGEMENT

I wish to thank Professor H.H.E. Leipholz and the Organizing Committee for the opportunity of being here and meeting with many experts of this subject area. In addition, I wish to acknowledge the encouragement of M.P. Gaus, S.C. Liu and C.A. Babendreier of the National Science Foundation, whose encouragement and support enable me to make several exploratory studies in this direction. Especially, I appreciate receiving valuable advice from Dr. Gaus in this regard during the past decade.

REFERENCES

[1] YAO, J.T.P., "Passive and Active Control of Civil Engineering Structures," presented at the ASCE Convention, Boston, MA., 2-6 April, 1979.

[2] ZUK, W., "The Past and Future of Active Structural Control Systems," General Lecture, presented at the IUTAM on Structural Control, University of Waterloo, Waterloo, Ontario, Canada, 4-7 June 1979.

[3] ZUK, W., "Kinetic Structures," *Civil Engineering*, ASCE, December 1968, pp. 62-64.

[4] WRIGHT, R.N., "Active Systems for Increased Structural Resistance to Exceptional Loads," Private Communications, January, 1968.

[5] NORDELL, W.J., "Active Systems for Blast-Resistant Structures," *Technical Report R-611*, Naval Civil Engineering Laboratory, Port Hueneme, CA, February 1969.

[6] YAO, J.T.P., "Adaptive Systems for Seismic Structures," *Report on NSF-UCEER Earthquake Engineering Research Conference*, University of California, Berkeley, CA, March 27-28, 1969, pp. 142-150.

[7] YEH, H.Y. and YAO, J.T.P., "Response of Bilinear Structural Systems to Earthquake Loads," presented at the Vibrations Conference, Philadelphia, Pennsylvania, March 30-April 2, 1969, *ASME Preprint No. 69-VIBR-20*.

[8]   YAO, J.T.P., "Concept of Structural Control," presented at the ASCE National Structural Engineering Meeting, Baltimore, MD, Preprint No. 1360, April 1971; Also, *Journal of the Structural Division*, ASCE, Vol. 98, No. ST7, July 1972, pp. 1567-1574.

[9]   GAUS, M.P., Private Communication, 1972.

[10] Wirsching, P.H., "A Monte Carlo Study of Design Concepts for the Improvement of Reliability of Siesmic Structure, Ph.D. Dissertation, Department of Civil Engineering, the University of New Mexico, Albuquerque, NM, June 1970.

[11] WIRSCHING, P.H. and YAO, J.T.P., "Safety Design Concepts for Seismic Structures", *Computers and Structures*, Vol. 3, 1973, pp. 809-826.

[12] MASRI, S.F., "Effectiveness of Two-Particle Impact Dampers," *The Journal of the Acoustical Society of America*, Vol. 41, No. 6, 1967, pp. 1553-1554.

[13] GUPTA, Y.P. and CHANDRASEKARAN, A.R., "Absorber System for Earthquake Excitations," *Proceedings of the Fourth World Conference in Earthquake Engineering*, 1969, pp. 139-148.

[14] SHINOZUKA, M., Private Communications, December, 1970.

[15] GREEN, N.B., "Flexible First-Story Construction for Earthquake Resistance", *Transactions*, ASCE, Vol. 100, 1935, pp. 645-674.

[16] FINTEL, M. and KHAN, F.R., "Shock Absorbing Soft Story Concept for Multi-Story Earthquake Structure," *ACI Journal*, Title No. 66-29, May, 1969.

[17] MATSUSHITA, K. and IZUMI, M., "Studies on Mechanisms to Decrease Earthquake Force Applied to Buildings," *Proceedings of the Fourth World Conference on Earthquake Engineering*, 1969, pp.

[18] PENZIEN, J. and HANSON, R.D., *The Gediz Turkey Earthquake of 1970*, National Academy of Sciences, Washington, D.C., 1970.

[19] KELLY, J.M., SKINNER, R.I. and HEINE, A.J., "Mechanisms of Energy Absorption in Special Devices for Use in Earthquake Resistant Structures," *Bulletin, New Zealand National Society for Earthquake Engineering*, Vol. 5, No. 3, Septmeber, 1973.

[20] SKINNER, R.I., BECK, J.L. and BYCROFT, G.N., "A Practical System for Isolating Structures from Earthquake Attack," *Earthquake Engineering and Structural Dynamics*, Vol. 3, 1975, pp. 297-309.

[21] KELLY, J.M. and TSZTOO, D.F., *The Development of Energy-Absorbing Devices for Aseismic Base Isolation Systems*, Report No. UCB/EERC-78/01, Earthquake Engineering Reserach Center, University of California at Berkeley, January 1978.

[22] KELLY, J.M. and TSZTOO, D.F., "Earthquake Simulation Testing of a Stepping Frame with Energy-Absorbing Devices," *Bulletin, New Zealand National Society for Earthquake Engineering*, Vol. 10, No. 4, December 1977, pp. 196-207.

[23] SKINNER, R.I., HEINE, A.J. and TYLER, R.G., "Hysteretic Dampers to Provide Structures with Increased Earthquake Resistance," *Proceedings, Sixth World Conference on Earthquake Engineering*, New Dehli, India, January 1977.

[24] SKINNER, R.I., BYCROFT, G.N., and McVERRY, G.H., "A Practical System for Isolating Nuclear Power Plants from Earthquake Attack," *Nuclear Engineering and Design*, Vol. 36, 1976, pp. 287-297.

[25] TYLER, R.G., "Dynamic Tests on Laminated Rubber Bearings," *Bulletin, New Zealand National Society for Earthquake Engineering*, Vol. 10, No. 3, September 1977, pp. 143-150.

[26] ROBINSON, W.H. and TUCKER, A.G., "A Lead-Runner Shear Damper", *Bulletin, New Zealand National Society for Earthquake Engineering*, Vol. 10, No. 3, September 1977, pp. 151-153.

[27] EIDINGER, J.M. and KELLY, J.M., *Experimental Results of an Earthquake Isolation System Using Natural Rubber Bearings*, Report No. UCB/EERC-78/03, Earthquake Engineering Research Laboratory, University of California at Berkeley, 1978.

[28] JOLIVET, F. and RICHLI, M.H., "Aseismic Foundation System for Nuclear Power Stations", Paper No. K9/2, *Proceedings of the 4th International Conference on Structural Mechanics in Reactor Technology*, San Francisco, August 1977, pp.

[29] SPENCER, P., ZACKAY, V.F., and PARKER, E.R., *The Design of Steel Energy Absorbing Restrainers and Their Incorporation into Nuclear Power Plants - Volume 1, Summary Report*, No. UCB/EERC 79/07, Earthquake Engineering Research Laboratory, University of California at Berkeley, February 1979.

[30] KELLY, J.M. and SKINNER, M.S., *The Design of Steel Energy Absorbing Restrainers and Their Incorporation into Nuclear Power Plants for Enhanced Safety; Volume 4 - Review of Current Uses of Energy Absorbing Devices*, Report No. UCB/EERC 79/10; Earthquake Engineering Research Laboratory, University of California at Berkeley, February 1979.

[31] BHATTI, M.A., PISTER, K.S. and POLAK, E., *Optimal Design of an Earthquake Isolation System*, Report No. UCB/EERC 78/22, Earthquake Engineering Research Laboratory, University of California at Berkeley, October 1978.

[32] CRANDALL, S.H., LEE, S.S. and WILLIAMS, J.H., Jr., "Accumulated Slip of a Friction-Controlled Mass Excited by Earthquake Motions", *J. Appl. Mech.*, Vol. 41, No. 4, December 1974, pp. 1094-1098.

[33] CRANDALL, S.H. nad LEE, S.S., "Biaxial Slip of a Mass on a Foundation Subjected to Earthquake Motions," *Ingenieur-Archiv*, Vol. 45, 1976, pp. 361-370.

[34] NAMAT-NASSER, S., Private Communications, 1 June 1978.

[35] KLEIN, R.E., CUSANO, C. and STUKEL, J.J., "Investigation of a Method to Stabilize Wind Induced Oscillations in Larger Structures," presented at the ASME Winter Annual Meeting, New York, N.Y., November 1972.

[36] KLEIN, R.E., "The Potential for Application of Closed-Loop Control Concepts in Structures," presented at the ASCE National Convention, Boston, April 1979.

[37] CHANG, M.I. and SOONG, T.T., "Optimal Control Configuration for Control of Complex Structures," presented at the IUTAM Symposium on Structural Control, University of Waterloo, Waterloo, Ontario, Canada, 4-7 June, 1979.

[38] SOONG, T.T., Private Communications, 13 April 1977.

[39] "Tuned Mass Dampers Steady Sway of Skyscrapers in Wind," *Engineering News Record*, 18 August 1977, p. 28-29.

[40] "Engineer's afterthough Sets Welders to Work Bracing Tower," *Engineering News Record*, 17 August 1978, p. 11.

[41] PETERSEN, N.R., "Design Considerations of Large-Scale Tuned Mass Dampers for Structural Motion Control," presented at the ASCE National Convention, Boston, April 1979.

[42] YAO, J.T.P. and TANG, J.P., *Active Control of Civil Engineering Structures*, Technical Report No. CE-STR-73-1, School of Civil Engineering, Purdue University, W. Lafayette,

[43] YANG, J.N. and YAO, J.T.P., *Formulation of Structural Control*, Technical Report No. CE-STR-74-2, School of Civil Engineering Purdue University, W. Lafayette,

[44] YANG, J.N., "Application of Optimal Control Theory to Civil Engineering Structures," *Journal of the Engineering Mechanics Division*, ASCE, Vol. 101, No. EM6, December 1976, pp. 819-838.

[45] SAE-UNG, S., "Active Control of Building Structures," Ph.D. Dissertation, School of Civil Engineering, Purdue University, W. Lafayette, May 1976,.

[46] SAE-UNG, S. and YAO, J.T.P., "Active Control of Building Structures," *Journal of the Engineering Mechanics Division*, ASCE, Vol. 104, No. EM2, April 1978, pp. 335-350.

[47] MARTIN, C.R. and SOONG, T.T., "Modal Control of Multistory Structures," *Journal of the Engineering Mechanics Division*, ASCE, Vol. 102, No. EM4, August 1976, pp. 613-623.

[48] CHANG, M.I.J. and SOONG, T.T., "Modal Control Design for Systems Having Complex Eignevalues," Private Communication, June 1979.

[49] ROORDA, J., "Tendon Control in Tall Buildings," *Journal of the Structural Division*, ASCE, Vol. 101, No. ST3, March 1975, pp. 505-521.

[50] SCHORN, G., "Feedback Control of Structures," Ph.D. Thesis, University of Waterloo, Ontario, Canada, 1975.

[51] YANG, J.N. and GIANNOPOULOS, F., "Active Control of Two Cable-Story Bridges," presented at the ASCE National Convention, Boston, April 1979.

[52] ABDEL-ROHMAN, M. and LIEPHOLZ, H.H., "Active Control of Flexible Structures," *Journal of the Structural Division*, ASCE Vol. 104, No. ST8, August 1978, pp. 1251-1266.

[53] ABDEL-ROHMAN, M. and LEIPHOLZ, H.H., "Structural Control by Pole Assignment Method," *Journal of the Engineering Mechanics Division*, ASCE, Vol. 104, No. EM5, October 1978, pp. 1159-1175.

[54] SCHORN, G. and LIND, N.C., "Adaptive Control of Design Codes," *Journal of the Engineering Mechanics Division*, ASCE, Vol. 100, No. EM1, February 1974, pp. 1-16.

[55] GOLDBERG, J.E., TANG, J.P. and YAO, J.T.P., "Reliability of Structures with Control Systems," *Proceedings of the International Symposium on Systems Engineering*, Purdue University, Vol. 2, 23-27 October 1972, pp. 153-155.

[56] EYKHOFF, P., *System Identification-Parameter and State Estimation*, John Wiley & Sons, 1974.

[57] SAGE, A.P. and MELSA, J.L., *System Identification*, Academic Press, 1971.

[58] HART, G.C. and YAO, J.T.P., "System Identification in Structural Dynamics," *Journal of the Engineering Mechanics Division*, ASCE, Vol. 103, No. EM6, December 1977, pp. 1089-1104.

[59] LIU, S.C. and YAO, J.T.P., "Structural Identification Concept," *Journal of the Structural Division*, ASCE, Vol. 104, No. ST12, December 1978, pp. 1845-1858.

[60] YAO, J.T.P., "Damage Assessment and Reliability Evaluation of Existing Structures," Invited Lecture, presented at the Symposium Honoring Professor T.V. Galambos, Washington University, St. Louis, MD, 17 April 1979.

[61] FU, K.-S. and YAO, J.T.P., "Pattern Recognition and Damage Assessment," to be presented at the ASCE EMD Specialty Conference at Austin, Texas, September 1979.

[62] ROORDA, J., "Experiments in Feedback Control Structures," General Lecture, presented at the IUTAM on Structural Control, University of Waterloo, Waterloo, Ontario, Canada, 4-7 June 1979.

[63] ZADEH, L.A., "Outline of a New Approach to the Analysis of Complex Systems and Decision Processes," *IEEE Transactions on Systems, Man and Cybernetics*, Vol. SMC-3, No. 1, January 1973.

STRUCTURAL CONTROL, H.H.E. Leipholz (ed.)
North-Holland Publishing Company & SM Publications
© IUTAM, 1980

THE PAST AND FUTURE OF ACTIVE STRUCTURAL CONTROL SYSTEMS

William Zuk

School of Architecture
University of Virginia
Charlottesville, Virginia, U.S.A.

1. INTRODUCTION

To many people, the concept of active structural systems is a radical new idea; yet its origins go back millions of years to a time when organic life first began. Perhaps the best way to present the case is by way of examples. I shall start with some relatively simple structures as found in very early nature; namely plants. We all know that plants, whose stalks and leaves can be seen as small cantilever structures, do indeed move and change shape, seemingly of their own accord. Actually, these movements, called tropisms, are in response to a variety of external stimuli or signals. There is phototropism (a response to light stimulus), heliotropism (a response resulting in twisting), geotropism (a response to gravity forces), hydrotropism (a response to water) and haptotropism (a response to touching). A branch growing up or out toward the direction of the sun is because of phototropism. A sunflower plant twisting to follow the orbital path of the sun illustrates heliotropism. Roots that grow downward are geotropic and roots that grow toward a source of water are hydrotropic. A dramatic example of haptotropism is the sudden closing of the

Venus fly trap plant when certain of its hairlike sensors are touched.

The mechanisms causing these controlled movements are ingeniously simple mechanically, although quite complex chemically. For a stalk or root to turn a corner, the plant merely adds more cells (through plant hormone chemistry) on one side over a small distance and the expansion of this side causes the structure to bend at that position. Much more rapid action, as the wilting of a mimosa leaf when touched is generated by fluid osmatic pressure. Water escapse from certain pressurized thin walled cells into intercellular voids; thereby causing drooping much like a balloon which has lost its air pressure.

Proceeding up the evolutionary ladder, creatures as insects, fish, animals and humans have other interesting structural control systems. Movements in creatures are achieved primarily by muscular action, although gas and fluid pressures are also used in some circumstances. An example of the latter is the arm of a starfish which is extended by increasing its internal fluid pressure and retracted by decreasing the pressure. A splendid example of muscular action can be found by looking no further than to ourselves. The movements of our hands, arms and legs are actuated by muscles (basically tension members) reacting against our bones (compression members). The human body contains 696 major muscles and 206 major bones. In our hip joint alone there are nineteen different sets of muscles enabling movement about the joint in almost any direction. More importantly, we know how to control with great precision these movements. Should we wish to move our leg a small distance in a given direction, our eyes (acting as sensors) focus on our leg, sending a message to the brain (a control centre) which would send a signal to the appropriate tendons to contract (an energizing action), thereby moving the leg the set amount.

2. STRUCTURAL CONTROL

I speak of man as an example, but obviously man did not invent his own bio-system. However, man did invent many other kinds of active control systems. His early attempts several thousand years ago resulted in the development of simple machines, such as Archimedes' water pump and Ktesibios' catapult. These were interactive machine-man systems where the machine performed the operation and man exercised the control. Through the years, these machine-man control systems gradually became more complex, and in the early 19th century machines were devised that started to control themselves in a number of ways. Pumps started and stopped automatically by a float control according to the water level to be maintained and steam engines automatically controlled their speed by mechanical governors. The concept of sensor feedback came into being as scientists developed sensing instruments able to detect both tangible and intangible quantities as velocity, acceleration, light, heat, pressure, magnetism and current; and as engineers developed sophisticated machines to perform tasks of greater complexity.

The early 20th century brought with it control systems in the field of electricity and electronics, stimulated in part by the need for accurate control of weaponry. Aside from all the hardware development associated with this field, as television, communication satellites, and guided missiles, this subject was responsible for the mathematical basis of control theory. Control of electrical and electronic apparatus is so rapid that unprecedented reliance had to be placed upon understanding of the feedback behaviour through mathematics. Mathematics thus permitted a "slow look" at what was happening or could potentially happen in complex circuitry.

Civil engineers, for the most part, ignored what was happening all around them in active control development. It was not until 1960 that it occurred to Eugene Freyssinet (a pioneer developer of prestressing of concrete bridges and buildings) that prestressing tendons have the potential of active control, much as tendons do in one's arm or leg. A few years later, Lev Zetlin (a most creative structural engineer and a good friend of mine) independently conceived the same idea and actually designed several tall buildings (a 1,000 ft. high one in Milwaukee and a 2,500 high one in New York) using active systems to control sway. The concept was not unlike that used in the human body. A set of vertical cables were fixed to the building frame, with the cables attached to hydraulic jacks at their lower ends. Sensors to detect movement at the top of the structure signaled a control device, which directed the action of the jacks. Unfortunately, neither structure was built.

3. APPLICATION OF STRUCTURAL CONTROL

I might mention that just in the last few years, two tall buildings with *partial* active structural control systems have been built. The first was the Citicorp Building in New York and the second was the John Hancock Building in Boston. Both structures, designed by the engineer William Le Messurier have at their top a large tuned mass damper. In strong wind conditions, the large mass is automatically lifted off its pad by hydraulic fluid and allowed to "float", reacting laterally against pneumatic springs which dissipate the buildings' vibrational energy through oil dashpots.

Because of the thousands of years of past history of buildings and bridges, almost totally limited to passive structures, as well as the natural conservatism to accept anything new, I do not believe that full realization and widespread acceptance of active control systems in large scale structures is likely until the 21st Century; as much as I should like to see it occur sooner. So, I shift my attention from the past to the future. I skip the present, because you gentlemen here at this symposium are the present and I am not intrepid enough to intrude upon your exceptional authority and expertise. I know, however, that your work and contributions in this emerging field are absolutely essential if there is to be a future. With full confidence that the future will be an exciting one, allow me to suggest some of the unique potentials of active structural control systems.

I see two basic long range applications of active structural control systems. The first is to make flexible objects stiffer, stronger or more stable. This condition generally involves small motions. The second is to generate controlled movements; generally involving large motions. Let me elaborate on these catagories with some examples, starting with the first. Modern materials are steadily getting stronger as researchers are improving the quality of old materials and producing new high strength ones. Unfortunately the increase in the stiffness moduli of these materials generally lags behind the increases in strength. This results in structures that are quite strong but very flexible and vibration prone. The new World Trade Towers in New York deflect laterally several feet in a strong wind as contrasted with the old Empire State Building which deflects only several inches. The new Verranzano Narrows suspension bridge in New York built of high strength steel deflects a total of 35 feet vertically under extreme conditions of load and temperature. A clear need thus exists for active control systems which come into play to limit excessive deformations. Flexure, torsion and even column buckling can be controlled to whatever degree required. The control movements from any source whatever can potentially be controlled with the proper active system. This includes normal dead and live loads, wind loads, seismic forces, general vibrational forces and temperature.

It may be said that additional passive systems as bracing and dampers can also suppress movements, but active systems can do certain things that passive systems just cannot do. Active systems can reduce deflections to absolutely zero; not just to a small amount, but to absolute zero. Active systems can alter the behaviour of columns so that the buckling load can actually exceed the generally accepted theoretical limit as derived by Euler by a factor of nine. What we have in active structural control systems is a basic change in kind, not just in amounts; as different as an automobile is to a wagon.

Buildings and bridges are by no means the only applications. A partial list of other uses would be for towers, aircraft and missile structures, ship and submarine hulls, fixed sea structures (as drilling rigs), communication structures (particularly antennas) and space stations. High strength and stiffness to weight ratios of these structures are very important. Using energy to replace mass is a promising way to achieve these ends.

4. KINETIC STRUCTURES

Although much could be said of active ways to reduce movement, let me move on to active systems as used to generate movements. At first thought, this subject may seem to be somewhat outside the scope of this symposium, but I believe it will become an important part of our future, opening up a new area of active structural control. In my own research on the subject, I have chosen to refer to these as kinetic structures. Conceptually, structures could be seen as machines which are designed to change and adapt; which are time-dependent rather than static. I visualize all manners of buildings as being able to change shape, modifying themselves to suit ever changing forces and functions. Was it not Hericlitus, the ancient Greek philosopher, who said that everything changes but change itself? Buildings could be compactly pre-packaged in a factory and conveniently transported to the site. At the site it would be energized, causing it to self-deploy or erect itself because of a predesigned control system. I have experimented with both hard and soft systems to perform these kinematics. Nature abounds with examples of this kind. A small compact acorn, when properly stimulated grows into a hugh oak tree, following a prearranged genetic control code. If structures can be self-erecting, they can be self-collapsing. Reversible buildings of this kind can be easily transplantable to new locations as site and use conditions change. Bridges, roads and utilities could all be designed for transplantability as well. The potentials for a kinetic city designed to physically change as human needs change are almost mind boggling.

At this point, let me outline some technical problems as I see them, that require future research and development. Basically these are mathematical analysis, hardware development, and kinematic structure research. In the area of mathematics, more must be learned of the behaviour of structures under different excitation and control conditions. We need to be absolutely certain when a system can go unstable. We need optimized solutions for practical implementation, considering both materials and energy. We need further analysis of fail-safe systems, should the primary active system fail.

In the area of hardware, three main units need physical development. The first deals with sensors. These must detect with reliability parameters as displacement, velocity, acceleration, strain and frequency. All of these sensors already exist in one form or another, but they need development for application in large scale structures under all kinds of field conditions. Perhaps the structures themselves may have to be modified to accommodate certain sensors. I am thinking of things as long sealed laser-beam tunnels or radar guides that may have to be part of a bridge or a tower. Other sensors as accelerometers and electrical strain gage can be attached with hardly any modification.

Signals from the sensors have to be transmitted to a brain as on on-board mini-computer. These, too, are already in existence, particularly on missiles and large aircraft. Concerning the new U. S. space shuttle "Columbia", the vehicle is so unstable that landing controls are activated by computers, not by the pilot. Reaction times of humans are too slow for the necessary complex operations required. Again, as in sensors, reliable computer equipment must be tailored for structural use.

The third bit of hardware, that of force actuators or energizers, requires the most work. Although many machines or vehicles have energizing hardware, that require for structures (particularly large ones as buildings and bridges) must be capable of producing rather large forces. I am drawn to seeing force generators as a system of tendons laced throughout a structure, analogous to muscles in a human body. These tendons could be pulled by any number of devices, as jacks, reels, levers, or possibly even thermal expansion or electromagnatic devices. However, other possible force generators are also possible, not necessarily related to tendons. These are gyroscopes, active mass dampers, air or gas pressure, or even controlled explosions. After all, explosions are how combustion engines and jet engines for aircraft produce their energy.

In regard to kinetic structures, there too, much development is needed. I divided this into two categories; one dealing with movements of deformable solids, and one dealing with defined articulated movements of essentially nondeformable solids. How does one control movements of deformable solids in a time-dependent way? One deals basically with three factors, namely forces, material properties and configuration of the material. A possible approach is to modify the forces as by use of controlled force generators previously described. Another possibility is to alter the shape of the structure. For instance, flexing a straight beam into an arch form will strengthen it; or folding a flat sheet of paper into corrugations will stiffen it. These changes in form could conceivably be done in a time-dependent way through various mechanisms. The third possibility of time-dependently modifying the material properties is also possible, but little used. Many factors do indeed alter the strength and stiffness of materials. To name several; there is temperature, chemicals (including water) electricity, radiation and the state of stress. The latter factor often puzzles students as they see a material as steel as strong and putty as weak; but if putty were triaxially confined, its strength becomes enormous. So, another alternative to controlling deformations is to actually control the properties of the materials themselves; ridigizing or plastisizing it electrically, thermally or chemically.

In the realm of articulated kinematics of solids, there are only two categories; linear translation and rotation. All other solid body motions whether planar or spacial are but combinations of these two. However, to implement these simple statements into structural connections as hinges, slide tracks, gears and free releases devices suitable for large scale structures subject to large forces and minimum maintenance is a notable challenge. Large structures which have utilized articulated kinematics do exist (chiefly as moveable bridges of the lift, swing and bascule variety) but I hesitate to suggest them as models for the future. Much more development must be done along these lines.

The following are a few examples of some actual projects, that my students and I have experimented with, as indicators of what might be done, particularly with buildings, in actively controlling enclosed space through structural manipulation. (See Figures 1 - 14).

The most fitting way to end my presentation is to underscore the obvious that much remains to be done in virtually all areas of this most exciting and potentially revolutionary concept of active control. I look forward to learning of the efforts and discoveries of all of those devoted to structural control.

Figure 1 - Three Dimensional Lattice Frame for Multistory Building. Frame is prefabricated and shipped to site in a collapsed state. At the site, it is opened as shown.

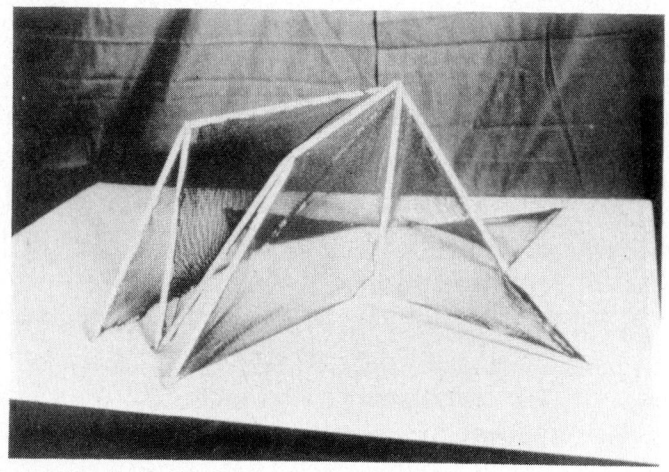

Figure 2 - Articulated Frame with Flexible Covering. Form of structure may be altered in a variety of ways to match changing spatial needs.

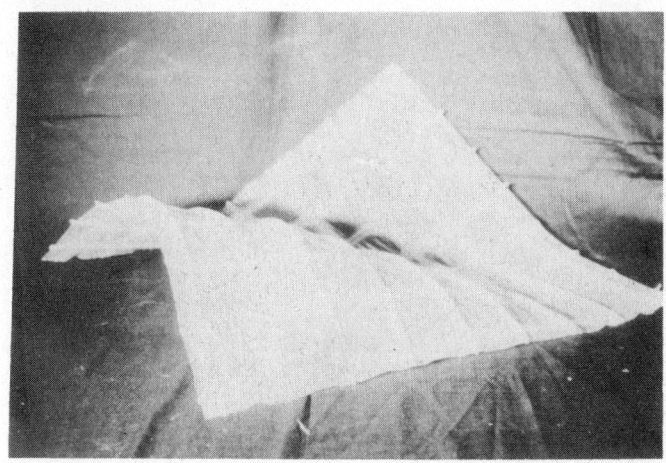

Figure 3 - Soft Shell Structure. The system of cross ribs can be rigidized or plastisized (chemically or thermally) to change the shape of the structure as required.

Figure 4 - Auditorium Structure, shown in a folded compact configuration.

Figure 5 - *Auditorium Structure, shown partly deployed. The folded plates function as structures as well as envelope.*

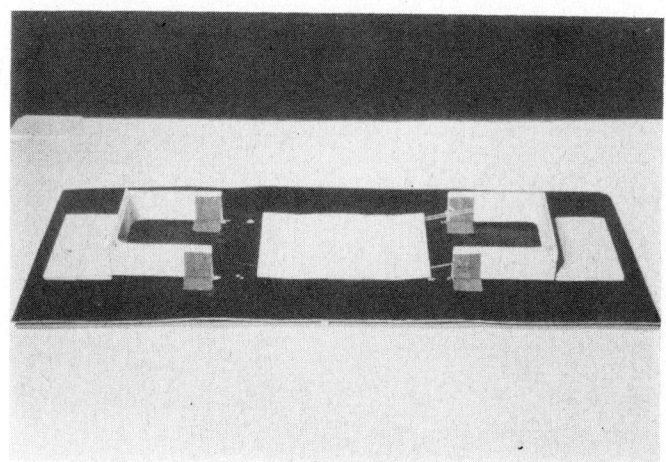

Figure 6 - *Self-Erecting Frame, shown in position ready for erecting. A set of cables are threaded continuously through the units.*

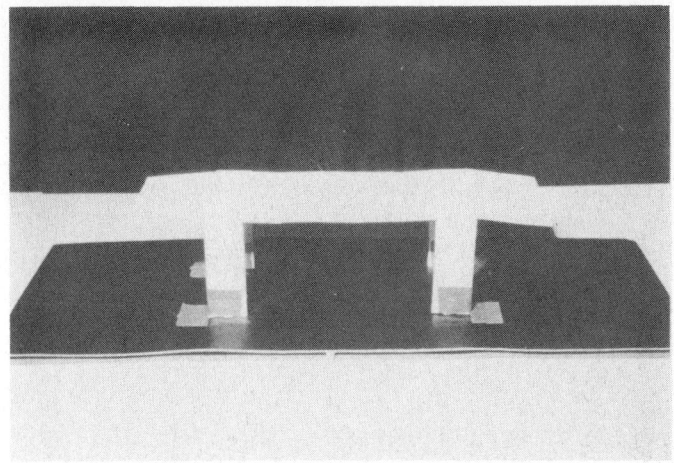

Figure 7 - Frame in Figure 6, shown erected and prestressed. By pulling the cable threaded through the units, the structure erects itself and post-tensions the units together.

Figure 8 - Three Story Dwelling shown Collapsed.

Figure 9 - Dwelling shown Erected. Rotation of the drive shaft at ground level causes the structure to rise and assume the desired form.

Figure 10 - Two Folding Shelters. One is shown erected and the other is shown laid out prior to erection. Simple hinges join the triangular panels.

Figure 11 - *Folding Shelter seen Partly Erected. Note the method of erection by rotation. The hinges allow for rapid erection as well as provide the connections for the transmission of forces in the structure.*

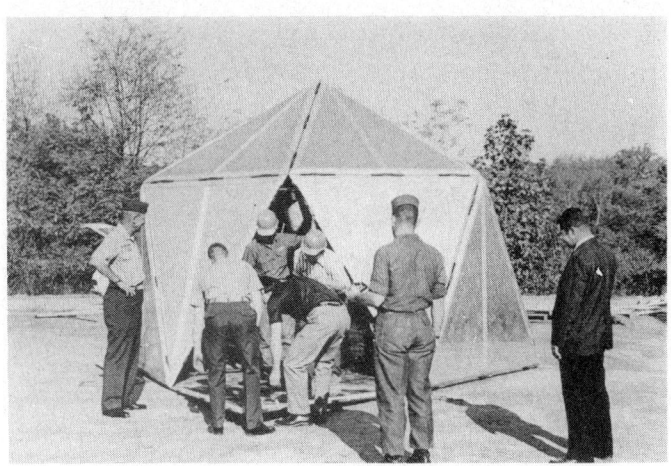

Figure 12 - *Full Scale Folding Shelter. Prototype panels are aluminum honeycomb sandwiches with extruded aluminum edge members. The hinges are high strength steel.*

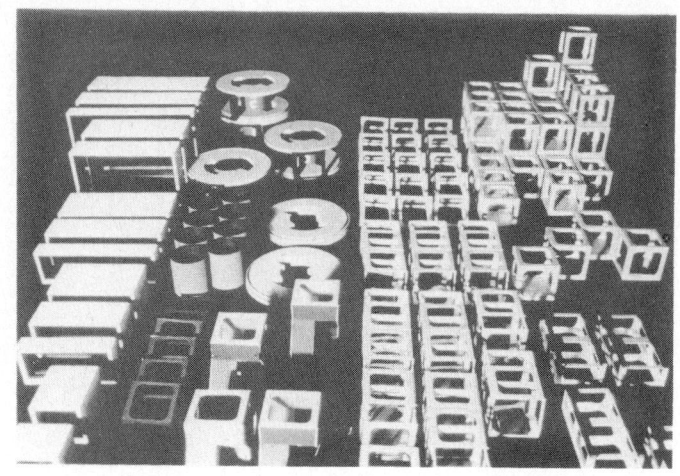

Figure 13 - A Kit of Modular Parts for a University Classroom Building.

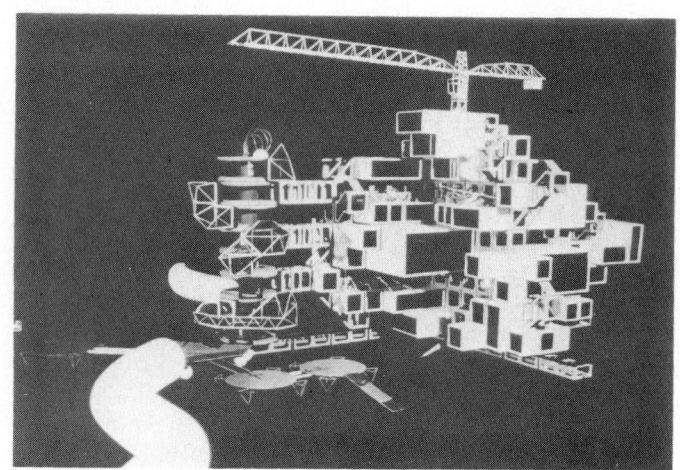

Figure 14 - Classroom Building Assembled. The permanent crane delivers modular units to any part of the structure. Units are interchangeable with reversible connectors so that the building can adapt to a wide variety of forms, changing with need and time, enlarging or contracting.

STRUCTURAL CONTROL, H.H.E. Leipholz (ed.)
North-Holland Publishing Company & SM Publications
© IUTAM, 1980

OPTIMAL STRUCTURAL DESIGN OF FLEXIBLE BEAMS WITH
RESPECT TO CREEP RUPTURE TIME

M. Życzkowski and W. Świsterski

Politechnika Krakowska
ul. Warszawska 24
31-155 Kraków, Poland

1. INTRODUCTORY REMARKS

The time factor appears in structural optimization mainly under dynamic loading conditions. However, the time factor may also be introduced by rheological properties of the material and such quasi-static problems will be considered in the present paper.

Optimal structural design in creep conditions was initiated a decade ago almost simultaneously by M.I. Reytman [1], W. Prager [2], M. Życzkowski [3] and Yu.V. Nemirovsky [4]. From the point of view of the time factor the problems of optimal design may be divided into two basic types: either this factor is inessential and may easily be eliminated, or it cannot be eliminated and the whole process of creep should be considered. The first, much simpler case takes place, as a rule, if we consider steady creep and neglect geometry changes of the structure: then, making use of Alfrey's analogy for physically linear bodies or Hoff's extension of that analogy to physical nonlinearity we may replace a viscoelastic structure by the corresponding elastic one and the time factor is eliminated. Such problems may be called "apparent creep optimization" though, of course, the solutions obtained give in most cases a correct answer for creep conditions. The above mentioned papers by Prager [2] and Nemirovsky [4] consider problems of this type; many effective solutions are given by Gajewski [5] and optimal design of pipe-line cross-section investigated by Życzkowski and Rysz [6] belongs also to the first case. The range of application of the "apparent creep optimization" is even broader: for example, Wojdanowska [7] considered optimal design of imperfect columns allowing for geometry changes (but described by the linearized theory) and in this case - under certain restrictions concerning the initial deflection line - the elimination of the time factor was also possible.

The present paper deals with a problem of the second type, where the time factor cannot be eliminated and the whole process of creep is studied.

Namely, we consider optimal shapes of geometrically nonlinear beams within the range of finite deflections as described by exact equations of creep bending with normal force allowing for stretchability of the beam axis.

The optimization problem is formulated as follows: minimal volume of the beam is the design objective; width $b = b(X)$ or depth $H = H(X)$ are the design variables the cross-sectional shape being assumed (X denotes here the Lagrangian coordinate measured along the axis of the beam); an assumed life-time of the structure t_k under the given loading yields the optimization constraint; the state equations are based on classical creep laws, (Maxwell, Voigt-Kelvin, Norton) as applied to small strains but finite deflections.

The above-mentioned optimization constraint will now be discussed in detail. The life-time of a structure in creep conditions is, in most cases, determined by creep rupture or creep buckling phenomena. However, even in certain problems of creep buckling the life-time may be determined by creep rupture (Życzkowski and Zaborski [8]). Hence we restrict the constraint to creep rupture only, assuming the simplest theory of brittle creep rupture as described by Kachanov's phenomenological equation [9].

L.M. Kachanov introduces the "continuity function" ψ (or the "damage function" $D = 1-\psi$) governed by the following differential equation

$$\frac{d\psi}{dt} = -C\left(\frac{\sigma}{\psi}\right)^\nu, \qquad (1)$$

where $\psi = A/A_o$, A and A_o are elementary cross-sections, current (diminished as a result of microcracks) and initial respectively, C and ν are material constants, t denotes time. This equation should be integrated with the initial condition $\psi(0) = 1$, and the condition $\psi = 0$ determines the onset of a macrocrack, identified here with the life-time of the structure, (Kachanov considered also the second period, namely the propagation of macrocracks up to complete failure, but this period is very short, as a rule, and will not be considered here).

Integrating (1) we may describe the life-time t_k by the equation

$$\int_o^{t_k} \sigma^\nu dt = \frac{1}{C(1+\nu)}. \qquad (2)$$

The beams which satisfy (2) for tensile outer fibres of each cross-section are called here "the beams of uniform creep strength". Similarly as in the elastic range we suppose that the optimal beam may be found among the beams of uniform creep strength (some exceptions may take place in the case of a considerable beam-column effect, but such cases will not be dealt with here). Since we confine our considerations to one design variable only, $H = H(X)$ or $b = b(X)$, condition (2) applied to particular sections X determines fully that variable and no further optimization is possible.

2. THE METHOD OF OPTIMIZATION

The process of creep of the beams under consideration will be described by nonlinear partial differential equations; these equations together with the integral condition (2) determine the optimal shape of the beam. The problem is rather complicated and neither analytical nor direct numerical procedures seem to be applicable. Hence we apply numerical integration of the governing equations combined with iterative determination of the optimal shape of the beam. The iterative procedure will be, to a certain degree, similar to the "growing reforming procedure", proposed by Umetani and Hirai [10], but applied by them to optimization of elastic beams only.

The basic steps of the procedure look as follows. First we determine an optimal elastic beam with finite deflections taken into account (geometrically nonlinear elastic beam of uniform strength); it is described by a system of ordinary differential equations. The stress in outer fibers is assumed to be

$$\sigma_o = \left[\frac{1}{C(1+\nu)t_k}\right]^{1/\nu}, \qquad (3)$$

corresponding to the life-time t_k under constant stresses. The creep process is then studied and the integral (2) for tensile stresses in outer fibers $\bar{\sigma}(X,t) = \sigma(X,H,t)$ of particular cross-sections X is evaluated. It should be constant along the beam, but it is not. Denote, for subsequent iterations i, the function $g_i(X)$ as follows

$$g_i(X) = \int_o^{t_k} [\bar{\sigma}_i(X,t)]^\nu dt, \qquad (4)$$

and denote the constant $1/C(1+\nu)$ by g_o. The next iteration of the shape of the beam is then described by the condition

$$d_{i+1}(X) = d_i(X) \left[\frac{g_i(X)}{g_o}\right]^m , \qquad (5)$$

where d denotes, generally, either b or H, and the exponent m depends on ν and on the design variable adopted. For example, if the width b of the beam is to be determined, then it is justified to assume $m = 1/\nu$, since $\bar{\sigma}$ is proportional to b^{-1} and $\bar{\sigma}^{-\nu}$ to $b^{-\nu}$; if the depth h is the design variable, then we assume rather $m = 1/2\nu$, since in pure bending $\bar{\sigma}$ is proportional to H^{-2}.

For each iteration the following integral is calculated

$$G_i = \frac{1}{\ell} \int_o^\ell [g_i(X) - g_o]^2 \, dX , \qquad (6)$$

and the iterative procedure is finished if G_i is sufficiently small.

3. GOVERNING EQUATIONS

We consider a flexible beam under bending with normal force (shear effects being neglected), and hence, to obtain a consistent theory, we allow for the stretchability of the beam axis and distinguish the Lagrangian coordinate along the axis X from the current length of the axis $s = s(X)$.

Further dependent variables are denoted as follows:

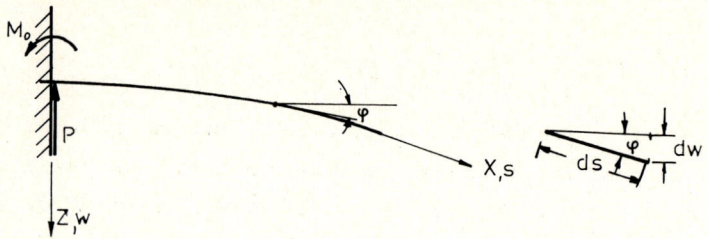

Figure 1

$M(X,t)$ - bending moment,
$N(X,t)$ - normal force,
$\phi(X,t)$ - angle of deflection (slope),
$w(X,t)$ - deflection,
$\sigma(X,Z,t)$ - stress distribution in the beam,
$\varepsilon(X,Z,t)$ - strain distribution in the beam.

Confining our considerations to cantilevers loaded by a concentrated force P we may write the equilibrium equations in the form:

$$\left. \begin{aligned} \frac{dM}{dX} &= P \frac{ds}{dX} \cos \phi , \\ N &= P \sin \phi . \end{aligned} \right\} \qquad (7)$$

Bernoulli's hypothesis of plane cross-sections describes the strain distribution as follows:

$$\varepsilon = \varepsilon(X,Z,t) = -\frac{d\phi}{dX} Z + \frac{ds}{dX} - 1 . \qquad (8)$$

Equations (7) and (8) together with the constitutive equation of creep, e.g.,

$$f(\varepsilon,\dot{\varepsilon},\sigma,\dot{\sigma},t) = 0 \ , \tag{9}$$

(where dots denote derivatives with respect to the time t) and with the integral formulae for M and N,

$$M = \int_A \sigma Z dA \ , \qquad N = \int_A \sigma dA \ , \tag{10}$$

determine six unknown functions: ε, σ, M, N, s and ϕ. Then the fibre stress $\bar{\sigma} = \bar{\sigma}(X,t)$ is known, the function $g_i(X)$ may be evaluated, (4), and the iterative procedure (5) may be employed. Moreover, the deflections w may be found from the equation (Figure 1),

$$\frac{dw}{dX} = \frac{ds}{dX} \sin \phi \ ; \tag{11}$$

the knowledge of w is not necessary when considering cantilever beams, but becomes necessary for simply supported or statically indeterminate beams.

Detailed calculations are carried out separately for physically linear and for physically nonlinear materials.

For a physically linear material the integrals (10) may be calculated in a general form, for arbitrary shape of the cross-section. Assume the constitutive equation in the form:

$$B_1(t)\varepsilon + B_2(t)\dot{\varepsilon} = B_3(t)\sigma + B_4(t)\dot{\sigma} \ , \tag{12}$$

which presents a generalization of the Prager-Hohenemser (standard) model to time-variable moduli B_i. Of course, Maxwell's and Voigt-Kelvin's bodies are also described by (12). Substituting (8) into (12) and integrating over the cross-sectional area A we obtain without multiplication by Z

$$B_1(t)A\left(\frac{ds}{dX} - 1\right) + B_2(t)A\frac{ds}{dX} = B_3(t)N + B_4(t)\dot{N} \ , \tag{13}$$

and with multiplication by Z

$$-B_1(t)J\frac{d\phi}{dX} - B_2(t)J\frac{d\dot\phi}{dX} = B_3(t)M + B_4(t)\dot M, \qquad (14)$$

where $J = J(X)$ denotes the moment of inertia of the cross-section about the axis Y. Both A and J contain the design variables: for example, for a rectangular section of the width b and the depth 2H we have $A = 2bH$, $J = \frac{2}{3}bH^3$, Figure 2. In this case four equations (7), (13) and (14) determine four unknowns M, N, s and ϕ; subsequently we can calculate $\bar\sigma$ and use the iterative procedure (5).

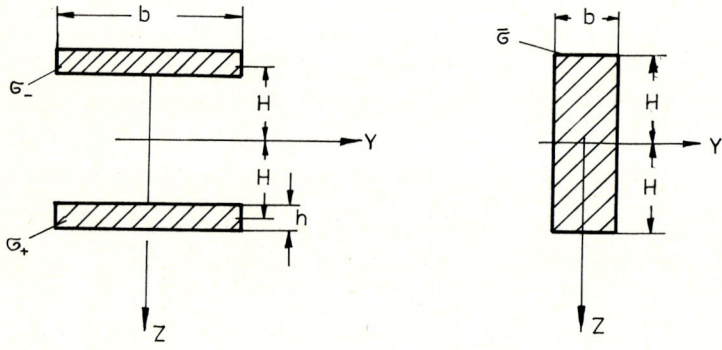

Figure 2

For a physically nonlinear material the integrations (10) cannot be performed in a general, analytical form. Hence, in this case, we confine our considerations to a theoretical I-section, Figure 2, and replace the integration by simple summation.

Denote by the indices + and - the stresses and strains in the lower flange $Z = +H$ and in the upper flange $Z = -H$, respectively. In view of (8) differentiated with respect to the time t, we may write

$$\left.\begin{aligned}\dot{\varepsilon}_- &= \frac{d\dot{\phi}}{dX} H + \frac{d\dot{s}}{dX} ,\\ \dot{\varepsilon}_+ &= -\frac{d\dot{\phi}}{dX} H + \frac{d\dot{s}}{dX} .\end{aligned}\right\} \quad (15)$$

The integrals (10) are here replaced by

$$\left.\begin{aligned}M &= (\sigma_+ - \sigma_-)bhH ,\\ N &= (\sigma_+ + \sigma_-)bh ,\end{aligned}\right\} \quad (16)$$

and hence

$$\left.\begin{aligned}\sigma_- &= -\frac{M}{2bhH} + \frac{N}{2bh} ,\\ \sigma_+ &= \frac{M}{2bhH} + \frac{N}{2bh} .\end{aligned}\right\} \quad (17)$$

The strain rates (15) are related to the stresses (17) by the constitutive equations. Assume, for example, the Norton-Bailey power creep law allowing for elastic strains

$$\dot{\varepsilon} = \frac{\dot{\sigma}}{E} + k\sigma^n , \quad (18)$$

where E denotes Young's modulus and k,n are material constants. In this case we may write, having solved (15) with respect to the derivatives $d\dot{\phi}/dX$ and $d\dot{s}/dX$,

$$\left.\begin{aligned}\frac{d\dot{\phi}}{dX} &= \frac{1}{2H}\left[\frac{\dot{\sigma}_- - \dot{\sigma}_+}{E} + k(\sigma_-^n - \sigma_+^n)\right] ,\\ \frac{d\dot{s}}{dX} &= \frac{1}{2}\left[\frac{\dot{\sigma}_- + \dot{\sigma}_+}{E} + k(\sigma_-^n + \sigma_+^n)\right] .\end{aligned}\right\} \quad (19)$$

Six equations (7), (17) and (19) contain six unknowns M, N, σ_+, σ_-, s and ϕ and describe the creep process; the deflections w may additionally be found from (11). In the case of a negative bending moment (as in a typical cantilever), the maximal tensile stress occurs in the upper flange, and hence we substitute σ_- into the integral (4).

4. NUMERICAL EXAMPLES

4.1 *Cantilever Beam, Norton-Bailey's Creep Law, width b as the Design Variable*

Numerical calculations are carried out in dimensionless quantities the number of dimensionless parameters being kept as low as possible. To this end, we introduce the dimensionless independent variables, $\xi = X/\ell$, $0 \leq \xi \leq 1$, and $\tau = kE^n t$, the following dimensionless dependent variables: $\chi = s/\ell$, $\zeta = w/\ell$, $\mu = \sigma/E$, $m = M/P\ell$, $\beta = 2Ehb/P$, and the dimensionless parameters $\lambda = H/\ell$, $\mu_o = \sigma_o/E$.

Substituting the second formula (7) for N we may now describe the elastic optimal design for the given stress μ_o by the following system of ordinary differential equations:

$$\left. \begin{aligned} \frac{d\phi}{d\xi} &= -\frac{m}{\beta\lambda^2}, \\ \frac{d\chi}{d\xi} &= 1 + \frac{\sin\phi}{\beta}, \\ \frac{dm}{d\xi} &= \frac{d\chi}{d\xi} \cos\phi, \\ \mu_- &= -\frac{m}{\beta\lambda} + \frac{\sin\phi}{\beta} = \mu_o = \text{const}. \end{aligned} \right\} \quad (20)$$

These equations, together with the boundary conditions $\phi(0) = \chi(0) = m(1) = 0$, determine the optimal elastic beam in terms of two parameters: life-time t_k (hidden in μ_o) and a certain slenderness parameter λ. The shape $\beta = \beta_1(\xi)$ may serve as the first approximation for the creep optimization. However, it turns out that $\beta_1(1)$ is too small, since creep brings larger deflections and an increase in the normal force at the free end (where the bending moment vanishes). Hence it is reasonable to add to the function $\beta_1(\xi)$ the following correction

$$\Delta\beta = \left(\frac{1}{\omega\mu_o} - \frac{\sin\phi}{\mu_o}\right)\xi^r, \qquad (21)$$

where ω and r are certain constants. The value $\omega = 1$ corresponds to the design of the free section $\xi = 1$ as subject to pure tension by the force P as a whole. In numerical calculations the values $\omega = r = 1$ were assumed and the shape of the beam $\beta_1 + \Delta\beta$ was used as the first approximation. The elastic deflections of the modified beam were found from (20), the last equation being dropped. The dimensionless stresses are given by

$$\left.\begin{array}{l} \mu_- = -\dfrac{m}{\beta\lambda} + \dfrac{\sin\phi}{\beta}, \\[2mm] \mu_+ = \dfrac{m}{\beta\lambda} + \dfrac{\sin\phi}{\beta}. \end{array}\right\} \qquad (22)$$

The creep process is governed by the constitutive equations (19) which take now the form

$$\left.\begin{array}{l} \dfrac{d\dot\phi}{d\xi} = \dfrac{1}{2\lambda}(\dot\mu_- - \dot\mu_+ + \mu_-^n - \mu_+^n), \\[2mm] \dfrac{d\dot\chi}{d\xi} = \dfrac{1}{2}(\dot\mu_- + \dot\mu_+ + \mu_-^n + \mu_+^n), \end{array}\right\} \qquad (23)$$

the third equation (20) and equations (22) remaining unchanged. The initial conditions are given by the elastic solution and the boundary conditions remain without change. The stress μ_- is substituted into the integral (4) and the iterative procedure is applied several times. Convergence of this procedure depends on the parameters λ and μ_o, and particularly on the life-time t_k assumed, "hidden" in μ_o. For larger life-times we obtain smaller μ_o and smaller deflections and then the iterative procedure is well convergent; smaller life-times result in larger deflections, larger redistribution of bending moment and normal force in time, and hence the convergence gets slower.

Figures 3 and 4 show the deflections and the shape of the optimal beam for the following input data: $P = 10$ kp, $\ell = 2000$ mm, $H = 20$ mm, $h = 1$ mm, $E = 1,7 \cdot 10^4$ kg/mm^2, $k = 0,16 \cdot 10^{-8}$ (kp/mm^2)$^{-n}$h^{-1}, $n = 3,3$, $C = 0,17 \cdot 10^{-6}$ (kp/mm^2)$^{-\nu}$h^{-1}, $\nu = 2,3$, $t_k = 5 \cdot 10^4$h. The procedures of numerical integration used were as follows: the Runge-Kutta 4 procedure along ξ with 100 basic steps but with additional subdivisions near the free end (128 steps in total); the Euler procedure along τ with 50 steps or 100 steps (no major differences were found).

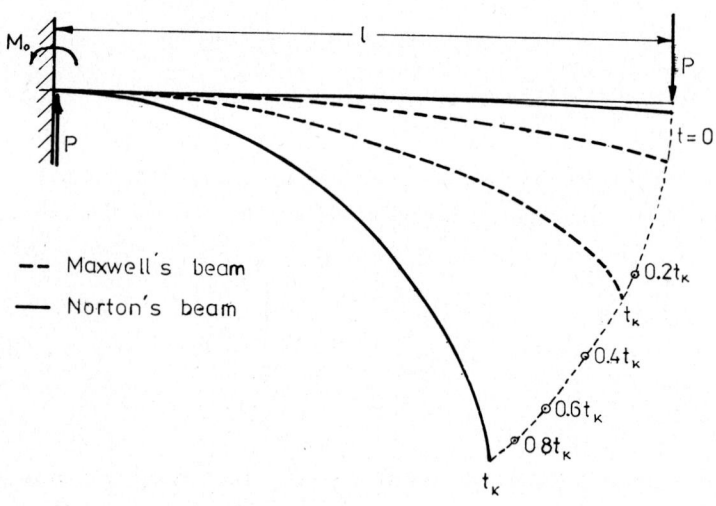

Figure 3

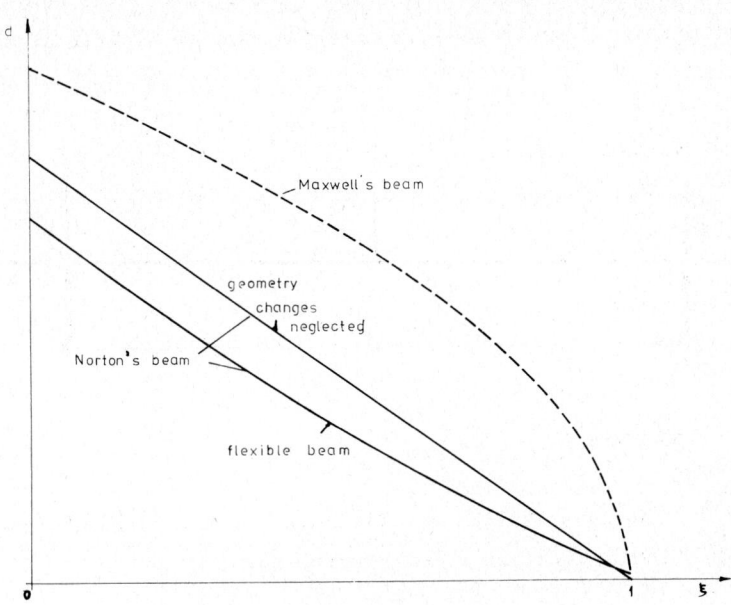

Figure 4

4.2 *Cantilever Beam, Maxwell's Creep Law, depth H as the Design Variable*

A similar procedure was applied to a rectangular Maxwell's cantilever beam with the depth H as the design variable. The results are also shown in Figures 3 and 4.

4.3 *Fully Clamped Beam, Norton-Bailey's Creep Law, width b as the Design Variable*

An attempt of solving statically indeterminate beams has also been made. Figure 5 shows a beam with full end fixity. In this case not only the equilibrium equations (7) are subject to change, but the deflections ζ have to be calculated as well as the horizontal displacements. Because of the symmetry and antisymmetry conditions it is sufficient to consider only one-fourth of the beam.

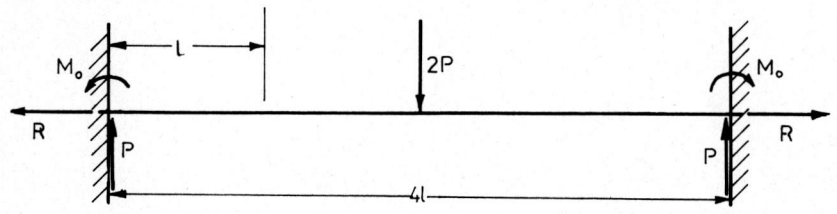

Figure 5

The constitutive equations (23) remain without change; the equilibrium equation takes now the form:

$$\frac{dm}{d\xi} = \frac{d\chi}{d\xi} (\cos\phi - r\sin\phi) , \qquad (24)$$

and the stresses are given by

$$\mu_{-} = -\frac{m}{\beta\lambda} + \frac{\sin\phi}{\beta} + \frac{r\cos\phi}{\beta} ,$$

$$\mu_{+} = \frac{m}{\beta\lambda} + \frac{\sin\phi}{\beta} + \frac{r\cos\phi}{\beta} , \qquad (25)$$

where the dimensionless longitudinal reactive force $r = R/P$ is unknown as yet. The deflections may be found from (11) expressed in the dimensionless form

$$\frac{d\zeta}{d\xi} = \frac{d\chi}{d\xi} \sin\phi . \qquad (26)$$

The system of equations given above needs four boundary conditions to determine the four integration constants and an additional

condition to determine r. They are as follows:
$\phi(0) = \chi(0) = \xi(0) = 0$ and $m(1) = 0$; the additional condition eliminates axial displacement at $\xi = 1$ and has the form

$$\int_0^1 \frac{d\chi}{d\xi} \cos\phi \, d\xi = 1 \; . \tag{27}$$

Numerical calculations are much longer here since we have three boundary conditions at the left-hand end and two boundary conditions at the right-hand end of the interval. Final results will not be quoted for this example.

5. FINAL REMARKS

The shape of the beams considered has been assumed constant during the whole creep process. Such an assumption corresponds to "passive" structural control in creep problems. Active structural control may be achieved if we can optimally shift the material from one place to another during the creep process.

Such an approach, theoretically possible, might also be realized in practice e.g., if the life-time of the structure is assumed for many years; however it exceeds the scope of the present paper and will not be dealt with here.

REFERENCES

[1] REYTMAN, M.I., "On the Theory of Optimal Design of Structures Made of Plastics with the Time Factor taken into Account", (in Russian), *Mekhanika Polimerov*, Riga, 1967, pp. 357-360.

[2] PRAGER, W., "Optimal Structural Design for Given Stiffness in Stationary Creep", *ZAMP*, Vol. 19, 1968, pp. 252-256.

[3] ŻYCZKOWSKI, M., "Optimal Structural Design in Rheology", *Proc. 12th Int. Congress Appl. Mech.*, Stanford, 1968; also *Trans. ASME*, Vol. E38, 1971, pp. 39-46.

[4] NEMIROVSKY, Yu.V. and REZNIKOV, B.S., "Beams and Plates of Uniform Strength in Creep Conditions", (in Russian), *Mashinovedenye*, Moskva, 1969, pp. 58-64.

[5] GAJEWSKI, A., "Optymalne Kształtowanie Wytrzymałościowe w Przypadku Materiałów o Nieliniowości Fizycznej", Zeszyty Naukowe Politechniki Krakowskiej, 12, Kraków, 1975.

[6] ŻYCZKOWSKI, M. and RYSZ, M., "Optimal Design of a Thin-Walled Pipe-Line Cross-Section in Creep Conditions", *Proc. Symposium on Mechanics of Inelastic Structures*, Warsaw, 1978, (in print).

[7] WOJDANOWSKA, R., "Optimal Design of Weakly Curved Compressed Bars with Maxwell-Type Creep Effects", *Archives of Mechanics*, Vol. 30, 1978, pp. 845-851.

[8] ŻYCZKOWSKI, M. and ZABORSKI, A., "Creep Rupture Phenomena in Creep Buckling", *Proc. IUTAM Symposium on Mechanics of Visco-Elastic Media and Bodies*, Göteborg, 1974, Springer, 1975, pp. 283-290.

[9] KACHANOV, L.M., "On the Rupture Time in Creep Conditions", (in Russian), Izv. Akad. Nauk SSSR, Otd. Tekhn. Nauk, 1958, pp. 26-31.

[10] UMETANI, Y. and HIRAI, S., "Shape Optimization for Beams Subject to Displacement Restrictions on the Basis of the Growing-Reforming Procedure", *Bull. JSME*, Vol. 21, 1978, pp. 1113-1119.